Inland Fisheries Management
in North America

Cover illustration by Beth A. Loftin

Inland Fisheries Management in North America

is a special project of the

**Education Section and Fisheries Management Section
American Fisheries Society**

Support for publication of this book
was provided by

**U.S. Fish and Wildlife Service
Exxon Company, U.S.A.
ARCO Foundation
National Park Service
Shell Oil Company
Conoco Inc.
U.S. Forest Service
U.S. Bureau of Land Management
Halliburton Company
James D. Woods**

Inland Fisheries Management in North America

Edited by

Christopher C. Kohler

Fisheries Research Laboratory
and Department of Zoology
Southern Illinois University

and

Wayne A. Hubert

Wyoming Cooperative Research Unit
Department of Fish and Wildlife
University of Wyoming

American Fisheries Society
Bethesda, Maryland, USA
1993

Suggested Citation Formats

Entire Book

Kohler, C. C., and W. A. Hubert, editors. 1993. Inland fisheries management in North America. American Fisheries Society, Bethesda, Maryland.

Chapter within the Book

Krueger, C. C., and D. J. Decker. 1993. The process of fisheries management. Pages 33–54 *in* C. C. Kohler and W. A. Hubert, editors. Inland fisheries management in North America. American Fisheries Society, Bethesda, Maryland.

Library of Congress Catalog Card Number 93-071888

ISBN 0-913235-83-0

Address orders to

American Fisheries Society
5410 Grosvenor Lane, Suite 110
Bethesda, Maryland 20814-2199, USA
Telephone (301) 897-8616

Contents

FISHERY ASSESSMENTS

5   Dynamics of Exploited Fish Populations

MICHAEL J. VAN DEN AVYLE

6   Practical Use of Biological Statistics

JOHN J. NEY

7   Socioeconomic Benefits of Fisheries

A. STEPHEN WEITHMAN

HABITAT MANIPULATIONS

8   Watershed Management and Land-Use Practices

THOMAS A. WESCHE

9 Stream Habitat Management

DONALD J. ORTH AND RAY J. WHITE

10 Lake and Reservoir Habitat Management

ROBERT C. SUMMERFELT

11 Maintenance of the Estuarine Environment

WILLIAM H. HERKE AND BARTON D. ROGERS

COMMUNITY MANIPULATIONS

12 Management of Introduced Fishes

HIRAM W. LI AND PETER B. MOYLE

13 Stocking for Sport Fisheries Enhancement

ROY C. HEIDINGER

14 Management of Undesirable Fish Species
ROBERT W. WILEY AND RICHARD S. WYDOSKI

15 Endangered Species Management
CLARENCE A. CARLSON AND ROBERT T. MUTH

16 Managing Fisheries with Regulations
RICHARD L. NOBLE AND T. WAYNE JONES

COMMON MANAGEMENT PRACTICES

17 Coldwater Streams
J. S. GRIFFITH

18 Warmwater Streams

C. F. Rabeni

19 Large Rivers

R. J. Sheehan and J. L. Rasmussen

20 Small Impoundments

S. A. Flickinger and F. J. Bulow

21 Natural Lakes and Large Impoundments

D. B. Hayes, W. W. Taylor, and E. L. Mills

22 The Great Lakes Fisheries

D. J. JUDE AND J. LEACH

23 Anadromous Stocks

J. R. MORING

Contributors

Frank J. Bulow (Chapter 20): Department of Biology, Tennessee Technological University, Box 5063, Cookeville, Tennessee 38505, USA.

Clarence A. Carlson (Chapter 15): Department of Fishery and Wildlife Biology, Colorado State University, Fort Collins, Colorado 80523, USA.

Daniel J. Decker (Chapters 2 and 3): Department of Natural Resources, Cornell University, 206 D Fernow Hall, Ithaca, New York 14853, USA.

Stephen A. Flickinger (Chapter 20): Department of Fishery and Wildlife Biology, Colorado State University, Fort Collins, Colorado 80523, USA.

J. S. Griffith (Chapter 17): Department of Biology, Idaho State University, Pocatello, Idaho 83209, USA.

Daniel B. Hayes (Chapter 21): Department of Fisheries and Wildlife, Michigan State University, East Lansing, Michigan 48824, USA. *Present address*: National Oceanic and Atmospheric Administration, National Marine Fisheries Service, 166 Water Street, Woods Hole, Massachusetts 02543, USA.

Roy C. Heidinger (Chapter 13): Cooperative Fisheries Research Laboratory, Southern Illinois University, Carbondale, Illinois 62901, USA.

William H. Herke (Chapter 11): Department of Forestry, Wildlife, and Fisheries, Louisiana State University, Baton Rouge, Louisiana 70803, USA.

Wayne A. Hubert (Coeditor): Wyoming Cooperative Fish and Wildlife Research Unit, University of Wyoming, Box 3166, Biological Sciences Building, Laramie, Wyoming 82071, USA.

T. Wayne Jones (Chapter 16): Route 3, Box 321 E, Nashville, North Carolina 27856, USA.

David J. Jude (Chapter 22): Center for Great Lakes and Aquatic Sciences, University of Michigan, 3116 Institute of Science and Technology, Ann Arbor, Michigan 48109, USA.

Christopher C. Kohler (Coeditor): Fisheries Research Laboratory, Southern Illinois University, Carbondale, Illinois 62901, USA.

Charles C. Krueger (Chapters 2 and 3): Department of Natural Resources, Cornell University, 206 D Fernow Hall, Ithaca, New York 14853, USA.

Berton L. Lamb (Chapter 4): U.S. Fish and Wildlife Service, National Ecology Research Center, 4512 McMurray Avenue, Fort Collins, Colorado 80525-3400, USA.

Joseph H. Leach (Chapter 22): Lake Erie Fisheries Station, Rural Route 2, Wheatley, Ontario N0P 2P0, Canada.

Hiram W. Li (Chapter 12): Oregon Cooperative Fishery Research Unit, Oregon State University, Corvallis, Oregon 97331, USA.

Edward L. Mills (Chapter 21): Cornell Biological Field Station, Rural Delivery 1, Bridgeport, New York 13030, USA.

John R. Moring (Chapter 23): Maine Cooperative Fish and Wildlife Research Unit, University of Maine, 313 Murray Hall, Orono, Maine 04469, USA.

Peter B. Moyle (Chapter 12): Department of Wildlife and Fisheries Biology, University of California, Davis, California 95616, USA.

Robert T. Muth (Chapter 15): Department of Fishery and Wildlife Biology, Colorado State University, Fort Collins, Colorado 80523, USA.

John J. Ney (Chapter 5): Department of Fisheries and Wildlife, Virginia Polytechnic Institute and State University, Blacksburg, Virginia 24060, USA.

Larry A. Nielsen (Chapter 1): Department of Fisheries and Wildlife, Virginia Polytechnic Institute and State University, Blacksburg, Virginia 24061-0321, USA.

Richard L. Noble (Chapter 16): Department of Fisheries and Wildlife Sciences, North Carolina State University, Campus Box 7646, Raleigh, North Carolina 27695, USA.

Donald J. Orth (Chapter 9): Department of Fisheries and Wildlife, Virginia Polytechnic Institute and State University, Blacksburg, Virginia 24061-0321, USA.

Charles F. Rabeni (Chapter 18): Missouri Cooperative Fish and Wildlife Research Unit, University of Missouri, Stephens Hall, Columbia, Missouri 65201, USA.

Jerry L. Rasmussen (Chapter 19): U.S. Fish and Wildlife Service, 608 East Cherry Street, Columbia, Missouri 65201, USA.

Barton D. Rogers (Chapter 11): Department of Forestry, Wildlife, and Fisheries, Louisiana State University, Baton Rouge, Louisiana 70803, USA.

Robert J. Sheehan (Chapter 19): Cooperative Fisheries Research Laboratory, Southern Illinois University, Carbondale, Illinois 62901, USA.

Robert C. Summerfelt (Chapter 10): Department of Animal Ecology, Iowa State University, Room 124, Science II Building, Ames, Iowa 50011, USA.

William W. Taylor (Chapter 21): Department of Fisheries and Wildlife, Michigan State University, East Lansing, Michigan 48824, USA.

Michael J. Van Den Avyle (Chapter 6): Georgia Cooperative Fish and Wildlife Research Unit, University of Georgia, Athens, Georgia 30602, USA.

Allen S. Weithman (Chapter 7): Missouri Department of Conservation, Post Office Box 180, Jefferson City, Missouri 65101, USA.

Thomas A. Wesche (Chapter 8): Department of Range Management and Wyoming Water Resources Center, University of Wyoming, Post Office Box 3354, Laramie, Wyoming 82071, USA.

Ray J. White (Chapter 9): 320 12th Avenue N, Edmonds, Washington 98020, USA.

Robert Wiley (Chapter 14): Wyoming Game and Fish Department, 528 South Adams, Laramie, Wyoming 82070, USA.

Richard S. Wydoski (Chapter 14): U.S. Fish and Wildlife Service, Division of Federal Aid, Post Office Box 25486, Denver Federal Center, Denver, Colorado 80225, USA.

List of Fish Species

The colloquial names of many fish species have been standardized in *Common and Scientific Names of Fishes from the United States and Canada* (4th edition, 1980; 5th edition, 1990) and *World Fishes Important to North Americans* (1991), published by the American Fisheries Society. Throughout this book, species listed in those publications are cited only by common name except when a fuller identification is important. The respective scientific names of these species follow.

Alewife .... *Alosa pseudoharengus*
American eel .... *Anguilla rostrata*
American shad .... *Alosa sapidissima*
Apache trout .... *Oncorhynchus apache*
Arctic char .... *Salvelinus alpinus*
Arctic grayling .... *Thymallus arcticus*
Ash Meadows Armagosa pupfish .... *Cyprinodon nevadensis mionectes*
Ash Meadows killifish .... *Empetrichthys merriami*
Ash Meadows speckled dace .... *Rhinichthys osculus nevadensis*
Atlantic croaker .... *Micropogonias undulatus*
Atlantic salmon .... *Salmo salar*
Atlantic sturgeon .... *Acipenser oxyrhynchus*
Atlantic whitefish .... *Coregonus canadensis*

Bigmouth buffalo .... *Ictiobus cyprinellus*
Black bass .... *Micropterus* spp.
Black bullhead .... *Ameiurus melas*
Black crappie .... *Pomoxis nigromaculatus*
Blackfin cisco .... *Coregonus nigripinnis*
Bloater .... *Coregonus hoyi*
Blue catfish .... *Ictalurus furcatus*
Blue pike .... *Stizostedion vitreum glaucum*
Blue tilapia .... *Tilapia aurea*
Blueback herring .... *Alosa aestivalis*
Bluegill .... *Lepomis macrochirus*
Bonneville cutthroat trout .... *Oncorhynchus clarki utah*
Bonytail .... *Gila elegans*
Bowfin .... *Amia calva*
Brook trout .... *Salvelinus fontinalis*
Brown trout .... *Salmo trutta*
Buffalo .... *Ictiobus* spp.
Bull trout .... *Salvelinus confluentus*
Bullhead catfish .... *Ameiurus* spp.
Burbot .... *Lota lota*

Chain pickerel .... *Esox niger*
Channel catfish .... *Ictalurus punctatus*
Char .... *Salvelinus* spp.
Chinook salmon .... *Oncorhynchus tshawytscha*

Chum salmon .... *Oncorhynchus keta*
Cisco .... *Coregonus artedi*
Coho salmon .... *Oncorhynchus kisutch*
Colorado River cutthroat trout .... *Oncorhynchus clarki pleuriticus*
Colorado squawfish .... *Ptychocheilus lucius*
Common carp .... *Cyprinus carpio*
Common shiner .... *Luxilus cornutus*
Creek chub .... *Semotilus atromaculatus*
Creek chubsucker .... *Erimyzon oblongus*
Cui-ui .... *Chasmistes cujus*
Cutthroat trout .... *Oncorhynchus clarki*

Deepwater cisco .... *Coregonus johannae*
Deepwater sculpin .... *Myoxocephalus thompsoni*
Desert pupfish .... *Cyprinodon macularis*
Devils Hole pupfish .... *Cyprinodon diabolis*
Dolly Varden .... *Salvelinus malma*

Emerald shiner .... *Notropis atherinoides*
Eulachon .... *Thaleichthys pacificus*
European perch .... *Perca fluviatilis*

Fathead minnow .... *Pimephales promelas*
Flathead catfish .... *Pylodictis olivaris*
Florida largemouth bass .... *Micropterus salmoides floridanus*
Freshwater drum .... *Aplodinotus grunniens*

Gar .... *Lepisosteus* spp.
Gila topminnow .... *Poeciliopsis occidentalis occidentalis*
Gila trout .... *Oncorhynchus gilae*
Gizzard shad .... *Dorosoma cepedianum*
Golden shiner .... *Notemigonus crysoleucas*
Goldfish .... *Carassius auratus*
Grass carp .... *Ctenopharyngodon idella*
Green sunfish .... *Lepomis cyanellus*
Greenback cutthroat trout .... *Oncorhynchus clarki stomias*
Gulf menhaden .... *Brevoortia patronus*

Herrings ....Family Clupeidae
Humpback chub .... *Gila cypha*
Hybrid striped bass (white bass x striped bass) ... *Morone chrysops x M. saxatilis*

Inland silverside .... *Menidia beryllina*

Kendall Warm Springs dace .... *Rhinichthys osculus thermalis*
Kiyi .... *Coregonus kiyi*
Kokanee .... *Oncorhynchus nerka*

Lahontan cutthroat trout .... *Oncorhynchus clarki henshawi*
Lake herring .... *Coregonus artedi*
Lake sturgeon .... *Acipenser fulvescens*
Lake trout .... *Salvelinus namaycush*
Lake whitefish .... *Coregonus clupeaformis*

Largemouth bass .... *Micropterus salmoides*
Las Vegas dace .... *Rhinichthys deaconi*
Little Kern golden trout .... *Oncorhynchus aguabonita whitei*
Longear sunfish .... *Lepomis megalotis*
Longjaw cisco .... *Coregonus alpenae*
Longnose sucker .... *Catostomus catostomus*

Menhaden .... *Brevoortia* spp.
Minnows .... Family Cyprinidae
Mohave tui chub .... *Gila bicolor mohavensis*
Mosquitofish .... *Gambusia affinis*
Mountain whitefish .... *Prosopium williamsoni*
Mozambique tilapia .... *Tilapia mossambica*
Mullet .... Family Mugilidae
Muskellunge .... *Esox masquinongy*

Nile perch .... *Lates niloticus*
Northern pike .... *Esox lucius*
Northern squawfish .... *Ptychocheilus oregonensis*

Owens pupfish .... *Cyprinodon radiosus*
Owens sucker .... *Catostomus fumeiventris*
Owens tui chub .... *Gila bicolor snyderi*

Pacific salmon .... *Oncorhynchus* spp.
Paddlefish .... *Polyodon spathula*
Pahranagat roundtail chub .... *Gila robusta jordani*
Pahranagat spinedace .... *Lepidomeda altivelis*
Pahrump killifish .... *Empetrichthys latos latos*
Pahrump Ranch killifish .... *Empetrichthys latos pahrump*
Paiute cutthroat trout .... *Oncorhynchus clarki seleniris*
Peacock bass .... *Cichla ocellaris*
Pecos gambusia .... *Gambusia nobilis*
Pickerel .... *Esox* spp.
Pink salmon .... *Oncorhynchus gorbuscha*
Prickly sculpin .... *Cottus asper*
Pumpkinseed .... *Lepomis gibbosus*

Quillback .... *Carpiodes cyprinus*

Rainbow smelt .... *Osmerus mordax*
Rainbow trout .... *Oncorhynchus mykiss*
Raycraft Ranch killifish .... *Empetrichthys latos concavus*
Razorback sucker .... *Xyrauchen texanus*
Red drum .... *Sciaenops ocellatus*
Red porgy .... *Pagrus pagrus*
Red shiner .... *Cyprinella lutrensis*
Redbreast sunfish .... *Lepomis auritus*
Redear sunfish .... *Lepomis microlophus*
Redhorse sucker .... *Moxostoma* spp.
Redside shiner .... *Richardsonius balteatus*
Rock bass .... *Ambloplites rupestris*

Round goby .... *Neogobius melanostomus*
Ruffe .... *Gymnocephalus cernuus*
Ruff .... *Centrolophus* spp.

Sacramento blackfish .... *Orthodon microlepidotus*
Sand seatrout .... *Cynoscion arenarius*
Sauger .... *Stizostedion canadense*
Saugeye (sauger x walleye hybrid) .... *Stizostedion canadense x S. vitreum*
Sculpins .... Family Cottidae
Sea lamprey .... *Petromyzon marinus*
Shortnose cisco .... *Coregonus reighardi*
Shortnose sturgeon .... *Acipenser brevirostrum*
Shovelnose sturgeon .... *Scaphirhynchus platorynchus*
Slimy sculpin .... *Cottus cognatus*
Smallmouth bass .... *Micropterus dolomieu*
Smallmouth buffalo .... *Ictiobus bubalus*
Snail darter .... *Percina tanasi*
Snook .... *Centropomus* spp.
Sockeye salmon .... *Oncorhynchus nerka*
Speckled dace .... *Rhinichthys osculus*
Spiny dogfish .... *Squalus acanthias*
Splake (brook trout x lake trout hybrid) .... *Salvelinus fontinalis x S. namaycush*
Spoonhead sculpin .... *Cottus ricei*
Spottail shiner .... *Notropis hudsonius*
Spotted bass .... *Micropterus punctulatus*
Spotted seatrout .... *Cynoscion nebulosus*
Squawfish .... *Ptychocheilus* spp.
Steelhead .... *Oncorhynchus mykiss*
Striped bass .... *Morone saxatilis*
Sturgeon. .... Family Acipenseridae
Suckers .... Family Catostomidae
Sunfish .... *Lepomis* spp.
Swordtail .... *Xiphophorus* spp.

Threadfin shad .... *Dorosoma petenense*
Threespine stickleback .... *Gasterosteus aculeatus*
Tiger muskellunge
(northern pike x muskellunge hybrid) .... *Esox lucius x E. masquinongy*
Tilapia .... *Tilapia* spp.
Tubenose goby .... *Proterorhinus marmoratus*
Tui chub .... *Gila bicolor*

Utah chub .... *Gila atraria*

Virgin River roundtail chub .... *Gila robusta seminuda*

Walleye .... *Stizostedion vitreum*
Warmouth .... *Lepomis gulosus*
White bass .... *Morone chrysops*
White crappie .... *Pomoxis annularis*
White perch .... *Morone americana*

White sturgeon .... *Acipenser transmontanus*
White sucker .... *Catostomus commersoni*
Winter flounder .... *Pleuronectes americanus*
Woundfin .... *Plagopterus argentissimus*

Yellow bass .... *Morone mississippiensis*
Yellow bullhead .... *Ameiurus natalis*
Yellow perch .... *Perca flavescens*

Preface

Inland Fisheries Management in North America was developed by fishery scientists associated with the American Fisheries Society, particularly the Education and Management Sections, to provide a description of the conceptual basis and current management practices being applied to manipulate freshwater and anadromous fisheries of North America. The text is devised for use in introductory university courses in fisheries management for juniors, seniors, or graduate students. It is assumed that students have a general background in ecology, limnology, ichthyology, and college-level mathematics. The book is designed as an educational tool, not a reference work; however, each chapter is fully referenced to allow readers access to the specific references and to advanced, more detailed information. Accordingly, the book should be of use to the practicing fisheries manager. The text emphasizes concepts of management and the decision-making process, while not overlapping information in *Fisheries Techniques* and *Methods for Fish Biology*. Both sport and commercial fisheries are discussed with emphasis on sport fisheries management.

The text is divided into five sections. The first section (Introduction) provides a historical overview of inland fisheries management in North America followed by chapters detailing the decision processes, communications, and legal considerations in fisheries management. The second section (Fishery Assessments) includes chapters that describe how biological statistics, including models of population dynamics, are used to assess the status of fisheries. A chapter on socioeconomic assessment is also included. In the third section (Habitat Manipulations) riverine, lacustrine, and estuarine environments are described along with specific measures that can be taken to maintain and improve the integrity of the habitats. Community manipulation is the theme of the fourth section. Chapters cover the management of introduced species, use of stocking in fisheries management, controlling undesirable species, managing endangered species, and manipulating sport fisheries through the use of regulations. The fifth section (Common Management Practices) concludes the book with chapters on coldwater streams, warmwater streams, large rivers, farm ponds and small impoundments, natural lakes and large impoundments, the Great Lakes, and finally, anadromous stocks.

The editors are indebted to the following individuals for assistance in developing the text outline: David H. Bennett, Frank J. Bulow, Robert F. Carline, James S. Diana, Charles C. Krueger, Stephen P. Malvestuto, John R. Moring, Peter B. Moyle, Donald J. Orth, Charles G. Scalet, Richard J. Strange, Michael J. Van Den Avyle, and Jimmy Winter. Although an attempt was made to develop a comprehensive textbook, the editors recognize that not every aspect of inland fisheries management can be covered within the confines of a single volume.

Each chapter was externally peer reviewed, and the criticisms were thoroughly considered by and constructive to the authors and editors. For the time and care they invested as peer reviewers, we thank: Joe G. Dillard, Peter A. Larkin, and Richard H. Stroud (Chapter 1); James R. Fazio and R. Ben Peyton (Chapters 2 and 3); Harvey R. Doerksen and Robert T. Lackey (Chapter 4); Robert M. Jenkins, Stephen P. Malvestuto, and David K. Whitehurst (Chapter 5); David K. Stevenson and William E. Ricker (Chapter 6); Kendall Adams, Richard C. Bishop and Lynne B. Starnes (Chapter 7); Burchard Heede and Keith McLaughlin (Chapter 8); Don A. Duff and John Orsborn (Chapter 9); Bobby G. Grinstead, Anthony A. Nigro, and Charles G. Scalet (Chapter 10); James B. Reynolds, Robert R. Stickney, and Alejandro Yáñez-Arancibia (Chapter 11); Walter R. Courtenay, Jr., Larry B. Crowder, and John J. Magnuson (Chapter 12); Brian R. Murphy, William L. Shelton, and Paul J. Wingate (Chapter 13); Mark B. Bain, Vern Hacker, and Fred P. Meyer (Chapter 14); Richard J. Neves, Edwin P. Pister and James A. Williams (Chapter 15); John G. Baughman, Thomas Gengerke and F. Joseph Margraf (Chapter 16); Richard A. Cunjak and Robert L. Hunt (Chapter 17); Vaughn L. Paragamian and Thomas R. Russell (Chapter 18); William D. Pearson and James C. Schmulbach (Chapter 19); James T. Davis, William D. Davies, James K. Mayhew, and William M. Lewis (Chapter 20); Eugene Maughan and David W. Willis (Chapter 21); Stephen B. Brandt, Jon G. Stanley, and Roy A. Stein (Chapter 22); and Lauren R. Donaldson, Arthur D. Hasler, Boyd Kynard, and Anthony A. Nigro (Chapter 23).

Several authors also wish to acknowledge the help they received as they created their respective chapters. Charles Krueger and Daniel Decker thank the following individuals for providing reviews of Chapters 2 and 3: G. A. Barnhart, T. L. Brown, A. Creamer, R. E. Eshenroder, D. McElveen, B. A. Knuth, L. A. Nielsen, P. Pister, C. P. Schneider, R. J. Wattendorf, and P. J. Wingate.

Berton Lamb and Beth Coughlan received helpful suggestions for Chapter 4 from E. R. Mancini, J. B. Gabbert, E. F. Tritschler, T. DeYoung, A. Locke, C. W. B. Stubbs, N. P. Lovrich, and P. Eschmeyer.

Steve Weithman appreciates reviews for Chapter 7 by R. O. Anderson, E. K. Brown, and D. J. Witter.

Thomas Wesche appreciates the critical reviews of Chapter 8 by Q. Skinner and J. F. Orsborn.

Donald Orth and Ray White thank C. A. Dolloff and P. L. Angermeier for thorough reviews of Chapter 9.

Robert Summerfelt is grateful to L. R. Mitzner and G. R. Ploskey for advice on Chapter 10.

William Herke and Barton Rogers thank the numerous respondents to their requests for information while preparing Chapter 11. An earlier draft was much improved as a result of critical reviews by H. Austin, D. Fruge, C. Levings, F. Perkins, R. Shaw, and T. Shirley. The same is true of reviews by students at Iowa State University, Louisiana State University, Montana State University, and the University of Massachusetts: M. Adkinson, C. Butson, Y. Lee, J. Mack, M. Mullins, S. Murphy, J. Serrano, and four who wished to remain anonymous. Special thanks to A. Yáñez-Arancibia for providing information for Mexico, and appropriate references to the Spanish literature. The criticisms and suggestions of the above were most helpful, but not all were accepted in the final version; therefore, errors or omissions must remain the responsibility of the authors and editors.

Hiram Li and Peter Moyle thank K. Currens and D. Buchanan for review comments on Chapter 12.

Roy Heidinger appreciates the critical reviews of Chapter 13 by M. Marcinko and P. Shafland.

Robert Wiley and Richard Wydoski appreciate the helpful suggestions on Chapter 14 by J. W. Mullan and W. F. Sigler.

Chapter 15 by Clare Carlson and Robert Muth is contribution 39 of the Colorado State University Larval Fish Laboratory. The authors extend gratitude to H. Rolston and R. J. Behnke for review comments and to O. Bray, J. Hamill, L. Lamb, and N. Kaufman for expert advice and unpublished material.

Richard Noble and Wayne Jones thank R. O. Anderson, F. A. Harris, C. R. Inman, and L. C. Redmond for providing review comments on Chapter 16.

Jack Griffith thanks D. W. Chapman, T. W. Hillman, and T. R. Angradi for reviewing Chapter 17.

Steve Flickinger and Frank Bulow thank the numerous respondents to their requests for information while preparing Chapter 20, the Alabama Division of Game and Fish for demonstrations used in some of the photographs, and T. C. Modde for supplying the basic drawing for the pond distribution map.

Dan Hayes, Bill Taylor, and Ed Mills thank S. Marod, J. Kocik, and R. Brown for their reviews of early drafts of Chapter 21. Assistance in literature searches by S. Marod are also gratefully acknowledged.

Chapter 22 by David Jude and Joseph Leach is contribution 513 from the Center for Great Lakes and Aquatic Sciences. The authors also extend thanks to T. Wells, J. Christie, D. Ryder, and R. Eshenroder for review comments. The seemingly endless draft versions of the chapter were typed by B. McClellan. The authors are in her debt.

John Moring thanks the following individuals for providing review comments on Chapter 23: D. Buchanan, H. Lorz, N. McHugh, M. Stratton, D. Locke, S. Pierce, E. Baum, N. Dubé, A. Meister, W. Hauser, D. Jones, D. Sidelman, G. Thomas, D. Misitano, A. Knight, F. Ayer, R. Barnhart, J. Delabbio, B. Glebe, G. Allen, L. Flagg, T. Squiers, and S. Ellis.

The editors gratefully acknowledge Teresa Cavitt and Nellore Collins for typing the entire volume on a single word processing program. This greatly facilitated the editorial process. We thank Allory Deiss, Thomas Lund, Carol Stevens, and Elizabeth Ono-Rahel with the graphics art shop at the University of Wyoming for the preparation of technical drawings.

We also appreciate the help provided by Beth Staehle and Amy Wassmann of the Society's headquarters staff who contributed editorial improvements in the final manuscripts and saw the book through production.

Funding for the book was provided by the U.S. Fish and Wildlife Service; Exxon Company, U.S.A.; ARCO Foundation; the National Park Service; Shell Oil Company; Conoco Inc.; U.S. Forest Service; U.S. Bureau of Land Management; Halliburton Company; and James D. Woods.

Any trade names and product vendors mentioned in this book are done so as a convenience to the readers. Specific citations do not imply endorsements by the American Fisheries Society, the authors and editors, or the employers of the book's contributors.

CHRISTOPHER C. KOHLER
WAYNE A. HUBERT

INTRODUCTION

Chapter 1

History of Inland Fisheries Management in North America

LARRY A. NIELSEN

1.1 INTRODUCTION

Fisheries management is a young profession. Its history—at least the part of greatest practical interest—can be recounted through the personal experience of older fisheries managers still working. Woody Seaman, after 44 years as a fisheries biologist, recently described himself as "a 'starter' of new things. . . . As you know, I was the first Chief of Fisheries, West Virginia Conservation Commission. Prior to this, it was only a hatchery program. I staffed up a group of fish biologists who tackled the fish management problems of the state" (Seaman 1988). That was in 1946, a time when many states and provinces were also just beginning their conservation programs and when many other starters like Woody Seaman were pushing the profession forward. They quickly developed most of the strategies and techniques used in fisheries management today. This rapid evolution of fisheries as a profession coincides with the accelerating development of the North American continent and with our increasing concern about the long-term consequences of human action.

The precedents of fisheries management, however, extend back to the origins of North American settlement. As is common with all subsets of human activity, the events and attitudes of society have affected fisheries and fisheries managers profoundly. The observation that fisheries management has a political and sociological context, as well as an ecological basis, is as accurate for earlier centuries as it is for today. Understanding fisheries management, therefore, requires a familiarity with the ideas and events that have shaped the North American personality and landscape.

This chapter has two purposes. First, it connects fisheries management and societal development. It illustrates that fisheries are part of the larger society, validating the obvious fact that fisheries management is, and always will be, subject to the desires of the public and societal leadership. Second, the chapter introduces many of the concepts developed in detail in later chapters. It displays the rich variety of approaches and techniques that make up fisheries management.

The chapter in total provides a conceptual definition of fisheries management. A dictionary-style definition might read, "the manipulation of aquatic organisms,

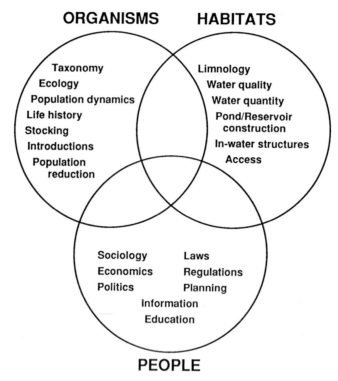

Figure 1.1 Fisheries management depicted as three overlapping circles which represent concern for aquatic organisms, aquatic habitats, and people.

aquatic environments, and their human users to produce sustained and ever-increasing benefits for people.'' This conception is often illustrated as three overlapping circles that represent the three principal components of fisheries: organisms, habitats, and people (Figure 1.1). Each is important, each affects the other two, and each presents opportunities for enhancing the value of fisheries resources.

1.2 HISTORICAL BASIS FOR FISHERIES MANAGEMENT

Fisheries in North America are public resources. State, provincial, and federal governments hold the resources in trust for the general use of their citizens. Although this system is different from the private ownership of fisheries in much of Europe, public ownership in North America derives directly from early English practices.

Ownership of fisheries resources has changed through time to follow the pattern of government. When the feudal system of governance spread across Europe during the Middle Ages, feudal kings and land barons declared themselves owners of the land. They claimed ownership of wildlife as well, not necessarily because wildlife was valuable, but because they wished to keep weapons out of the hands of the people. Later, as the feudal system in England evolved to a national monarchy, royalty expanded their ownership to include all wildlife and fisheries, exercising their assumed right to assign both exclusive and nonexclusive franchises (permission to use) to private individuals (Bean 1977). This system also governed fisheries use and ownership in North America up to the time of the American Revolution.

After the United States became a sovereign nation, the courts struck down the exclusive rights previously given by English royalty. They reasoned that exclusive franchises violated the principle that the king held fisheries as a public trust. After the revolution, all the rights and responsibilities of being the trustee of public fisheries were transferred to the new governments, specifically the state governments. Since that time, state governments in the United States have been the principal guardians of fisheries resources, vested with the major responsibilities for protecting and managing inland fisheries (Bean 1977).

In Canada, the responsibilities for managing inland fisheries were originally held by the provinces, operating as independent governments. Since the Confederation Act of 1867, however, when the Canadian provinces agreed to join under a federal government, fisheries have been shared by provincial and federal governments. The provinces own most inland fisheries, holding them in trust and managing the distribution of fisheries benefits. The federal government holds the responsibility to protect inland fisheries. This dual authority over Canadian fisheries management, which has produced a complicated legal history, has evolved through time; most responsibility now lies with the provincial governments (Thompson 1974). Management of Mexican resources is conducted primarily by the federal government, which assumed control of wildlife resources in the 1910 revolution (Beltran 1972).

The question of how a public trustee should treat fisheries has also been answered by a tradition as old as government itself: fisheries are common property. Common properties are those resources owned by the entire populace, without restriction on who may use them and, at least in earlier times, on how they may be used. The principle of common property was established formally in 1608, when the Dutch statesman Hugo Grotius proffered the doctrine of freedom of the seas. The open ocean, including its fishes, was declared the property of all people. This seemingly idealistic principle was actually just a statement of fact. In 1608, no one could control the oceans because defining boundaries and then protecting them was essentially impossible. Ownership was not necessary in any case because the wealth of the oceans was believed to be inexhaustible—the oceans held more fish than anyone could ever imagine needing or being able to catch (Nielsen 1976). The doctrine of freedom of the seas has been modified throughout history, but has remained the principle for the management of fishes, both marine and inland, as common property to this day. 1608

The common property principle is a good one—under certain conditions. As long as productivity of a fishery exceeds the demand for its products, common property provides an efficient and equitable allocation system. When demand gets too high, however, the productive capacity of the resource declines, and everyone suffers. For centuries, fisheries did supply all the food that people sought, and little or no attention was paid to managing the resource. Within the past century, however, the demand for fish has exceeded the available supply in place after place around the world.

Fisheries management was born of a need to balance the supply–demand equation. The modern history of fisheries is basically a chronicle of individual and governmental attempts to control the exploitation of common property fisheries.

Over the past century, scientists and public officials have struggled to develop suitable objectives for managing fisheries—objectives that preserve the time-honored ideal of free access to fisheries and that preserve the productive capacity of fish populations. In doing so, they have also developed the technical capacity to enhance fisheries productivity and to reduce the influence of other human activities on fisheries resources.

1.3   THE PRELUDE TO FISHERIES MANAGEMENT

North America truly was a new world. The continent was populated at such a low density and was so naturally productive that early European colonists faced a land entirely beyond their experience. Resources were seemingly limitless. Of the Chesapeake Bay, Robert Beverly wrote in 1705:

> As for fish, both of fresh and saltwater, of shellfish, and others, no country can boast of more variety, greater plenty, or of better in their several kinds. . . . In the spring of the year, herrings come up in such abundance into their brooks and fords to spawn that it is impossible to ride through without treading on them. . . . Thence it is that at this time of the year, . . . the rivers . . . stink of fish. (quoted in Wharton 1957)

It is little wonder that the principle of common property and an attendant lack of concern for conservation were the standards for conduct.

European settlers approached the new continent aggressively. The untamed landscape was viewed as an enemy that had to be subdued in order to provide a suitable human environment. Various explanations of this aggressiveness have been offered, including the Judeo-Christian ethic and the rise of democracy, but the prevailing forces must surely have been necessity and opportunity. The colonists were strictly utilitarian, and their early laws were developed to govern only the commercial aspects of natural resource use (Kawashima and Tone 1983).

Native Americans, in contrast, lived more harmoniously with nature. Natural resources, including fish, were the base of their existence, and virtually all Native Americans lived largely by subsistence fishing, hunting, and gathering (Swanton 1946). Their mythology often ascribed human qualities to natural objects, thus compelling an ethical treatment of their environment. Where Native American populations were dense, as in coastal California, they restricted their fish harvests in ways that sustained high productivity for centuries (McEvoy 1986). In most areas, however, their population density was low; at the time of European colonization, fewer than 12 million humans occupied North America. The sparse population, along with the low-technology forms of agriculture and fish harvest they practiced, would never make significant impacts on the continent's re-sources.

The aggressiveness of the colonists could not be sustained long without local problems. Local resource depletion and environmental problems, like those long experienced in Europe, developed rapidly in North America. Pastures were overgrazed, forests were overharvested, streams were overburdened with wastes, and fisheries were overexploited. Laws to correct and prevent local degradation became common in colonial governments, beginning with a 1652 Massachusetts

Figure 1.2 The westward colonization of North America used rivers as liquid highways. The Ohio River was perhaps the most important highway because its westward flow carried settlers and cargo across the rugged Appalachian Mountains. (Drawing courtesy of the U.S. Army Corps of Engineers, Huntington, West Virginia.)

law restricting fish catches (Stroud 1966). By the time of the American Revolution, hundreds of statutes restricted the times, places, and mechanisms for harvesting fish, precursing the variety of similar laws in force today.

The answer to resource depletion and environmental degradation, however, was not yet to be found in good management. The United States' answer was immortalized in the words of Horace Greeley, ''Go West, young man!'' Westward expansion of the new nation was a dominant goal of the United States' leaders, most notably Thomas Jefferson. Jefferson envisioned a nation of small landowners—hardworking, stable, and dedicated to preserving the nation. Federal land policy, from the Louisiana Purchase in 1803 to the Homestead Act of 1862, reflected the Jeffersonian doctrine of expanding, subdividing, and settling the frontier (Cox 1985).

Water played a dominant role in western settlement, as a liquid highway transporting people and goods across the eastern mountains. The Ohio River was the primary route of immigrants embarking at Pittsburgh for destinations in the Northwest Territory (Figure 1.2). River travel was difficult and dangerous, interrupted by portages around rapids, strandings in shallow water, and damage from floating debris. Waterway improvements, therefore, also became national priorities, and the United States' first federally supported public works were for improved navigation on the Ohio River (Smith 1971). By 1824, under the guise of improving national defense, the U.S. Army Corps of Engineers was actively and continuously modifying the Ohio and Mississippi rivers by digging canals, removing snags, and deepening the channels. Since then, water development projects have become a dominant feature in U.S. domestic policy and have modified the fisheries of virtually every major U.S. river system (see Chapter 19).

Fisheries in the 19th Century were primarily subsistence and commercial. The subsistence use of fisheries is poorly recorded, but it certainly provided substantial food supplies for local communities. Commercial fisheries, which had been based on marine fishes in colonial times, began to grow in fresh waters. The Great Lakes commercial fisheries were well established by 1835, with expectations of much greater possible harvests as the century progressed (Whitaker 1892). The state of Iowa reported a commercial fish catch of more than 1.8 million kilograms in 1886, or close to 1.4 kilograms per person (Carlander 1954). The Mississippi River was well used for commercial and subsistence fishing, although not everyone agreed that the river offered either palatable food or enjoyable sport (see Box 1.1).

Giving the western lands to eastern migrants also meant taking it away from Native Americans. During the first half of the 19th Century, scores of treaties between Indian nations and the U.S. government traded millions of hectares desired by settlers for more distant lands currently not desired. As the frontier pushed westward, so did the treaty lands, removing Native Americans progressively farther from their homes and disrupting their ties with the land and its resources. The continuing efforts of the U.S. government to accommodate western settlement progressively stressed relocation, separation, and assimilation of Native Americans, none of which improved their lifestyles or protected their rights. During this time, however, a body of Indian law developed and a series of treaty-based promises accumulated. Modern interpretation of these laws and promises shows that in many cases Native Americans retain ownership of the natural resources, including fish, on their aboriginal and treaty lands (Busiahn 1985). Recognition of those rights is reshaping the governance and use of many fisheries today and will continue to do so for the foreseeable future (see Chapter 4).

1.4   THE BIRTH OF FISHERIES MANAGEMENT

The completion of the U.S. transcontinental railway in 1869 signified both practical and symbolic ends to the frontier ethic. The continent had been tamed by the railroad, which allowed fast and convenient transportation across the formerly treacherous landscape. The question was no longer *if* the land would yield to human development, but *how* that development would proceed. The last half of the 19th Century represents the period when North Americans began to exert their influence broadly on natural resources, including fisheries, and when they developed a unique New World character tied closely to the character of their environment.

1.4.1   The Cultural Basis for Natural Resource Management

The wilderness of the New World had a deep impact on its settlers. While carving homesteads out of the wilderness, the pioneers also developed a kinship with the expansiveness and wildness of the land. After 1850, that kinship flowered into a North American personality, reflected in many aspects of human endeavor. The culture emphasized the natural environment, represented in philosophy by the transcendentalism of Ralph Waldo Emerson; in art of the Hudson River school of landscape artists, most notably Thomas Cole; in literature by romanticists such as William Cullen Bryant (Figure 1.3). The worldwide recognition of this cultural

Box 1.1 Life on the Mississippi

The Mississippi River was a dominant force in the lives of riverside communities. Harriet Bell Carlander (1954) chronicled the history of the river's fisheries, quoting examples of the wonder, inspiration, and indignation of river watchers.

Thomas Jefferson was an early champion of the Mississippi, as he recorded in his "Notes on Virginia":

> The Mississippi will be one of the principal channels of future commerce for the country westward of the Allegheny. . . . This river yields turtle of a peculiar kind, perch, trout, gar, pike, mullets, herrings, carp, spatula fish of fifty pound weight, catfish of one hundred pounds weight, buffalo fish and sturgeon.

Estwick Evans swam in the Mississippi in 1818 and noted the sediment in the water with mixed emotion:

> It is, however, not very unpalatable and is, I think, not unwholesome. The fish in the river are numerous and large; but they are too fat to be delicate.

Mark Twain quotes an English sea captain who visited the Mississippi in 1837 and was less ambivalent:

> It contains the coarsest and most uneatable of fish such as catfish and such genus. . . . There are no pleasing associations connected with the great common sewer of the western America which pours out its mud into the Mexican Gulf, polluting the clear blue sea for many miles beyond its mouth.

Charles Lanman, a New Englander who wrote about the river in 1856, also was dubious about catfishes:

> ... this fish is distinguished for its many deformities and is a great favorite with all persons who have a fancy for muddy water. In the Mississippi they are frequently taken weighing upwards of one hundred pounds...but it has always seemed to us that it requires a very powerful stomach to eat a piece from one of the mammoths of the western waters.

The editor of the *Muscatine Journal* assessed fishing in 1869:

> Fishing parties are fashionable now-a-days. For our part we should prefer lighter employment, such as sawing wood or carrying a hod to the seventh story of a building.

The truth about the Mississippi, however, is probably best recorded by Mel Ellis, writing in the *Milwaukee Journal* in 1949:

> If you haven't fished Ol' Man Mississip, forget about any preconceived notions you may have as far as rivers are concerned. Because Ol' Man River isn't really a river at all. In fact, he's a hundred rivers and a thousand lakes and more sloughs than you could explore in a lifetime. He is creeks, bayous, ditches, puddles, and thousands and thousands of impenetrable lotus beds that break big yellow flowers out above green pads.

Figure 1.3 The harmony between the North American personality and its wild natural environment is the subject of this 1849 painting by Asher Durand. Entitled ''Kindred Spirits,'' it depicts Thomas Cole, the foremost natural landscape artist of the time, and William Cullen Bryant, a naturalistic poet and journalist, in a setting like those that inspired them. (Photograph courtesy of the New York Public Library.)

growth legitimized the place of nature in the arts, and it welded an appreciation for the natural environment to the growing surge of land development.

This linkage between the human and the natural landscapes was first tied to resource management through the writings of George Perkins Marsh. Marsh's 1849 book, *Man and Nature,* showed how human activity affected the physical environment and how both could be improved by applying scientific and aesthetic principles. He denounced the destructive impacts of human exploitation and anticipated the development of true natural sciences (Nash 1987). His work aimed landscape design in a new naturalistic direction, one that would be implemented in full form through the genius of Frederick Law Olmsted, the architect of New York's Central Park and of the first U.S. national parks.

While Marsh was changing landscape architecture, a transformation also occurred in natural science. Until the mid-1800s, natural science had been an avocation. Amateur scientists collected bizarre and grotesque animals or spent their lives classifying animals, plants, and geological specimens. The application of physical and biological sciences to human endeavors, however, was like the touch of flame to oil. The pace and magnitude of human affairs exploded. As depicted in the United States' 1876 Centennial Exposition in Philadelphia, the industrial revolution unfolded in an array of machinery and engines. Natural science radiated into useful specialties. The application of zoology to agriculture was mandated in the United States by the creation of Land Grant Colleges in 1863 (these same universities house many of the U.S. fisheries programs today). The science of ecology began to take form at about the same time, as the encyclopedic observations of Canadian and U.S. naturalists provided the basis for theoretical development (Fry and Legendre 1966). Charles Darwin's *Origin of Species,* published in 1859, was the precursor to ecology, a term formally defined by Ernst Haeckel in 1866 (Egerton 1976).

Fisheries and other natural resources were also subject to the modernization brought by the industrial revolution. Modifications of the landscape were more massive and ubiquitous, multiplying and merging the local environmental impacts of previous generations. Power dams were built on many rivers; on the Connecticut River, they had blocked the spawning migrations of Atlantic salmon and American shad by 1849. The combination of mill dams, deforestation, and pollution had virtually exterminated Atlantic salmon from the St. Lawrence system by the mid-1800s. Throughout the continent, the use of waterways for power, transportation, mining, and waste disposal and the use of watersheds for rapid exploitation of timber, minerals, and crops were destroying the capacity of aquatic environments to sustain fish populations.

1.4.2 First Steps for Fisheries Conservation

Fisheries exploitation had always been a local industry, but the industrial revolution of the late 1800s allowed rapid expansion of exploitation for an ever-growing market. The demand for fish increased as the population grew and as spreading railroad lines made possible rapid transportation of fish to distant cities. Fish processors developed better canning and refrigeration techniques, which led to massive exploitation of concentrated fish stocks. On the Sacramento River in California, for example, the chinook salmon fishery expanded from 2,500 cases of canned fish in 1874 to 200,000 cases in 1882, caught by 1,500 boats and processed in 21 canneries. Fishers responded by improving their own technology. Steam replaced sail as the power for fishing vessels in the 1880s, permitting larger vessels and more reliable fishing schedules. The effectiveness of fishing equipment advanced regularly, for example, with the development of the otter trawl and the deep trap net in the late 1860s.

Concern for the decline of fisheries was building throughout this period. Along with the societywide interest in natural resources, fishers, scientists, and government officials began to doubt that fisheries were inexhaustible, in either marine or fresh waters. The Sacramento River salmon, for example, responded to heavy exploitation with regularly declining spawning runs; after 1882, the fishery declined as well. The United States institutionalized its concern in 1871 by creating the U.S. Fish Commission, for the express purpose of investigating the decline in commercial fisheries (Allard 1978). Similar actions were underway around the world, with national investigations in Russia, Germany, and England during this time (Nielsen 1976). Provinces and states began establishing their own fisheries agencies, beginning with Quebec in 1857, New York in 1868, and California in 1870. A small group of fish culturists in 1870 formed the American Fish Culturists' Association (now the American Fisheries Society) to promote fish culture as a cure for the widespread destruction of fisheries (Thompson 1970).

Although the modern techniques of fisheries management were used in rudimentary form during the late 1800s, most people believed that the future belonged to fish culture. In its second year of operation, the U.S. Fish Commission was given the added task of raising fish and distributing them throughout the United States for the promotion of commercial fisheries. Spencer Baird, the first U.S. Fish Commissioner, hired Livingston Stone in 1872 to carry American shad to the West Coast and, once there, to establish salmon culture stations for Pacific salmon eggs to be transported back east.

These early fish culture operations were enormously successful, and they spawned an era of unrestrained enthusiasm for raising and stocking fish throughout the continent (Regier and Applegate 1972). Striped bass were shipped from New Jersey to California in 1879, developing into a commercial fishery by 1889 (Craig 1930). Rainbow trout were first distributed into the eastern United States in 1880, and by 1896 many states east of the Rocky Mountains boasted selfsustaining rainbow trout populations (Wood 1953). European introductions were equally popular, with common carp imported from Germany in 1877 and brown trout in 1883. Missouri built its first hatchery in 1881, for the production and distribution of common carp (Callison 1981). The belief that fish culture could sustain commercial fisheries was the established position of the U.S. government throughout this period. A 1900 law, for example, required fishers in Alaska to establish sockeye salmon hatcheries on every river they fished (Roppel 1982). The technical improvements of the time also enhanced fish stocking activities. Long-distance movements of fish were accomplished in specialized railway cars, first used in 1873 and finally retired in 1947 (Leonard, no date). The initial transfers used milk cans to carry eggs, but soon the cars were outfitted with specialized fish-holding cans (Fernow pails) and later with tanks carrying ice and aeration devices (Figure 1.4).

Fish rescue operations provided another major source of fish for stocking and redistribution, especially in the midwestern United States. Annual flooding of the Mississippi River and its major tributaries stranded multitudes of fish as the waters receded. State officials reasoned that the death of these fish reduced fish abundance in the river. In 1876, therefore, Iowa began an annual program of rescuing fish, a program that continued until 1930 (Carlander 1954). Most of the fish were returned to the main channel of the river, but game fishes were kept for distribution into other waters throughout the region. The work was difficult—

Figure 1.4 Railroad cars, like this one, were specially adapted to transport fish and fish eggs across the continent. Long-distance movements by rail continued well into the 1940s. (Photograph courtesy of the U.S. Fish and Wildlife Service.)

weeks of seining in shallow, silt-bottomed ponds in hot weather. The death of one young worker after several years of fish rescue work prompted an infamous epitaph on his Iowa gravestone (Figure 1.5).

Fishing regulations proliferated during the latter 19th Century, generally in response to the declining fisheries and the desire to protect stocked fishes. Most laws regulated either the seasons or methods of commercial fishing. Closed seasons were implemented to protect spawning fish, under the implicit belief that spawners were needed to assure future yields. Regulation of fish-catching methods, however, was usually politically motivated, designed to restrict the effectiveness of some fishers while enhancing the effectiveness of others (Nielsen 1976). Such regulations were usually ineffective for fish conservation, riddled with loopholes and contradictions. Early commercial fishing laws also provided no means for enforcement, other than the efforts of ordinary police. The few regulations that did exist for sport fishing were even less likely to be enforced. Among the first law enforcement officers specifically hired for natural resource work were those in California, authorized in 1876, and in New York, authorized in 1883.

At the end of the 19th Century, fisheries were emerging from the worst conditions in history. Fish populations were badly depleted throughout the developed world. Environmental degradation tied to industrial frenzy, and overfishing tied to insatiable demand, had affected freshwater and anadromous fisheries to the point that some commercially exploited fishes were extinct, and many were economically useless. But a new attitude was pervading the continent. The respect for nature that was intrinsic in the North American heritage was being directed towards fisheries. Unlimited exploitation had been benign on the frontier, but people recognized that the closure of the frontier necessitated a conservative attitude. Attempts to control exploitation were largely political and insufficient, but the scientific basis for more rational management was beginning to develop. The new century would harbor a new science and new approaches to the use and conservation of natural resources.

Figure 1.5 The dreadful truth about fisheries work is shown on this gravestone standing in Manchester, Iowa. The young man who died was not really a fish hatchery worker, but a member of fish rescue crews that seined fish along the Mississippi River.

1.5 THE SCIENTIFIC MANAGEMENT OF FISHERIES

Most historians mark the beginning of the North American conservation movement with the Governor's Conference of 1908. United States President Theodore Roosevelt invited governors of all states and territories, federal legislators, the Supreme Court, cabinet officers, and selected authorities to the White House to discuss the "conservation of natural resources." The conference serves well as a milestone depicting the transition of the conservation movement from a collection of separate ideas and actions into a national priority.

Natural resource conservation received this attention partly because of the acknowledged need for better management, but mostly because it was swept along in a new national outlook. The 20th Century began with an obsession for efficiency in the conduct of human affairs—the progressive movement. Born of the waste and corruption of the concluding century, the progressive movement presumed that direct manipulation of human affairs, often aided by the best scientific expertise, could produce a better society—even a perfect one. The philosophy was the force behind the work of Gifford Pinchot, Chief of the U.S.

Forest Service under President Roosevelt. Known in conservation as the founder of modern forestry, Pinchot was a zealous proponent of efficiency in all aspects of resource use and societal development—so zealous that his forceful rhetoric has been called "the gospel of efficiency." He popularized the notion that natural resources should be used for the long term, conserving their capacity to produce human value indefinitely. The other part of that philosophy, however, held that fully using the annual production of renewable resources was an obligation in the service of the present human population, as he observes here for coal:

> The first principle of conservation is development, the use of the natural resources now existing on this continent for the benefit of the people who live here now. There may be just as much waste in neglecting the development and use of certain natural resources as there is in their destruction. We have a limited supply of coal, and only a limited supply. Whether it is to last for a hundred or a hundred and fifty or a thousand years, the coal is limited in amount, unless through geological changes which we shall not live to see, there will never be any more of it than there is now. But coal is in a sense the vital essence of our civilization. If it can be preserved, if the life of the mines can be extended, if by preventing waste there can be more coal left in this country after we of this generation have made every needed use of this source of power, then we shall have deserved well of our descendants. (Pinchot 1910)

Fish populations under the progressive philosophy served the primary purpose of providing food and the secondary purpose of providing economic value, much like agricultural crops (Bower 1910). For example, at Roosevelt's 1908 Governors Conference, only two speakers mentioned fisheries, both extolling the virtues of commercial fisheries as thriving industries that provided a wholesome food supply (Smith 1971). The idea that natural resources were crops to be planted, managed, and harvested would later evolve into the founding principle of wildlife management (Leopold 1933), and would dominate the thinking of fisheries scientists for the first half of the 20th Century.

 The efficient use of fish populations became known as maximum sustainable yield, or MSY. During the early 20th Century, the MSY concept was developed independently several times. E. S. Russell, a British scientist studying marine fisheries, presented it most effectively in his classic book, *The Overfishing Problem* (Russell 1942). He related the present abundance of a fish population to additions via growth, recruitment, and immigration, and to losses via natural mortality, fishing mortality, and emigration. Russell's simple mathematics showed that the greatest long-term yield of fish was achieved by allowing small fish to grow before harvesting them. The "vital statistics" of a fish population—growth, recruitment, natural mortality, fishing mortality, immigration, and emigration rates—became the standard descriptors of fish population dynamics and remain so to this day (see Chapter 5).

Logical arguments like Russell's and political philosophies like Pinchot's were powerful, but they also needed a strong scientific basis. The science of ecology answered that need. Ecology, which began in earnest in the late 1800s, developed rapidly during the 1920s and 1930s (McIntosh 1976). Much of the early work in ecology centered on aquatic environments, relating the lives of organisms to the physical and chemical characteristics of the habitat. Significant in that early work was Stephen A. Forbes, Director of the Illinois State Laboratory of National History in the 1880s. His classic paper, "The Lake as a Microcosm," explains the attraction of aquatic ecosystems and the need for comprehensive information:

> It forms a little world within itself—a microcosm within which all the elemental forces are at work and the play of life goes in full, but on so small a scale as to bring it easily within the mental grasp. Nowhere can one see more clearly illustrated what may be called the *sensibility* of such an organic complex, expressed by the fact that whatever affects any species belonging to it, must have its influence of some sort upon the whole assemblage. He will thus be made to see the impossibility of studying completely any form out of relation to the other forms; the necessity for taking a comprehensive survey of the whole as a condition to a satisfactory understanding of any part. If one wishes to become acquainted with the black bass, for example, he will learn but little if he limits himself to that species. He must evidently study also the species upon which it depends for its existence, and the various conditions upon which *these* depend. He must likewise study the species with which it comes in competition, and the entire system of conditions affecting their prosperity; and by the time he has studied all these sufficiently he will find that he has run through the whole complicated mechanism of the aquatic life of the locality, both animal and vegetable, of which his species forms but a single element. (Forbes 1925)

1.5.1 Science for Fisheries Management

Ecology and fisheries science grew in mutualistic fashion, with ecology supplying hypotheses and principles and fisheries science supplying natural laboratories for testing them. Fisheries science grew exponentially in this exciting environment, with early scientists concentrating on gathering information in three areas critical to fisheries management.

First, scientists needed to describe and survey the fish and invertebrate animals in important waters before the management of fisheries could even be considered. Building on classical studies by naturalists such as Louis Agassiz, many prominent ichthyologists, including Tarleton Bean, David Starr Jordan, Barton Evermann, and J. R. Dymond conducted faunal surveys for state, provincial, and federal agencies in the early 1900s. Conservation agencies and universities carefully documented those surveys in elaborate reports complete with taxonomic keys and full-color drawings. The first volume of the *Roosevelt Wild Life Annals,* for example, published by the New York State College of Forestry, contains a 300-page report of the fishes and fishery of Oneida Lake (Figure 1.6; Adams and Hankinson 1928). This style of reporting was largely replaced by the next generation of zoologists, who prepared standardized taxonomic references for the continent or for large ecological regions. Works by Carl Hubbs and Karl Lagler concerning fishes of the Great Lakes (1941), and by Ronald Pennak concerning freshwater invertebrates of North America (1957), for example, became major references for biological surveyors.

The second type of information desired by fisheries scientists related to the physicochemical characteristics of important waters. Work in this area had developed a strong theoretical base through the work of European scientists like

Figure 1.6 Agency reports were the first textbooks on fisheries management and ecology. Drawings like this, from an intensive survey of Oneida Lake, New York, provided scientifically accurate records of fish species.

Francois Forel, who named the new field "limnology" in 1869 (Egerton 1976). Based on the ecological link between organisms and their environments and on the concern for the devastating pollution of the early 1900s, physical and chemical limnology became important components of fisheries management (Osburn 1933). Realizing the need to survey the conditions of lakes and streams, researchers blanketed the continent with meters, water bottles, and nets. The most famous field workers were Edward Birge and Chancey Juday, limnologists for the University of Wisconsin who surveyed waters throughout North America in the early 1900s (Beckel 1987). Their efforts were matched in Canada by Donald Rawson, who used his extensive field experience to create indices predicting fish production based on lake characteristics (Northcote and Larkin 1966). Large-scale physicochemical surveys are seldom conducted by fisheries managers today, and are now handled by specialists in water management. Nevertheless, the idea remains firmly planted in fisheries management that water quality determines the type and productivity of a fishery.

The third set of information desired by early fisheries scientists concerned fish life history and ecology. Scientists recognized that management of fish populations required knowledge of the critical elements in a fish's life and of the factors that altered the vital statistics of the population. Therefore, management agencies began to directly sponsor research on fish ecology (Larkin 1979). The Canadian government opened the Pacific Biological Station at Nanaimo, British Columbia, in 1908 to advance both marine and freshwater research. Similar laboratories were founded throughout Canada and the United States, providing an institutional home for fisheries research. Fisheries ecology grew in this environment so massively that it defies a brief synopsis. The list of outstanding scientists working in this area is a virtual who's who of fisheries (Box 1.2).

This growing wealth of facts and theories regularly confirmed the experiences of fishers and government officials that habitat destruction and unlimited fishing could have monumental impacts on fish populations. The corroboration of

Box 1.2   A Fisheries Hall of Fame

Fisheries science developed so rapidly during the mid-1900s that a complete history would fill many volumes. A partial story, however, may be told in the accomplishments of those scientists honored to receive the American Fisheries Society's Award of Excellence. (Biographical information was compiled by Yanin Walker of the American Fisheries Society.)

1969 - William E. Ricker   -   population dynamics theory and quantitative
                                   fisheries computation methods

1970 - Stanislas F. Snieszko   -   fish diseases, especially bacteriological, and fish
                                   health management

1971 - F. E. J. Fry   -   dynamics of exploited fish populations and
                                   environmental effects on fish

1972 - Ralph O. Hile   -   population dynamics of Great Lakes fishes and
                                   fisheries statistics

1973 - Carl L. Hubbs   -   taxonomy, distribution, ecology, life history,
                                   and evolution of fishes

1974 - Clarence M. Tarzwell   -   pollution biology, water quality requirements of
                                   fish, and toxicity testing

1975 - Robert R. Miller   -   taxonomy and evolution of freshwater fishes and
                                   protection of endangered fishes

1976 - Alfred W. H. Needler   -   marine mollusk culture and management of
                                   marine fisheries

1977 - Arthur D. Hasler   -   fish ecology, fish behavior, and environmental
                                   effects on fishes

1978 - Peter Doudoroff   -   toxicity testing of fishes and fish physiology and
                                   energetics

1979 - Kenneth D. Carlander   -   life history of freshwater fishes and fish
                                   population management

1980 - Reeve M. Bailey   -   taxonomy, distribution, ecology, and
                                   nomenclature of freshwater fishes

1981 - Gunnar Svardson   -   genetics of northern European freshwater fishes,
                                   especially whitefish

1982 - Douglas G. Chapman   -   population dynamics of fish and marine mammals

1983 - Peter A. Larkin   -   population dynamics of Pacific salmon and fish
                                   ecology

1984 - James L. McHugh   -   dynamics of exploited fish populations and
                                   management of North Atlantic fishes

1985 - Lauren R. Donaldson   -   culture of Pacific salmon, especially nutrition
                                   and genetics

1986 - William B. Scott   -   taxonomy, distribution, and ecology of
                                   freshwater and marine fishes

1987 - David Cushing   -   dynamics of exploited marine fishes and
                                   productivity of the sea

1988 - Clark Hubbs   -   fish ecology in relation to environmental
                                   conditions and endangered fish management

1989 - John Cairns, Jr.   -   environmental toxicology and biological
                                   monitoring

1990 - John A. Gulland   -   population dynamics of marine fishes

1991 - Kenneth Wolf   -   fish virology, cell culture, fish parasitology

1992 - Henry A. Regier   -   impacts of use and development on fishery
                                   resources and rehabilitation of damaged
                                   ecosystems

practical knowledge by the new sciences of ecology and limnology thus provoked the flowering of fisheries management to a form much as we see it today. Most of the techniques and approaches used today in fisheries management were developed during the first half of the 20th Century.

1.5.2 Management of Fish Populations

The exploration of population dynamics principles by Russell (1942) and others for marine fishes was readily transferred to inland fisheries. Population dynamics became a fundamental concern to scientists and managers with developments occurring rapidly throughout North America. In the Pacific Northwest, concern for Pacific salmon fisheries fueled the pioneering work of William Ricker and Earle Forester at the Nanaimo laboratory. They demonstrated the individuality of salmon stocks in different rivers and the relationships between the abundance of spawning fish and the subsequent generation (Ricker's stock–recruitment relationship; Ricker 1958). Ricker also invented and published computational techniques for measuring the vital statistics of fish populations; Chapter 5 in this book relies directly on Ricker's 6 decades of population dynamics research.

In the southeastern United States, population dynamics was developing from an empirical approach. Homer Swingle, an extension entomologist at Auburn University in Alabama, began work in the 1930s on the possibility of raising fish effectively in farm ponds (Figure 1.7). His work expanded quickly into a continuing series of pond experiments relating the species, sizes, and densities of fish to the sustained quality of fishing. Swingle called his concept of sustainable quality fishing the "balance" of the fish population, monitored by a series of ratios comparing the relative abundance of predators and prey (Swingle 1950). Swingle's ideas have guided two generations of fisheries managers, and like Ricker's, form the basis for the practical fisheries statistics described in Chapters 6 and 20.

The mideastern and midwestern areas of the continent have contributed many management innovations. Because of the abundance of natural waters and the high density of human populations, fisheries have always been important recreational and commercial resources in these regions. The dominant fishes of these waters, however, often have not been the species desired by anglers. Managers in these areas, therefore, pioneered techniques for the removal or control of undesired fishes. Commercial netting to reduce the abundance of unwanted fishes was used commonly in many states. In Minnesota, large-scale netting began in the 1920s and continued for several decades (Johnson 1948). Commercial fishing is still used successfully today to control undesirable fishes in several situations, but it has never produced the large increases in desired fishes that were expected. Fisheries managers, therefore, turned increasingly to chemical fish poisons. First used in 1913, chemical poisons became common tools after World War II as a larger variety and supply of chemicals became available (Cummings 1975). The use of fish poisons is now greatly restricted, but a few poisons still are important parts of specific management plans (see Chapter 14).

Fish stocking changed in important ways in the mid-1900s. Stocking remained a favorite tool for fisheries managers, but the promiscuous introductions of the late 1800s were supplanted by a more scientific, or at least conscientious, approach. Increasing knowledge of fish ecology and an increasing body of stocking experience allowed fisheries managers to choose stocking locations

Figure 1.7 Homer Swingle pioneered the empirical approach to fisheries management. Working with first a few ponds, and later with hundreds, Swingle and his colleagues at Auburn University discovered combinations of species, sizes, and densities of fish that provided quality pond fishing.

where growth and survival were probable (Wood 1953). This knowledge was first employed systematically in 1927 by George Embody, a New York fishery scientist, who published tables relating trout stocking densities to stream characteristics. Stocking eggs and larval fish was largely abandoned in the 1920s; instead, fish were raised in hatcheries for extended periods, to be stocked when larger and more likely to survive. This idea was rapidly expanded to include stocking fish at a size desired by anglers, creating the "put-and-take" fisheries so common today (Chapter 13).

1.5.3 Management of Habitats

Since earliest times, fisheries managers have modified aquatic habitats. Among the earliest laws governing fisheries in Europe and the United States were those that restricted dams. Dams that blocked entire rivers in order to catch fish, called "fixed engines" in England, were first prohibited in the Magna Carta (Nielsen 1976). Interest in habitat improvements in North America followed closely on the investigation of life history requirements of fishes and the development of the ecological concept of "carrying capacity." Once these habitat-related limits on the abundance or size of fish were outlined, managers began developing ways to remove them.

Initial attempts to increase carrying capacity focused on the living space available for fish. In 1938 Carl Hubbs and Ralph Eschmeyer published their handbook, *The Improvement of Lakes for Fishing*, the first comprehensive description of lake habitat management (Hubbs and Eschmeyer 1938). They emphasized the addition of artificial structures, especially brush piles, as a way to increase the standing crop of fishes. Managers had long been using analogous measures in coldwater streams, placing various low dams and shoreline obstructions where they would diversify the depth and velocity of water (Cooper 1970). Although the effectiveness of such structures has been evaluated only rarely, the manipulation of habitat diversity by adding structures is enormously popular, and, therefore, continuously employed (see Chapters 9 and 10).

Where natural lakes were scarce, the preferred method for expanding fisheries habitat was to build artificial lakes. Farm ponds and other small impoundments have always graced the human landscape, but the environmental problems of the early 20th Century greatly stimulated their construction. When the extended droughts of the 1920s settled in on the Great Plains, farming practices had already - produced a land stripped of its natural protection from erosion and with little water available for irrigation. In 1934, the U.S. Soil Erosion Service (now the Soil Conservation Service) was created to rectify the causes of the Dust Bowl. It developed a cost-sharing program for building farm ponds. The purpose was to raise the water table, but the ponds also provided fisheries habitat. As an incentive to build ponds, the U.S. Fish and Wildlife Service supplied farmers with free fish for stocking. Thus began nationwide programs of farm pond construction and fish stocking that have produced over 5 million small fishing ponds in the United States. Agency-sponsored fish stocking is greatly reduced today, but small impoundments are still primary opportunities for fishing (see Chapter 20).

Reservoir construction had similar beginnings. Fewer than 100 reservoirs (larger than 200 hectares) existed in North America in 1900. Under the impetus of burgeoning human development, the U.S. government accelerated the construction of large reservoirs in the 1930s (Figure 1.8). The development of the Tennessee River valley, for example, began in 1933 as part of a federal program to revitalize the region's economic and social well-being through natural resource conservation. By 1941, the Tennessee Valley Authority had built seven major impoundments, providing navigation, flood control, hydroelectric energy—and fishing (TVA 1983). More than 1,300 large reservoirs had been built in the United States by 1970, mostly in the Southeast and upper Midwest (Jenkins 1970). As the most desirable sites for reservoir construction have been used, the rate of construction has slowed markedly, and few new reservoirs are being built today. Although fishing was generally a secondary objective for most reservoir construction, the nearly 4 million hectares in major impoundments are intensively managed for fisheries today (Chapter 21).

1.5.4 Management of Fisheries Users

Early in the 20th Century, the importance of recreational fisheries began to grow. This expansion caused demands for regulations on the competing harvests of commercial fishers. Exhortations for governmental restrictions on environmental degradation and on overfishing became common after World War I. In

Figure 1.8 Reservoir construction has provided millions of hectares of fishing water in North America. Most construction occurred between 1930 and 1960, when agencies like the Tennessee Valley Authority built hundreds of major impoundments. (Photograph courtesy of the Tennessee Valley Authority.)

response to massive depletion of inland fish stocks for food during the war, the U.S. government passed the Black Bass Act of 1926. This act, which prohibited interstate movement of bass taken in violation of state laws, effectively eliminated most large commercial markets for freshwater predatory fish in the United States. Since that time, commercial fishing in freshwaters has been regularly reduced and is now a minor part of most inland fisheries management programs.

Regulation of recreational fishing, in contrast, has fluctuated widely. The earlier regulatory emphasis on closed seasons and areas was supplemented with new techniques based on studies of fish population dynamics. Once managers began to understand the value of higher growth rates and lower mortality rates, they tried to improve them through regulations. Highly restrictive fishing regulations, including minimum size limits, fishing equipment restrictions, and daily catch (creel) limits, were implemented broadly and uniformly by state agencies (Redmond 1986). In the early 1940s, however, the regulatory pendulum swung in the other direction, as several new studies showed the inappropriateness of restrictive regulations in many waters. State agencies responded by liberalizing regulations, again in a largely uniform manner. The pendulum has reversed direction again, and stricter fishing regulations have returned. Because regulations represent one of the few sure methods of decisively manipulating fisheries, they are popular among managers (see Chapter 16).

The decades from 1900 to 1950 might well be called the "Golden Age of Fisheries Management." The continent had discovered the value of recreational fisheries and had backed the research and management to improve them. Scientific information grew rapidly and, when combined with the experience of previous years, provided fertile ground for the proliferation of management techniques. Maximum sustainable yield was firmly established as the goal for fisheries management, whether for commercial or recreational fisheries. Managers stocked and poisoned fish, they built and modified waterbodies, and they regulated fish harvest with the single aim of providing the greatest sustained quantity of fish. With a well-developed kit of tools, the significant changes after 1950 would come in the purposes, rather than the techniques, of fisheries management.

1.6 MODERN FISHERIES MANAGEMENT

The modern era of fisheries management began after the end of World War II. State and provincial agencies grew rapidly by enlarging management, fish culture, and law enforcement staffs. Spurred on by the trend for higher education and by the increasing use of fisheries resources for recreation, governments invested generously in education, research, and management. Along with this investment came an outpouring of data and ideas that has allowed fisheries management to develop as the multifaceted profession it is today.

Perhaps the greatest stimulus to inland fisheries management in the United States was the passage in 1950 of the Federal Aid in Sport Fish Restoration Act, popularly known as the Dingell–Johnson Act (D-J). Patterned after the earlier Pittman–Robertson Act for wildlife, the Dingell–Johnson Act created a 10% excise tax on specified fishing equipment. The tax was dispersed to state fisheries agencies to support the creation and improvement of recreational fisheries. From its inception through 1985, the D-J program had dispersed over $480 million for fisheries development and research. In 1985, the Dingell–Johnson program was expanded significantly, increasing the range of items taxed, adding marine fuel taxes to the program, and authorizing development of marine, as well as freshwater, recreational fisheries. This important legislation, known as the Wallop–Breaux act, has more than doubled the annual funds available, with $332 million dispersed in 1992 alone.

The initial decades of the 20th Century had been dominated by the concept of fisheries as crops, with the single objective of achieving the highest physical yield, or MSY. Since 1950, the focus on MSY has been challenged repeatedly, with a continual expansion of the possible objectives for management. This has not been a sequential process, with each new idea following neatly behind the previous. For most modern viewpoints, the antecedents extend well back into the last century. Their rise to widespread acceptance, however, has followed a somewhat identifiable path.

The initial challenge to MSY was the idea that producing physical yields was really secondary to the more universal objective of producing economic value. Biologists had learned that physical yield could be maximized (theoretically) by regulating the total fish catch via a quota or similar means. Economists now reasoned that if the effort used to catch those fish was so great that the cost equalled the revenue, no economic value (profit) would exist. This idea was expressed eloquently by Michael Graham, a leading British fisheries scientist, in his classic essay *The Fish Gate*: "Fisheries that are unlimited become unprofitable" (Graham 1943). The result of such economic inefficiency was a different type of common property dilemma: although the natural resource was preserved, the dominant human benefit—economic gain—would be lost if fishing effort was unrestricted.

In the 1950s, fisheries economists began to point out that MSY should be replaced by the concept of maximizing profit, alias maximizing net economic revenue (MNER). Management for MNER required a different approach. Scientists such as Milner Schaefer, who analyzed the Pacific tuna fisheries, and Anthony Scott, working in British Columbia, stated that not only must the fish harvest be regulated, but also the effort used to acquire that harvest must be limited. The fundamental economic premise was that the fish harvest must be

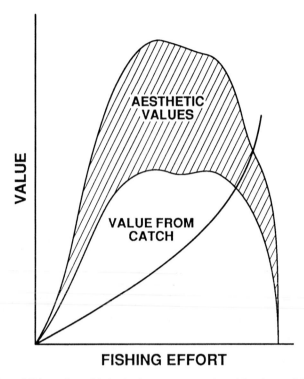

Figure 1.9   The addition of nonbiological concerns to the objectives of fisheries management is shown in this figure from James McFadden's keynote address to the American
Fisheries Society in 1968.

allocated to particular fishing units—individual fishers, corporations, nations—in
order to keep the fishing effort low. The concept, which is usually called "limited
entry," grants some persons the right to fish while excluding everyone else. This
economic principle is not universally accepted, but it is being increasingly used for
important commercial fisheries and is the basis for the 200-mile limit that most
coastal nations now claim over their marine fisheries.

While economic concerns were challenging MSY in commercial fisheries, MSY
was also being challenged as a sufficient objective for recreational fisheries. The
quality of recreational fishing had been measured traditionally as the number and
size of harvested fish. Nevertheless, anglers had always admitted that other parts
of the fishing experience were at least as important as the catch itself. People
value companionship and pleasant surroundings in their fishing experiences, and
their preference for catch may vary from one large fish to many small fish. The
idea that such sociological concerns should be part of fisheries management
gained popularity in the 1960s as public opinion became more important in
directing government decisions. The addition of aesthetic values to the relationship between fishing effort and fish harvest was formalized by James McFadden
(1969) in a modern classic of fisheries literature (Figure 1.9). Since then, the
development of socioeconomic principles for fisheries management has been a
high priority for management agencies (see Chapter 7).

The third major addition to the objectives of fisheries management was the result of continuing advances in ecological science. Because fisheries are the primary human uses of the productivity of aquatic ecosystems, ecological research has continuously provided ideas for the better understanding of fisheries management. In the 1970s, for example, ecologists supplied the notion that the management of single fish species must be replaced by multiple-species management. The yield of predatory fish (e.g., largemouth bass) depends on the condition of the food base; when the food base is also exploited (e.g., bluegill), the pair of species must be managed together, not separately. The idea that fisheries must be thought of as communities, or at least as interacting groups of populations, has become firmly rooted through a series of symposia focusing on multispecies fisheries (e.g., Clepper 1979). Since the days of Forbes, ecology has been and will continue to be the basic science that contributes most to the understanding of fisheries management.

The accretion of additional concerns—economic, sociological, ecological—into the management of fisheries forced MSY off its throne in the 1970s (see Box 1.3). It has been replaced by a new guiding principle: management for optimum sustainable yield, or OSY. Optimum sustainable yield was formalized in a 1975 symposium (Roedel 1975) that assessed management from a variety of viewpoints. The basic tenets of OSY are that the appropriate goal for fisheries management includes a broad range of considerations (not just maximizing physical yield), and that a unique management goal exists for each fishery. Optimum sustainable yield thus greatly complicates fisheries management. Defining the OSY for a fishery is much more difficult than defining the MSY, because fishery-specific information is needed about biological, ecological, economic, and sociological aspects of fishery use. Optimum sustainable yield, however, is also much more realistic in that it recognizes the diversity of aquatic ecosystems and the diversity of human needs in relation to them.

The job of the fisheries manager has evolved rapidly. What the manager does today has been conditioned as much by the events of the past 2 decades as by the accumulated history of fisheries. The remainder of this chapter describes some of the substantive areas of fisheries work today, specifically as they relate to later chapters in the book.

1.6.1 Habitat Management

The growth of the environmental movement in the 1960s changed the world of the fisheries manager. United States law had required fisheries managers to assess the impacts of federal development projects since the 1930s (the Fish and Wildlife Coordination Act), but recent decades have greatly expanded and defined that general charge (see Chapter 4). Of greatest significance were the U.S. National Environmental Policy Act of 1969, which created the process of developing environmental impact statements, and the 1970 amendments to the Canadian Fisheries Act, which broadly prohibits harm to aquatic habitats (Pearse et al. 1985). Fisheries managers are now required to review thousands of government-sponsored developments each year, and many states have passed legislation requiring impact analysis for state-regulated projects.

The need to predict the relationship between habitat modification and fisheries impacts has given rise to entirely new specialties concerned with habitat analysis (see Chapter 9). Habitat analysis has been concentrated in the western half of the

---

Box 1.3   R.I.P. for MSY

Maximum sustainable yield (MSY) served the fisheries profession well for many decades. In the 1970s, however, the concept of optimum sustainable yield (OSY) supplanted MSY as the dominant fisheries objective. In the keynote address at the 1976 American Fisheries Society meeting, Peter Larkin (1977) laid MSY to a gentle rest:

> Whatever lies ahead in the development of new concepts for harvesting the resources of the world's fresh waters and oceans, it is certain that the concept of maximum sustained yield will alone not be sufficient. The concept has served an important service. It arrived just in time to curb many fisheries problems. To appreciate what MSY has done, we need only ask what the world's fisheries would have looked like today if the concept had not been developed and advocated with such fervor. The fish, I'm sure, would shudder to think of it. Like the hero of a western movie, MSY rode in off the range, caught the villains at their work, and established order of a sort. But it's now time for MSY to ride off into the sunset. The world today is too complex for the rough justice of a guy on a horse with a six-shooter. We urgently need the same kind of morality, but we also need much more sophistication.

Accordingly, I tender the following epitaph:

M. S. Y.
1930s–1970s
Here lies the concept, MSY.
It advocated yields too high,
And didn't spell out how to slice the pie.
We bury it with the best of wishes,
Especially on behalf of fishes.
We don't know yet what will take its place,
But hope it's as good for the human race.
R. I. P.

---

continent, where water supplies are low and where fisheries compete directly with consumptive uses of water. The most significant efforts have helped define the water flow conditions needed to sustain fish life (usually called "instream flow"). Scientists have created methods for expressing the relations between instream flow and habitat conditions and have determined the specific habitat conditions needed by dozens of fish species.

Environmental laws have also allowed fisheries managers to address the ecological principle that land uses throughout a watershed affect the well-being of fish (see Chapter 8). Fisheries managers have advanced well past the point of merely predicting changes. They now prescribe remedies for the anticipated problems in the design, construction, and operation of development projects (Swanson 1979). Mitigation, as this process is called, allows the manager to participate in the planning process for land and water projects. The manager can then help developers make good choices—eliminating avoidable damages by redesign or improved operation, and gaining suitable restitution for the damages that are unavoidable.

The legal stature of fisheries, along with their increasing economic and social value, has also allowed for the small- and large-scale rehabilitation of habitats. Stream reconstruction in the West and wetland reconstruction in the East are widely practiced by public agencies and private companies. Earlier techniques of stream improvement have been augmented so that now entirely new stream channels can be constructed, kilometers of stream bottoms can be removed, washed, and replaced, and existing dams and other structures can be removed (see Chapter 9). Although lake habitat is not as easily remodeled as stream habitat, the methods available far exceed those of previous decades. New reservoirs are constructed with outlets designed to enhance fisheries, and a seemingly endless supply of devices is available for improving water quality and fisheries habitat (see Chapter 10).

1.6.2 Organism Management

As ecological knowledge has increased, the breadth of interest in management of fish and other aquatic organisms has similarly increased. The concentration on a limited number of predatory fish has been replaced by the recognition that the entire aquatic community is valuable and needs management attention.

Endangered and rare species are now managed primarily by fisheries and wildlife professionals. Although the addition of this responsibility to their more traditional roles has been resisted by some managers, fisheries agencies are the natural homes for endangered species concerns. Fisheries managers have the skills and experience to manipulate ecosystems to enhance endangered species, just as they manipulate other fisheries organisms. Today the management of endangered species is recapitulating the entire course of fisheries management, rapidly moving through the stages of life history descriptions, distribution studies, and habitat assessments to the stage of prescribing how to remove species from the danger of extinction (see Chapter 15).

Management interest has also broadened to incorporate the nonharvested species of a typical aquatic environment. Attention is now being directed toward smaller fishes that provide the food for predators and toward competing fishes that reduce the reproduction, growth, and survival of desired predators (see Chapters 12 and 13). Concern has also developed for other species that may not enter directly or indirectly into a fishery. Conservation biology, as this interest is called, concentrates on managing species that are seldom abundant but are broadly distributed. Conservation of such resources, the aquatic equivalent of songbirds, is stronger in Europe than in North America (Maitland 1974), but significant interest is developing here. For example, the preservation and reconstruction of native fish faunas is now a priority in the management of U.S. national parks.

Aquaculture remains one of the most valuable management techniques in fisheries today (Stroud 1986). It still serves traditional purposes of stocking fish for angling, but it also serves faunal restoration efforts and endangered species management. As the use of natural waters for commercial fishing has declined, aquaculture has developed rapidly as the source for food fish. Without question, aquaculture will continually increase in importance in fisheries management. Scientists and managers, therefore, will be forced to confront many significant issues related to aquaculture, including genetic engineering, the effects of hatchery-reared fish on wild stocks, and the use of public waters for commercial fish farming.

1.6.3 People Management

The human component in fisheries management, so long de-emphasized, is now a major element in all management activities. Optimum sustainable yield demands that the needs and desires of fisheries users be discovered and incorporated into management plans. Consequently, creel surveys and attitude assessments are now continuing activities of state and federal agencies. Methods for assessing user demands and for comparing the relative value of various fisheries are still rudimentary, but they are evolving rapidly in the face of both professional and legal demands for information (see Chapter 6).

Now that fisheries agencies know more about the demand for fisheries, they have changed their notions about how to serve society. Fisheries programs now typically include urban fisheries and fisheries accessible to handicapped anglers. Aquatic education, which opens fishing to people without a family tradition of angling, is increasingly important in fisheries agencies. A more representative balance also is being achieved in the allocation of fish between commercial and recreational fisheries. Comanagement of fisheries, by which traditional regulatory agencies relinquish some authority to local governments, Native American groups, and conservation organizations, is beginning and seems highly likely to increase in the future. The utility of socioeconomic information to guide management will continue to expand as techniques improve and as the success of incorporating such information is established.

As fisheries management has become more complex—more quantitative, more political, more scientifically sophisticated, and more diverse—the need for effective planning of fisheries programs has become apparent. Beginning in the 1970s, the concept of comprehensive planning has become increasingly popular as a way to anticipate the future. An increasing number of state, provincial, and federal agencies are functioning under strategic plans (which describe what could be done) and operational plans (which describe how to do it). The emphasis is now placed on setting specific objectives for fisheries management and monitoring the achievement of those objectives (see Chapter 2).

1.7 CONCLUSION

And just what is this profession we call "fisheries management?" Viewpoints about management are as diverse as the collection of human activities that we call "fisheries." As the profession has evolved and radiated in recent decades, the precepts that fisheries management was applied biology, applied ecology, or even applied economics have all proven too restrictive. A dictionary definition is not really important; a principle for guiding the management of fisheries, however, is necessary. Most fisheries professionals would agree that the principle of fisheries management is to provide people with a sustained, high, and ever-increasing benefit from their use of living aquatic resources. In the pursuit of that principle, fisheries managers manipulate all aspects of the natural and human ecosystem. Where people and water meet, fisheries exist; where people and water could meet, potential fisheries exist; wherever fisheries, real or potential, exist, fisheries management can make them better.

1.8 REFERENCES

Adams, C. C., and T. L. Hankinson. 1928. The ecology and economics of Oneida Lake fish. Roosevelt Wild Life Annals 1(34):235–548, Roosevelt Wild Life Forest Experiment Station, Syracuse, New York.

Allard, D. C., Jr. 1978. Spencer Fullerton Baird and the U.S. Fish Commission. Arno Press, New York.

Bean, M. J. 1977. The evolution of national wildlife law. U.S. Council on Environmental Quality, Washington, D.C.

Beckel, A. L. 1987. Breaking new waters. A century of limnology at the University of Wisconsin. Transactions of the Wisconsin Academy of Sciences, Arts and Letters, Special Issue, Madison, Wisconsin.

Beltran, E. 1972. Programs for renewable natural resources in Mexico. Transactions of the North American Wildlife and Natural Resources Conference 37:4–18.

Bower, S. 1910. Fishery conservation. Transactions of the American Fisheries Society 40:95–100.

Busiahn, T. R. 1985. An introduction to native peoples' fisheries issues in North America. Fisheries (Bethesda) 9(5):8–11.

Callison, C. 1981. Men and wildlife in Missouri. Missouri Department of Conservation, Jefferson City.

Carlander, H. B. 1954. History of fish and fishing in the upper Mississippi River. Upper Mississippi River Conservation Committee, Rock Island, Illinois.

Clepper, H., editor. 1979. Predator–prey systems in fisheries management. Sport Fishing Institute, Washington, D.C.

Cooper, E. L. 1970. Management of trout streams. American Fisheries Society Special Publication 7:153–162.

Cox, T. R. 1985. Americans and their forests. Romanticism, progress, and science in the late Nineteenth Century. Journal of Forest History 29:156–168.

Craig, J. A. 1930. An analysis of the catch statistics of the striped bass (*Roccus lineatus*) fishery of California. California Department of Fish and Game, Sacramento, California.

Cummings, K. B. 1975. History of fish toxicants in the United States. Pages 5–21 *in* P. H. Eschmeyer, editor. Rehabilitation of fish populations with toxicants: a symposium. American Fisheries Society, North Central Division, Special Publication 4, Bethesda, Maryland.

Egerton, F. N. 1976. Ecological studies and observations before 1900. Pages 311–351 *in* B. J. Taylor and T. J. White, editors. Issues and ideas in America. University of Oklahoma Press, Stillwater.

Forbes, S. A. 1925. The lake as a microcosm. Illinois Natural History Survey Bulletin 15:527–550.

Fry, F. E. J., and V. Legendre. 1966. Ontario and Quebec. Pages 487–519 *in* D. G. Frey, editor. Limnology in North America. University of Wisconsin Press, Madison.

Graham, M. 1943. The fish gate. Faber and Faber, London.

Hubbs, C. L., and R. W. Eschmeyer. 1938. The improvement of lakes for fishing. Michigan Department of Conservation, Institute for Fisheries Research, Bulletin 2, Lansing.

Jenkins, R. M. 1970. Reservoir fish management. American Fisheries Society Special Publication 7:173–182.

Johnson, R. E. 1948. Maintenance of natural population balance. Proceedings, Convention of the International Association of Game, Fish and Conservation Commissioners 38:35–42.

Kawashima, Y., and R. Tone. 1983. Environmental policy in early America: a survey of colonial statutes. Journal of Forest History 27:168–179.

Larkin, P. A. 1977. An epitaph for the concept of maximum sustained yield. Transactions of the American Fisheries Society 106:1–11.

Larkin, P. A. 1979. Maybe you can't get there from here: foreshortened history of research in relation to management of Pacific salmon. Journal of the Fisheries Research Board of Canada 38:98–106.

Leonard, J. R. No date. The fish car era. U.S. Fish and Wildlife Service, Washington, D.C.

Leopold, A. 1933. Game management. Scribner, New York.

Maitland, P. S. 1974. The conservation of freshwater fishes in the British Isles. Biological Conservation 7(1):7–14.

McEvoy, A. F. 1986. The fisherman's problem. Ecology and law in California fisheries 1850–1980. Cambridge University Press, Cambridge, United Kingdom.

McFadden, J. G. 1969. Trends in freshwater sport fisheries of North America. Transactions of the American Fisheries Society 98:136–150.

McIntosh, R. P. 1976. Ecology since 1900. Pages 353–372 in B. J. Taylor and T. J. White, editors. Issues and ideas in America. University of Oklahoma Press, Stillwater.

Nash, R. 1987. Aldo Leopold's intellectual heritage. Pages 63–90 in J. B. Callicott, editor. Companion to A Sand County Almanac. University of Wisconsin Press, Madison.

Nielsen, L. A. 1976. The evaluation of fisheries management philosophy. U.S. National Marine Fisheries Service Marine Fisheries Review 38(12):15–23.

Northcote, T. G., and P. A. Larkin. 1966. Western Canada. Pages 451–486 in D. G. Frey, editor. Limnology in North America. University of Wisconsin Press, Madison.

Osburn, R. C. 1933. Some important principles of fish conservation. Transactions of the American Fisheries Society 63:91–97.

Pearse, P. H., F. Bertrand, and J. W. MacLaren. 1985. Currents of change. Environment Canada, Inquiry on Federal Water Policy, Final Report, Ottawa.

Pinchot, G. 1910. The fight for conservation. Doubleday, New York.

Redmond, L. C. 1986. The history and development of warmwater fish harvest regulations. Pages 186–195 in G. E. Hall and M. J. Van Den Avyle, editors. Reservoir fisheries management: strategies for the 80's. American Fisheries Society, Southern Division, Reservoir Committee, Bethesda, Maryland.

Regier, H. A., and V. C. Applegate. 1972. Historical review of the management approach to exploitation and introduction in SCOL lakes. Journal of the Fisheries Research Board of Canada 29:683–692.

Ricker, W. E. 1958. Handbook of computations for biological statistics of fish populations. Fisheries Research Board of Canada Bulletin 119.

Roedel, P. M., editor. 1975. Optimum sustainable yield as a concept in fisheries management. American Fisheries Society Special Publication 9.

Roppel, P. 1982. Alaska's salmon hatcheries, 1891–1959. Alaska Historical Commission, Studies in History 20, Juneau.

Russell, E. S. 1942. The overfishing problem. Cambridge University Press, Cambridge, UK.

Seaman, E. A. 1988. Silent letters released. Seven decades of one life. Vantage Press, New York.

Smith, F. E. 1971. Land and water, 1492–1900. Chelsea House Publications, New York.

Stroud, R. H. 1966. Lakes, streams, and other inland waters. Pages 57–73 in H. E. Clepper, editor. Origins of American conservation. Ronald Press, New York.

Stroud, R. H., editor. 1986. Fish culture in fisheries management. American Fisheries Society, Fish Culture Section and Fisheries Management Section, Bethesda, Maryland.

Swanson, G. A., technical coordinator. 1979. The mitigation symposium: national workshop on mitigating losses of fish and wildlife habitats. U.S. Forest Service General Technical Report RM-65.

Swanton, J. R. 1946. The Indians of the southeastern United States. Smithsonian Institution, Bureau of American Ethnology, Bulletin 137, Washington, D.C.

Swingle, H. S. 1950. Relationships and dynamics of balanced and unbalanced fish populations. Alabama Agricultural Experiment Station, Auburn University, Bulletin 274.

TVA (Tennessee Valley Authority). 1983. The first fifty years: changed land, changed lives. TVA, Knoxville.

Thompson, P. C. 1974. Institutional constraints in fisheries management. Journal of the Fisheries Research Board of Canada 31:1965–1981.

Thompson, P. E. 1970. The first fifty years—the exciting ones. American Fisheries Society Special Publication 7:1–12.

Wharton, J. 1957. The bounty of the Chesapeake: fishing in colonial Virginia. University Press of Virginia, Charlottesville.

Whitaker, H. 1892. Early history of the fisheries of the Great Lakes. Transactions of the American Fisheries Society 21:163–179.

Wood, E. M. 1953. A century of American fish culture, 1853–1953. Progressive Fish-Culturist 15:147–162.

Chapter 2

The Process of Fisheries Management

CHARLES C. KRUEGER AND DANIEL J. DECKER

2.1 INTRODUCTION

Over the last century, fisheries management in North America has evolved institutionally, socially, and conceptually. Modern management incorporates not only specific concerns about fish and their habitats but also considers economics, aesthetics, user attitudes and desires, and the interests of the general public. Considerable efforts are being made by management agencies to integrate these concerns, both conceptually and in practice, into a comprehensive approach to management. For the purposes of this chapter, we define fisheries management as the integration of ecological, economic, political, and sociocultural information into decision making that results in the implementation of actions to achieve goals established for fish resources.

The process of management has evolved to depend on many types of information, often scientifically based, as an essential input to decision making. Effective management requires an information base about the society that uses fish as well as about fish and their habitats. Unfortunately, natural resource agencies do not have the comprehensive information needed to make perfect decisions for optimal management of a fishery. Yet every day decisions must be made by agencies based on whatever information is available. The best management approaches are those that permit careful observation of the effects of today's decisions so that new information can be acquired and used in future decisions.

Who manages fisheries or fish populations? Who makes the decisions and implements the actions? Many government agencies have authority to implement specific actions to provide for the wise use and conservation of aquatic resources. As a general rule, in most areas of North America either the state or provincial governments have direct authority over fishery resources. Some situations exist, however, where authority is vested in federal or tribal governments (see Chapter 4).

Institutional complexity surrounding fisheries management increases when natural resources are shared between countries along international boundaries. For example, state, provincial, tribal, and federal agencies each have authority over specific aspects of Great Lakes fisheries management (see Chapter 22). To facilitate cooperation among these agencies, the Great Lakes Fishery Commission was created in 1955 by an international treaty between the federal governments of Canada and the United States. The complications caused by multiple agencies

33

sharing responsibility for management increase when agencies have different
goals and different internal organizational structures. Coordination of each
agency's actions is essential if effective management is to occur. All cases demand
that managers effectively communicate and cooperate with each other (see
Chapter 3).

Fisheries management is not only the concern of governmental agencies; it
requires active involvement by members of the public. Historically, management
agencies focused solely on the desires of recreational anglers or commercial
fishers who sometimes formally organized into associations or lobby groups to
communicate their interests more effectively to fisheries agencies. In the last 3
decades more specialized groups (e.g., bass anglers and trout anglers) have
emerged from the traditional users, dramatically increasing the number of public
interest groups that interact with agencies. These groups often represent interests
distinct from each other, and sometimes their interests overlap. In addition, new
groups representing environmental interests have formed which are different from
those of the traditional fishing public. For example, nonconsumptive resource
users can have environmental values (e.g., preservationist) that may conflict with
those of sport and commercial users. Also, in some locations in North America,
Native Americans represent a group with special privileges as consumptive users,
often outside the jurisdiction of non-Indian management agencies. Thus, resource
agencies have the complicated and challenging task of managing fisheries for an
array of users who represent different sets of values. In some management
jurisdictions, public groups directly participate in agency decision making.
Regardless of the situation, a key to effective management is mediating conflict
and focusing on common interests among the groups (Fisher and Ury 1981).

What are the products of fisheries management? What are the goals that
management attempts to achieve? Is it the capture of a cutthroat trout from a
Rocky Mountain stream? The successful restoration of Atlantic salmon to streams
in the maritime provinces? The sustained commercial harvest of channel catfish
from the Mississippi River? Successful fisheries management can be responsible
for all of these events. In a sense, these events may be thought of as the products
of effective decision making in a management agency and reflect the achievement
of specific goals. Products of management may not only be the catching of fish but
more fundamentally can be thought of as simply the opportunity to go fishing.
People enjoy knowing such opportunities exist close to where they live even if
they can only rarely participate. In the case of restoring fish populations, such as
Atlantic salmon, many people can appreciate when a species has regained a
secure, stable status even if they never see or fish for it themselves. Goals and
products of management are closely related concepts that are derived from the
interests of the public. Goals are the focus of fisheries management activities.

Success in achieving fisheries management goals requires meeting the challenges posed by a deficient information base, as well as by the complexity of the
governmental and public sectors discussed earlier. Meeting the challenge depends
on the willingness of fisheries managers to adopt a comprehensive approach to
fisheries management. The concept used must be general enough to be applicable
to management of an urban fishery in New York City or a wilderness alpine lake
in the Cascade Mountains. The success of a fisheries professional depends on a
thorough understanding of the general management process, and also on an

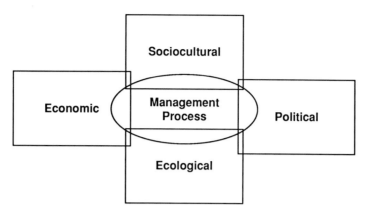

Figure 2.1 Representation of the management environment. The ecological, economic, political, and sociocultural components categorize the forces that shape management.

individual's ability to communicate effectively with other fisheries professionals and relevant public sectors (see Chapter 3; Pringle 1985).

In this chapter we present the concept of fisheries management as being a cyclic, self-correcting, goal-oriented process. The chapter is organized to describe the environment that management must function within, the team of individuals who conduct management, and the elements of the management process. The process presented here is idealized in that it does not necessarily portray fisheries management as it has occurred in the past; rather the portrayal provides a conceptual approach to fisheries management of the future.

The intent of the chapter is to describe the entire fisheries management process. Many elements of the process are similar to past descriptions of natural resource planning (e.g., Anderson and Hurley 1980; Crowe 1983) and business management (e.g., Odiorne 1979), and closely follow an earlier general description of natural resource management (Krueger et al. 1986). However, the concepts described here include not only the steps that are often written down in a strategic plan for management of a fishery but also the actual implementation of such a plan.

2.2 MANAGEMENT ENVIRONMENT: THE FORCES THAT SHAPE MANAGEMENT

The combination of ecological, political, economic, and sociocultural factors that influence the management process can be thought of as the special "environment" where fisheries management is conducted (Figure 2.1). The ecological characteristics of this environment control potential fish production and yield from a water resource, such as the number of walleye caught from a lake. Similarly, the sociocultural aspects of this environment will limit the types of actions that a manager can take. For example, it may be socially unacceptable for state government to condemn private property for improvement of stream watersheds. Understanding each component of the management environment is essential for effective fisheries management.

The ecological component of the environment includes the ecosystem within which fish and other species of interest complete their life cycles. This component includes biotic and abiotic environmental factors, such as forage availability, abundance of predators, and quality and quantity of habitats required for different life stages. Manipulation of these environmental factors has been a traditional focus of fisheries management. However, the ecological component also includes genetic factors such as traits that affect migration, habitat preferences, and survival. Increased concern by agencies for the genetic aspects of fisheries management has resulted in recognition of exciting new opportunities to manage fishery resources more effectively (e.g., Billingsley 1981; Ryman and Utter 1987). Both environmental and genetic factors control the annual production of fish and other aquatic organisms within bodies of water. Managers who understand the potential for fish production can realistically assess public fishing opportunities and identify ways to enhance or protect fishery resources.

The economic component of the management environment includes the market and nonmarket forces that influence monetary valuation of fishery resources. The economic component has long been an important consideration in making decisions about commercial fisheries (e.g., Christy and Scott 1965; Clark 1976), and has been increasingly recognized as important in sport fisheries management (e.g., Talhelm 1987; Chapter 7). This component also includes the economics of fisheries agencies such as license fee structures (e.g., nonresident angler fees, trout stamps) and annual budgets available to conduct management activities. Agency budgets limit the magnitude and type of management actions that may be considered for implementation. Before making a decision a manager should study the entire economic component to analyze the costs and benefits to society of different fishery policies (Bishop 1987).

The political component of the management environment includes both established laws and policies of government and the personal values of government employees who enact, enforce, or interpret those laws and policies. Laws and policies of government are often well defined by constitutions, statutes, administrative codes, and published agency management plans. Laws and policies establish, in general terms, the mission of an agency. In contrast, the internal politics of government are constantly changing as employee responsibilities and personal attitudes shift, and as new employees are hired. Both aspects of the political component must be considered if current management practices are to change. For example, before a new program in environmental protection can be proposed, managers must first determine whether their agency has legal jurisdiction to enforce the program and whether the political climate in their agency or government is "right" for the new proposal. The political component includes the branches of government that surround a fisheries agency (executive, legislative, judicial), and these can have an important decision-making function for the agency. If fishery-related conflicts occur that are not resolved by the manager or the agency, then this component will determine a political solution to the conflict, sometimes through judicial process.

The sociocultural component includes traditions, values, norms, religions, and philosophies of the society within which management is conducted. This component provides the primary motivation for fisheries management to occur. Fisheries management is conducted only because the end products of the process are believed to have value to society. Study of this component will help managers understand, and possibly influence, fishery-related values held by the public at individual, community, and regional levels. These social values influence choice of management goals (Brown and Manfredo 1987), especially for typically high-value species such as walleye or lake trout (see Chapter 14). In some instances, the fisheries profession works to change or influence society's values about fishery resources. For example, the American Fisheries Society published a book about common carp to enhance the public's awareness of the potential value of this species as a fishery resource (Cooper 1987). As fisheries managers, our interpretation of ecological, economic, and political components is influenced strongly by our identification with particular values and philosophies within the sociocultural component (Nielsen 1985). Our personal values influence what we do and how we choose to do it. Understanding social values will help managers understand the needs and motivations of different public sectors, and help managers to communicate effectively with the public.

Separation of the management environment into ecological, economic, political, and sociocultural components provides a useful categorization of the forces that shape management. Clearly the components of the management environment described here are not independent of each other. For example, the economic and political components are derivatives of the sociocultural component but are important enough to management to warrant their separate consideration. Managers faced with a pressing fishery problem should first consider how each component of the management environment may affect the problem and its possible solutions, rather than plunging headlong into a favorite mode of thinking, such as pursuing an ecological solution. Often, factors other than ecology are more important. For example, decision making for fisheries management of chinook salmon stocks in the Pacific Northwest has been influenced more strongly by sociocultural and political factors than by ecological considerations, despite the presence of adequate biological data and analytical capabilities (Fraidenburg and Lincoln 1985).

2.3 MANAGEMENT TEAM APPROACH

Concerns related to each component of the management environment should be considered during agency decision making (Figure 2.2). Fisheries managers are required to predict the economic and cultural effects of proposed new regulations as well as the effects of regulations on fish populations. It is unlikely that every fisheries manager could be educated sufficiently in all disciplines needed to make such predictions. Instead, a team approach should be used to provide input to decision making (Harville 1985). The team should include individuals who have general fisheries science backgrounds, and those having a special focus in one or more of the core areas of ecology, economics, government, or social science. Every fisheries manager should have a sufficiently broad educational background to communicate effectively with team members trained in areas outside their specialty. The team should also include representatives from every agency sharing jurisdiction for a fishery. Leaders from some key public sectors often constitute important additional members of a management team. Public participation can produce especially helpful insights on potential economic and cultural effects that may result from new management programs. Representation on the

P.O. BOX 818
WILSON, NY 14172

**Toward Total Economic Valuation of
Great Lakes Fishery Resources**

RICHARD C. BISHOP

*Department of Agricultural Economics, University of Wisconsin
Madison, Wisconsin 53706, USA*

KEVIN J. BOYLE

*Department of Agricultural and Resource Economics, University of Maine
Orono, Maine 04469, USA*

MICHAEL P. WELSH

*Department of Agricultural Economics, Oklahoma State University
Stillwater, Oklahoma 74078, USA*

Figure 2.2   The fisheries management team includes members representing the different components of the management environment.

team from the disciplines of ecology, economics, government, and social sciences; agencies that have jurisdiction; and the public will help ensure proper guidance of the fisheries management process.

Based on our definition, management teams comprise individuals with a wide variety of backgrounds and who will likely approach the challenges of management quite differently. This diversity is the strength behind using the team approach to fisheries management, but can also be its downfall (Miller 1984). Team members must be tolerant and understanding of a diversity of viewpoints if the strengths of a team are to be channelled into innovative management strategies.

The team approach to fisheries management is valid at all scales of management, from local areas within a state or province to international fisheries. For example, a local manager may be solely responsible for a set of lakes and streams in the region. Managers in this case must recognize that the concerns of disciplines outside their own training are equally relevant and must be considered in decision making. Managers should not hesitate to seek advice from individuals who have the needed expertise. Such expertise may be found within agencies and universities, or may be located by referral through professional organizations such as the American Fisheries Society. At international levels of fisheries management, teams of individuals representing several disciplines may be assembled on location to work on specific management problems. Public involvement on teams may occur at all levels.

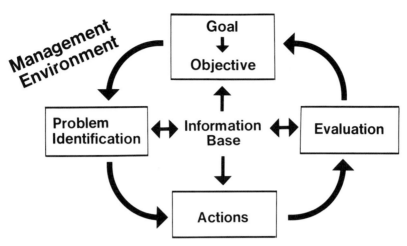

Figure 2.3 The steps of the fisheries management process as they cycle around the information base.

2.4 THE FISHERIES MANAGEMENT PROCESS

2.4.1 Process Summary

The management process has five steps: (1) choice of goals, (2) selection of objectives, (3) identification of problems, (4) implementation of actions, and (5) evaluation of actions (Figure 2.3). The steps of the process require the use of an information base that contains knowledge about each component of the management environment. Managers use and contribute to the information base as each step is executed. Goals provide long-term statements about what fisheries programs are to achieve. Objectives specify measurable expected outcomes that indicate achievement of goals, and state the date when achievement is expected. Problem identification determines what factors impede achievement of goals and objectives. Actions are the activities chosen and implemented to overcome the problems. Evaluation includes the measurement of the responses of the management environment to the actions implemented in terms of the levels specified in the objectives. Management program revision occurs based on assessment of the information gained in the evaluation step. The term management program refers to the application of the process to achieve goals related to a fishery (e.g., the management program to rehabilitate lake trout populations in Lake Ontario). After completion of the evaluation step, management then returns to the next most appropriate step in the process—to redefine goals, choose new objectives, identify new problems, and implement alternative actions. Below, we provide an actual example of this process, followed by descriptions of the conceptual details of each step.

2.4.2 Example of Management Process

Lake trout was originally one of the most important species native to the Great Lakes. Over the last 60 years, lake trout populations have declined sharply due to predation by the exotic sea lamprey, overfishing, and habitat degradation (see Chapter 22). In Lake Ontario, lake trout were presumed to be extinct around 1960.

In 1983, a formal plan for lake trout rehabilitation in Lake Ontario was established between fisheries agencies in the United States and Canada (Schneider et al. 1983). The plan documented the past and desired future direction of lake trout management. Implementation of this plan illustrates the fisheries management process as described above (Figure 2.3). All block quotations are from Schneider et al. 1983.

> The goal of the lake trout management program in Lake Ontario is: To rehabilitate the lake trout population of Lake Ontario such that the adult spawning stock(s) encompasses several year classes, sustains itself at a relatively stable level by natural reproduction, and produces a usable annual surplus.

This goal clearly and concisely states that the intent or purpose of this fishery program is to restore naturally reproducing lake trout populations to Lake Ontario in sufficient numbers and quality to permit a fishery.

An interim objective was then formulated to measure achievement of the rehabilitation goal stated above.

> *Interim Objective:* By the year 2000, demonstrate that rehabilitation is feasible by developing a Lake Ontario lake trout stock consisting of 0.5 to 1.0 million adult fish with adult females that average 7.5 years of age and produce 100,000 yearlings annually.

This objective specifies measurable expected outcomes that can later be evaluated and also defines a specific date (year 2000) when this objective is to be achieved. The development of this objective required knowledge from the information base about the productive capacity of Lake Ontario and an understanding of lake trout population dynamics. An ultimate objective was also established that defines and quantifies the final achievement of the goal.

> *Ultimate Objective:* To develop a lake trout population in Lake Ontario of 0.5 to 1.0 million adults that produce 2 to 3 million yearlings annually and provide 450,000 kg of usable surplus.

This objective is unusual because a date is not specified for its achievement. The agencies involved decided that a realistic date could not be specified until achievement of the interim objective had occurred. The time required to achieve the interim objective would provide an important piece of information for setting a completion date.

Several problems were identified that would prevent achievement of these objectives, for example:

- sea lamprey predation on lake trout
- over-exploitation of lake trout
- environmental degradation of Lake Ontario
- lack of optimal procedures for stocking lake trout.

Management actions were then developed and implemented to solve the problems. Strategies were identified (e.g., to reduce lake trout mortality) and then specific actions were chosen to accomplish the strategy. For example:

1. Sea lamprey control. As agents of the Great Lakes Fishery Commission, the Department of Fisheries and Oceans (Canada) and the U.S. Fish and Wildlife Service have maintained an active sea lamprey control program since 1971 primarily through the application of selective lampricides into Lake Ontario tributary streams (strategy: increase lake trout population by reduction of mortality).

2. Sport fishery regulation. The state of New York instituted more restrictive angler regulations in 1993 to reduce the fishing mortality of lake trout that have reached spawning size since being stocked into the lake (strategy: increase lake trout population by reduction of mortality).

3. Water quality legislation. Agencies in both countries continue to support legislation to reduce the pollution of Lake Ontario (strategy: protect and restore lake trout habitat).

4. Lake trout stocking. Approximately 1.5–2.0 million lake trout of several genetically different strains are marked and stocked into Lake Ontario annually. The marking program has permitted determination of the strains that survive best and identification of improved stocking procedures (strategy: increase lake trout population by improved stocking methods).

Evaluation has been ongoing since the program began; efforts include field studies and annual assessment meetings by the fisheries agencies of both countries. For example, the natural production of yearlings, and the size and age of the adult spawning stock are evaluated annually by trawling and gill netting conducted by the U.S. Fish and Wildlife Service, New York State Department of Environmental Conservation, and Ontario Ministry of Natural Resources. Assessment occurs by comparing the results of the field studies to the program's objectives during annual meetings of the Lake Ontario Committee organized by the Great Lakes Fishery Commission. This committee includes representatives of the management agencies that are directly involved in lake trout management. The management program is annually revised and fine-tuned based on discussions at the meetings.

2.4.3 Goals: Why We Manage Fisheries

The first step in management is to determine the goals that are to be achieved (Figure 2.3). Goals are long-term, broad statements of intent about management that define the purpose of management but not how the goals will be achieved. Goals are the mission statements that explain why agencies manage a particular fishery resource. Consequently, goals set the frame of reference in which management actions are chosen and subsequently implemented.

Goals that are specified within agency programs will often be multiple and hierarchically organized. The highest goal level may be determined legislatively. For example, Wisconsin state statutes establish that conservation law and the Department of Natural Resources were instituted "...to provide an adequate and flexible system for the protection, development, and use of forests, fish and game, lakes, streams, plant life, flowers and other outdoor resources..." (Wisconsin State Statutes 23.09). Under this statement is an administrative code which specifies that "...the goal of fish management is to provide opportunities for the optimum use and enjoyment of Wisconsin's aquatic resources, both sport and commercial..." (Wisconsin Administrative Code NR 1.01[2]). Although such legislative mandates are useful in establishing management authority for an agency, they are usually so general they only help to define goals lower in the hierarchy. Nested within these broad statements are goals that direct specific management programs. The goal statement described earlier for lake trout in Lake Ontario is an example. Goals at this level are more specific and give real direction to fisheries management programs.

Determination of goals is usually a difficult, complex process that must be done carefully because of the long-term effect that goals have on future direction of management programs. Both fisheries professionals and representatives from the public should be involved in discussions leading to the choice of goals. Senior-level agency administrators responsible for policy must always be included in the process of deciding goals (Crowe 1983). Their involvement will help them to understand a program so they can later provide the essential internal agency support needed for program implementation. Inclusion of regional or district managers in the process will help ensure that goals will be realistic with respect to the resource and the public. In some cases, fisheries scientists from other governmental agencies or universities may be helpful in providing specialized information or viewpoints.

The public should always be involved in the goal-determination process (Anderson and Hurley 1980). If programs are to succeed, they must reflect the values and interests of society (Yarbrough 1987). Participation ensures that management will address the public's interests. Participation also provides the opportunity for agencies to communicate new ideas and information which will help guide future program direction. Fisheries programs that include goals chosen with public input are more likely to be successful because the public will better understand program direction, and therefore politically support the management efforts of the agency. Management programs that result from public participation are politically defensible, especially when the goal-determination process has been documented and is available for review (e.g., meeting minutes). Forums for this type of public involvement include task forces, steering committees, advisory councils, review committees, surveys, and public hearings (Anderson and Hurley 1980).

A major challenge associated with public participation is effective communication between the fisheries agency and public groups (see Chapter 3). Various media, communication, and extension specialists can provide substantial assistance in bridging the gap between fisheries professionals and the public. Discussions about goals are often difficult to initiate with either agency personnel or the public because there is a natural tendency to discuss management actions (e.g., stock more fish) rather than to ensure consensus about what management is to achieve (Giles 1981). To counteract this tendency, the management process should be reviewed at the start of discussions to help participants conceptualize the significance of the definition of goals.

Administrators, regional biologists, resource scientists, and public groups will provide important linkages to the information base (Figure 2.3). Ecological information will help define which goals are biologically feasible. Economic information will help predict the effect on businesses, jobs, markets, and the standard of living which might result from the achievement of different goals. Political information will identify which goals are legally appropriate for the agency to pursue. Sociocultural information can identify which groups of people will be affected by programs designed around different goals or how different groups may react to a set of goals. Fair consideration should be given to age classes, health, wealth, residence, race, and sex of potential users affected by different goals.

Procedures to determine goals that are optimally compatible with sociocultural values remain to be defined for fisheries management. Philosophers and social scientists need to provide systems of thought that could be used by fisheries managers to define, interpret, and understand the values of various groups that make up the public sector. These definitions would help managers to identify where values among groups overlap. In addition, a process is needed to identify the values of greatest importance to society in general. The identification of these critical values would then provide a definition of the ''public interest'' which fisheries agencies are often required by legislation to serve. This type of value analysis could then be combined with decision theory, such as is used in business management, to help choose goals for fisheries management. The process of goal determination should be a top priority for study by philosophers and scientists interested in fisheries management (see Scarnecchia 1988).

Realistically, management of fishery resources cannot wait until a decision theory that includes analysis of cultural values is developed and tested. Management goals must be chosen. The key questions to answer are, Where are we? and Where do we want to go? (Crowe 1983). These questions are equally valid for setting management objectives. The management team must combine and intuitively use all types of information to reach a fair decision about goals. Public participation will provide a valuable guide to how public sectors will respond to a change in policy. Though the goal-determination process may be imperfectly defined, fisheries managers should choose to be advocates for the long-term well-being of fishery resources and the use of those resources by future generations. These positions are, at times, not fully represented in discussions about management goals. Occasionally, the well-being of fishery resources is assured and yet conflicts over resource allocation occur among user groups. In this case, the manager may assume a more neutral role as a facilitator to resolve conflicts and to fairly allocate the resource. Managers, however, should not simply be mediators among special interest groups without regard for the resource (see Behan 1978 for an alternative viewpoint).

2.4.4 Objectives: Management Targets with Deadlines

The second step in the management process is to choose objectives that will define progress towards achievement of goals (Figure 2.3). Objectives have two important characteristics: they are measurable and they have a specified time period in which to be accomplished. Objectives are the criteria by which agencies can determine their progress toward achieving goals over time. Objectives provide a measurable definition of successful management.

**Box 2.1   Measures that Could be Used as Objectives for Fisheries
           Management.**

- Catch per unit effort
- Number of fish caught
- Number of angler hours
- Number of angler trips
- Proportion of anglers catching limits
- Measures of variation related to species, size, or number caught per trip
- Number of trophy fish caught
- Measures of angler satisfaction
- Value of landed commercial catch (maximum economic yield)
- Kilograms of fish per year (maximum sustained yield)
- Spawning stock abundance (species restoration management)
- Year-class recruitment index (species restoration management)
- Number of eggs deposited (species restoration management)

Choice of the measures used to define objectives requires several considerations. First, measures must relate to the management goal, and are usually
related to the anticipated products of management (see Box 2.1). Second, they
must be realistic or attainable. Similar to goal setting, an assessment of ecological,
economic, political, and sociocultural information is required to ensure that the
objectives are reasonable relative to the management environment. Third, the
feasibility, both logistically and economically, of conducting studies to measure
the progress towards achievement of the objectives must be assessed. Such
measurement is a requirement for the fifth step in the management process—
evaluation (Figure 2.3). Money from the agency's budget will have to be available
for these studies. Fourth, if variables are chosen, then they must be measurable
with sufficient precision (and sometimes accuracy) to permit statistical comparison during the evaluation step. The degree of precision attainable by the
evaluation studies will be determined by available technology and budget.

Management objectives related to a commercial fishery may be defined more
easily than those related to a recreational fishery. For commercial fisheries a body
of theory has been devoted to the definition and measurement of maximum
sustainable yield and maximum economic yield (see Chapters 5 and 7), concepts
which serve both as goals and measurable objectives for commercial fisheries
management (e.g., Gulland 1977). Measurable objectives that gauge the satisfaction of recreational anglers are more difficult to define (Larkin 1980). Objectives
include measures of resource use such as the proportion of anglers who catch their
legal bag limit of walleyes. Alternatively, recreational satisfaction could be
measured directly through sociological surveys of the users. Objectives that
quantify goals related to population reestablishment or endangered species
management would most likely be measures of population characteristics such as
those previously illustrated for lake trout in Lake Ontario. Similar to the
goal-setting process, the process of choosing valid measures for recreational
fisheries objectives needs to be further developed by the fisheries profession. As

the management process moves through time, new experience and knowledge will be gained that can be used to modify objectives to improve the measurement of goal achievement.

Objectives should be defined with the help of fisheries professionals from several disciplines. Applied ecologists, sociologists, economists, statisticians, planners, and budget analysts either from within or external to an agency may be helpful. This team must also include the field managers who are responsible for implementation of the management program. Typically, senior-level administrators need not be involved as intimately as they were in the goal-determination process (Crowe 1983). Sociocultural studies can provide the public input needed to help guide managers in their selection of parameter values that best define goals.

2.4.5 Problem Identification: What Will Prevent Success

The next step in fisheries management is identification of problems that could prevent achievement of objectives, and hence goals. If the managers have answered, Where do we want to go? then they should ask, What will prevent us from getting there? For example, sea lamprey predation was considered to be an important obstacle to achieving the lake trout restoration goal in Lake Ontario (Figure 2.4). To develop a list of possible problems, questions need to be asked, such as: Is stream habitat degradation reducing production of trout? If we stock more salmon will there be adequate forage? Do markets exist to accommodate expansion of commercial harvest of catfish, and thus increase the value of the catch? Is fishing mortality too high for the muskellunge population to sustain itself? Is tournament fishing affecting largemouth bass populations? Is acid precipitation affecting brook trout reproduction? Will fishing effort by anglers stay the same if striped bass populations increase? Do anglers know about the fishing opportunities available? Does current fisheries management need to be changed? The group of problems identified will be the focus for the next step—management actions (Figure 2.3).

Problem identification should be conducted by a group of specialists who represent disciplines related to the measures specified in the objectives. This group must also include regional or local field managers responsible for program implementation. Resource scientists external to the agency can provide helpful expertise not represented by agency personnel. The development of this part of the lake trout program for Lake Ontario included state, provincial, and federal agency personnel, and scientists from universities specializing in hatchery production, sea lamprey control, population dynamics, and genetics. Involvement of specialists in identifying problems ensures that the information base will be properly used to analyze the problems. Public involvement may also occur at this step, especially when problems related to user conflicts need to be identified. The discussions that occur within the group should be conducted in an open, creative atmosphere where all ideas are at least initially considered valid no matter how ridiculous they may seem to some. Evaluation of the problems identified will occur later in the next step.

2.4.6 Actions: Solutions to Problems

The fourth step in the management process is to identify strategies and implement actions required to solve problems (Figure 2.3). Actions are the

Figure 2.4 Lake trout mortality caused by sea lampreys is considered an important problem preventing reestablishment of lake trout populations in Lake Ontario. (Photo courtesy of U.S. Fish and Wildlife Service.)

specific tactics or techniques used by an agency to solve problems that prevent achievement of goals and objectives. If we have answered the question, Where do we want to be? (goals and objectives) and have identified, What will prevent us from getting there? then this step answers the question, How will we get there? (Crowe 1983). Sea lamprey control and angler regulation are examples of actions chosen and implemented to speed lake trout restoration in Lake Ontario (Figure 2.5). Actions or tactics (also sometimes called techniques) are unfortunately what many public groups and some fisheries professionals envision as management. Management as discussed in this chapter includes the entire process set within the management environment (Figures 2.1 and 2.3).

The problems identified earlier first need to be examined to determine which problems have the greatest effect on the objectives and offer the most potential for the agency to solve. Limited agency budgets usually prevent all problems from being addressed initially. A final subset of problems needs to be established; however, all the problems identified should be documented because the agency may need to reconsider them later. At this point, a careful analysis of the subset should occur to determine those strategies upon which actions should focus.

Strategies define the purpose and state the desired outcome of actions. It is important to determine strategies before implementing actions. Often specific problems are part of a larger, more general problem. In this case, strategies can be developed that address the general problem and join together many specific problems. For example, the reduction of lake trout mortality is a strategy that links and solves the problems of sea lamprey predation and angler overharvest. Most importantly, the identification of strategies often results in the identification of additional, less obvious solutions or actions. For example, a manager might

Figure 2.5 Stocking of yearling lake trout and regulation of angler harvest were two important actions chosen and implemented as a part of lake trout management in Lake Ontario.

respond to an overharvest problem by choosing the obvious action of reducing the creel limit imposed on anglers. However, if it is decided that the strategy will be to increase the fish population by reducing all forms of mortality, then actions could include creel limits, size limits, seasons, fish refuges, reduction of predation and other natural causes of mortality, and the establishment of a voluntary catch-and-release program. Once strategies are established, actions should be chosen and implemented.

Actions may address any component of the management environment—ecological, economic, political, and sociocultural. Ecologically related actions include tactics often thought to be the traditional domain of fisheries management. Traditional actions by fisheries agencies include regulation of resource use (e.g., season or bag limits; Chapter 16), population manipulation (e.g., stocking or chemical control of sea lamprey; Chapters 13 and 14), and habitat manipulation (e.g., stream improvement; Chapter 9). However, management actions include a much broader array of tactics than these examples. Economic solutions for a commercial fishery could include the establishment and administration of monthly catch quotas to prevent temporary market gluts that drive product prices down. Political solutions initiated by a fisheries agency are often related to new legislation designed to assist some other action. Examples include the establishment of laws that give authority to an agency to manage fisheries through limited entry or to establish the annual regulations for angling seasons and creel limits. Sociocultural actions may be related to education of resource users. Such actions include programs to teach users outdoor ethics or how to harvest underused resources (e.g., how to catch and cook common carp). These actions could also focus on encouraging landowners to provide anglers access to their lands or to encourage them to adopt wise land-use practices to reduce soil erosion and stream siltation.

Choice of management actions can follow either a qualitative approach based on experience or a quantitative analytical process or a combination of both. Often the local fisheries manager will have an intuitive sense based on past experience of how the resource and its users (the management environment) will respond to a particular action. This experience can be some of the most valuable and applicable information available to help in choosing a new tactic. This experience sometimes can be effectively augmented by having public groups suggest actions that would be acceptable to them. In other cases, sufficient information will be available to construct a simulation model of the management environment to assist in the decision process (see Walters 1986). Such models can have input variables that are management actions and output indicators that include the measures specified in the objectives. The model is used to predict output indicators for several different combinations of actions to identify the best set of tactics to achieve objectives and goals. Regardless of the process used, usually all tactics cannot be implemented because of dollar and personnel constraints. Actions should be based on cost-effectiveness and the greatest potential for solving problems.

Individuals involved in selecting management actions will be similar to those who identified problems in the previous management step. It is important to include local managers who are responsible for program implementation and have experience with the specific resource and its users. Others who know the agency's conservation laws and represent law enforcement will be helpful, if actions related to regulation of users are to be implemented. If simulation models will be used as input to the decision process, then people with resource modeling expertise should be included. Public involvement often can be helpful at this stage. In some situations, several actions may be equally effective in addressing a particular strategy. Public input can help managers choose those tactics most socially acceptable. For example, after several meetings with the public, the New York State Department of Environmental Conservation chose size-limit and season regulations over more restrictive creel limits as the regulations most acceptable to the public for the reduction of angler-induced lake trout mortality in Lake Ontario.

The group involved in this step may conclude that the problems related to the achievement of goals are insurmountable. As a result, the goals and objectives must be redefined. Alternatively, the group may decide that the problems and associated strategies cannot be clearly identified without more information, and they may recommend as the first management action that studies be initiated for problem definition.

After management actions are chosen, they must be implemented. Fish must be propagated and stocked. New laws that govern anglers must be passed. Extension publications must be written and distributed. Implementation of actions requires the allocation of money and personnel from agency budgets. Control of implementation is often termed operational or tactical planning (Crowe 1983). Each year a tactical plan for Lake Ontario is developed by state, provincial, and federal agencies that determines, for example, which tributary streams will experience chemical control of sea lampreys, and how many lake trout will be produced by hatcheries and where they will be stocked. After money has been allocated, the actions can be implemented and the management environment observed for its response.

2.4.7   Evaluation: Information for Program Improvement

The fifth step is evaluation of the management environment's response to the actions implemented, and an assessment of that response in terms of the entire

Figure 2.6 The R/V *Seth Green* operated by the New York State Department of Environmental Conservation is used to measure the parameters stated in the management objectives for lake trout in Lake Ontario.

management process (Figure 2.3). This step answers the question, Did we make it relative to our goals and objectives (Crowe 1983)?. This step has four parts: (1) measurement of response, (2) comparison to objectives, (3) assessment of the comparison, and (4) revision, continuation, or termination of management programs.

First, the expected outcomes chosen as objectives must be measured and used as indicators of the management environment. For example, studies are annually conducted to determine how many yearling lake trout are produced naturally in Lake Ontario (Figure 2.6). Careful statistical design of such studies is essential to detect differences between parameter estimates. Such measurements are the focus of many of the field studies conducted by state, provincial, and federal fisheries agencies, and often occur simultaneously with implementation of actions rather than just on the date specified in the objectives. These measurements throughout the course of program implementation provide continuous feedback to guide management.

Second, the values measured above must be compared to those stated in the objectives. For instance, the question must be answered as to whether 100,000 yearling lake trout were produced the past year by natural reproduction in Lake Ontario (see the objectives stated in section 2.4.2).

Third, an assessment of the management program must follow, especially when objectives were not achieved. If objectives were achieved then the agency must decide whether to maintain, expand, reduce, or terminate the current management program. Failure to achieve objectives is probably the most informative result for guiding future management. We often learn more from our mistakes than from our successes. Assessment of the most probable causes for failing to achieve objectives is conducted next. If few or no yearling lake trout are captured in Lake Ontario, it must be determined why. Is the interim objective unrealistic? Were spawning substrates unsuitable? Were adults unable to locate spawning reefs? Were stocking and sea lamprey control ineffective? Assessment of the Lake

Figure 2.7 Assessment of annual evaluation activities occurs at the Lake Ontario Committee meetings attended by New York, Ontario, and the two federal fisheries agencies. The meetings are organized by the Great Lakes Fishery Commission. (Pictured here: Carlos Fetterolf, Jr., former Executive Secretary, Great Lakes Fishery Commission.)

Ontario program occurs at an annual meeting of all the relevant agencies (Figure 2.7).

Fourth, management program revision is conducted based on the above assessment. Management should now cycle to the next most appropriate step. Goals may need to be reviewed and revised. Objectives may need to be redefined with different time frames. New problems may have arisen that need identification. New strategies may need to be developed and different actions may need to be implemented to solve problems.

The evaluation step is critical to the management process. This step provides the information required for program redirection. Evaluation is the step that gives management the capability to correct earlier errors and to adapt to changes in the management environment that have occurred since the program began. Evaluation permits analysis of the past so that management can improve in the future. To choose to conduct the evaluation step means that limited resources within agency budgets will become more scarce. However, the long-term benefits of evaluation will more than counter-balance budget and personnel expenditures and the temporary delay in implementation of other management actions.

2.4.8 Information Base: Resources for Effective Decision Making

The information base provides knowledge to assist decision making and is central to the fisheries management process (Figure 2.3). The information base is used and enhanced at each step of the process. The evaluation step will especially cause the information base to grow. Information includes the undocumented experience of fisheries managers and the public, and the published results of scientific research. Ecological, economic, political, and sociocultural knowledge is included in the information base. For example, information could include anecdotal observations by a hatchery manager that lake trout yearlings reared at high densities in raceways appear in poor condition, or a published account of the economic benefits caused by lake trout rehabilitation. A conscious effort must be made to use all available information not just ecological topics. Agencies, universities, private foundations, and industries contribute to the information base available for fisheries management.

2.5 DEFENSIBILITY: KEY TO PROGRAM SURVIVAL

Management programs must survive long enough to have a positive effect on fishery resources. Programs may be canceled prematurely by legislative actions on agency budgets or due to lack of public support. Therefore, defensibility is an important characteristic for program survival. Use of a goal-oriented approach to management will help ensure that a sound rationale exists for program development. Fisheries managers with effective communication skills will then be able to explain the rationale behind past and planned agency decisions and subsequent actions.

Defensibility is especially enhanced when public involvement has occurred throughout the evolution of the program. In this chapter, we have identified some key steps in the management process where direct public involvement will assist and improve the task of fisheries management. In addition, agency personnel need to communicate continually with the public about the current status of management programs. An informed public is the most effective ally when an agency needs political support to maintain programs.

2.6 ADAPTIVE MANAGEMENT: LEARNING FROM PAST ACTIONS

The management process described above is based on the fundamental assumption that insufficient information exists for optimal management to occur, and various levels of uncertainty exist for every decision made. Evaluation as a step in the process was emphasized because it permits learning about the responses of the management environment. The characteristics of this process are similar to the concept of "adaptive management" presented by Walters (1986). Walters and his colleagues view management actions as primary tools for whole-system study of the management environment. This management style is in contrast to the more typical incremental changes implemented by many agencies (Lindblom 1959). Instead, the adaptive management concept promotes aggressive experimentation with actions coupled with careful evaluation. The primary purpose of evaluation is to understand or "adaptively learn" about the full range of environmental responses that are possible as various actions are implemented. The predictability of these responses becomes an important aid to decision making in management.

Styles similar to adaptive management will rapidly provide the information needed for optimal management. This is especially true when compared to the traditional incremental approach used by agencies or the reductionist approach of many scientists who conduct narrowly focused research. However, aggressive experimentation with actions incurs the real risk of negatively affecting fishery resources (at least temporarily) which could upset users and result in political liability for a program. To avoid such risk, an agency should first seek public approval to use an adaptive management approach.

2.7 CONCLUSION

Adoption of a goal-focused, cyclic process for fisheries management will prevent haphazard development of programs. Budgets are always limited, yet our fishery resources demand more management attention than in the past as our human populations increase. Society simply cannot afford the luxury or potential waste that can be caused by the unfocused evolution of fisheries management programs. To stay within the process, the fisheries manager must always evaluate each decision in terms of a program's goals and objectives and the strategies that actions are to address.

The conduct of management demands day-to-day decision making. The fisheries manager must recognize that past decisions will often affect the types of choices or opportunities that will be available in the future. Each decision must be made with an eye to its potential effects on future decisions. For example, the decision to introduce a species may be irrevocable. The species may establish natural populations or a new group of anglers may develop who specialize in catching a species that must be continuously stocked.

The personal wisdom and skill required for effective decision making in fisheries management is not something gained simply by completion of a course or reading a textbook. The aspiring fisheries professional must recognize that management effectiveness requires practical experience in an agency and a lifelong commitment to study the forces that affect management. The professional must also work hard to develop skills to communicate effectively with fellow professionals and the public. Citations within this chapter have been given to assist the reader in the continued study of the processes of management.

Fisheries management offers an exciting profession to those who enjoy direct involvement with management of a portion of the Earth's natural resources. Decisions made today by a manager may ensure conservation and wise use of fish for the future. This aspect of direct involvement in management decisions helps distinguish the fisheries manager in an agency from the fisheries scientist in a university. The scientist's role is primarily to discover, interpret, and provide information. Managers within agencies are responsible for making decisions that actually affect fishery resources.

The foci of the management process are fish and people. Fish, and the aquatic communities of which they are a part, provide delightful and intriguing subjects for study and management. Equally fascinating is understanding the economic, political, and sociocultural forces caused by the many interactions between fish and people. These interactions inextricably weave fish and people together within the process of fisheries management. Thus, the best fisheries managers are those who enjoy, understand, and work effectively with both fish and people.

2.8 REFERENCES

Anderson, K. H., and F. B. Hurley, Jr. 1980. Wildlife program planning. Pages 455–471 *in* S. D. Schemnitz, editor. Wildlife management techniques manual, 4th edition. The Wildlife Society, Bethesda, Maryland.

Behan, R. W. 1978. Political dynamics of wildlife management: the Grand Canyon burros. Transactions of the North American Wildlife and Natural Resources Conference 43:424–433.

Billingsley, L. W., editor. 1981. Stock concept international symposium. Canadian Journal of Fisheries and Aquatic Sciences 38:1457–1921.

Bishop, R. C. 1987. Economic values defined. Pages 24–33 *in* D. J. Decker and G. R. Goff, editors. Valuing wildlife - economic and social perspectives. Westview Press, Boulder, Colorado.

Brown, P. J., and M. J. Manfredo. 1987. Social values defined. Pages 12–23 *in* D. J. Decker and G. R. Goff, editors. Valuing wildlife - economic and social perspectives. Westview Press, Boulder, Colorado.

Christy, F. T., Jr., and A. Scott. 1965. The common wealth in ocean fisheries. Johns Hopkins University Press, Baltimore, Maryland.

Clark, C. W. 1976. Mathematical bioeconomics: the optimal management of renewable resources. Wiley-Interscience, New York.

Cooper, E. L., editor. 1987. Carp in North America. American Fisheries Society, Bethesda, Maryland.

Crowe, D. M. 1983. Comprehensive planning for wildlife resources. Wyoming Game and Fish Department, Cheyenne.

Fisher, R., and W. Ury. 1981. Getting to yes - negotiating agreement without giving in. Penguin Books, New York.

Fraidenburg, M. E., and R. H. Lincoln. 1985. Wild chinook salmon management: an international conservation challenge. North American Journal of Fisheries Management 5:311–329.

Giles, R. H. 1981. Assessing landowner objectives for wildlife. Pages 112–129 *in* R. T. Dumke, G. V. Burger, and J. R. March, editors. Wildlife management on private lands. The Wildlife Society, Wisconsin Chapter, Madison.

Gulland, J. A. 1977. Goals and objectives of fishery management. FAO (Food and Agriculture Organization of the United Nations) Fisheries Technical Paper 166.

Harville, J. P. 1985. Expanding horizons for fishery management. Fisheries (Bethesda) 10(5):14–20.

Kennedy, J. J., and P. J. Brown. 1976. Attitudes and behavior of fishermen in Utah's Uinta Primitive Area. Fisheries (Bethesda) 1(6):15–17,30–31.

Krueger, C. C., D. J. Decker, and T. A. Gavin. 1986. A concept of natural resources management: an application to unicorns. Transactions of the Northeast Section of the Wildlife Society 43:50–56.

Larkin, P. A. 1980. Objectives of management. Pages 245–262 *in* R. T. Lackey and L. A. Nielsen, editors. Fisheries management. Blackwell Scientific Publications, Cambridge, Massachusetts.

Lindblom, C. E. 1959. The science of "muddling through." Public Administration Review 19:79–88.

Miller, A. 1984. Professional collaboration in environmental management: the effectiveness of expert groups. Journal of Environmental Management 16:365–388.

Nielsen, L. A. 1985. Philosophies for managing competitive fishing. Fisheries (Bethesda) 10(3):5–7.

Odiorne, G. S. 1979. MBO II, a system of managerial leadership for the 80s. Fearon-Pitman Publishers, Belmont, California.

Pringle, J. D. 1985. The human factor in fishery resource management. Canadian Journal of Fisheries and Aquatic Sciences 42:389–392.

Ryman, N., and F. Utter, editors. 1987. Population genetics and fishery management. University of Washington Press, Seattle.

Scarnecchia, D. L. 1988. Salmon management and the search for values. Canadian Journal of Fisheries and Aquatic Sciences 45:2042–2050.

Schneider, C. P., D. P. Kolenosky, and D. B. Goldthwaite. 1983. A joint plan for the rehabilitation of lake trout in Lake Ontario. Great Lakes Fishery Commission, Lake Ontario Committee, Ann Arbor, Michigan.

Talhelm, D. R., editor. 1987. Social assessment of fisheries resources proceedings. Transactions of the American Fisheries Society 116:289–540.

Walters, C. 1986. Adaptive management of renewable resources. Macmillan, New York.

Yarbrough, C. J. 1987. Using political theory in fishery management. Transactions of the American Fisheries Society 116:532–536.

Chapter 3

Communication: Catalyst for Effective Fisheries Management

DANIEL J. DECKER AND CHARLES C. KRUEGER

3.1 INTRODUCTION

Communication is a process that each of us engages in daily—at school, on the job, and at home. We are all highly experienced in the use of communication, either as the originators of information or recipients of information that bombards us from every direction. Think about it for a second. Every person is involved in hundreds to thousands of communication events daily. Our routine exposure to communications includes conversations, telephone calls, letters, meetings, television commercials and newscasts, radio programs, newspapers and magazines, posters, billboards, "junk" mail, graffiti, and even course lectures and textbooks like this one, all of which are intended to communicate certain messages to us in the hopes that we will buy, believe, or behave in a certain way. Some are very effective, some are not effective at all. Ever wonder why? Communication professionals have amassed a wealth of information on the communications process that can be useful to fisheries managers. Effective communication among staff of a natural resources agency, and between staff and the public, is a key catalyst for the fisheries management process presented in Chapter 2.

In this chapter we will use a term borrowed from program evaluation specialists to refer to all people interested in fisheries management—stakeholders. As the name implies, a stakeholder is someone who has a stake in an issue or program that addresses a problem, interest, or concern of that person. We feel this term is more appropriate as a general descriptor than others commonly used to describe people interested in fisheries management; terms such as constituent, which really means a supporter, although not all those interested in management are supporters; or client, which really means someone who receives a service, usually for a fee, although not all those interested in management pay for it; or user, which means someone who actually uses the resource, like an angler, although not all those interested in management personally use the resource; or public, which from a public relations standpoint is accurate, but lacks intuitive appeal. Thus, we have settled on stakeholder as the best general term for people with an interest in fisheries management.

Stakeholders in fisheries management include a wide variety of people, such as direct users, people who benefit from others' use (e.g., business operators who sell supplies and services to anglers), people who incur costs because of fishery resources (e.g., riparian landowners who suffer trespass problems from anglers), staff of related natural resource agencies, people who might not even realize initially that they have a stake in the resource (e.g., local government officials who will eventually have to deal with providing expanded services in an area with a developing recreational fishery, such as occurred in conjunction with the Great Lakes salmonid fishery), and even people who are opposed to sport fishing. These and other stakeholders, as individuals and groups, may all be involved in any single fisheries management issue. Effective communication must occur with them to develop and implement fisheries management programs. This represents a real challenge for the fisheries manager, but can contribute to an interesting and exciting career!

The primary purpose of this chapter is to increase your awareness of the communication process and to help you understand its role in fisheries management. Communication is viewed broadly in this chapter. The discussion of the management process (Chapter 2) should have started you thinking about the role of communication in fisheries management. Aspiring fisheries managers need to become aware of the direct relationship between effective communication and effective fisheries management early in their academic training, while they still have an opportunity to enroll in courses that can help develop and hone communication skills. These skills may become the most important fisheries management techniques used by a professional manager. Some reasons why this is the case will be discussed. Minimally, aspiring fisheries professionals should recognize that management is as much working with people as with fish. Dealing effectively with people is achieved by the skillful use of communication.

3.2   FISHERIES MANAGEMENT IN TRANSITION

Fisheries management is in a period of transition that began about 20 years ago (Harville 1985). Two major changes associated with the transition have heightened the importance of communication in fisheries management. First, the broad goals of fisheries management have moved toward optimum sustainable yield (OSY). In the OSY philosophy some optimal set of stakeholders' preferences and other economic and sociocultural benefits are to be achieved (see Chapter 1; Roedel 1975). Such benefits are defined by people in terms of their wants, desires, values, preferences, and satisfactions. For fisheries managers to set goals to produce economic and sociocultural benefits, they must engage in communication that will give them a clear understanding of the kinds of benefits desired by different user groups and other stakeholders in the fishery. Managers also need to be able to communicate to stakeholders their plans to provide the benefits desired. As a result, the changing management philosophy has redefined and expanded the roles and responsibilities, and therefore the academic training and expertise required, of contemporary fisheries managers.

The second major change associated with the transition in fisheries management is greater public involvement in the management decision-making process. The management process presented in Chapter 2 reflects this philosophy. Though outside the context of inland fisheries management, the U.S. Fishery Conservation and Management Act of 1976 created a mandate to include active public participation throughout the planning and decision-making process for marine fisheries management. This notion of including various groups of interested stakeholders in decision making has spilled over into the philosophy and conduct of fisheries management in general. Fisheries managers cannot operate without public involvement in policy and management decision making. As Harville (1985) remarked, "While enlightened fish and wildlife agencies long have recognized the need to involve their user-publics in review and dissemination of agency policy, . . . our constituency now demands active public participation in the formulation of that policy." This participation requires effective, two-way communication with stakeholders, particularly resource users.

In addition to the changes in fisheries management, the identities of the stakeholders have changed, further complicating communication. They are more diverse in their interests and more sophisticated in their ecological understanding and political activities. Today's stakeholders have greater expectations for clear and complete communication, which are certain to challenge a manager's abilities.

3.3   FUNDAMENTALS OF COMMUNICATION

Every natural resource professional, whether employed by a university, consulting firm, foundation, or agency should understand the fundamentals of communication. Understanding communication is a *requirement* for managers. Texts on this subject have been written especially for the natural resource professional. For example, the book *Public Relations and Communications for Natural Resource Managers* (Fazio and Gilbert 1986) is full of practical information on communication theory and practice and should be in every manager's personal library. If managers develop a working understanding of this subject, they will be able to facilitate the kinds of communication needed to apply the concept of fisheries management described in Chapter 2. This section introduces fisheries managers to the general process of communication and addresses the ways that managers can use communication in fisheries management.

3.3.1   Communication: The Process of Interaction

The term communication is used in a variety of ways and may mean different things to different people. A definition that reflects the notion of communication used in this chapter is: the interchange of information (e.g., data, beliefs, insights) by written and spoken words, visual illustrations, and actions for the purpose of developing mutual understanding (but not always agreement) between the sender and receiver about a situation, concept, or event. A principal element of this definition is the concept of interchange among people. There should be two-way communication between a fisheries management agency and its stakeholders. Listening to stakeholders' opinions and ideas relative to a management issue is as important as telling them the agency's viewpoints. Building effective feedback, or "listening" mechanisms, is essential to two-way communication needed for meaningful public involvement in decision making by agencies. A simple model of the communication process has five basic components: communicator, message, channel, audience, and feedback (Figure 3.1). To simplify the discussion below,

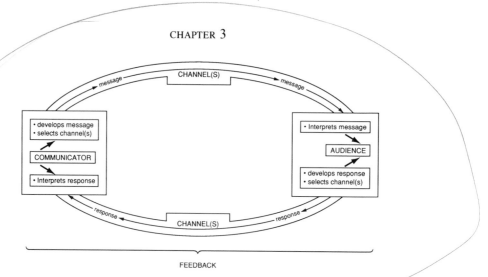

Figure 3.1 Model representing the five basic components of the process of communication.

fisheries managers can be thought of as the "communicators" and stakeholders as the "audience"; however, in two-way communication these roles often switch ("feedback" arrow in Figure 3.1).

The communicator (i.e., source of a message) needs to be perceived as being technically sound and credible if the message is to be accepted and acted upon by the audience (Figure 3.1). Communication starts with a "climate" of belief on the part of the audience. This climate is created by the credible past performance of the communicator. Trust and honesty are particularly important components of credibility. Such performance reflects an earnest desire to serve the audience and helps the audience to have confidence in the communicator (Cutlip et al. 1985). When a fisheries manager or agency does not hold the confidence of stakeholders, the desired communication will not take place.

Managers as communicators should know their audience—its preferences, beliefs, attitudes, communications habits, and other traits pertinent for effective interchange. Good communicators have skills in selecting and structuring messages, selecting and using channels, and soliciting and interpreting feedback. Fisheries managers, not the public, are primarily responsible for effective communication.

Possibly the most difficult task for a communicator is that of developing meaningful ways to express messages. This task is challenging because to do it well requires that the manager transform fisheries management ideas into words and illustrations that can be correctly interpreted by the public. Managers have to recognize that the text and illustrations they use to convey their message may not be interpreted in the manner they intended. Fisheries managers cannot simply create messages that are clear to them and be confident that they have executed their communication responsibilities adequately. Instead, they must carefully and skillfully develop ways to express their messages so that their audiences will understand.

The message is the information that the manager wants the audience to understand, usually to accept, and often to act upon (Figure 3.1). Consequently messages should be clearly interpreted and easily understood by the recipients and perceived as being useful to them. Two aspects of messages are important (Cutlip et al. 1985):

1. Content. The message must have meaning for the recipients and must be relevant; people tend to listen to messages that offer them something tangible and useful. Usually the content of a message largely determines the audience (i.e., those who will listen).

2. Clarity. The message must be put in simple terms. Words must mean the same thing to the audience as they do to the communicator. Complex fisheries issues must be simplified and clarified.

Basically, a "good" message should be: (1) matched—to the mental, social, economic, and physical capabilities of the audience; (2) significant—economically, socially, or aesthetically—to the needs, interests, and values of the audience; (3) concise—containing no irrelevant material; (4) timely—especially when seasonal factors are important and issues are current; (5) balanced—by factual material covering both sides of an issue; (6) applicable—so the audience can use the recommendations presented; and (7) manageable—so the message can be handled skillfully within the constraints imposed by the time and resources available for communication.

The channel is the way in which a message reaches the intended audience (Figure 3.1). Channels of communication can be thought of as bridges over which the message travels connecting the communicator and the audience. Some examples of channels commonly used by fisheries managers are meetings (and other face-to-face situations), letters, newspapers, newsletters, magazines, radio announcements and broadcasts, and reports. Established channels of communication should be used as much as possible, particularly channels that the audience uses frequently and trusts. Although new channels may be difficult to create, at times fisheries managers may find it necessary to do so because existing channels to which they have access are inadequate or inappropriate.

Surveys can be used to identify the channels used by different groups. For example, a fisheries manager may want to reach charter boat captains with a message. The manager may discover through a survey of readers that the agency magazine used in the past only reaches 5% of the captains. As a result, a special newsletter may have to be developed for direct mailing just to this group to be sure that messages reach them in an effective and timely fashion.

Fisheries managers should use multiple channels to reach an audience, rather than rely entirely on a single channel for most communication purposes. Use of channels in "parallel" rather than in "series" is advisable. In other words, use several channels that reach the audience directly instead of one channel to reach another channel that then reaches the audience. The disadvantage of the latter approach is that information "gatekeepers" usually exist within channels who can alter the message in such a sequence. Using several channels has multiple advantages, such as the reinforcement of ideas caused by the effect of receiving a message more than once, added credibility associated with receiving a message from more than one channel, and the increased likelihood of reaching more people.

Overall the selection and use of channels should be approached with attention to the following considerations: specific goal of the message, content of the message, characteristics of the audience, channels available that will reach the audience, personnel available with the right skills needed to use a particular channel, channels that can be combined and used simultaneously, channels in series, cost of channel use relative to its effectiveness, and time constraints related to the urgency of the message.

Most of the important considerations relative to the audience have been presented in the discussion of the other elements (Figure 3.1). Given that a message reaches the intended audience (and they read or listen to it), possibly the most important consideration is the ability of the audience to interpret the message as the communicator had anticipated. If the message was developed with this essential purpose in mind, then communication will probably be effective and the audience will gain a new or better understanding of the situation through correct interpretation. However, understanding a message does not mean the audience will agree with it.

Feedback relates to either being receptive to or actively seeking the audience's reactions and responses to the message (Figure 3.1). Obtaining accurate and comprehensive feedback is important in fisheries management for two reasons. First, the manager needs confirmation that the message was received. Second, reactions to the message need to be assessed so decisions can be made about the need for additional communication efforts. Feedback can be thought of as either intentional or unintentional. Intentional feedback is the type that stakeholders engage in to make certain the agency knows their interpretation of the content of the message they received and their opinions about that message. Unintentional feedback is information the manager receives about stakeholders' reactions through observations, informants, and other means. This kind of feedback may be revealing and useful for assessing the need for additional communication efforts.

Fisheries managers should not leave the feedback element to chance. Obtaining feedback is simply too important. Rather, they should develop mechanisms to solicit feedback from the audience. This approach has two distinct advantages. First, managers are more likely to obtain accurate, timely, and representative information if the feedback mechanisms are developed carefully (e.g., a scientific survey of the angling public). Second, seeking input has an important public-relations value. By doing so, managers are telling their stakeholders that they are not merely being "talked at" but that their opinions are valued enough to be sought. Mechanisms that are used to encourage feedback may be as simple as the address of an agency contact person on a brochure, or as formal as surveys and public hearings.

Contemporary fisheries managers need to have a basic understanding of the communication process and a willingness to develop and fine-tune their communication skills. The ability to speak well in public or to write interesting media releases must be combined with a good understanding of communication as a process to apply these kinds of skills effectively. When contemplating communication, managers should keep the entire process in mind to improve the probability of engaging in an effective interchange.

3.3.2 External Communication: Requirement for Good Public Relations

Experience has shown that good public relations is a prerequisite to success in managing fishery resources. A definition of public relations applicable to fisheries

---

Box 3.1   Principles of Public Relations

Principle 1   Every agency action makes an impression on its publics.

Principle 2   Good public relations is a prerequisite to success in agency programs.

Principle 3   The public is actually many different groups of people.

Principle 4   Truth and honesty are essential to credible public relations.

Principle 5   Proactive is more effective than reactive.

Principle 6   Communication is the key to good public relations.

Principle 7   Planning comprehensive communication strategy is essential.

---

managers is the planned effort to influence public opinion through good character and responsible performance, based upon mutually satisfactory two-way communication (Fazio and Gilbert 1986). This definition of public relations may give a slightly narrower view of the communication function than that proposed for managers in this chapter. The goal of communication in the context of the fisheries management process is more than the development of a favorable public opinion climate for the agency; it is also the development of informed groups who can contribute to decision making through public involvement processes. Developing and maintaining good public relations is a responsibility of all agency personnel involved in fisheries management. Fazio and Gilbert (1986) discussed in detail seven principles of public relations which have been modified and outlined in Box 3.1.

External public relations must be carefully planned and executed to be effective. External communication can be multipurpose: (1) to listen to stakeholders' desires for fishery resources, (2) to inform them of the status of the management environment and what future options appear available, (3) to assess their preferences, (4) to gain their support for proposed actions, and (5) to evaluate their reactions to management actions. Consequently, ways to accomplish effective communication vary with the purpose of the communication and the particular characteristics of stakeholders who are the audience. Methods also depend on the particular situation. For example, if a message needs to reach a group of anglers in a short time, the channel of choice may be first-class mail, telephone, teleconference, or a special meeting, depending on the geographic distribution of the target audience and the amount of funds available to contact them. On the other hand, if there is no urgency about the message, some combination of a news release, article in an agency magazine, or an agenda item on a regularly scheduled meeting may suffice.

A number of barriers to communication may exist in any management situation, but four in particular have needlessly persisted in fisheries management and should be avoided. These barriers are ''assumptions'' that a few fisheries managers have accepted and have used as reasons for not working more diligently on communication. They are presented here so you can recognize and avoid them.

1. Assumption: Anglers have little concern for either the resource base or the future of the fishery.

> Response: Most fisheries managers who work to communicate effectively with anglers and other users discover that anglers care about the resource base and, given an informed choice, will choose the side of conservation (Pringle 1985; Walters 1986).

2. Assumption: The public is unable or unwilling to understand the ecological concepts underlying fisheries management deliberations.

> Response: This assumption reflects an attitude of professional elitism that can seriously impede sound resource management. Although many people may not be able to easily understand some aspects of fisheries management, the opinion leaders of stakeholder groups are typically both intelligent and highly motivated to learn. If just these people understand the basis for management and lend their support, others who may not fully understand the alternative often will trust the leader and also be supportive of management. Thus, fisheries managers should communicate the ecological bases for their management deliberations.

3. Assumption: The public will oppose almost any fisheries management program the agency suggests.

> Response: This assumption emanates from an "us versus them" attitude that sees management as a contest of wills between the agency and the public, rather than a team approach to solve problems held in common between the agency and the public (Fisher and Ury 1981). Managers who understand that stakeholders care about fishery resources will find many concerns they share in common.

4. Assumption: The public has nothing to offer the management process; fisheries managers are professionally trained people who know best how to manage the resources.

> Response: The public is made up of many individuals and groups of stakeholders having diverse and extensive experience. Many people are astute observers who often have important insights and may be able to teach managers valuable lessons. To gain these helpful insights, managers must have a receptive attitude about open interchange with their stakeholders.

Fisheries managers need to remember that "contact" with the public is not equivalent to effective communication. In other words, simply reaching stakeholders with a message does not mean that they heard or saw it, interpreted it correctly, or put it in the context of management as managers had intended. Managers should not be satisfied that a piece of information simply gets into the hands of the people who they want to have it. Rather, managers should want people to be able to interpret, understand, and use that information to participate meaningfully and responsibly in the fisheries management process. In essence, managers should view themselves as educators of their stakeholders.

3.3.3 Internal Communication: Interchange of Information Within Agencies

Professionals inside one's own agency are sometimes disregarded, particularly by inexperienced managers, when communication outside the agency is planned. Inevitably, the importance of good internal relations and attendant communication become apparent. In some instances, internal communication is more important and challenging than external communication.

A principle that managers soon learn is that good internal relations among professionals is a prerequisite for good external public relations (Decker 1976). Many examples of resource management difficulties and failures can be traced to poor communication within agencies that led to external public relations problems. The individual fisheries manager must be mindful not to work independently within an agency. Support for a fisheries management program is essential from administrators, supervisors, other managers, and technicians. Program support by other agency personnel results from understanding the elements and rationale of current management.

Poor internal communication can have disastrous results for a program. The effort devoted to developing the support of a skeptical outside interest group can be undone by someone else in the agency making a disparaging remark about the program at the wrong time. The reaction of the outside skeptic is: If they can't even agree on the merits of this program, why should we support it? In a brief moment the public relations work of weeks or possibly months may be negated. There is no guarantee that managers can avoid this scenario by engaging in purposeful internal communication; however, they can significantly reduce the probability of such happenings if they plan and execute communication to establish broad agency support for proposals.

Managers should be careful not to overlook any group of people potentially important to a program when considering internal communication (e.g., conservation law enforcement personnel). In Chapter 2, important agency personnel were identified for each step in the management process. Several general categories of agency personnel exist, including administrators, law enforcement officers, researchers, office staff, seasonal staff, licensing agents, commissioners, and advisory boards (Fazio and Gilbert 1986). Field staff represent one category of individuals with whom it is absolutely essential to develop continuous communication. Many fisheries management agencies have a decentralized organizational structure where most managers and technicians are located in regional offices. What commonly develops is tension between field staff and central office staff. This counterproductive atmosphere occurs largely because of poor communication. Regardless of the cause of this situation, if a program proposal is to move ahead, be accepted, and have a reasonable chance for success, the manager has to pay careful and continuous attention to internal communication. To ignore this reality of management is to court failure.

Internal communication can be pursued in several ways. Seeking input from all relevant groups within an agency prior to establishing policies can contribute substantially to the development of internal support. Proposals must be explained clearly and reactions to them need to be addressed. Some suggestions will be incorporated and some will not, but even people whose ideas are not fully accepted will feel better if they know their suggestions have been considered honestly and rejected for understandable reasons. The process of explaining why an idea was not adopted will help the person(s) who offered the idea learn even more about the plan. If handled skillfully, these learning episodes can help establish strong internal support for a management program.

A special type of communication is that which occurs among agencies. This type of "internal" communication is essential if effective fisheries management is to occur where jurisdiction is shared by two or more agencies. Communication among agencies often verges on diplomacy. Special kinds of sensitivities must be considered when representatives of agencies that share jurisdictions interact with one another. Especially important is developing a thorough understanding of each agency's jurisdiction over fishery resources. Agencies often jealously guard their authority over resources. Communication strategies must be designed to facilitate the cooperation needed. Remember that communication with other agencies is actually communication with individuals. Be sensitive to them as people and fellow professionals. Correcting a communication blunder is much more difficult and time-consuming than doing the planning necessary ahead of time to avoid problems.

3.3.4 Innovation-Adoption: Idea Acceptance by Individuals

Innovation-adoption theory proposes that new ideas or practices (i.e., innovations) are seldom adopted as the result of a single decision by an individual. Rather, a person usually goes through a process that can be thought of as a series of stages, each stage having a characteristic set of information-seeking behaviors and decisions, and each leading to a progressively advanced stage in the process. The final stage is to adopt the idea or reject it as being inappropriate for adoption at the time. Thus, the process is a series of stages where: (1) a person becomes aware of the idea, (2) develops an interest in it, (3) evaluates it for personal application, (4) tries it on a limited basis, and (5) finally adopts the idea or, alternatively, rejects the idea as being inappropriate (Figure 3.2; Box 3.2).

At each stage in the process, information is important for an individual's continued progress to the next stages. Research has shown that people rely on different kinds and sources of information at different stages in the process. For example, mass media are of greatest importance in the awareness and

Five stages of adopting an innovation:

Figure 3.2 Innovation-adoption in the angler's world.

Box 3.2 Adoption of Downrigger Trolling by Anglers

The innovation-adoption process can be illustrated by examining the use of new technology by anglers. Let us go through the process as it has occurred probably thousands of times since the early 1970s as the Great Lakes salmonid fishery developed. An angler with no previous experience in fishing for trout and salmon in a large body of water hears from other anglers at a tackle shop or reads in a fishing magazine about the tremendous catches of salmon and trout being made using moderate-sized boats from 16 to 24 feet in length equipped with fish-finders and downriggers (awareness stage). The stories and pictures are alluring but uncertainty about necessary skills and equipment slows enthusiasm for the potential of this new activity. This person begins reading more fishing magazines and going to exhibitions and boat shows to learn more about boats, motors, downriggers, and sonar (interest stage). After accumulating additional knowledge about this type of fishing, and talking to people who have tried it, the angler soon forms some opinions about its suitability for his or her own enjoyment. It is likely that the angler will try to find someone with whom to go fishing a few times, or will perhaps go out on a charter boat to gain first-hand experience (evaluation stage). If this experience-evaluation period is positive and reinforces the notion that this type of fishing experience is "right" (including being affordable), the angler might next invest in the minimum equipment to get a fishing boat together and use this limited-investment outfit for a season or two (trial stage). If the angler is successful and enjoys this new kind of fishing, and anticipates that its rewards will continue, he or she may invest more into the existing outfit or upgrade to an even more elaborate outfit that suits his or her needs as they have developed (adoption stage).

---

interest stages, opinions of friends and neighbors are most influential in the evaluation and trial stages, and personal experience gained during the trial stage has the greatest bearing on the ultimate decision to adopt or reject. Knowledge of the particular information needs and preferred sources of information will help a manager in determining message content and in selecting communication channels as stakeholders go through each stage of the adoption process.

3.3.5 Adoption-Diffusion: Idea Acceptance by Society

The adoption-diffusion theory is an explanation of how an idea gains acceptance throughout a social system. This theory can help fisheries managers understand how proposals for a new regulation, such as the establishment of a slot limit for lake trout in Lake Ontario, gains acceptance among anglers. The adoption-diffusion theory also helps managers have realistic expectations for the rate and extent of acceptance of a new idea. Basically, the theory explains that a new idea is adopted, or accepted, at differential rates by people in a social system. Five categories of people will be discussed: innovators, early adopters, early majority, late majority, and nonadopters (Table 3.1; Figure 3.3).

The first people to adopt a new practice are the experimenters, or in terms used by sociologists, the innovators. These people tend to be more highly educated,

---

Box 3.3   Adoption-Diffusion of Sonar by Anglers

The adoption-diffusion process operates constantly among boat anglers. The last 10 years have witnessed a proliferation of technological advancements in equipment to aid anglers in their pursuit of salmonids, largemouth bass, walleye, and many other fish. With the development of each new product such as sonar or fish finders (a relatively new product in the 1960s), the experimenters are the first to purchase the gadget, regardless of cost. The early adopters tend to wait a while and see what kinds of refinements are made to the equipment. They are assessing the applicability and advantages of this new development compared to the old way of fishing. Given a positive assessment, the early adopters will obtain the new fish-finder and will readily communicate their successful use to members of their peer group. The early majority will eventually be convinced to make similar purchases to increase their catch rate and to be in step with the new wave of technologically advanced anglers. The late majority will follow suit, and typically have access to a greater variety of models and price ranges as the market grows and more manufacturers are competing for a share of the market. Some anglers will be very late in adopting the now not-so-new technology or may never adopt it (nonadopters). Their reasons for not adopting may be as pragmatic as the inability to afford it or as value-oriented as wanting to continue a "traditional" approach to fishing (e.g., sonar provides an unfair advantage; anglers should spend time "learning" a lake without this equipment).

---

more affluent, and more willing to take risks. Usually, they are not the local opinion leaders in a social system, but the opinion leaders keep an eye on the innovators to identify "good" ideas. Innovators tend to be information seekers; they are willing and able to access primary sources of information, such as agency personnel and university researchers.

The early adopters are very influential at the local level, operating within a variety of social groupings, such as fishing associations. These people do not necessarily hold elected offices in organizations. Nevertheless, they are looked to and typically sought out for their opinion regarding new ideas. Like the innovators, early adopters are well educated and informed. They constantly seek information from a variety of sources, so they usually are the willing recipients of any communication offered about management.

The early majority is made up of many informal leaders. They tend to be older and more cautious than early adopters. These people are considered wiser, more conservative, and are similar in many respects to the majority of people in the social system. They have considerable influence when it comes to the adoption of ideas by the majority. Members of the early majority may hold offices in state and local groups, such as angling clubs and associations of charter boat operators. Thus, these people must be kept informed of any developments relative to fisheries management. Opinion leaders of the early majority are the people who are sought by individuals during the evaluation stage of the innovation-adoption process. Opinion leaders of both the early adopters and early majority represent those people most likely to be directly involved in the process of fisheries management.

Table 3.1 Characteristics of different categories of adopters. (Adapted from Fazio and Gilbert 1986.)

Adopter category	Salient characteristics	Personal characteristics	Communications behavior	Social relationships
Innovators (First 2.5% to try out idea)	Venturesome; willing to accept risks	Youngest age; highest social status; largest and most specialized operations; largest income	Closest contact with scientific information sources; most interaction with other innovators; greatest use of impersonal sources	Some opinion leadership; very cosmopolitan
Early adopters (Next 13.5%)	Respectable; regarded by many others as the social role model	High social status; large, specialized operations	Greatest contact with local agents of change	Greatest opinion leadership of a category in most social systems; very local
Early majority (Next 34%)	Deliberate; willing to consider innovations only after peers have adopted them	Above-average social status; average-sized operation	Considerable contact with agents of change and early adopters	Some opinion leadership
Late majority (Next 35%)	Skeptical; overwhelming pressure from peers needed before adoption occurs	Below-average social status; small operation; little specialization; small income	Secure ideas from peers who are mainly late majority or early majority; less use of mass media	Little opinion leadership
Nonadopters (Last 15%)	Traditional; oriented to the past	Little specialization; lowest social status; smallest operation; lowest income; oldest	Neighbors, friends, and relatives with similar values are main information source	Very little opinion leadership; isolated

The late majority are the people who take "a wait and see" posture before making a commitment to adopting a new idea. They rely on the early majority for much of their information and for their formation of opinions about an idea. These people tend to be the followers, the rank-and-file members of angler organizations or the people who never affiliate with any organization. For sake of efficiency, a minimum amount of time should be spent targeting communication with the late majority because the payoff for the effort will be low. Energy is better spent on the early adopters and early majority who will be the key influencers of this category of people anyway. This is not to suggest that the late majority should be ignored, but the amount of attention they receive should be tempered by the knowledge of how they react to and use information emanating directly from an agency.

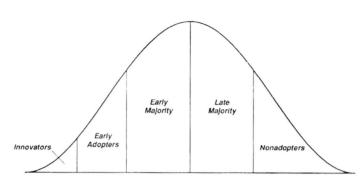

Figure 3.3 Adoption-diffusion: relative frequency of different types of people regarding adoption of an innovation or idea.

The nonadopters are people who either do not adopt or are exceptionally slow to adopt new ideas. These people depend almost entirely on peers for their information. From a strictly pragmatic standpoint, it is nearly a waste of time to try to reach these people. From an ethical standpoint, managers have a responsibility to offer them access to information about fishery resources and their management. Again, knowledge of this category of people suggests that the best way to reach them, for all parties involved, is through their peers in the other categories; thus, this provides another reason to do an effective job in educating the latter groups.

This discussion of innovation-adoption and adoption-diffusion (Box 3.3) is a brief introduction to a body of thinking (Rogers and Shoemaker 1971; Rogers 1983) that can serve the manager in planning comprehensive communication efforts in support of fisheries management. Although the concepts were presented in terms of communication with external sectors, they apply also to internal agency communication. Both theories were developed in the context of voluntary behavior, so their applicability to some areas of management, such as acceptance of new regulations, may not be valid.

3.4 PUBLIC INVOLVEMENT IN DECISION MAKING

The trend toward greater public participation in management decision making brings the importance of effective communication "front and center." Public involvement ". . . will generally produce better decisions and increased acceptance, but, with few exceptions, will likely make the process no easier" (Peyton and Talhelm 1984). Public involvement is not an easy out for managers; it will not make a decision for the manager (Heberlein 1976). Nevertheless, communication between stakeholders and managers will ensure that the manager understands societal goals for the fishery resource and interprets those goals correctly as specific objectives and actions are developed. Similarly, communication between managers and stakeholders is essential to improve the latter group's understanding and support of how managers are translating the goals into objectives and objectives into actions.

Managers will find that communication required for public involvement can be difficult. However, without effective communication fisheries management can

---

Box 3.4 Public Involvement and Communication

Seven basic elements of a comprehensive public-involvement program can be identified (Peyton and Talhelm 1984).

1. A staff that understands the dynamics of public behaviors, attitude formation, and group decision making.

2. A staff that is aware of existing public preferences, values, and systems of belief pertaining to fishery resources.

3. A continuous monitoring system to anticipate and detect changes in public behaviors, needs, preferences, values, and perspectives.

4. An ongoing information system to develop and maintain adequate public awareness.

5. A structured public-involvement system to provide opportunity for appropriate public participation in management.

6. A coordinated, credible public image.

7. A network of stakeholders who are capable of maximizing the agency's attempts to facilitate public involvement in fisheries management.

---

become a frustrating, unrewarding, and often unproductive undertaking. Poor communication about management programs can result in confrontations between stakeholders and agency representatives during meetings and the pursuit of political solutions by the public. Ultimately, uninformed or misinformed stakeholders may circumvent the management process (from the fisheries manager's perspective) through nonparticipation, litigation, and various levels of noncompliance with regulations required to meet program goals. The probability and severity of these unfortunate outcomes can be greatly diminished through carefully planned and executed communication efforts designed as public involvement mechanisms (Box 3.4).

3.4.1 Four Functions of Public Involvement

Identifying the purpose or purposes of public involvement is important to the development of effective communication mechanisms. Four functions of public involvement may be described as follows (Heberlein 1976):

1. Informational—Disseminating information to and obtaining it from stakeholders.

2. Interactive—Working together on an issue or problem with information going back and forth between managers and stakeholders freely and rapidly.

3. Assurance—Making sure that a group of stakeholders knows that its views have been considered in the program-planning process.

4. Legalistic—Involving stakeholders to satisfy legal requirements or social

norms and to allow input from any sources that may have been inadvertently overlooked.

The goal is to convince others (i.e., besides the key stakeholders) that there have been open mechanisms for public involvement. This type of public involvement should be documented (Box 3.5).

3.4.2 Integration of Public Involvement Into Management

Public involvement efforts will vary depending on an agency's purpose and type of management program. We identify three categories of management and each requires a different approach to public involvement (from Peyton and Talhelm 1984).

1. Mandated Management. Some management programs are mandated by legislation, administrative codes, commission rulings, or court orders. In this case, the political or legal component of the management environment is dominant in decision making. When these mandates are clear and not a matter of interpretation, the public education system of the agency must simply communicate these mandates to the public. When options are provided or the mandates must be interpreted, involvement of the affected stakeholders is necessary. The agency should also be monitoring changes in public perceptions, needs, and preferences which may lead to changes in or creation of mandates.

2. Ecological Management. Many fisheries management decisions are con-strained by ecological or biological considerations. Sometimes, the survival of the fishery resource requires decisions based on biological aspects. Little flexibility for alternative approaches may be available to the manager. The ecological component of the management environment is the dominant force in decision making. In this case, stakeholders must be educated to under-stand the ecological constraints imposed on management. Care must be taken not to treat an issue as this type when in fact it is really the next type listed below. Managers must be alert not to attempt to defend decisions by overstating or misrepresenting the biological aspects of an issue. To do so could quickly lead to diminished credibility and trust among important stakeholders.

3. Issue Resolution Management. Most decisions facing a fisheries manage-ment agency involve issue resolution management. They are not clearly or precisely mandated, nor are they ecologically constrained to the point where no flexibility or consideration of options are possible. The sociological and economic components of the management environment strongly influence decision making. Issue resolution management situations are confounded by conflicting pressures from stakeholders with disparate views about the best management objectives and approaches. Managers typically have to deal with considerations of different priorities among stakeholders, inadequate data, and insufficient time lines for decisions and action. Carefully designed public involvement can help resolve conflicts among user groups, and between user groups and the agency.

Box 3.5 Possible Outcomes of Public Involvement

Peyton and Talhelm (1984) report that public involvement can be used:

1. to educate the public in the processes of effective public participation and about fishery resources and management;

2. to change public behaviors and attitudes;

3. to supplement staff resources and provide additional expertise or volunteer efforts;

4. to prevent anticipated conflict by involving expected opponents in the decision-making process;

5. to use existing public-opinion leaders and their influence to get broader public acceptance of decisions;

6. to legitimize agency decisions among stakeholders;

7. to present an image of citizen participation without compromising agency autonomy (tokenism);

8. to resolve conflict (e.g., in matters of resource allocation);

9. to achieve better representation of various stakeholder and resource interests in decision making.

Peyton and Talhelm noted that although some of the outcomes listed above may seem to represent more noble purposes (e.g., 9) than others (7), an ethical case could be developed for the judicial balance of all of the above in a public-involvement effort related to a fisheries management program.

It is often difficult for the fisheries manager to determine the most effective means to obtain information about public needs and requirements for public involvement. Many possible methods exist, but the decision about which one or ones to employ is seldom straightforward. Each can be rated in terms of their effectiveness in serving the four functions of public involvement (Table 3.2; Fazio and Gilbert 1986). In every case, the conscientious manager should determine why public involvement is needed and what result is desired from public involvement (see Box 3.5; Heberlein 1976).

3.4.3 Benefits from Public Involvement

What can public involvement bring to the decision-making process to benefit management? Involvement of stakeholders in management is intended to lead to better decisions and decisions that stakeholders support. Specifically, public involvement can result in the following benefits (from Fazio and Gilbert 1986):

1. New factual data may emerge about the resources involved.
2. Preferences, conflicts, and unknown social complexities will be clarified.

Table 3.2 Effectiveness of various methods in serving the four functions of public involvement. (Adapted from Fazio and Gilbert 1986.)

| Form of public involvement | Function of public involvement | | | | | |
| | Information | | Interactive | Assurance | Legalistic | Effectiveness |
	To give	To get				
Open public meetings	Good	Poor	Poor	Fair	Yes	Poor
Workshops (small)	Excellent	Excellent	Excellent	Excellent	Yes	Potentially good
Presentations to groups	Good	Fair	Fair	Fair	Yes	No clear assurance
Ad hoc committees	Good	Good	Excellent	Excellent	Yes	Potentially good
Advisory groups	Good	Good	Excellent	Excellent	Yes	Potentially good
Key contacts	Excellent	Excellent	Excellent	Excellent	No	No clear assurance
Analysis of incoming mail	Poor	Good	Poor	Poor	Yes	Poor
Direct mail from agency to public	Excellent	Poor	Fair	Good	No	Potentially good
Questionnaires and surveys	Poor	Excellent	Poor	Fair	Yes	Potentially excellent
Behavioral observation	Poor	Excellent	Poor	Poor	No	Potentially excellent
Reports from key staff	Poor	Good	Poor	Poor	No	No clear assurance
News releases and mass media	Good	Poor	Poor	Poor	Yes	Potentially fair to good
Analysis of mass media	Poor	Fair	Poor	Poor	Yes	Potentially fair to good
Day-to-day public contacts	Good	Good	Excellent	Fair	No	Poor
Nominal group process	Poor	Excellent	Poor	Excellent	Yes	Potentially good
Delphi technique	Fair	Excellent	Fair	Excellent	Yes	Potentially excellent

3. People will develop respect and confidence for an agency that openly invites their participation in management.
4. Involvement reduces the likelihood that management decisions will be reversed in court, by new legislation, or by superiors.

The ultimate result desired from public involvement is the development of the best possible fisheries programs to serve the public interest and ensure the long-term well-being of fishery resources.

3.5 STUDIES OF STAKEHOLDERS: A SPECIAL TYPE OF COMMUNICATION

Identifying values, attitudes, satisfactions, and preferences of stakeholders often requires systematic studies of stakeholder groups. Comprehensive studies that use in-depth interviewing and observation techniques in combination with attitude and opinion surveys can effectively yield valuable information about stakeholders. Studies of stakeholders are a form of communication in two ways. First, such studies will systematically obtain information for an agency from a targeted group of stakeholders such as steelhead anglers who use drift boats in Oregon, or largemouth bass anglers in Florida. Improved knowledge of the varying tastes, preferences, expectations, beliefs, economic characteristics, and life-styles of stakeholders helps managers understand them. The effective communicator, whether an educator, diplomat, or fisheries manager, learns to tailor messages and select methods to deliver information based on key characteristics of the intended audience. Second, the process of conducting a study communicates to those contacted and others who learn of the effort that fisheries managers are interested in their opinions, observations, and preferences. This concept is an important message to send to stakeholders, and it can become even more powerful if the people in the study observe that the information they provided was used as input to decision making (Figure 3.4).

3.6 CONCLUSION

Managers can learn much about communication that will facilitate the process of fisheries management. Managers need to: (1) understand the elements of the communication process; (2) understand the process individuals go through in adopting or accepting new ideas, and the kinds of communications that people use in each stage of the process; (3) understand the way new ideas gain acceptance throughout a social system, and the influence that other people have in forming opinions about and accepting new ideas.

Fisheries managers should keep in mind several specific ideas.

1. Contact does not mean communication.

2. Communication is a two-way interchange leading to mutual understanding.

3. The public comprises many individuals and groups of people having a stake in fisheries management. Communication must occur with professionals

Figure 3.4   Examples of survey forms used to gain information from anglers—a critical step in fisheries management.

inside agencies and externally with the public. Understanding each group's characteristics is important for effective communication to occur.

4. Public relations is important for support of agency programs, and effective communication is the vehicle of public relations. Credible performance in fisheries management is the foundation of good public relations and necessary for effective two-way communication with public groups.

5. Public involvement is usually a requisite of contemporary fisheries management and requires effective use of communication skills to realize its full benefit.

Managers should expect that regardless of how hard they try to avoid conflicts with other people over management, they will occur. Knowing this beforehand will allow fisheries managers to deal rationally with conflict situations and manage them to positive resolutions. Courses and textbooks are available that describe techniques to help managers resolve conflicts (e.g., Fisher and Ury 1981).

Finally, communication with all individuals and groups external and internal to agencies is a primary responsibility of the fisheries manager. Ability to communicate effectively will be directly related to the effort put into learning more about the concepts discussed in this chapter and cumulative experience in applying them. Skill in communication is a requirement for a professional to be effective in the fisheries management process. Clearly, fish are not the only focus of the management process. A fisheries manager must communicate effectively with many types of people.

3.7 REFERENCES

Cutlip, S. M., A. H. Center, and G. M. Broom. 1985. Effective public relations. Prentice-Hall, Englewood Cliffs, New Jersey.

Decker, D. J. 1976. The influence of internal communication on the development of the Bureau of Wildlife's public image in relation to deer management in the peripheral Adirondack region of New York State. Master's thesis. Cornell University, Ithaca, New York.

Fazio, J. R., and D. L. Gilbert. 1986. Public relation and communications for natural resource managers, 2nd edition. Kendall/Hunt, Dubuque, Iowa.

Fisher, R., and W. Ury. 1981. Getting to yes - negotiating agreement without giving in. Penguin Books, New York.

Harville, J. P. 1985. Expanding horizons for fishery management. Fisheries (Bethesda) 10(5):14–20.

Heberlein, T. A. 1976. Principles of public involvement. University of Wisconsin, Cooperative Extension Programs, Rural and Community Development, Madison.

Peyton, A. B., and D. L. Talhelm. 1984. Expanding public involvement programs in the Michigan Department of Natural Resources: a recommendation of the Great Lakes Fisheries Advisory Committee. Michigan Department of Natural Resources, Fisheries Division, Lansing.

Pringle, J. D. 1985. The human factor in fishery resource management. Canadian Journal of Fisheries and Aquatic Sciences 42:389–392.

Roedel, P. M., editor. 1975. Optimum sustainable yield as a concept in fisheries management. American Fisheries Society, Special Publication 9.

Rogers, E. M. 1983. Diffusion of innovations, 3rd edition. The Free Press, New York.

Rogers, E. M., and F. F. Shoemaker. 1971. Communication of innovations: a cross-cultural approach. The Free Press, New York. Walters, C. 1986. Adaptive management of renewable resources. Macmillan, New York.

Chapter 4

Legal Considerations in Inland Fisheries Management

BERTON L. LAMB AND BETH A. K. COUGHLAN

4.1 THE ROLE OF LAW IN NORTH AMERICAN SOCIETIES

Sharon Douglas is a fictitious mid-level manager with an agency of the United States. In this capacity, she supervises fisheries and wildlife biologists who are routinely and frequently called upon to consult with other federal and state agencies, as well as private citizens, on the economic development of water resources. These consultations are often neither easy nor pleasant. The biologists must tell citizens, agency representatives, and decision makers that their favored projects are unwise as advocated, because they are harmful to the aquatic life of a stream. Sometimes the projects must be modified or even abandoned in order to protect society's aquatic resources. Douglas has found that her staff biologists are frequently not well prepared for consultations which require not only scientific knowledge, but a healthy measure of diplomacy.

The limitations that Douglas has identified in her associates reflect their lack of understanding of how law works to resolve conflicts in the political atmosphere of regulatory (and other) processes. Although Douglas is a fictitious supervisor, the problem she faces is real. Often managers in fish and wildlife agencies must build an understanding of the law and how it works without formal training. The law is more alive than most professionals recognize and an understanding of the law grows with experience and observation.

4.1.1 How Government Works

Understanding law begins with an appreciation of government. People first introduced to the natural sciences are often attracted by the rigor of analysis and comprehensive perspective provided by the "laws of science." Later they realize that even these scientific laws are never absolute. Experience teaches the scientist to have a healthy skepticism and to be careful about accepting any claim. The same can be said for understanding government, politics, and law.

It is important to recognize that each nation has its own particular culture that broadly defines how government works; this is called political culture. Consider political culture as the environment in which political institutions live and evolve. In North America, nations divide their political institutions into three parts: executive, legislative, and judicial. Beyond this similarity, there are vast differ-

ences among the nations in terms of political history, beliefs about appropriate
political behavior, and the nature of lawmaking. For example, in the United States
a person convicted of a crime may appeal that decision to the highest court in the
land. Once that court decides, it sets a precedent for future cases. Lower courts
must follow the precedent. This is not the system in Mexico, where each case is
decided solely on the basis of the written statute as passed by the legislature. Each
nation has its own view of law depending on its political culture, and that culture
is based, in large measure, on the history of the nation.

In contrast to Mexico, Canada and the United States have histories that reflect
many similarities. Both were colonies of other nations and inherited the British
legal system; both have strong democratic traditions. But these nations also differ
in ways that profoundly affect how natural resource management is conducted.
Not the least of these differences is found in the executive part of their political
institutions. In the United States, with its presidential system, the executive
branch consists of the President, the executive office of the President (including
the Council on Environmental Quality),[1] the President's cabinet, and the departments. For inland fisheries management, the important departments are the
Department of the Interior, Department of Commerce, Environmental Protection
Agency, Department of Defense, and Department of State.

In Canada there is no separate executive branch. The Canadian government is
a parliamentary system in which executive functions are an extension of the
legislature. Of course, there are bureaucracies, led by ministers who are also
elected members of the legislature. The federal ministries that affect inland
fisheries management are the Department of Fisheries and Oceans and Environment Canada. Like the United States, Canada and Mexico have federal systems
that provide for responsibilities in these areas to be shared between central and
regional or local governments. In Mexico and the United States, these responsibilities are shared with several states. The United States suffered a major civil war
to determine the role of the states versus the federal government and Mexico's
system was shaped, in part, by its revolution of 1917. Canada has not experienced
the trauma of civil war, and is still evolving a system in which lawmaking as well
as enforcement are shared with the provinces. Canada's Constitution is embodied
in a series of documents beginning with the British North America Act (1867
[U.K.] R.S.C. 1970, Appendix II, No. 5) with which the British implemented a
plan for confederation that had been worked out in a series of conferences (Hogg
1985). The process of developing a constitution was completed in 1982 when the
Act was adopted with minor adjustments along with the Charter of Rights and
Freedoms (Constitution Act, 1982). The federal system is continuing to develop in
all three countries.

Among democracies, the functions of the executive branch are probably
nowhere more fundamentally different than in North America. Mexico has had a
strong one-party government where the Institutional Revolutionary Party (PRI)
has virtually complete control. This party selects a candidate for President, and
that candidate is elected. The PRI has not lost a national election since 1929. The

---

[1]In May 1993 the U.S. Congress was considering legislation that would disband the
Council on Environmental Quality, transfer responsibility for the National Environmental
Policy Act to the Environmental Protection Agency (EPA), and raise the EPA to cabinet
rank.

opposition parties have experienced success in recent state elections but their lack of a national victory means that while Mexico is technically a democracy, political power is strongly centralized.

The United States exhibits a rather unique political system with its two-party competition and "checks-and-balances" relationship among the three branches of government. The President is very powerful, but that power is limited by the other branches. Canada's parliamentary system is radically different in terms of the chief executive: the Prime Minister is an elected member of the parliament chosen by the majority party to organize a government. Thus, the executive and legislative duties are formally interconnected, giving the Prime Minister substantial control of the legislature. These three patterns are essentially repeated in the organization of each country's state and provincial executives.

The legislative branch is made up of the most important lawmakers in North America. Although the legislature is not the only source of law, a statute created by a legislature is often equated with the law. In Canada and Mexico the majority party formulates policy, and because it controls the legislature, it ensures the passage of statutes reflecting party preferences. In the United States, the majority party is frequently unable to guarantee this sort of legislative success, and statutes are passed through a process of bargaining among members of the legislature.

In Mexico, courts—those denizens of the judicial branch—do not interpret the law, they apply it literally. Comparing the countries illustrates one of the common misunderstandings of law in Canada and the United States—the importance of the heritage imparted by English common law. This is a history Mexico does not have because of its Spanish and French colonial past. In Canada and the United States, courts use both statutory and common law. In those nations, the law grows and develops with each decision a court makes, as well as with legislative changes. Law in Canada and the United States has three sources: statute, court opinion, and common practice. The law in Mexico is much more restricted to statutes created by the legislature.

4.1.2 The Role of Law in Conflicts

Many beginning professionals in fish and wildlife biology see the law as a set of static principles held in place for all time, unless superceded by new laws. This view ignores the dynamic nature of law in resolving conflicts. In fact, it is this conflict-resolution process in which most resource managers uncomfortably find themselves.

Learning that there is little court interpretation of statutes in Mexico might lead to the conclusion that enforcement is simple. Nothing could be further from the truth. Law is applied according to political culture. In Mexico, that culture requires bargaining and compromise among the many agencies and organizations that statutes designate to enforce the laws. The political culture of the United States and Canada also puts a premium on compromise. The three countries perform this bargaining differently, but all use negotiation in protecting the natural environment.

It is in the context of negotiation that law serves as a conflict resolver. Courts provide a formal remedy when bargaining fails. More than that, law is a guide to negotiation and promotes resolution of conflict. Where the law says little about how society should operate, conflict may be resolved by resorting to force. Happily, violence is rarely a part of fisheries management.

4.1.3 The Impact of Law on Fisheries Management

There are many ways in which law impacts fisheries management. Four of the most important are treated here: (1) prescribing rules of conflict, (2) balancing the powers of government branches, (3) defining the powers of the central government, and (4) describing the boundary between legal and political issues.

The first impact of law is in prescribing rules for conflicts. If Sharon Douglas' employees have a poor understanding of law, they probably believe they will end up in court whenever they are involved in a dispute. No matter whether Ms. Douglas is supervising an office in Mexico, Canada, or the United States, this is probably a wrong assumption. Her employees *will* participate in negotiations and enforcement actions outside the courtroom. Law does more than just guide conflict into the judicial system, it also guides the behavior of agencies, sets their missions, constrains their actions, and defines their powers.

The second impact concerns balancing the power of legislative and executive branches. In the United States, the balance between the legislative and executive branches is a kind of dynamic equilibrium that works itself out over time. Which of the two branches decides the priorities of fisheries protection? Who plans? Who budgets? Which branch sets the standards for successful protection? All these questions must be answered by either the executive branch, legislative branch, or both.

The third impact of law on fisheries management is by defining the powers of the central government as opposed to the states and provinces. The first level of law in this regard is the Constitution. Each nation has a Constitution defining basic government powers. The Constitution clarifies such matters as who has the power to regulate trade between the states or provinces. Who owns the land, water, or air—and who owns the fish—may be set forth in the Constitution, statutes, or court cases. All of these questions are subject to litigation and may be answered by court cases. For example, how far can a Canadian province go in regulating the harvest of inland fish populations? Is this a question for the federal or provincial government?

The fourth impact of law is by deciding what is a legal question and what is political. In democracies, almost everything is open for debate. However, some things are regarded as beyond partisan politics. In the United States, the individual states are understood to own the fish and wildlife populations within their boundaries; this is no longer a subject for serious debate. What is debated is the degree of control that comes with ownership. For example, treaties and federal statutes can limit state prerogatives. In another example, it was once thought that the states also owned the water, but this has come under question where the federal government controls the protection of habitat for endangered species, the management of water on national forests and Indian reservations, and interstate transfers.

4.1.4 Who Owns Fish Under the Law?

Most fisheries managers are trained to think in terms of rules and regulations governing the taking of fish, e.g., size limits, closed seasons, closed areas, and permissible equipment (see Chapter 16). In these cases, the agencies are enforcers of clear rules. The result of this perspective is that legal problems may be seen as cut and dried.

Certainly, fish are owned by some jurisdiction, such as a nation, a state, or a province. In that ownership capacity, there is a trust responsibility to protect the fishery for the people. In the United States, that responsibility can result in curtailing water development, commercial harvests, or land use in order to preserve the fishery. Ownership also means that the government can regulate the taking of sport fish and enforce civil or criminal penalties for illegal activity. The police power of states and provinces means that they can abridge the rights of private property owners in order to protect the fishery resource. The inland fishery is owned by the federal government in Mexico and Canada; it is owned by the individual states in the United States.

4.2 ALLOCATING WATER TO FISHERIES PURPOSES

4.2.1 Control of Water Allocation Versus Control of Fishing

The control of fishing and the enhancement of fisheries through hatchery propagation is not enough to provide the public with an adequate inland fishery resource. Inland fisheries management requires more than the regulation of fishing especially in the face of extensive land and water development. What is needed is the protection of habitat. Suitable habitat is a necessity for maintaining fish populations, and while fishing pressure can eliminate a population at a particular site, lack of adequate habitat can eliminate an entire fishery.

Protection of habitat goes beyond the question of controlling harvest and enters the realm of environmental values. In this field, water development is the major impediment to maintenance of fish habitat. Economic development may take the form of diverting water for agriculture, hydroelectric power production, municipal water supply, or impoundments for navigation. Because the states and provinces own the water, they may regulate its use. Generally, water development cannot proceed without water rights. The right to use water is granted by the states and provinces. However, in recent years the federal governments of Canada and the United States have exerted increasing authority over these water allocations. This development has raised the specter of conflict over which jurisdiction has the power to regulate the use of water. Controlling the use of water means controlling the quality of fish habitat. In both the United States and Canada, two basic legal doctrines control the allocation of water.

4.2.2 Riparian Doctrine

Where rainfall is plentiful, it is common to find a water rights doctrine reflecting that abundance. The idea of the doctrine is very simple: persons owning land that abuts a body of water have the right to use that water; persons whose land is not contiguous to a body of water have no such right. Typically, these nonriparian landowners must rely on groundwater. The simplicity of the riparian doctrine belies the confusion that arises when water becomes scarce or many users compete for a limited supply.

When water is scarce, the riparian doctrine applies one of two cardinal principles. The first of these principles is known as the natural flow rule, where a landowner is entitled to use the water in the stream or lake so long as the water body remains substantially undiminished in either quantity or quality. This rule seems to maintain habitat because water right holders cannot diminish the flow of a stream. However, because the doctrine is laxly enforced this protection is minimal. The natural flow rule developed in connection with domestic water uses and an industry powered by water wheels. Such uses either consumed very little water or operated on the basis of run of the river. When other uses began competing for water, the natural flow rule lost its appeal.

The other principle for dealing with scarcity in riparian law jurisdictions is the reasonable use rule. Under this principle, a riparian land owner may use water as long as that use is reasonable. This means that a reasonable use might completely dewater a stream. What is a reasonable use of water? That question is answered in a relative fashion—reasonableness depends on other uses on the stream. If there is only one riparian owner on the stream, almost any use is reasonable. Between two competing users, a court must weigh each claim and determine which is relatively more reasonable. Should a third user arrive on the stream, the determination must be made again—also by a court weighing the relative merits of the three uses.

This system does not work well for the fisheries manager because to regulate this water rights system requires special environmental, water regulation, or fisheries management legislation, or a court finding that environmental protection is somehow required. Typically, there is so much water in riparian doctrine jurisdictions that fish habitat needs have not been an issue and few statutes exist. As a consequence, there is generally little legal leverage for the fisheries manager under the riparian doctrine.

Laws protecting fish habitat have often stemmed from environmental protection statutes empowering agencies to promulgate regulations to protect stream environments. An example can be found in Iowa, where state agencies set streamflow levels below which water users may not divert. When streamflows approach the protected low level, the state informs riparian owners to cease or reduce diversions so that the flow levels are maintained (see generally, Trelease and Gould 1987).

4.2.3 Appropriation Doctrine

The maintenance of instream flows for fish habitat is also cumbersome under the appropriation doctrine, although it is more common. The appropriation doctrine developed as a way to effectively allocate a chronically scarce resource. The heart of the doctrine entails a water right holder understanding that the first person to put water to a "beneficial use" on a stream has the best right to use that water. The second person to apply water to a beneficial use has the second best water right, and so on; "first in time is first in right." The measure of a right is not land ownership as in the case of the riparian doctrine; rather, it is sustained beneficial use. Each jurisdiction has defined what constitutes a beneficial use: they include diversion of water for agriculture, industry, municipalities, mining, and other developmental purposes. These water rights are private property that may be sold and moved. Historically, the concept of beneficial use excluded environmental protection. However, beginning in 1959 instream flows began to be recognized as beneficial uses of water under the appropriation doctrine (Lamb and Doerksen 1990).

Instream flows are accommodated under the appropriation doctrine in one of three ways: (1) appropriations of water are allowed for instream uses, (2) an amount of water is set aside that cannot be used for water rights, or (3) special requirements are placed on diversionary water rights.

First, it is possible for a province or state to declare instream flows for fish habitat to be a beneficial use and allow appropriation of water for that purpose. Such a program involves either a water management agency or a fish and wildlife agency appropriating water for use between two points in a stream, for a certain amount of flow, and for preferred species of fish. This water right would be administered in time of priority like any other right. An example of this system is found in Colorado, where the Water Resources Conservation Board, on the recommendation of the Division of Wildlife, can file for and hold an instream flow right. Because instream flow programs began late in the 1970s, these water rights are very new. Old water rights are said to be "senior," reflecting their more secure status, while newer rights are "junior" because they are less frequently allowed to take water. However, under a provision of the appropriation doctrine a junior user is entitled to have the stream remain in the condition in which it was found. Therefore, a junior instream flow water right holder can prevent senior rights from being moved upstream.

Second, a way to protect aquatic habitat is through the reservation of water. This is very much like a water right. A model reservation system is found in Washington State, where the Department of Ecology sets aside a base flow of water from further appropriation. This agency set-aside is managed as a water right, but it can be revised by administrative action, whereas a water right is held in perpetuity.

Third, some jurisdictions (such as California) allow the fish and wildlife agency to review water rights applications and to suggest adjustments. These conditions usually require that the right holder not use water when flow levels fall below a specified volume (see generally, Trelease and Gould 1987).

4.2.4 Mixed Doctrine Jurisdictions and Special Arrangements

Some jurisdictions, including California and Nebraska, use a mixture of the appropriation and riparian doctrines. One essential difference between the two doctrines is that the riparian doctrine, which governs based on land ownership, cannot be lost by nonuse, whereas an appropriative right can be lost if not persistently applied to a beneficial use. This raises the specter of an instream flow appropriative right being injured without recourse by a long dormant riparian right.

4.2.5 The Public Trust Doctrine

Under English common law as it has developed in the United States and Canada, governments hold fish and wildlife in trust for the people. The government may allow the use and taking of fish and wildlife so long as this management is mindful of the trust responsibility. The public trust doctrine is the "bedrock of modern wildlife regulation" (Veiluva 1981). It is under this doctrine that state legislatures have developed laws to control the use of the resources held in trust for the public.

Similarly, the government has a trust responsibility for other natural resources, including water. Acknowledgment of the public trust for water resources is growing slowly, but in some jurisdictions has come to include the protection of fish and wildlife habitat, coastal access, and aesthetic characteristics. The test for the public trust doctrine is whether the government is cognizant of its trust responsibilities when making decisions about allocating resources to private uses. Failure to consider the public trust may result in a court reversing a natural resource management decision even years after the decision was made.

4.2.6 The Taking of Private Property

The Constitution of the United States (Amendment XIV, Section 1 [1868]) provides that private property cannot be taken by the government without just compensation under due process of law. Thus, when the government wishes to build a highway through a person's land, the government must pay the full market price for the land. This is called "a taking" of private property. However, when a government creates a statute or regulation to promote the public's welfare, safety, or health, property rights may be infringed upon without constituting a taking.

For example, if a state were to pass legislation protecting wetlands from draining and filling in order to preserve the natural environment, landowners might not be able to sell wetlands to developers. This would mean a loss of value to the landowner, but it would not be a taking so long as the landowner still has some economic use of the land. This remaining economic use might be for activities as low in value as growing wild cranberries. Inasmuch as legislation protecting beaches, wetlands, rivers, endangered species, and other environmental values is becoming commonplace, the principle of taking is in a state of flux.

4.2.7 Federal Reserved Rights

Reserved water rights are only an issue in the United States. Beginning with Indian tribes, the U.S. Supreme Court created a doctrine enabling the federal government to obtain water rights in states that follow the appropriation doctrine even though those water rights may have lain dormant for many years.

In 1908, the Supreme Court found that when Congress reserved the land of an Indian reservation, it reserved with that land enough water to accomplish the purposes of the reservation. Succeeding cases have extended the reserved rights doctrine to include other federal enclaves such as national parks and monuments, national forests, national wildlife refuges, recreation areas, and military reservations. The amount of water reserved is the minimum amount necessary to ensure that the purposes of the reservation are not entirely defeated. In a national park, this may mean the entire natural flow of all rivers and streams. In national forests, it may mean that amount of water flow necessary to maintain stream channels. On refuges, it may mean a minimum flow or lake level to propagate fish and waterfowl. In any event, the amount of water that is reserved depends entirely on the purposes of the reservation.

This doctrine is often controversial because it can cause long-time users of water to lose senior standing. If the date of the reservation predates senior water rights, the reservation will become the senior user, adversely affecting the state-granted water rights. Such a result causes a great deal of consternation among water management interests and has led to a long series of convoluted state and federal court cases. The final shape of this doctrine is still unknown.

4.2.8 Interstate Compacts

Under the federal system in the United States, each state administers water within its borders. However, rivers have a way of crossing state boundaries. When rivers flow between and among states, administration of water resources becomes complicated by the plethora of jurisdictions controlling water use. When this happens, states may enter into interstate compacts, which are authorized under the U.S. Constitution (Article I, Section 10, Clause 3). Each one operates as a treaty between the states and requires the consent of the Congress. These compacts have been used to settle disputes over navigation, boundaries, fishing rights, consumptive use of water, and water quality (Doerksen and Wakefield 1975; Goldfarb 1988).

Not all conflicts between the states are resolved by compact. Other techniques include congressional apportionment and judicial apportionment. In disputes between the states, the U.S. Constitution (Article III, Section 2, Clause 2) provides that the Supreme Court has original jurisdiction. The Supreme Court has also held that Congress can itself apportion water among the states contiguous to an interstate navigable stream (Arizona v. California, 29 U.S. 558 [1936]).

4.2.9 Canadian Federal Policy

The Department of Fisheries and Oceans in Canada has determined that "fish habitats constitute healthy production systems for the nation's fisheries and, when the habitats are functioning well, Canada's fish stocks will continue to produce economic and social benefits throughout the country" (DFO 1986). Under the federal Fisheries Act of 1970 (R.S.C. 1970, C.F-14), habitat is defined as those parts of the environment on which fish depend either directly or indirectly for their life processes. The objective of the federal policy is an increase in the natural productive capacity of habitats for the nation's fisheries resources to benefit present and future generations of Canadians (DFO 1986). This objective leads to fish habitat conservation based on the guiding principle of "no net loss of the productive capacity of habitats." The policy also envisions rehabilitating productive habitat in selected areas, and creating habitats where there is some social or economic benefit (DFO 1986). In implementing the Fisheries Act, the Department of Fisheries and Oceans works closely with other agencies, including Environment Canada, the Department of Indian and Northern Affairs, and the provincial governments, to ensure compliance in matters related to habitat destruction, pollution control, habitat improvement, and fisheries research.

4.2.10 Mexican Federal Policy

The focus of the federal government in Mexico has been on economic development (Juergensmeyer and Blizzard 1973). In developing countries it is common to focus on the economy and not give much attention to environmental protection. However, this has begun to change in recent years because of a growing awareness in Mexico of the importance of sound environmental policies in improving the long-term outlook for the national economy (Mumme et al. 1988). In 1988, the federal government enacted legislation greatly increasing the strength of the nation's environmental programs. This statute, the General Law of Ecological Equilibrium and Protection of the Environment, became effective on March 1, 1988 (Diario Oficial de la Federación, January 28, 1988) and is administered by the Secretariat of Urban Development and Ecology. The Ecological Equilibrium statute provides greater authority to prevent and control environmental disruptions and allows state and municipal governments to share with the federal government in resource conservation. The law encourages joint approaches to solving environmental problems. Important provisions of the statute call for protecting natural areas and conserving plants and animals, along with conservation education in elementary schools.

4.2.11 U.S. Federal Policy

In the United States, as in Mexico and Canada, there are many federal statutes relating to management of inland fisheries resources. These include the Endangered Species Act, Anadromous Fish Conservation Act, Clean Water Act, National Environmental Policy Act, Federal Power Act, and the Fish and Wildlife Coordination Act. Among these, the Fish and Wildlife Coordination Act sets forth the national policy probably as well as any statute. The Endangered Species Act is stronger, the National Environmental Policy Act is more comprehensive, and the Clean Water Act has many important provisions, but the Fish and Wildlife Coordination Act sets the tone of federal policy. This act requires that agencies undertaking water development activities—either through construction or issuing permits—consult with the Fish and Wildlife Service, National Marine Fisheries Service, and state fish and wildlife agencies. It does not require that the action agency follow the recommendations of the wildlife agencies. Quite often, however, the views of the fish and wildlife agencies prevail in decision making and projects are revised to take account of their concerns (Hamilton 1980; Clarke and McCool 1986).

4.3 FEDERAL REGULATION OF WATER DEVELOPMENT AND QUALITY

4.3.1 Environmental Regulation in Canada and Mexico

Numerous Canadian federal statutes include water pollution control. The most powerful of these is the Fisheries Act of 1970. Section 33 states that "no person shall deposit or permit the deposit of a deleterious substance of any type in water frequented by fish or in any place under any conditions where such deleterious substance or any other deleterious substance that results from the deposit of such deleterious substance may enter any such water." Judicial rulings have found that the stirring up of silt and sediment, which endangers fish eggs, is covered under this provision. The deleterious substance need not be discharged in harmful quantities to be prohibited; all that must be shown is that a substance is harmful to fish in some concentration.

The Fisheries Act gives the Canadian federal government broad powers in controlling water pollution and other development activities in a stream, although federal regulation must be tied to the harmful effects on fish. Approval of the Department of Fisheries and Oceans must be obtained for any project or undertaking that may possibly harm fish habitat. Proponents of a project submit a statement that includes a description of the proposed undertaking, any probable effects on fish habitat, and the plans for mitigation and compensation of harmful effects. The Department of Fisheries and Oceans can either approve the project, impose conditions on approval, or reject the proposal. In practice, the negotiations can be long and arduous. The Act is administered jointly by the Department of Fisheries and Oceans and Environment Canada. In addition, a federal Environmental Assessment and Review Process was established by the Guidelines Order (S.O.R./84-467). Under this Order the application for specific permits triggers an environmental assessment that must investigate all impacts within federal jurisdiction (Elder and Thompson 1992).

The Canada Water Act of 1970 (R.S.C. 1970, 1st Supp. C.5) authorizes the establishment of federal–provincial committees to consult on water resource issues, including designing specific projects for conservation, development, and use of any waters in which management is of significant national interest. The Act also authorizes studies, quality tests, recommendation of water quality standards, and the design, construction, and operation of wastewater facilities by federal-provincial committees or designated provincial agencies. The Canada Water Act sets forth the "polluter must pay" philosophy, emphasizing that the costs of pollution control should be considered part of production costs. It is one of several statutes governing the pollution of inland waters in Canada (see generally, Cram 1971).

The Mexican federal government has predominant jurisdiction over environmental issues. The Constitution of 1917 declares the ownership of all waters within the nation's boundaries to be vested in the nation. Environmental laws are also issued by the federal government and administered by secretaries (secretariats) at the federal level. These secretariats are equal to U.S. departments and Canadian ministries. Only minor roles are delegated to the state or local governments in Mexico. In the administration of all environmental legislation, enforcement has not been a priority (Mumme et al. 1988).

Environmental protection and pollution control in Mexico are addressed by the Federal Law for the Prevention and Control of Environmental Pollution of 1971 (Diario Oficial de la Federación, March 23, 1971), as amended in 1984, 1987, and updated by the massive Ecological Equilibrium Act of 1988. These laws establish the framework for legislative control and regulation of the environment including pollution control, management, exploitation, and education. The Ecological Equilibrium Act is a detailed ecological plan covering the rational use of water, the aquatic ecosystem, and living resource (DuMars and Beltran del Rio 1988). No authority for state or local governments was originally defined, although all levels of government, along with the Secretariats of Agriculture and Hydrologic Resources, Industry and Commerce, Fisheries, Oceans, and Urban Development and Ecology are involved.

The regulation of pollution in Mexico is targeted at urban areas and the larger lakes, principally Lake Chapala and Lake Patzcauro (see for example, Faudon et al. 1983). Federal involvement is usually in the form of investments in wastewater treatment facilities and encouraging industries to voluntarily comply with existing regulations (Juergensmeyer and Blizzard 1973; Cabrera Acevedo 1978).

Water allocation is an important federal responsibility in Mexico. Several priority uses are defined in the Ecological Equilibrium Act, including urban public service, livestock, irrigation, electrical power for the public, industry, aquaculture, electrical power for industry, landwashing and flood irrigation, domestic, and other uses. The National Water Plan was prepared in 1975 by the Secretariat of Agriculture and Hydrologic Resources to organize and encourage the development of policies for the socioeconomic development of Mexico's water resources. The goal of the plan was to develop water for economic purposes.

In Canada, the Constitution Act of 1867 (R.S.C. 1870, Appendix II, No. 5, see also Department of Justice, Canada, *A Consolidation of the Constitution Acts, 1867-1982* [1982]) grants the provinces jurisdiction over local works and undertakings (Section 92 [10]). These works include the generation and distribution of energy. A provincial agency, such as Alberta's Department of the Environment, is responsible for approving hydroelectric development plans. Federal approval is required under the Fisheries Act for any undertaking that may have harmful effects on fish, and under the Constitution Act for projects on navigable waters (Hogg 1985). The generation of electrical power is exclusively an activity of the federal government in Mexico, and any fisheries protection is worked out by the central government.

Every country needs to build water supply projects for irrigation and municipal needs. A number of books have detailed how the U.S. government has subsidized these projects. Reisner's *Cadillac Desert* (1986), Hundley's *Water and the West* (1975), and Nadeau's *The Water Seekers* (1950) detail how the great cities of the American West—Denver, Los Angeles, San Francisco, Salt Lake City, Phoenix—strove for and found water in pristine valleys, future national parks, wilderness areas, and productive farmlands, moving it hundreds of miles to slake the thirst of city dwellers. Also chronicled are stories of how federal agencies and Congress aided in the development of this water, not only for cities, but for farmers and ranchers. These books argue that subsidizing water development with federal tax dollars is no longer appropriate. Whatever the suitability of some water development, it is clear that the combined constituencies of city and farm are powerful influences on Congress when it considers water projects.

Mexico has been involved in similar projects. Beginning in 1947, river basin commissions under the control of the Secretariat of Agriculture and Hydrologic Resources were established in several regions to develop and administer large-scale irrigation projects. These commissions also became involved in flood control, hydropower development, comprehensive regional planning, and the coordination of the activities of other secretariats and agencies within their region (Barkin and King 1970).

4.3.2   Statutes in the United States

The U.S. statute that seeks to integrate environmental protection policies, including fisheries management, is the National Environmental Policy Act (NEPA) of 1969 (42 U.S.C. Sections 4321-4361). The statute has been effective in bringing federal agencies together to consider the environmental consequences of their actions by requiring that federal agencies prepare environmental impact statements for "major federal actions significantly affecting the quality of the human environment." The courts have interpreted this to mean that federal agencies must give "at least as much automatic consideration to environmental factors" as they do to economic factors (Zabel v. Tabb, 430 F 2nd 199 [1970]). Furthermore, individuals can bring suits against agencies to force them to comply with NEPA (Sierra Club v. Morton, 405 U.S. 727 [1972]). The watchdog of the NEPA process is located in the executive office of the President, at the Council on Environmental Quality (see footnote 1, page 78).

The Fish and Wildlife Coordination Act is often referred to as FWCA (16 U.S.C. Sections 661-666). This act was promulgated in response to the general deterioration of rivers caused by major urban development (Veiluva 1981). The idea is that fisheries will be preserved by forcing federal agencies to work together on construction projects. The FWCA specifies how the studies and recommendations of the Fish and Wildlife Service, National Marine Fisheries Service, and state fish and wildlife agencies will be reported to and considered by the Army Corps of Engineers, Bureau of Reclamation (as well as other agencies of the Department of the Interior), Department of Agriculture (especially the Soil Conservation Service), and the Congress as they plan impoundments, diversions, dredging, and navigation projects. In short, the Act requires that construction agencies consult with fish and wildlife agencies in order to determine how to compensate or mitigate for losses to fish and wildlife resources resulting from the project. The construction agencies are to give full consideration to the reports of these fish and wildlife agencies. Veiluva (1981) argues that FWCA is an ineffective mechanism for protecting fisheries because, at least into the late 1970s, federal action agencies continued to issue permits and build projects over the objections of fish and wildlife agencies. More recently, the Department of the Interior, through the Fish and Wildlife Service, has developed joint working agreements with the Army Corps of Engineers and the Environmental Protection Agency to improve the consultation process.

The Fish and Wildlife Coordination Act has an impact on federal hydropower licensing in the United States. The Federal Energy Regulatory Commission issues licenses to private parties who wish to develop hydroelectric power facilities. Under a number of authorities, including the Federal Power Act (16 U.S.C. 791a-828 [1982]), the Commission may approve hydropower projects and must take into consideration the fish and wildlife impacts of these developments. The FWCA is a further reinforcement of the requirement to consult with the fish and wildlife agencies before issuing a license to anyone wishing to generate electric power. The Commission has routinely given weight to the opinions of the Fish and Wildlife Service, National Marine Fisheries Service, and state fish and wildlife agencies. This does not mean that the task is easy for fisheries managers faced with evaluating hydropower projects. In fact, the large number of hydropower license applications, dearth of information on fish species life requirements, and the federal incentive program encouraging hydropower projects has meant that fish and wildlife agencies are often overwhelmed.

The Clean Water Act (33 U.S.C. Section 1251 et seq.) has a long and fascinating history in the United States. It has evolved as the nation's understanding of pollution has developed incrementally. Together with the other environmental statutes—FWCA, NEPA, and aspects of the Federal Power Act—the Clean Water Act helps weave together a fabric of protections. The history of this extensive statute can be traced from the Water Quality Act of 1965 (33 U.S.C. Section 1151) to the present statute. This law redefines pollution from what was considered conventional wisdom in 1965. In Section 502, it is defined as the "man-made or man-induced alteration of the chemical, physical, biological, and radiological integrity of water." Goldfarb (1988) argues that what Congress meant by the "integrity of water" is its ecological stability. Before 1972, the public did not have a right to clean water, only to water not polluted beyond an individual state's standards. "Pollution was defined as excessive discharge" (Goldfarb

1988). The goals of the Act today include elimination of the discharge of pollutants into U.S. waters, protection of aquatic life, prohibition of the discharge of pollutants, financial assistance for the construction of wastewater treatment plants, and research funding.

The Clean Water Act is an ambitious statute. Perhaps the most important and controversial aspect for fisheries management is Section 404 because it affects state water allocation practices and influences a vast array of nonfederal water developments. This section requires developers to acquire a permit from the Army Corps of Engineers before discharging dredge or fill material into the "waters of the United States." Much broader than navigable waters, this description of the jurisdiction of the Corps takes in almost all water courses. Goldfarb (1988) points out that intrastate streams, freshwater wetlands, drainage ditches, mosquito canals, and intermittent streams are all covered by Section 404. This has led to a pitched battle between environmentalists and developers over who really controls water allocation, the states or federal government. On the advice of fish and wildlife agencies, the Corps has denied permits under this statute, strictly on ecological grounds. This may have the effect of superceding state-granted water rights.

Another similar statute is Section 10 of the Rivers and Harbors Act of 1899 (33 C.F.R. 320 et seq.). This section prohibits the unauthorized construction in or alteration of any navigable water of the United States. Under the statute, the Corps issues permits for "the construction of any structure in or over any navigable water . . . the excavation from or depositing of material . . . or the accomplishment of any other work affecting the course, location, condition, or capacity of (navigable) waters" (see generally, Goldfarb 1988). Of course, Section 10 is administered within the consultation requirements of the Fish and Wildlife Coordination Act.

The Fish and Wildlife Service (FWS) also operates under the Anadromous Fish Conservation Act (16 U.S.C. Section 757 [1965]) and a host of other statutes, including the Federal Aid in Fish Restoration Act (16 U.S.C. Sections 777–777k as amended). The Anadromous Fish Conservation Act provides that the FWS can plan and conduct research and develop programs affecting anadromous species whenever two or more states have a common interest in any river basin. Programs can be designed to conserve, enhance, or develop anadromous fishery resources nationwide. The Federal Aid in Fish Restoration Act is often referred to as the Dingell–Johnson Act after its main sponsors and as Wallop–Breaux (P.L. 98-369 [1984]) after the sponsors of the latest amendment. The Act is one of several providing federal aid to states for management and restoration of fish. Funds for this aid come from taxes on the sale of fishing equipment, boats, motors, and motorboat fuel.

Whenever a U.S. federal agency undertakes an action, such as building a water resources project or issuing a permit for construction, dredging and filling, or hydropower production, it must ask the Fish and Wildlife Service if any threatened or endangered species are present at the project site. This is the main element of the Endangered Species Act (16 U.S.C. Section 1531 et. seq.). The act has four approaches to protect endangered species: listing of species, agency consultation, mandatory agency responsibility to conserve endangered species, and a ban on any taking of endangered species. The first step is to list endangered species, and no action can be taken under the Act until this has been completed.

Once a species is listed, it cannot be taken either through collection or harassment; federal agencies are prohibited from actions adversely affecting the species. When the Fish and Wildlife Service makes a tentative determination that an endangered species is present at a project site, the action agency must submit a biological assessment to the FWS. If the assessment identifies potential adverse impacts, the FWS prepares a biological opinion setting forth whether or not the action will likely jeopardize the species. The opinion may recommend no project, mitigation, or conservation measures. If differences over a project's effects on the species cannot be resolved, a federal committee may grant an exception to the otherwise absolute ban against injuring endangered species.

This is a stringent law with controlling power over many projects. It is not administered as prohibitively as might be expected, however. Yaffee (1982) has demonstrated that throughout the history of the Act, the FWS has bargained rather than regulated the impacts on endangered species. Chapter 15 deals extensively with the implementation of this Act.

4.4 ENVIRONMENTAL PROTECTION: STATE AND LOCAL REGULATION

4.4.1 State and Provincial Environmental Protection Statutes

In a recent count, 30 U.S. states had "little NEPAs" or state Environmental Protection Acts. These statutes are similar to the federal NEPA and are applicable to state and local projects. Typically, they require a state environmental impact statement. These little NEPAs are operative for state projects where the conditions placed on the projects by the state do not conflict with the paramount jurisdiction of the United States under a law passed by Congress. Generally speaking, when Congress has acted rightfully under the Constitution, the states may not contradict Congressional intent. A good example of this would be where a state attempts to prevent construction of a hydroelectric project that has a permit from the Federal Energy Regulatory Commission. The state is likely to lose this battle because Congress has authorized the Commission, under the Commerce Clause of the Constitution, to regulate hydropower (see generally, Trelease and Gould 1987).

The authority of Canadian provinces to enact environmental protection legislation is based on the British North American Act of 1867. These policies were carried forward in the Canadian Constitution of 1982, giving the provinces jurisdiction over property and civil rights. Pollution control legislation exists in all the provinces in the form of a Water Resources, Pollution Control, or Public Health Act. These statutes are administered by a provincial water, pollution control, or health board. Under these statutes, activities that could cause pollution to waters within the province are regulated and licensed. The jurisdiction over and liability for activities that cause pollution in another province or in interprovincial waters has not been clearly determined (Morgan 1970; Hogg 1985).

State and local governments in Mexico received the authority, under a 1982 revision of the Law to Prevent and Control Environmental Pollution (Diario Oficial de la Federación, March 23, 1971), to enact environmental legislation. The actual power to take action, however, is restricted due to lack of funding. In 1987, the federal government provided some modest funding for awareness campaigns and educational programs. The Ecological Equilibrium Act of 1988 extends some of these programs, but implementation remains problematic.

4.4.2   Health Codes

In the United States, individual states have enacted legislation for the protection of public health. These laws are enacted under state "police powers" and range from wide-reaching to ineffective in their impact on the protection of the natural environment. Control of pesticides and toxic substances have important ramifications for fisheries management. Just as in the United States, Canadian municipalities may construct and operate water and wastewater treatment facilities. Public health acts provide the provincial departments of health with the power to control unsanitary conditions.

Mexico, too, has these health codes. The Secretariat of Agriculture and Hydrologic Resources is responsible for the control of all wastewater discharges and the regulation of water quality for industrial use. The Secretariat of Health and Welfare has the authority under the Federal Law on Environmental Protection (Diaro Oficial de la Federación, January 11, 1982), which combines several older statutes to protect the public health. However, in 1975, only 50% of Mexico's population was served by municipal water supplies, while sewage treatment facilities were available to only 25% of the population. Disposal of toxic wastes is not regulated (Juergensmeyer and Blizzard 1973; Mumme 1985; Mumme et al. 1988). All the nations of North America have been painfully slow in recognizing the problem of toxic waste.

4.4.3   Land-Use Planning, Special Districts, and Commissions

One of the most bewildering aspects of fisheries management is the plethora of state and local boards, districts, and commissions that have some sort of impact on fish habitat. These can range from the municipal planning board, to water conservation districts, and even to intergovernmental councils. There are many organizations that develop water or help local governments plan for economic and natural resource development. Examples in the United States are councils of government in which several cities and counties may be represented in planning efforts. Examples in Canada are the Canadian Council of Resource and Environmental Ministers, which was formed to enhance interprovince communications, as well as the Prairie Provinces Water Board which helps allocate the water of interprovincial rivers. There are literally tens of thousands of such organizations, most of which have public participation programs, planning, and rule-making authority.

4.4.4   State and Provincial Fish and Wildlife Agencies

State fish and wildlife agencies in the United States enact their own bag limits, sell fishing licenses, designate special fishing areas, and enforce fish and game laws. In some states, game wardens are deputized law officers. States operate fish hatcheries and stocking programs, conduct research, and manage nongame populations. Some state agencies manage commercial fishing. All of this is done under the state's trust responsibility to protect the wildlife resources of the state, and state legislatures pass statutes to effect this purpose.

Less well understood is the state fish and wildlife department's work in commenting on other state and federal programs. These programs range from Federal Energy Regulatory Commission licenses—where the state must make a Clean Water Act Section 401 certification that any discharge will comply with state standards—to recommending streamflow levels under the state's instream flow program (Lamb and Doerksen 1990).

In Canada, fishing regulations established by the provinces are enacted as federal regulations under the Fisheries Act. These regulations are administered by provincial fisheries agencies in Ontario, Manitoba, Saskatchewan, Alberta, Quebec, and British Columbia. The federal government directly manages the fisheries of Newfoundland, New Brunswick, Nova Scotia, Prince Edward Island, and the Northwest and Yukon Territories.

4.4.5 State and Provincial Endangered Species Protection

State fish and wildlife agencies provide information to the Fish and Wildlife Service, petition for changes in the status of endangered species, and help enforce the provisions of the federal Endangered Species Act. More than two-thirds of the states have some form of endangered species law or regulation. Most of these state statutes are copies of the federal law. Many states also have cooperative agreements with the Fish and Wildlife Service. Yaffee (1982) summarized the situation when he observed:

> The state agencies, whose historic constituency is hunters and fishermen, are by and large game animal agencies. Their professional traditions are those of conservation and management of wildlife to produce huntable surplus. Most of their programs are financed principally by hunting and fishing license fees.
>
> Because of the fear of their game animal bias, the states were not given a larger scale in the implementation of the ESA.

Endangered species in Canada are protected by provincial legislation such as that of Ontario, which lists species of plants and animals that are considered endangered and makes it an offense to kill, injure, or destroy the habitat of these species. Such legislation is administered by provincial fish and wildlife departments (Burton 1984).

4.5 REGULATING THE TAKE

4.5.1 The Connection Between Regulating and Managing

Regulating is a tool of management (see Chapter 16). When to regulate, what to regulate, and how to regulate are all policy questions. In the United States, Canada, and Mexico, the power to regulate is derived from some type of ownership of the resource, plus some type of authority to control the harvest in order to protect the public. In the United States, this authority is referred to as the "police power," the power to control individuals for the good of society, and it must be balanced against the rights of individuals—including property rights, religious freedom, and free speech. However, the power to regulate is not an end in itself. How the resource is regulated depends on management philosophy. The right to regulate may not change, but how states and provinces carry out their regulatory responsibilities depends on political culture and management decisions.

Traditionally, fish and wildlife departments in the United States have concentrated on producing enough fish and game to provide adequate hunting and fishing experiences. The environmental movement has led to more emphasis on nongame species and management for ecological protection, but the primary emphasis has remained service to a hunting and fishing constituency.

Competition occurs between environmental and harvest (conservation) uses, but there is international competition as well. All managers pay attention to international treaties as well as the impact of high-seas harvests of anadromous stocks by domestic and foreign commercial fishing enterprises. Foreign fishing interests are not regulated by the states or provinces when fishing is outside a country's territorial limits, but the impact must be counted in domestic regulations. Thus, regulation of fishing is guided by a management philosophy. That philosophy must be based on the law, and reflect good management practices in the face of many competing uses of the resource.

4.5.2 Regulating Sport Fishing

State and federal government in the United States share power over fishery resources. This sharing is an evolving, complicated concept arising out of conflict over states rights and federal powers. There is little doubt that the states do have the right to regulate the taking of wildlife within their borders (see generally, Bean 1977). A classic case setting forth this right to regulate is Geer v. Connecticut (161 U.S. 519 [1896]), in which the U.S. Supreme Court found that the states have the right "to control and regulate the common property in game." Going further, the Court said that "the wild game within a state belongs to the people in their collective sovereign capacity." However, in the same case, the Court said that the states have this power "insofar as exercise may not be incompatible with, or restrained by, the rights conveyed to the federal government by the Constitution."

Three rights conveyed to the U.S. federal government are important in limiting state power. These are the treaty power, the commerce clause, and the property clause. Briefly, the federal government has the exclusive right to conclude treaties with foreign powers, and those treaties have the force of the Constitution; the Congress can regulate interstate commerce—from which comes most of the nation's environmental legislation; and Congress has the right to manage federal property such as national forests, parks, and refuges. Finally, the federal government is also trustee for American Indians. In this trust relationship, it must protect the various tribes. The states own the game within their boundaries, but the federal government can often tell the states how to behave.

A similar sort of shared power over fishery resources exists in Canada. Besides the authority to prohibit the discharge of deleterious substances into waters frequented by fish, the Fisheries Act grants the Minister of Fisheries and Oceans complete discretion to issue and determine conditions of fishing leases and licenses, as well as authority to regulate a wide range of activities relating to fishing. In practice, the administration of fishing regulation is shared with the provinces (VanderZwaag 1983). Where anadromous fisheries are concerned, the federal agencies do indeed fully manage these fish stocks and administer the

Fisheries Act. However, in all provinces, freshwater fisheries are managed by the provincial agencies and those agencies actually manage the federal Fisheries Act. For example, Alberta has passed fisheries regulations pursuant to the federal Fisheries Act. The province administers the act subject to annual federal approval of regulations. Provincial administration of the Fisheries Act is accomplished through what might be best termed an administrative understanding between Alberta and the federal government.

Mexico's fishing regulations are more centralized. The Federal Law on Fisheries Development was enacted in 1972, although fishery laws have existed since the 1920s (Torres García 1987). The Secretariat of Fisheries was established in 1976, and elevated to cabinet status in 1982. Known as SEPES, this Secretariat is responsible for the management and development of the nation's fishery resources. The Secretariat has been primarily concerned with the development of Mexico's marine fishing industry and is responsible for overseeing the operation of the national fishing industry. At the federal level, the Secretariat consists of four divisions: fishery development, infrastructure, aquaculture, and the National Institute of Fish. Thirty-one delegates represent the interests of the states. The Secretariat also coordinates the activities of the National Fishery Collectives Commission, which is an advisory board consisting of representatives from other cabinet-level departments and the fishing industry (Cicin-Sain et al. 1986).

4.5.3 Regulating Commercial Fishing

The government of Canada is also involved in the management of commercial fisheries. This too is shared informally with the provinces. Several acts affect federal financial involvement in the fishing industry. These acts, plus the Department of Fisheries and Ocean's complete discretion on fishing leases and licenses, form a strong regulatory base (VanderZwaag 1983).

The first significant legislation regulating fishing was enacted in 1857. This act dealt primarily with protecting the salmon fisheries of Canada's rivers and estuaries, and developing the marine fishery resource. The legislation defined the types of gear and fishing practices allowed, established closed seasons and licensing and leasing of fishing rights, and has provided the basis for the current policy for the regulation of fishing. The Minister of Fisheries and Oceans' licensing authority does not provide exclusive federal control of fishing because some provincial jurisdiction has been established by the courts under the constitutional role of the provinces to regulate property rights. As a result, the federal and provincial governments cooperate to regulate both commercial and sport fishing in inland waters. Provincial departments are responsible for developing regulations, although these must be enacted at the federal level (Scott and Neher 1981).

Mexico's history has been one of regulating and promoting the marine fishery. There are a number of important inland fisheries as well. An example of these is Lake Chapala, near Guadalajara. As is the case elsewhere, fishing cooperatives harvesting from this lake are regulated by the federal government. In the United States, individual states regulate the commercial take. An example of this is the paddlefish fishery in Iowa, where the state sets the times and means of fishing, monitors the commercial harvest, and makes adjustments to assure that the population remains viable. All the surrounding states also regulate the paddlefish commercial take in the same way. This means that without close interstate cooperation—which is a sometimes thing—regulating the population by control of timing and methods can be an uncertain process.

4.5.4 Regulating Native American Fishing

Indian fishing in the United States is an area of great uncertainty. Some tribes have a right to virtually unregulated subsistence fishing, as in Alaska, while others have treaty rights. Though states can regulate some aspects of tribal fishing, the federal government's trustee status makes it an important actor in this process.

The United States recognizes 507 tribes and Alaska native groups. These tribes and groups have a total population of 1.4 million. The federal government is the trustee for the lands and natural resources held by these tribes. The Bureau of Indian Affairs, within the Department of the Interior, is the government agency charged with representing Indian interests. In addition to this trust relationship with the federal government, Indian tribes are, to a certain extent, sovereign nations that have binding treaties with the United States. Even though the tribes are part of the states in which they reside, they are exempt via the treaties from many state laws, pay few state taxes, and may collect royalties for the extraction of natural resources.

In the area of fisheries management, states can and do regulate aspects of tribal fishing. Generally the tribes have treaties that protect some element of their sovereignty in regard to fishing. Tribes can control sport and commercial fishing on the reservations and under treaty may have relatively open access to fishing sites off the reservations. The limits that may be placed on the take by Indian fishers is a matter of controversy. States have a responsibility to conserve viable fish populations, tribes have sovereign rights under treaty to take fish, and the federal government has a trust responsibility to protect Indian natural resources. Any attempt to summarize the long line of court decisions that have tried to make sense out of this jumble would fail to be brief, but a fairly recent case is, perhaps, a good place to start. In Washington v. Washington State Commercial Passenger Fishing Vessel Association (99 S. Ct. 3055 [1979]), the Supreme Court seems to have upheld a kind of fair-share approach to Indian fishing rights off the reservation. The Court said that, at least for certain types of treaty tribes, a 50-50 split of the harvestable fish with nontribal fishers is a fair result. What is "harvestable" is in debate because it means leaving some fish in the stream for conservation and reproduction. Good introductions to these concepts are provided by Landau (1980), Miklas and Shupe (1986), and Shupe (1986).

Native rights in Canadian law stem from "aboriginal rights," the property rights of Native Peoples that are theirs by virtue of their occupation of certain lands from time immemorial. These rights were first recognized in the Royal Proclamation of 1763 ([U.K.], R.S.C. 1970, Appendix II, No. 1) when Britain forbade settlement on Indian lands and decreed that such lands be obtained only by purchase or cession. Treaties were also made in recognition of these rights, granting the right to unrestricted fishing and hunting at all times on all unoccupied government lands or any other lands to which Native People may have a right of access. As a result, all Native Peoples can claim fishing rights, either under a treaty or through aboriginal rights. It is unclear to what degree aboriginal rights provide protection of fishing and hunting in the absence of a treaty. Native rights are under federal jurisdiction and provincial laws can make only limited infringements on any rights granted under a treaty (Section 88, Indian Act). For example, certain sections of the Alberta Wildlife Act specify that while Native Peoples may hunt year-round for food, they cannot sell the meat or use automatic weapons.

At one time, the federal parliament could enact legislation limiting treaty rights. Such limitations of hunting rights were included in legislation and had been upheld by the courts (Cummings and Mickenberg 1972). However, as with other areas of law, this has changed in recent years as the duties of the federal government regarding Native Peoples have been defined. The Constitution Act (Sec. 35[1], 1982) provides protection of "existing aboriginal rights." The meaning of this and other provisions is still being worked out in the courts. A recent decision by the Supreme Court of Canada summarized the law as follows: "Federal power must be reconciled with federal duty and the best way to achieve that reconciliation is to demand the justification of any government regulation that infringes upon or denies aboriginal rights" (Ronald Edward Sparrow v. Her Majesty the Queen, et al., Supreme Court of Canada, File No. 20311, 3 November 1990).

4.5.5 International Treaties

International treaties are negotiated by national governments subject to the provisions of their constitutions. In the United States, this means negotiation by the President and subsequent ratification by a two-thirds vote of the Senate. There are several treaties in force between the United States and Canada affecting fisheries; three such treaties are also in force between the United States and Mexico (Rohn 1983). Important treaties regulate the boundary waters between nations and the take of marine fisheries. Because treaties become the supreme law of the land—equivalent to the Constitution—it is imperative that they be worked out carefully and that the states and provinces be cognizant of their provisions. To this end, treaties frequently establish commissions that are responsible for approving and carrying out projects affecting treaty-protected resources. An example of this is the International Joint Commission covering boundary water issues for the United States and Canada, and the Great Lakes Fisheries Commission (Dworsky 1986; Sadler 1986).

4.6 WHAT THE FISHERIES MANAGER MUST KNOW: THE BALANCING ACT

At the opening of this chapter, the problem faced by Sharon Douglas was presented. That fictitious problem might be summarized in this quote: "We were never trained to work in an arena where everything is controversial and everything is political" (quoted in Booth 1988). The biologist who said this might well have added that science is rarely the controlling factor in the decision-making process. Fisheries management is a bewildering mixture of the scientific, legal, and political, where the technically trained manager must integrate professional and pragmatic judgments about policy outcomes.

4.6.1 The Essential Points of Law

The essential points of law fall into six categories: constitutional, treaties, Native Peoples rights, fishing regulation, water allocation, and property rights. Constitutional law means discerning those powers given to the federal government from those reserved to the states or provinces. Another way of saying this is to ask, Who has jurisdiction? The answer to this question is evolving in every nation, but the trend seems to be toward central control.

Treaty making is a constitutional power, but it has evolved a new area of the law affecting fisheries management. The combination of Indian water rights and Indian fishing rights leads to complexity for any manager seeking to conserve and protect fish habitat and populations. The final answer on who controls this important resource is not fully known, and may not be known for some time. Treaties, of course, are also international. These formal international agreements are important considerations for the manager because their provisions may be paramount over other statutes.

The rights of Native Americans are a further intervening variable in fisheries management. These rights arise from treaties, aboriginal rights, and federal trusteeship. Native Americans have a right to fish for subsistence, religious, and commercial purposes and for maintenance of a moderate standard of living. These rights are to be balanced against state ownership responsibilities, including the duty to provide self-sustaining fish populations.

These rights may be abridged by federal statute in both the United States and Canada—this may happen in the case of an endangered species, for example—but is politically difficult to accomplish.

Fishing regulation is the responsibility of the states in the United States and the federal governments in Mexico and Canada. In practice, fishing regulation is shared between and among jurisdictions in each country. Treaties, federal trust responsibilities, and the regulation of interstate commerce gives the United States power to affect state regulation, while property rights make Canadian provinces strong forces in that country's fishing regulation.

Water allocation is important because it affects fish habitat. As one noted biologist has commented, "We've become so focused on preserving a species here and a species there that we are losing sight of the habitat" (quoted in Booth 1988). The water allocation systems of the North American continent historically have not considered habitat preservation as a beneficial use of water, worthy of legal protection. This situation has been slowly changing, but taking an ecosystem view of water rights is not on the legal horizon. Most jurisdictions still give water in perpetuity to the first user who invests money in developing the resource. Environmental uses of the water are struggling to achieve a strong place in that system.

Property rights stand for a host of legal doctrines and policies that essentially tell landowners what they can or cannot do with their property. The states have a police power to regulate unsafe practices and may legislate land uses to protect the public. The federal power to regulate interstate commerce has extended to the control of pollution. The public trust doctrine means that states cannot dissolve their ownership of public land or natural resources without first considering their obligation to the public to protect those resources. All of these are limits on private property.

4.6.2 The Essential Points of Politics

It is often remarked that "politics is the art of the possible." Learning what can be done and how to accomplish goals is the key to the political art. Four kinds of politics concern the fisheries manager.

First, electoral politics are of concern as they would be to any citizen. The process of selecting good leaders is important in a democracy. Second, legislative politics is sometimes important to fisheries management because the manager is called upon to report to the legislature, provide background for pending legislation, or provide expert advice on policy. It is important to understand these processes and to be attuned to the needs of the legislative branch in order to promote the democratic process. As public servants, managers have an obligation to be well-informed and efficient in carrying out legislative programs.

Third, judicial politics is uniquely frustrating and complex. The court system provides a formal means of resolving disputes, related to the rules of advocacy and fair play in debate. As such, there are many rules that seem foreign to the casual observer, but which serve to assure each side protection until a judge or jury can decide. This is a form of politics that the fisheries biologist will likely face in one of two capacities: expert witness or client. As an expert witness, the fisheries biologist is a servant of the court. The biologist will be called to testify by one side, based on special knowledge. Imparting that knowledge to the court means using expert scientific judgment to give the court enough information to decide. The adversary process can make this a formidable task. As a client, the biologist is the one who guides the preparation of the case. The client uses lawyers to make sense out of the trial process. It is the client whose interests are represented and who must instruct counsel on pertinent facts, agency policy, and procedures. It is based on this knowledge that lawyers can present an adequate case.

Fourth, implementation politics requires an additional skill. This political arena deals with the process for working out how statutes will be administered by the executive branch. The task of implementing statutes may sound straightforward, but it is not. Yaffee (1982) has demonstrated how even the most clear statute may require interagency bargaining at all phases of implementation. The fisheries manager must come to recognize these bargaining situations and plan for them. This means combining a knowledge of both the law and the past behavior of agencies to develop and carry through an effective bargaining position.

4.6.3 Expertise of Others

What should be evident from this discussion is that modern fisheries management is multidisciplinary. Certainly the manager—like the fictitious Sharon Douglas described at the beginning of this chapter—must have a strong knowledge base in a number of fields. But more than that, the manager must have a sense of how to seek out needed expertise in order to complete the administrative task.

It is imperative that the manager have an understanding of the relevant law and how it works. Knowledge of how law is made, interpreted, and enforced is crucial. When faced with a new legal problem, the manager must know where to turn to develop understanding of the history of the statute or case, its precedents, and relevant procedures. Building this knowledge will lead the manager beyond the law itself into the fields of hydrology, meteorology, chemistry, physics, and other technical subjects. Because management deals with people and agencies, however, problems almost always involve other disciplines, including public administration and applied policy analysis. It is in these areas that the manager comes to fully appreciate the legal considerations in fisheries management.

4.7 REFERENCES

Barkin, D., and T. King. 1970. Regional economic development: the river basin approach in Mexico. Cambridge University Press, London.

Bean, M. J. 1977. The evolution of national wildlife law. Praeger, New York.

Booth, W. 1988. Reintroducing a political animal. Science (Washington, D.C.) 241:156–159.

Burton, E. C. 1984. Enforcement of natural resources litigation. Carswell Company, Ottawa.

Cabrera Acevedo, L. 1978. Legal protection of the environment in Mexico. California Western International Law Journal 8:22–42.

Cicin-Sain, B., M. K. Orbach, S. J. Sellers, and E. Mansanilla. 1986. Conflictual interdependence: U.S.–Mexican relations on fisheries resources. Natural Resources Journal 26:769–792.

Clarke, J. N., and D. McCool. 1986. Staking out the terrain. State University of New York Press, Albany.

Cram, J. S. 1971. Water: Canadian needs and resources. Harvest House Ltd., Ottawa.

Cummings, P. A., and N. H. Mickenberg. 1972. Native rights in Canada. General Publishing, Toronto.

DFO (Department of Fisheries and Oceans). 1986. National policy on fisheries. Environment Canada, Ottawa.

Doerksen, H. R., and G. Wakefield. 1975. Columbia basin compact issues review. Pacific Northwest Regional Commission, Washington State Water Research Center, Pullman.

DuMars, C. T., and S. Beltran Del Rio. 1988. A survey of the air and water quality laws of Mexico. Natural Resources Journal 28:787–813.

Dworsky, L. B. 1986. The Great Lakes: 1955–1985. Natural Resources Journal 26:291–336.

Elder, P. S., and D. Thompson. 1992. Rivers and Canadian environmental impact assessment. Rivers 3:45–52.

Faudon, E. E., E. F. Tritschler, and G. J. E. R. Michel. 1983. Lago de Chapala; investigación actualizada 1983. Universidad de Guadalajara, Instituto de Geografía y Estadística, Guadalajara, México.

Goldfarb, W. 1988. Water law, 2nd edition. Lewis Publishers, Chelsea, Michigan.

Hamilton, M. S. 1980. Summaries of selected federal statutes affecting environmental quality. Colorado State University, Cooperative Extension Service, Fort Collins.

Hogg, P. W. 1985. Constitutional law in Canada. The Carswell Company, Toronto.

Hundley, N. 1975. Water and the west: the Colorado River compact and the politics of water in the American West. University of California Press, Berkeley.

Juergensmeyer, J., and E. Blizzard. 1973. Legal aspects of environmental control in Mexico: an analysis of Mexico's new environmental law. Pages 101–102 in A. Utton, editor. Pollution and international boundaries. University of New Mexico Press, Albuquerque.

Lamb, B. L., and H. R. Doerksen. 1990. Instream water use in the United States—water laws and methods for determining flow requirements. Pages 109–116 in 1987 National water summary. U.S. Geological Survey, Water-Supply Paper 2350:109–116.

Landau, J. L. 1980. Empty victories: Indian treaty fishing rights in the Pacific Northwest. Environmental Law 10:413–456.

Miklas, C. L., and S. J. Shupe, editors. 1986. Indian water 1985. American Indian Resources Institute, Oakland, California.

Morgan, F. 1970. Pollution—Canada's critical challenge. Ryerson Press and Maclean-Hunter, Toronto.

Mumme, S. P. 1985. Dependency and interdependence in hazardous waste management along the U.S.–Mexico Border. Policy Studies Journal 14:1601–68.

Mumme, S. P., C. R. Barth, and V. J. Assetto. 1988. Political development and environmental policy in Mexico. Latin American Research Review 27:7–34.

Nadeau, R. A. 1950. The water seekers. Peregrine Smith, Santa Barbara, California.

Reisner, M. 1986. Cadillac desert: the American West and its disappearing water. Viking Penguin, New York.

Rohn, P. H. 1983. World treaty index. ABC-Clio Information Services, Santa Barbara, California.

Sadler, B. 1986. The management of Canada–U.S. boundary waters: retrospect and prospect. Natural Resources Journal 26:359–376.

Scott, A., and P. A. Neher. 1981. The public regulation of commercial fisheries in Canada. Minister of Supply and Services, Ottawa, Canada.

Shupe, S. J. 1986. Water in Indian country: from paper rights to a managed resource. University of Colorado Law Review 57:561–592.

Torres García, F. 1987. El marco legal de la pesca en Mexico. Pages 185–193 in Sintesis pesquera, 1982–1987. Secretaria Pesca, Mexico City.

Trelease, F., and G. A. Gould. 1987. Water law case book. West Publications, Minneapolis, Minnesota.

VanderZwaag, D. L. 1983. The fish feud. Lexington Books, Toronto.

Veiluva, M. 1981. The fish and wildlife coordination act in environmental litigation. Ecology Law Quarterly 9:489–517.

Yaffee, S. L. 1982. Prohibitive policy: implementing the federal endangered species act. MIT Press, Cambridge, Massachusetts.

FISHERY ASSESSMENTS

Chapter 5

Dynamics of Exploited Fish Populations

MICHAEL J. VAN DEN AVYLE

5.1 INTRODUCTION

Knowledge about the dynamics of fish populations is essential for developing management plans and evaluating management success. In the context of fisheries management, population dynamics includes estimation of changes in population numbers, composition, or biomass. Population dynamics can be subdivided into two areas having different applications in fisheries management. The first area of application, termed population assessment, focuses on estimating characteristics of certain populations and comparing them with other populations or with the same population over time. Assessments are important for describing fluctuations in abundance, measuring responses to exploitation or other perturbations, and defining fishery components that could be successfully managed. The second area of application uses mathematical models to predict future trends. Models have been used to forecast changes in fish population size and harvest by assuming certain levels of fishing effort. Estimates of parameters in the models are obtained from assessment studies. The most common application of forecasting techniques has been to predict yield of commercial fisheries, but several models have proven useful for predicting trends for sport fisheries.

5.2 CHARACTERISTIC DYNAMICS OF INLAND FISH POPULATIONS

5.2.1 Population Biology and Regulation

Exploitation reduces the number of harvestable-size fish in a body of water, but the influences on long-term stability of the population are rarely obvious. Models developed for predicting effects of harvest are based on the premise that a population, even when not fished, remains at or near some equilibrium level of abundance. Populations that fluctuate randomly or cyclicly above and below a long-term equilibrium level can also be regarded as stable.

Factors or processes that regulate population size can be categorized as density dependent or density independent. A factor is said to be density dependent if its influence on a population varies with the density of that population. Density-dependent factors such as food availability, predation, cannibalism, diseases, parasites, and availability of spawning sites moderate extremes in population size.

For example, with increasing fish population density, food availability per fish declines, leading to slower growth and poorer condition of the surviving fish. These responses may, in turn, lead to increased vulnerability to predation or cause delayed maturation, which would cause a decline in population size because of reduced rates of survival and reproduction. A low population density could lead to rapid growth and maturation, relatively high survival and reproduction, and increased population size.

Density-independent factors have effects on populations that do not vary with population density. Water temperature, river flow, lake level, drought, and other features of the environment may affect a population in ways that are not influenced by the number of fish present.

The relative importance of density-dependent and density-independent factors in regulating population size can vary among ecosystem types and life stages of a species. In systems where the environment is relatively stable or undergoes recurring long-term cycles, density-dependent processes tend to produce an equilibrium level about which the population varies randomly because of density-independent environmental factors. Oceans and large lakes often support fish populations that are regulated principally by density-dependent factors. Ecosystems with relatively unstable and unpredictable physical characteristics have fish populations that are regulated to a greater extent by density-independent factors. Streams, rivers, and some reservoirs, which are subject to influence by storms, draw-downs, temperature changes, and other weather factors, are examples of ecosystems that may be regulated by density-independent factors.

For many freshwater fishes, the period of reproduction, including spawning and egg incubation, occurs during a relatively short period in spring, when weather conditions, water levels, turbidity, and other factors can be unstable. Hence, egg-hatching success and the number of young produced may be a function of density-independent factors. Survival of juveniles and adults is generally thought to be regulated primarily by density-dependent processes, though in many freshwater environments the most pronounced influence of increased population size is reduced growth. Reduced growth can lead to increased mortality from predation because fish remain at small, vulnerable sizes for longer periods. Older life stages are better capable of surviving or avoiding extremes of physical environmental factors than are eggs and larvae.

Most freshwater fish populations are characterized by considerable variation in the number of young produced annually. Such variation is most pronounced for species with brief spawning periods or those that spawn in variable, unpredictable environments. The extent to which variation in reproduction influences the adult population depends on the rate of survival of the group of fish spawned in a given year (termed a cohort or year-class) and the age structure of the population. The overall population abundance of a long-lived species will show less annual variation than the abundance of short-lived species subjected to the same annual variation of reproduction.

A key requirement for management of a fishery is knowledge of processes and factors that control survival of young fish to maturity or until they reach a desired size. The relative abundance of a year-class (termed year-class strength) at any early developmental stage may show no obvious relationship to either the abundance of the spawning population that produced it or the number of fish that eventually is added to the adult population. Knowledge of processes that regulate dynamics of year-classes is required to prescribe effective management programs.

The idea that year-class strength is established during some specific, relatively distinct phase of a species' life cycle has been applied to fish populations. The term critical period is defined as the time when natural regulatory factors determine the eventual abundance of a cohort. The concept that a critical period exists during the early life of fishes is consistent with the belief that natural processes of population regulation have their greatest influence on the youngest life stages, leading to fluctuations in abundance of year-classes rather than directly affecting survival of adult fishes.

The critical period has been postulated to occur during early larval development, at a time when the fish become reliant on exogenous food (Cushing and Harris 1973). Initially, larvae use energy contained in the yolk sac to develop functional mouth parts and become capable of swimming and foraging for food. When larvae begin to need external food, energy reserves are low, and the fish are vulnerable to weather extremes, food shortages, and predation. Hence, biologists have often concluded that the number of fish surviving to juvenile and adult stages is functionally determined during the larval stage.

It is possible that a critical period may occur at later development stages. In temperate inland waters, seasonal changes in food availability can lead to a critical period during the first winter of life. This has been demonstrated for juvenile largemouth bass in some reservoirs (Shelton et al. 1979). Young largemouth bass initially feed on zooplankton and small aquatic invertebrates but switch to larger invertebrates and small fish as they grow. Individuals in a cohort that grow quickly can realize a greater survival advantage because of their size and greater flexibility in selecting prey. During fall and winter, prey abundance can decline to an extent that food becomes limited, especially to the smallest individuals that have the least flexibility in prey selection. Consequently, more fish are lost to predators, disease, or other stresses. In this example, regulation is a function of density-dependent processes, but the intensity of their influence is expected to vary with severity of fall and winter weather conditions.

Knowledge of the existence and timing of a critical period during a species' life cycle can provide guidance for making management decisions. Management efforts in systems regulated like those described above should not be directed at increasing the number of young during the fall because survival is regulated by food availability during winter. A more effective approach might be to use procedures that would enhance growth in summer and fall, thereby increasing overwinter survival. Estimates of cohort abundance should be made after, not before, the critical period if one intends to use the data to forecast year-class strength in a fishery.

5.2.2 Effects of Fishing

Unexploited stocks are typified by a high proportion of old fish, slow individual growth rates, and low rates of total annual mortality (Clady et al. 1975; Goedde and Coble 1981). The presence of old fish in poor condition (termed senile fish) is often reflective of little or no exploitation. When unexploited populations are opened to fishing, length- and age-frequency distributions typically shift toward smaller and younger fish, mean age declines, and total mortality increases. For stocks that are naturally regulated by density-dependent processes, it is also expected that individual growth rates of surviving fish would increase after exploitation because of reduced intraspecific competition (Backiel and Le Cren 1967). At initial stages of exploitation, a population is usually relatively stable because the abundance of adult fish is not reduced to an extent that reproduction is affected. In fact, it is possible that the reduction in numbers of larger, older fish could lead to increased survival of young because of reduced cannibalism (Ricker 1954). Management objectives in these cases are usually directed at maximizing the recreational or economic benefit that can be obtained from each fish newly added to the population by reproduction. However, if harvest is further increased, the reproductive potential of the population may be reduced to an extent that the adult population declines substantially. At such times, management goals are adjusted to help assure adequate reproduction in the population.

Angler catch rates (number or weight of fish caught per unit effort) often are high for newly exploited stocks but decline rapidly thereafter. Redmond (1974) estimated that 39–66% of largemouth bass populations were removed by anglers during the first 3 days that five Missouri lakes (9 to 83 hectares in size) were opened to angling. Similarly, Goedde and Coble (1981) showed that one month's angling in a recently opened Wisconsin lake (5 hectares) reduced the number of harvestable-size pumpkinseeds, yellow perch, largemouth bass, and northern pike, by 74, 86, 53, and 46%, respectively. Such high initial exploitation rates are partly a function of intense angling pressure, but unexploited stocks may contain a large proportion of naive fish that are highly vulnerable to exploitation.

The effects of exploitation on the abundance of mature fish in a population are determined by the extent to which replacement by reproduction is altered. Because populations vary in age structure, fecundity, and relative importance of density-dependent and independent mortality factors, it is difficult to define any typical pattern of response. The various possibilities range from direct proportionality between abundance of adults and the number of fish that will be replaced by reproduction to total independence between the two levels of abundance. It is possible that a reduction of the adult population may cause an increase in the number of young that eventually reaches maturity.

Ricker (1954) developed generalized models of the relationships between abundance of adult fish (stock size) and the number of new fish surviving to eventually reach an exploitable size or age (termed recruitment) for stocks that are regulated by density-dependent factors prior to recruitment. Such stock–recruitment models have proven useful for predicting population responses to changes in exploitation and for estimating optimum levels of harvest, primarily for marine fisheries. Application of stock–recruitment models to freshwater fisheries has been less common because the greater amount of environmental variation typical of inland waters suggests that recruitment is regulated to a greater extent by factors independent of stock size.

5.2.3 Quantification of Dynamics

A stock is the biological unit of interest in studies of fish population dynamics. Stocks are expected to respond differently to exploitation because of differences in growth or mortality rates. They can often be defined as geographically isolated, and biologists generally attempt to gather information for distinct stocks and manage each separately. The term stock is almost synonymous with the term population, which is defined as a collection of interbreeding organisms having its own birth rate, death rate, sex composition, and age structure. The major distinction is that stock refers to the biological unit that is exploited; it may be a subset of a larger population or a collection of species that is exploited as a single unit.

Delineation of stocks and descriptions of their biological, behavioral, and genetic characteristics are important to fisheries management; such studies are necessary to define the extent to which management actions may influence a particular fishery. Stocks of many inland fish species are easily defined spatially— populations that occur in isolated lakes or are geographically distant obviously represent different stocks. Stock identification is often difficult for marine fisheries managers because extensive migrations lead to mixing of marine stocks. It is also difficult in some freshwater environments to know if a particular species within a body of water comprises one or more stocks. This is most common in large river systems, where the potential for evolution of distinct stocks is greater than in most lakes. Migratory behavior of adults may lead to aggregations of individuals from several stocks in one location. It would be important to know that management actions influencing the species' abundance and mortality in this location would affect several stocks simultaneously; conversely, management actions directed toward improving survival at one stock's spawning grounds might not have the expected influence on the overall abundance of adults because recruitment from other stocks is unaffected.

Managers need to recognize the genetic characteristics of stocks in order to make effective decisions about the creation of new fisheries and transfers of fish among water bodies. For example, establishment of several species of salmonids in the Great Lakes in the 1970s and 1980s was accomplished by introducing hatchery-reared fish produced from stocks native to Pacific coastal rivers of North America. The combination of differences in original genetic composition of the introduced fish and differences in habitat characteristics of the receiving waters has led to variation in growth and survival of the stocked fish.

Traditionally, the most important biological statistics of fish populations have been population size, total mortality rates at successive ages, natural mortality and fishing mortality, individual growth rates, recruitment rates, and the rate of surplus production (Ricker 1975). These statistics are needed to determine the greatest amount of biomass that can be harvested from a stock on a sustained basis.

For recreational fisheries typical of inland waters, management objectives are usually far more complicated than simply maximizing harvest, suggesting that managers need to collect information beyond the statistics listed above. Fisheries that illustrate this need emphasize catch-and-release, fishing-for-fun, or trophy angling, where success is measured in terms of recreational enjoyment rather than biomass harvested. Management objectives of inland fisheries also frequently address the need for maintaining balance in systems regulated by density-dependent processes. Here, the size distribution of fish available to anglers can be more important than biomass. In such cases, aesthetic and economic values of a given fishery may not be simply related to stock biomass, meaning that other measurements will be needed to monitor management success. These might include estimates of the number of trips or hours fished by anglers, economic benefits derived from a fishery, numbers of fish caught and released, numbers of fishing licenses sold, and various indices of condition of the fish population itself. Methods for conducting angler surveys to determine fishing pressure, catch rates, and harvest have been reported by Malvestuto (1983), and procedures for determining social and economic values are described in Smith (1983) and Weithman (1986).

5.3 ESTIMATING POPULATION PARAMETERS

5.3.1 Population Estimation

Estimates of population size often provide the information needed for making fisheries management decisions. To identify influences of environmental factors and human exploitation of a stock, research or survey programs track fluctuations in numbers of fish and ultimately are used to identify effective management strategies. As such, population monitoring activities often make up a significant percentage of a fisheries biologist's work load. This section introduces three commonly used methods of population estimation: counts on sample plots, mark–recapture, and decline in catch per unit effort. Otis et al. (1978), Seber (1982), White et al. (1982), and Brownie et al. (1985) contain information on more advanced models.

Population size can be estimated by determining the average density of animals per unit area in sample plots and multiplying this value by the total area covered by the population. Seber (1982) outlined three main steps in developing a sampling scheme of this type.

1. The size and shape of the sample area (plot) should be determined. This choice will be a function of the behavior of the animals to be evaluated, physical features of the habitat, and practical constraints associated with the sampling gear. Plots can cover a standardized area and be shaped like squares, circles, or rectangles, termed quadrats, or the plots could consist of nonoverlapping strips running through the population area, termed transects.

2. The number of plots to be sampled should be established in advance. Replication of counts from plots is necessary to estimate sampling variance, and the desired level of precision of the population estimate can be used to determine the number of plots required.

3. The sample plots should be located randomly so that valid statistical estimates of sampling error can be calculated.

This method of estimation is used primarily when all members of the target population within each sample plot can be counted with reasonable certainty. For example, plots could be established by using nets to block off sections of a small stream, and fish could be counted following removal with toxicants or electrofishing. Another example is the use of a seine to block a standardized quadrat along the shoreline of a lake or reservoir, followed by application of toxicants, such as rotenone, to facilitate removal and counting of the fish.

Counts of fish made per unit of time or volume can also be used to estimate population size, provided that the steps described above are followed. For example, counts of larval fish or plankton samples can be expanded to estimate population size provided that samples are collected randomly and have a standard sample volume.

An estimate of the population size (N) in an area can be calculated from the individual plot counts as follows:

$$\hat{N} = \frac{A}{a} \, \bar{n}, \qquad (5.1)$$

where A = total population area, a = the size of the plot (same unit of measure as A), and $\bar{n}$ = the average number of animals counted per sample plot. The estimated variance denoted as $V(\hat{N})$ of the population estimate is calculated as:

$$V(\hat{N}) = \frac{A^2}{a} \frac{V(n)}{s} \frac{(A - s \cdot a)}{A}, \qquad (5.2)$$

where $V(n) = \sum_{i=1}^{s} (n_i - \bar{n})^2 / (s - 1)$,

n_i = number of animals counted in the ith plot, and

s = number of plots used (Cochran 1977).

An approximate 95% confidence interval for the true population size can be calculated as $\hat{N} \pm t \, [\sqrt{V(\hat{N})}]$; t is calculated for $s - 1$ degrees of freedom and a probability (P) of 0.05.

In designing a study, it is important to predetermine a desired level of precision to be achieved for estimates of important parameters. A convenient way of expressing the precision is to calculate a coefficient of variation, CV, which is defined as the square root of the variance of an estimate divided by the estimate itself. Thus, CV is a unitless measure of the relative amount of variation about an estimate. When using counts from sample plots to estimate population size, we define the coefficient of variation to be $CV = \sqrt{V(\hat{N})}/\hat{N}$. A coefficient of variation of 0.20 or less is usually judged adequate.

The number of plots sampled is a principal determinant of the precision of population estimates obtained from simple random sampling designs. Prior to conducting the field work, a researcher should determine the number of plots that need to be sampled to achieve the target level of precision (Cochran 1977).

The above sampling procedures are termed simple random sampling when all potential sampling plots, transects, or intervals within their respective population areas or times have an equal chance of being included in the sample. Thus, every animal in the population has an equal chance of being included in the sample provided that the members of the population are randomly or uniformly distributed throughout the area. Fish populations, however, are rarely distributed randomly and more typically are aggregated in certain areas or times. In such cases, estimates from equation (5.1) generally are not biased but usually have poor precision because of extreme variability in counts among plots. If fish distribution patterns are known prior to conducting a population study, precision may be improved by subdividing the population area into zones, or strata, expected to have different fish densities and selecting sample plots at random within each stratum. This is termed stratified random sampling.

The simplest mark–recapture technique of population estimation requires one sample period in which fish are collected, marked, and released, and another period in which fish are collected and examined for marks. This method is the Petersen index (alternatively known as the Lincoln index), which is based on the assumption that the proportion of marked fish in the second sample estimates the proportion of marked fish in the total population. The estimator of population size is:

$$\hat{N} = MC/R, \tag{5.3}$$

where M is the number of fish initially marked and released, C is the number of fish collected and examined for marks in the second period, and R is the number of recaptures found in C. This estimate applies to the population present during the first sample period, not the recapture period.

The Petersen index can give biased estimates of population size when the number of fish sampled is low, but several modifications of equation (5.3) have been proposed to help correct this bias. Bailey's (1951) modification is

$$\hat{N} = \frac{M(C + 1)}{(R + 1)}, \tag{5.4}$$

with variance

$$V(\hat{N}) = \frac{M^2 (C + 1) (C - R)}{(R + 1)^2 (R + 2)}. \tag{5.5}$$

These are used in cases where sampling during the recapture period is conducted "with replacement," meaning that each fish is returned to (replaced in) the population after it is examined for marks and thus is eligible to be included in the sample again. Chapman (1951) recommended using

$$\hat{N} = \frac{(M + 1) (C + 1)}{(R + 1)} - 1, \tag{5.6}$$

with variance

$$V(\hat{N}) = \frac{(M + 1) (C + 1) (M - R) (C - R)}{(R + 1)^2 (R + 2)}.$$

This model is used when sampling during the recapture period is done "without replacement," as in cases when fish caught by anglers are examined for recaptures or when all fish collected in the recapture period are marked in a way different from the first mark and then released. Differences among population estimates obtained from equations (5.3), (5.4), and (5.6) would probably be of little significance in making fisheries management decisions if the number of fish recaptured (R) exceeds 7.

Several important conditions or assumptions must be met to obtain valid estimates using the Petersen index or its modifications: (1) marked fish do not lose their marks prior to the recapture period; (2) marked fish are not overlooked in the recapture sample; (3) marked and unmarked fish are equally vulnerable to capture in the recapture period; (4) marked and unmarked fish have equal mortality rates during the interval between the marking and recapture sample periods; (5) following release, marked animals become randomly mixed with the unmarked ones, or recapture effort is distributed in proportion to the number of animals in different parts of the population area; and (6) there are no additions to the population during the study interval.

Assuring that these conditions are satisfied is one of the most difficult aspects of estimating population size with the Petersen index. Any factor causing under-representation of marked fish in the second sample will lead to overestimation of the population size. This could result from poor mark retention, failure to recognize all recaptures in the second sample, impaired survival of marked fish, and immigration of new (thus unmarked) animals before the recapture sampling. Conversely, any factor leading to over-representation of marked fish in the second sample, caused perhaps by increased susceptibility of marked animals to capture, will result in underestimation of the true population number.

Several methods may be used to find confidence intervals for population estimates obtained by using the Petersen index. These have been thoroughly developed for the most common sampling design—one in which sampling is done without replacement during the recapture period. In these cases, the random variable is the ratio R/C, which estimates M/N for the population, and the distribution of R/C is hypergeometric. Unfortunately, neither tables nor explicit formulas are available for determining exact confidence intervals for the hyper-geometric distribution. Consequently, various approximations based on the binomial, Poisson, or normal distributions have been used, depending on the magnitude of R/C and the values of M, C, and R for a particular study (Seber 1982).

Precision and accuracy of Petersen estimates are affected by the numbers of fish marked and subsequently checked for marks. Charts prepared by Robson and Regier (1964) can be used to determine values of M and C required to produce Petersen population estimates that are expected to differ from the true population number by no more than 50%, 25%, or 10% at the 95% level of confidence. They recommend the 50% level for preliminary surveys, 25% for management studies, and 10% for research evaluations. Use of the charts requires an initial guess of population size. A particular combination of M and C can be chosen as a function of the relative costs associated with marking fish and sampling for recaptures.

Capture–recapture methods of population estimation that use two or more sample periods for marking animals are termed multiple-census procedures. The simplest of these was originally described by Schnabel (1938). Fish are collected from the population, marked, and released for a series of samples; the numbers of recaptures and unmarked fish collected in each sample are recorded. Assumptions of the method are identical to those of the Petersen index except that no mortality is allowed during the study. Because of this requirement, the multiple-census is most appropriate when sampling periods are closely spaced and restricted to a relatively short overall period so that the occurrence of mortality would not have a great influence on the validity of the population estimate. Table 5.1 illustrates typical data and computational procedures.

Table 5.1 Data records and calculations for Schnabel multiple-census population estimate.

Sample (t)	Number of fish captured			Total number of marked fish released prior to sample period (M)	C × M
	Marked (R)	Unmarked	Total (C)		
1	0	150	150	0	0
2	22	203	225	150	33,750
3	26	86	112	353	39,536
4	53	150	203	439	89,117
5	38	80	118	589	69,502
6	28	53	81	669	54,189
7	87	150	237	772	171,114
Total	254				457,208

The Schnabel population estimation formula is

$$\hat{N} = \frac{\sum_{t=1}^{n} C_t M_t}{\sum_{t=1}^{n} R_t},$$ (5.7)

where the subscript t refers to the individual sample period, and n is the number of periods. For our example,

$$\hat{N} = \frac{\sum_{t=1}^{7} C_t M_t}{\sum_{t=1}^{7} R_t} = \frac{457,208}{254} = 1,800.$$

Confidence limits are determined by first computing the variance of $(1/N)$ because the inverse of N is more normally distributed than is N itself:

$$V(1/\hat{N}) = \frac{\sum_{t=1}^{7} R_t}{(\sum_{t=1}^{n} C_t M_t)^2}.$$ (5.8)

We next determine a 95% confidence interval for $(1/N)$ as $(1/\hat{N}) \pm 1.96\ V(1/\hat{N})$ and compute the inverses of the limits to find the confidence interval of N itself. For our example, $1/\hat{N} = 1/1800 = 0.000556$, and

$$V(1/\hat{N}) = \frac{254}{(457,208)^2} = 1.215 \times 10^{-9}.$$

From these, the 95% confidence interval for $(1/N)$ is $0.000488 - 0.000624$, and by calculating the inverses of these limits, we obtain a confidence interval for N of 1,602–2,049.

In cases where the total number of recaptures in the study is small, say less than 25, we do not expect $(1/N)$ to be normally distributed, making it important to calculate confidence limits by alternative methods. This can be done by using tables of the Poisson distribution (see Ricker 1975) to determine 95% limits for R, and then substituting these values into equation (5.7) to determine limits for N.

Population size can be estimated from data on fishing effort and catch rates. Several estimators have been developed, all of which are based on the theory that the number of fish caught per unit of effort will progressively decline as members of the population are removed. The most common methods assume that (1) all members of the target population are equally vulnerable to capture, (2) vulnerability to capture is constant over time, and (3) there are no additions to the population or losses other than those due to fishing during the study interval. Additionally, one must be able to either quantify fishing or sampling effort or create a sampling situation in which equal effort is expended in consecutive sample periods. Examples of ways that effort could be quantified include hours spent electrofishing, angler trips, vessel days, seine hauls, or gill nets fished.

The Leslie and DeLury methods are used to quantify population size in cases where sampling effort varies among periods. These models have been most effective for large populations where there is low probability that an individual fish will be caught during a single unit of effort; typical applications have included commercial fisheries where data on sampling effort and catch are obtained by monitoring those who are fishing.

According to the Leslie method of estimation, the number of fish caught per unit of effort during some time interval, t, is proportional to the number of fish present at the beginning of the interval:

$$\frac{C_t}{f_t} = q\,N_t,\qquad(5.9)$$

where C_t = catch during period t,
  f_t = units of fishing effort during period t,
  N_t = number of fish present at beginning of period t, and
  q = the catchability coefficient.

Because the method assumes that the population is closed to additions or losses other than fishing, we can express N_t as a function of the original population size (N_0) minus the total number of fish caught and removed (K_t) prior to time t as:

$$N_t = N_0 - K_t.\qquad(5.10)$$

We can substitute this expression for N_t into equation (5.9) and obtain

$$\frac{C_t}{f_t} = qN_0 - qK_t,\qquad(5.11)$$

which is a linear relationship of the form $Y = a + bX$, where $Y = C_t/f_t$ and $X = K_t$. A plot of catch per effort (C_t/f_t) versus cumulative catch (K_t) will approximate a straight line with slope (actually, the absolute value of the slope) equal to q and intercept of qN_0. We can use linear least-squares regression methods to estimate the slope and intercept, and then estimate the original population size as:

$$\hat{N}_0 = \frac{\text{intercept}}{|\text{slope}|} = \frac{qN_0}{q}.\qquad(5.12)$$

When the fraction (q) of a population that is taken by a given unit of fishing effort is small, say less than 0.2 (2% of the population), DeLury's modification of the Leslie model is preferred. The DeLury method also is based on the premise that catch per effort is proportional to population size (equation 5.9) and assumes a closed population (other than losses due to removal), but uses a different expression of population decline,

$$N_t = N_0 e^{-qE_t}, \tag{5.13}$$

where E_t = cumulative total effort expended prior to period t, and other variables are defined as before. This implies that the population declines in proportion to total effort, whereas the Leslie method assumes that the decline is a function of the total catch.

Substituting the expression of N_t from equation (5.13) into equation (5.9), we get $C_t/f_t = qN_0 e^{-qE_t}$, and by taking the natural log of both sides, we obtain

$$\log_e (C_t/f_t) = \log_e (qN_0) - qE_t, \tag{5.14}$$

which is also of the form $Y = a + bX$, with $Y = \log_e (C_t/f_t)$ and $X = E_t$. So in this method, we can plot the natural log of catch per effort versus cumulative effort, and again use linear least-squares regression methods to estimate the slope and intercept. The estimate of population size is

$$\hat{N}_0 = \frac{e^{\text{intercept}}}{|\text{slope}|}. \tag{5.15}$$

Confidence limits for population estimates obtained from the Leslie and Delury methods are calculated from intermediate statistics obtained when performing the least-squares regression, and they may be determined using Ricker (1975). Because regression techniques are used, these methods of estimation require a minimum of three sample periods. Precision can be improved by increasing the number of sample periods, but the influences of immigration or natural mortality on accuracy of the estimates could become significant if the duration of the study is extended.

Removal methods of population estimation are also used in situations where the catchability of fish is high and equal effort is expended in each sample period. The most common example of this is sampling small streams, where sections are blocked off with nets, and fish are collected by making consecutive passes with electrofishing gear. Each pass represents one sample period. Fish captured during each period can be released outside the sample reach or marked and then released back into the sampling area. Marking in this case can be used to "remove" fish from consideration in subsequent samples.

A model for population estimation was described by Zippin (1956, 1958) as:

$$\hat{N} = \frac{C}{1 - \hat{p}^s}, \tag{5.16}$$

where C = total catch over all sample periods ($\sum_{t=1}^{s} C_t$),
    s = the number of sample periods, and
    $\hat{p}$ = probability that a fish escapes capture during a sample period (i.e., $\hat{p} = 1 - \hat{q}$, where $\hat{q}$ is the catchability coefficient as defined before).

Table 5.2 Summary of types of mortality rates.

Mortality rates	Symbol	Relationships
Instantaneous		
Total	Z	$A = 1 - e^{-Z}$
From fishing	F	$F = Z - M$
From natural causes	M	$M = Z - F$
Actual		
Total	A	$A = u + v$
From fishing (exploitation rate)[a]	u	$u = FA/Z$
From natural causes[a]	v	$v = MA/Z$

[a]Relationships shown are for cases where fishing mortality and natural mortality are in the same ratio throughout the year.

To calculate $\hat{N}$, we must first estimate $\hat{p}$ from experimental data using the equation:

$$\frac{\hat{p}}{\hat{q}} - \frac{s\hat{p}^s}{(1 - \hat{p}^s)} = \frac{\sum_{t=1}^{s} (t - 1) C_t}{C}. \tag{5.17}$$

Using an example from Seber (1982), we have the following catches made from three consecutive sampling efforts: $C_1 = 165$, $C_2 = 101$, $C_3 = 54$. First estimate $\hat{p}$ by iteratively solving equation (5.17);

$$\frac{\hat{p}}{\hat{q}} - \frac{3\hat{p}^3}{(1 - \hat{p}^3)} = \frac{(1 - 1)165 + (2 - 1)101 + (3 - 1)54}{320} = 0.65;$$

and find $\hat{p} = 0.58$. Then using equation (5.16), calculate $\hat{N} = 320/(1 - 0.58^3) = 400$. Charts originally published by Zippin (1956) and partially reproduced in Seber (1982) can be used to help solve equation (5.17).

Confidence intervals for N can be calculated as $\hat{N} \pm 1.96 \sqrt{V(\hat{N})}$, where

$$V(\hat{N}) = \frac{\hat{N}(1 - \hat{p}^s)\hat{p}^s}{(1 - \hat{p}^s)^2 - (\hat{q}s)^2 \hat{p}^{s-1}}. \tag{5.18}$$

For our example, $V(\hat{N}) = 691.1$, and the confidence limits are 348–452.

One limitation of the Zippin method is that a large proportion of the population must be sampled to obtain reasonably accurate and precise estimates. This is especially restrictive as the population size decreases; thus, for small populations, it may be necessary to mark fish to simulate removal or to otherwise hold the collected fish for eventual release in order to avoid depleting the population.

5.3.2 Mortality Rates

Sources of mortality in fish populations are usually placed into one of two categories: natural mortality, including losses to predation, diseases, and weather, or fishing mortality, which is mortality due to harvest (Table 5.2). The combined effects of natural and fishing mortality are termed total mortality. For exploited stocks, we usually regard the life span of the fish as having a pre-recruitment phase, when only natural mortality occurs, and a post-recruitment phase, when

Figure 5.1 Relationship between the number of fish alive and age for a population that has a stable age distribution and a constant proportion dying per unit of time.

fishing and natural mortality occur. In this context recruitment refers to the addition of fish to the exploited portion of the stock.

Methods used to estimate mortality rates vary in relation to assumptions that are made regarding temporal patterns of reproduction and survival rates in the stock. The simplest estimation procedures are based on three assumptions: (1) reproduction is constant from year to year; (2) survival is equal among all age groups; and (3) survival is constant from year to year. Under these assumptions, the population will be in steady state, that is, the number of fish added equals the number dying annually, and the population's age composition is stable. Hence, information on the population's age composition at any point in time would be representative of mortality rates over a much broader interval. A more realistic model for mortality of inland fish stocks would allow for annual variation in reproduction and survival rates. These variations would produce an unstable age structure in the population, meaning that samples would have to be collected annually for several years (at least) to estimate mortality rates.

For a population having a stable age distribution, the number of fish alive at any age could be described by the curve shown in Figure 5.1. The curve could also represent the decline in abundance of one year-class (or cohort) throughout its life, given the assumption of equal mortality among age groups. The curve in Figure 5.1 is based on a mathematical model that assumes a constant proportion (Z) of the population (N) dies per unit of time (t), where:

$$\frac{dN}{dt} = -ZN.$$

Upon integration, we obtain an exponential equation for predicting population size:

$$N_t = N_0 e^{-Zt} \qquad (5.19)$$

where N_t = number alive at time t,
$\quad N_0$ = number alive initially (at time t_0),
$\quad Z$ = force of total mortality (also known as the instantaneous total mortality rate), and
$\quad t$ = time elapsed since t_0.

Although any unit of time can be used with the above model, years are most commonly used so that the number of fish alive at any age t can be computed as a function of the number alive initially and the force of total mortality. If we let t = 1 year, the probability that a fish survives the year (S) can be expressed as

$$S = \frac{N_1}{N_0} = e^{-Z}, \qquad (5.20)$$

and the complement of survival, the annual mortality rate (A), is equal to $1 - S$ or $1 - e^{-Z}$.

In most applications, the exponential model is useful for fish that are age 1 and older; it sometimes is not applicable until age 2, 3, or older. Mortality rates of early life stages, including eggs, larvae, and juveniles, are typically much higher than those experienced by older fish. Additionally, because many management efforts are focused on the post-recruitment phase of the species' life span, the exponential model has traditionally been used to describe mortality beginning with the youngest age group that is exploited in the fishery. In these cases, the assumption of constant reproduction is modified to become an assumption that the number of fish alive at the age first considered in the model is constant from year to year. If this age coincides with the age at recruitment, the assumption is one of constant recruitment.

Total mortality rates are often estimated from the age composition of samples taken from a population. If we assume that recruitment and survival rates are equal across years, the population age structure would be stable, and a random sample taken from the population at any time would always show the same age composition (allowing for random variation among samples). Thus, relationships among numbers of fish at specific ages in the sample can be used to estimate mortality rates.

In practical applications, it is first necessary to assure that the sample is representative of the entire population. Most sampling gear, including gear used in commercial operations, are selective for certain sizes (and hence certain ages) of fish, and they would produce biased estimates. A typical problem is that young fish are underrepresented in the sample because they are either too small to be effectively sampled by the experimental gear or they occur in a different habitat than that sampled. In these cases, age groups that are not fully vulnerable to the sampling gear are excluded from the mortality computations.

The simplest methods of estimation assume that mortality is equal among ages (in addition to the assumptions of constant recruitment and mortality among

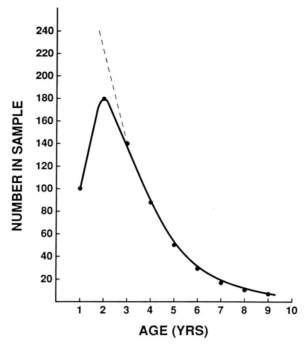

Figure 5.2 Plot of number of fish collected versus age, illustrating under-representation of age-1 and age-2 fish relative to numbers expected from an exponential model (dashed line).

years). Consider the following hypothetical sample, where we have data from a population where the annual survival rate (S) is 0.60.

Age	1	2	3	4	5	6	7	8	9
Number sampled	100	180	140	84	50	30	18	11	6

It is obvious that age-1 fish are underrepresented in the sample, and closer inspection also shows that age-2 fish were inadequately sampled (e.g., 140/180 exceeds the actual survival rate). Thus, ages 3–9 would be used in further computations.

A more realistic situation is one where we have the sample, but the survival rate is unknown (it is, of course, the object of the investigation). One way to evaluate gear selectivity in this case is to plot the number collected versus age. This plot should be similar to Figure 5.1, but in the example given above, we find that the youngest age groups fall below the line expected from the exponential model (Figure 5.2). An alternative method is to plot the sample results on semi-log paper (Figure 5.3). This plot is known as a catch curve. Because we assume that $N_t = N_0 e^{-Zt}$ (equation 5.19), a plot of the logarithms of the numbers sampled at each age (N_t) versus age (t) should be a descending straight line with slope = $-Z$. In Figure 5.3 we see that the numbers collected for fish age 3 and older form a straight line whereas the points for ages 1 and 2 lie below an extension of this line, indicating they were not adequately sampled. The catch curve can be used to

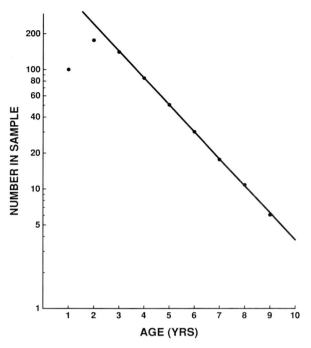

Figure 5.3 Catch curve illustrating the linear relationship between the number of fish collected and age, when plotted on a semi-log scale. Ages 1 and 2 were under-represented in the sample.

estimate mortality rates. By taking the natural log of both sides of equation 5.19, we obtain

$$\log_e(N_t) = \log_e(N_0) - Z(t),$$

which is of the form $Y = a + bX$, with $Y = \log_e(N_t)$ and $X = t$. Linear least-squares regression can then be used to estimate the annual instantaneous total mortality rate, Z. For example, assume we have collected the following data:

Age (t)	1	2	3	4	5	6
Number (N_t)	100	150	95	53	35	17

For ages 2–6, we regress $\log_e(N_t)$ versus t and obtain a slope of -0.54. Thus, $\hat{Z} = 0.54$, and $\hat{S} = 0.59$ (from $S = e^{-Z}$). By assuming constant recruitment and survival over time and equal survival among ages, we have determined that 59% of fish age 2–6 survive annually (hence, total annual mortality is 41%).

The accuracy and precision of this estimate are affected by the number of age groups included and the representativeness of the sample. Age groups having fewer than five fish in the sample usually are excluded from the regression because of the extreme variation that could be introduced by the collection of, or failure to collect, a few individuals. This typically occurs for the oldest age groups. In practice, the data set is truncated beginning with the youngest age group having fewer than five fish in the sample. Because regression techniques are used, data for at least three age groups that are fully vulnerable to the gear are required.

Confidence limits for Z are equal to the confidence limits for the slope of the regression; methods for computation are included in most statistical texts.

Other models that can be used to estimate survival in stable-age populations are based on ratios among numbers of the various ages of fish collected. A method described by Robson and Chapman (1961) estimates survival as

$$\hat{S} = \frac{T}{n + T - 1},$$

(5.21)

where n is the total number of fish in the sample, beginning with the first fully vulnerable age, and T is determined from the age distribution of the sample. First, the sample data are coded so that the number collected in the first fully vulnerable age is labeled N_0, the number in the next oldest group is N_1, etc., so that N_k is the number of the oldest age group in the sample. For our example, we have:

Age	2	3	4	5	6
Coded age (x)	0	1	2	3	4
Number (N_x)	150	95	53	35	17

Mathematically,

$$T = \sum_{x=0}^{k} x(N_x) = 0(N_0) + 1(N_1) + 2(N_2) + 3(N_3) + 4(N_4),$$

which, for the example, is

$$T = 0(150) + 1(95) + 2(53) + 3(35) + 4(17) = 374.$$

Hence, from equation 5.21, we have

$$\hat{S} = \frac{374}{350 + 374 - 1} = 0.52.$$

The variance of this estimator is

$$V(\hat{S}) = \hat{S}\left(\hat{S} - \frac{T - 1}{n + T - 2}\right),$$

(5.22)

which, for our example, is

$$V(\hat{S}) = 0.52\left(0.52 - \frac{374 - 1}{350 + 374 - 2}\right) = 0.0018.$$

Methods for estimating mortality rates in populations that do not have stable age distributions can require considerable amounts of data and may be mathematically complicated.

For exploited fish populations, it is important to separately account for the influences of natural and fishing mortality. The total instantaneous mortality rate (Z), can be partitioned into instantaneous rates of fishing mortality (F) and natural mortality (M) from the relationship $Z = F + M$, and equation (5.19) can be modified as:

$$N_t = N_0\, e^{-(F + M)t} = N_0\, e^{-Ft}e^{-Mt}.$$

(5.23)

Likewise, the actual total mortality rate (A), can be expressed as the sum of two components,

$$A = u + v, \tag{5.24}$$

where u is the expectation of death from fishing (also known as the rate of exploitation), and v is the expectation of death from natural causes. In cases where fishing and natural mortality are similarly apportioned throughout the year, it is possible to equate

$$\frac{Z}{A} = \frac{F}{u} = \frac{M}{v},$$

and upon rearrangement, we find that

$$u = FA/Z \text{ and} \tag{5.25}$$

$$v = MA/Z. \tag{5.26}$$

These relationships are useful if it can be reasonably assumed that the number of fish harvested in any given time period (perhaps a month) relative to the total number harvested annually is similar to the ratio of the number dying naturally in the same period divided by the total number dying from natural causes annually.

Separation of total mortality into fishing and natural components is accomplished usually by first estimating total and fishing mortality and then estimating natural mortality as the difference. The exploitation rate, u, can be calculated by estimating the population size at the beginning of some time interval and then counting the number of fish harvested by anglers in the ensuing period. Provided that there is no immigration or other additions to the population, the number harvested divided by the initial population size is an estimate of u. If the number harvested is monitored for 1 year, the value of u estimates the annual exploitation rate. The precision and accuracy of the estimate are a function of biases and variances associated with methods used in estimating the population size and determining the number harvested.

Another way to estimate u is to release a known number of tagged fish at one point in time and then determine the proportion of tagged fish harvested in a year. Counts of numbers harvested are often obtained from anglers who report their catches of tagged fish; in many cases, monetary rewards are offered by management agencies to encourage complete reporting of harvested fish. The validity of reward procedures is based on assumptions that tagging does not affect the fish's vulnerability to angling or other sources of mortality, that the tagged fish are representative of the target population, that tags are not lost by the fish, and that all harvest of tagged fish is reported. Precision of the estimate is primarily a function of the number of fish tagged and released.

A final approach for estimating mortality components is based on the relationship $Z = M + F$. The instantaneous fishing mortality rate (F) can be related to the amount of fishing effort (f) as:

$$F = qf, \tag{5.27}$$

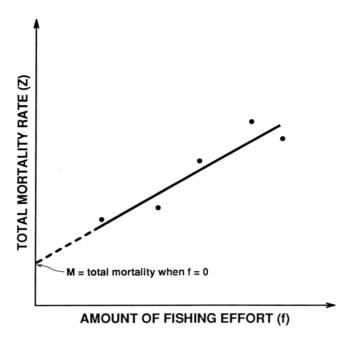

Figure 5.4   Illustration of a straight line (solid) fit to 5 years' estimates of fishing effort and total instantaneous mortality. The point at which an extension of this line (dashed) intercepts the vertical axis estimates the instantaneous natural mortality rate (M).

where q is a catchability coefficient. Because $Z = M + F$, we can substitute for F and obtain

$$Z = M + qf, \tag{5.28}$$

which is equivalent to a linear equation with slope q and intercept M. The relationship between Z and f is illustrated in Figure 5.4. Thus, if we have annual estimates of Z and corresponding values of annual fishing effort, it is possible to estimate M using least-squares regression techniques. The method requires several years of data (a minimum of three) and assumes that catchability coefficients are constant over time and among age groups. The time-specific annual estimates of Z could be obtained from age-specific catch data as described earlier in this section. The basis of this approach is that if natural mortality (M) is constant, any annual variation of total mortality (Z) will be due only to variation in fishing mortality (F), which in turn varies only in relation to effort (f).

5.3.3 Growth Rates

Growth of fish, as of many other poikilothermic animals, is indeterminant, meaning that individuals have no innate pattern of growth and can continue to increase in size throughout life. Because growth can be affected by food abundance, weather, competition, and many other factors, the measurement of growth rates is a common way fisheries biologists evaluate the effectiveness of management practices. Additionally, because the size of fish caught recreationally and commercially greatly affects aesthetic and economic values of the catch, an

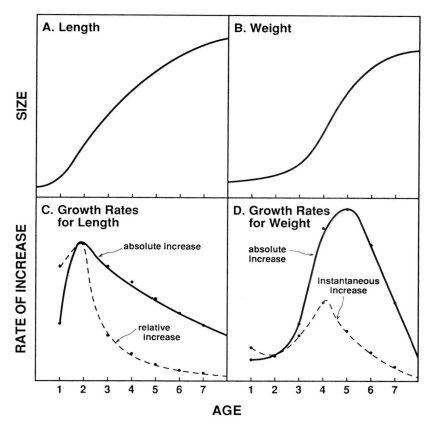

Figure 5.5 Typical trends of length (panel A), weight (B), and growth rates (C and D) throughout the life of a fish.

understanding of growth dynamics in populations is important to adequately predict fishery trends. Mathematical models of fish growth are often incorporated into models of population dynamics to predict changes in stock biomass resulting from various harvest strategies.

Growth may be measured in terms of length (l) or weight (w) and is expressed in several ways:

1. absolute increase per unit time: $l_2 - l_1$ or $w_2 - w_1$;

2. relative rate of increase per unit time: $(l_2 - l_1)/l_1$ or $(w_2 - w_1)/w_1$; and

3. instantaneous rate of increase per unit time: $\log_e l_2 - \log_e l_1$ or $\log_e w_2 - \log_e w_1$.

Estimates of these rates can be made from observations of length (or weight) of a cohort of fish at two or more points in time, from tagging studies, or from size-at-age data obtained from age-and-growth analyses.

These expressions should generally not be calculated for time periods exceeding 1 year because rates of growth typically vary with age. Trends of length and weight of a cohort throughout life usually show an early period of rapid growth and a subsequent period of more gradual increase (Figure 5.5). Likewise, absolute,

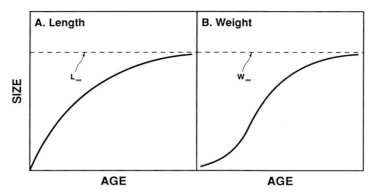

Figure 5.6 Growth trends predicted from von Bertalanffy models of length (panel A) and weight (panel b).

relative, and instantaneous growth rates computed from size at successive ages are initially low, increase to a maximum, and then decline with age (Figure 5.5).

One popular model that mimics this pattern of declining growth rate with age was originally described by von Bertalanffy (1938). The von Bertalanffy model is based on the theory that the rate of change in length per unit of time (dl/dt) will decline and approach zero as a fish nears its maximum possible size (L_∞). Mathematically,

$$\frac{dl}{dt} = K(L_\infty - l), \tag{5.29}$$

where K is a growth parameter (not a rate in the sense defined earlier) and l_t is the length at time t. Equation (5.29) shows that the rate of increase in length is a constant proportion (K) of the difference between the maximum size and present length ($L_\infty - l_t$). Upon integration of equation (5.29), we obtain a predictive relationship:

$$l_t = L_\infty [1 - e^{-k(t - t_0)}], \tag{5.30}$$

where t is time (or age) in years and t_0 is the time at which $l_t = 0$. A plot of this relationship shows a progressive increase in body size that asymptotically approaches L_∞ (Figure 5.6). Parameters of the model (L_∞, K, and t_0) are typically estimated from annual length-at-age data; procedures have been outlined by Gulland (1969) and Galluci and Quinn (1979).

In many stock analyses, it is more desirable to model growth in weight because stock biomass is the product of population size times mean weight. The von Bertalanffy model can be converted to an expression for weight by first using the allometric relationship:

$$W = aL^b, \tag{5.31}$$

where the parameters a and b describe the form of the weight–length relationship. Methods for estimating these parameters are outlined in Carlander (1969) and Ricker (1975). In many cases, b is near 3.0, and in practice, a value of $b = 3$ is often used and only the parameter a is estimated. This produces a simple form of the von Bertalanffy weight model:

$$w_t = W_\infty[1 - e^{-K(t - t_0)}]^3, \tag{5.32}$$

where $W_\infty = aL_\infty^3$, and the other parameters are the same ones used in the length model. A plot of w_t versus age shows an initial period of accelerating growth, an intermediate period when growth is approximately linear, and a final phase of decelerating growth as w_t approaches W_∞ (Figure 5.6).

5.4 PREDICTING FISHERY TRENDS

5.4.1 Goals of Predictive Models

A major activity of fisheries managers has been to predict the effects of different amounts of fishing effort on the numbers and sizes of fish obtained from a stock on a continuing basis. Models are developed within constraints imposed by data availability, types of predictions desired, and mathematical and computational complexity. Models are tools that should be as simple as possible while providing appropriate types of predictions—emphasis should be placed on making decisions regarding feasible management practices rather than impossible ones. Formulation of any mathematical representation of population dynamics will require certain assumptions, and the validity of the assumptions will affect the accuracy and meaning of predictions from a model. Evidence of failure of assumptions does not necessarily mean that the model is useless; rather, the manager should recognize the limitations of the model and determine if such limitations will have a significant impact on decisions or recommendations that will be made (Gulland 1983). Models often can be used to explore the influences of different management options on a specific fishery even if the predicted values are known to be only approximately proportional to the actual values.

Three general types of models have been used for fisheries predictions. Models of the first type, surplus production, consider trends in a population as a whole in relation to harvest. The influences of growth, mortality, and reproduction are combined into a model of overall population change (Shaefer 1968). The second type predicts the yield that would be obtained from a year-class of fish throughout its lifespan as a function of harvest practices. These are known as yield-per-recruit models because the predictions are usually expressed as the yield per fish that is newly recruited to the stock. Such models do not explicitly account for the effects of variable recruitment on yield. The most common of these models is known as the dynamic-pool yield model, which was first proposed by Beverton and Holt (1957). The third category of models treats each age group of a population separately and sums yield predictions among ages and years to obtain an overall estimate of long-term yield. Such predictions are used as a reference, or index, in determining optimal management policies (Walters 1969). Age-structured models require estimates of age-specific rates of reproduction, growth, and mortality, meaning that data requirements are greater than those of surplus production or yield-per-recruit models, which rely primarily on data that are readily available from commercial fishery landings.

Surplus production models can be used to predict yield for a fishery by using information on either stock abundance or fishing effort. Surplus production models do not explicitly consider the growth, reproduction, and mortality rates in a population; rather, they deal primarily with relationships among overall stock biomass, yield, and fishing effort. These models have been popular because their parameters can be estimated from commonly available statistics, such as annual records of commercial harvest and effort.

Biological assumptions of surplus production models are similar to those discussed for stock–recruitment relationships. Populations are assumed to be regulated by density-dependent processes that affect reproduction and control survival and growth early in life, prior to the size (or age) at which fish are harvestable. Survival after this time is assumed to be density independent and is not necessarily compensatory. An unexploited population would be expected to occur at some equilibrium level of abundance where the number of fish added each year equals the number dying. If the stock is subjected to a fixed level of fishing mortality annually, the population is expected to reach a new equilibrium abundance at which the number of recruits produced annually exceeds the number of adults required to produce them. This excess of new recruits, termed surplus production, is available for exploitation. Because of density-dependent regulation, it is reasonable to expect that the greatest harvestable "surplus" will occur at some level of stock abundance that is less than the primitive equilibrium because, by definition, there is no surplus production when recruitment exactly equals the stock size required for replacement.

Graham (1935) proposed a surplus production model based on the assumption that the annual change in biomass of a stock was proportional to the actual stock biomass and also to the difference between present stock size and the maximum biomass the habitat would support:

$$\frac{dB}{dt} = kB \frac{(B_\infty - B)}{B_\infty}, \qquad (5.33)$$

where B = stock biomass,
$\quad B_\infty$ = maximum biomass,
$\quad k$ = instantaneous rate of increase of stock biomass, and
$\quad t$ = time in years (Ricker 1975).

This equation describes how the stock biomass would increase (dB/dt) following a reduction of biomass from B_∞ to B. If the reduction in biomass to B has been brought about by human harvest, and if the level of fishing mortality (or effort) operates in a manner that stock biomass remains at B from year to year, the stock is regarded as being at equilibrium, and the yield that could be obtained annually from the fishery is dB/dt. Hence, equation (5.33) can be rewritten as:

$$Y = kB \frac{(B_\infty - B)}{B_\infty}, \text{ or} \qquad (5.34)$$

$$Y = kB - \frac{k}{B_\infty}(B^2) \qquad (5.35)$$

where Y = annual yield when the stock is at equilibrium biomass B.

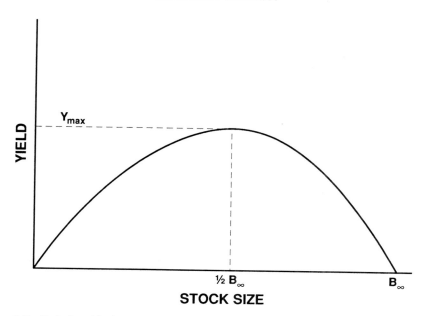

Figure 5.7 Relationship between yield and stock size for the surplus production model. The maximum sustainable yield (Y_{max}) theoretically occurs at a stock size equal to one-half of the maximum biomass (B_∞) that the habitat will support.

The primary purpose of developing the model usually is to determine maximum sustainable annual yield, the level of fishing effort required to obtain maximum sustainable yield, and the stock size at which this would occur. Equation (5.35) is a parabolic relationship (Figure 5.7), with the greatest yield (Y_{max}) at a stock abundance (B) equal to $B_\infty/2$. By substituting $B = B_\infty/2$ into equation (5.35), the maximum sustainable yield is found to be:

$$Y_{max} = \frac{kB_\infty}{4},$$

showing that estimates of k and B_∞ are needed to determine the maximum sustainable yield. Data on equilibrium yield obtained at two or more different levels of fishing effort are used to estimate these parameters (Ricker 1975). Shaefer (1954) outlined methods for estimating the parameters for nonequilibrium situations.

Dynamic-pool yield models are based on the premise that stock biomass varies with growth and mortality rates in the stock. Hence, yield is taken from a changing, dynamic "pool" of available biomass. Because biomass is the product of the average fish weight times the number of fish present, it is possible to predict yield (Y) as:

$$Y = \int_{t_c}^{t_l} F \, N_t W_t dt. \tag{5.36}$$

This indicates that total yield from a given cohort of fish in a stock is the integration of the force of fishing (F) times the number of fish alive (N_t) and the average weight per fish (W_t) for the period of life during which the cohort is exploited. This period begins at the age when fish are first exploited (t_c) and ends at the cohort's theoretical maximum age (t_l). If a stock is at a stable equilibrium, the total yield each year would equal the total harvest of a cohort during its life, so equation (5.36) also estimates the annual equilibrium yield (Shaefer 1968).

Dynamic-pool models are useful for predicting yield where fishing effort and the minimum size (or age) of capture can be regulated. Mesh size restrictions in a commercial fishery or minimum length limits in a recreational fishery would affect the age at which a cohort is first exploited. Regulations that influence fishing effort, such as restrictions on numbers of licenses issued, or those that otherwise influence fishing mortality, such as closed seasons or refuge zones, would affect F. By simulating the effects of different levels of F and age at first harvest, one can evaluate the extent to which yield is influenced and determine levels that produce maximum yield.

Implementation of the model requires mathematical expressions for changes in numbers and average weight of a cohort during its life span. The formulation most commonly used follows that originated by Beverton and Holt (1957):

$$Y = FRW_\infty \, e^{-M(t_c - t_r)} \sum_{n=0}^{3} \left[\frac{U_n}{F + M + nK} \, e^{-nK(t_c - t_0)} \right], \qquad (5.37)$$

where t_r = the age, in years, when fish reach a size at which they could be harvested;

t_c = the age at which fish actually are first harvested ($t_c \geq t_r$);

t_0 = a constant introduced upon integration of equation (5.36), usually interpreted as the age at which fish length is zero;

F = annual instantaneous rate of fishing mortality, assumed equal for all fish older than age t_c;

M = annual instantaneous rate of natural mortality, assumed constant over time and equal for all fish older than age t_r;

R = recruitment, the number of fish alive in a cohort at age t_r;

W_∞, K = parameters of a von Bertalanffy growth model, assumed constant over time; and

U_n = a mathematical term used to simplify equation (5.37), where $U_0 = 1$, $U_1 = -3$, $U_2 = 3$, and $U_3 = -1$.

The model includes three distinct periods during a cohort's life span. The first period is the pre-recruitment phase, when the fish are too young (small) to be harvested. This stage begins at age t_0 and ends at t_r when the fish reach a size that they could be harvested. The post-recruitment stage in the life cycle can include a pre-exploited phase as well as an exploited phase. The age at which the fish are first harvested (t_c) sometimes exceeds t_r because of economic or social factors. For example, fishing technology may be available to commercially harvest small fish, but dollar value per unit biomass may be low for these fish, and it thus would be more profitable to harvest fish at a larger size (older age). In a recreational fishery, t_r would correspond to the age at which fish reached a size that anglers

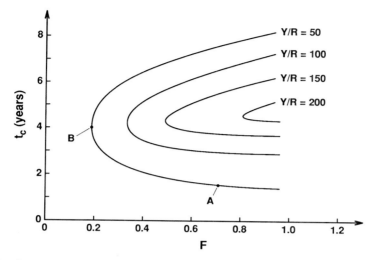

Figure 5.8 Summary of yield predictions obtained from the Beverton-Holt dynamic-pool yield model across a range of age at first capture (t_c) and fishing mortality (F). Lines are isopleths indicating combinations of t_c and F expected to produce the same yield per recruit (Y/R).

would harvest in the absence of length limits whereas t_c would correspond to the age at which fish reached a legally imposed size limit.

The rate of fishing mortality (F) and the age at first harvest (t_c) can be manipulated by fisheries managers, and the influences of these on yield can be predicted using equation (5.37). Estimates of the number of recruits entering the fishery are rare, and it has become customary to divide both sides of equation (5.37) by R so that the prediction of yield actually is yield per recruit rather than an estimate of total yield. Yield-per-recruit estimates (Y/R) can be used as an index of actual yield for stocks that have relatively stable (albeit unknown) levels of recruitment. Gulland (1983) suggested that yield-per-recruit models are best suited to providing guidance on how to manage specific cohorts in stocks that have variable, environmentally regulated recruitment levels. In these cases, data from some index of cohort abundance at the time of recruitment could be used to define appropriate timing and levels of harvest.

In practice, fisheries managers are often more interested in how the yield-per-recruit predictions would vary across a range of management options, as opposed to obtaining point estimates for specific combinations of parameters. In the dynamic-pool model, it is possible to have many combinations of F and t_c that would produce the same Y/R. After calculating yield for ranges of t_c and F in a given fishery, the results can be summarized by plotting isopleths that connect the same Y/R estimates (Figure 5.8). The plots then can be used to choose desirable levels of F and t_c for the fishery.

For example, suppose that the fishery represented in Figure 5.8 has been operating at point A, where $F = 0.7$, $t_c = 1.5$ year, and $Y/R = 50$. If our goal is to increase Y/R, it is clear that we could accomplish the most by altering t_c rather than increasing F. If F is maintained at 0.7, t_c should be increased to about 4.5 years to maximize yield. Thus, by delaying the age at first harvest by about 3 years, we would expect a three to fourfold increase in Y/R.

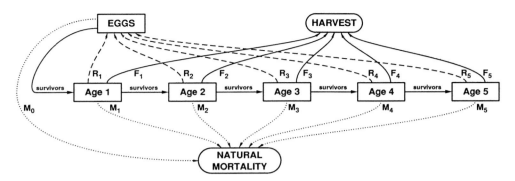

Figure 5.9 Representation of an age-structured population with maximum age of 5 years. Age-specific rates of reproduction (R_i), fishing mortality (F_i), and natural mortality (M_i) regulate survival from one age to the next.

If, however, the fishery was presently operating at point B on Figure 5.8, Y/R also equals 50, but $F = 0.2$ and $t_c = 4$. In order to increase Y/R from this point, the most effective approach will be to increase F, possibly by allowing or promoting increased effort or improved harvest techniques. An approximate fourfold increase in F would be required to approach the maximum yield.

Despite the inclusion of the adjective dynamic in the name dynamic-pool model, the suitability of using the yield predictions for making management decisions is dependent on several assumptions of stability or constancy of the stock's vital statistics. For example, the model assumes that natural mortality and growth rates remain constant over time and are independent of stock size. This may be a reasonable assumption for stable environments or for stocks where post-recruitment survival and growth are relatively constant from year to year. The model also assumes that the fishery has an equal influence on all ages of fish older than t_c and that rates of growth and natural mortality are not influenced by changes in the level of harvest.

The basic structural unit of a fish population is the age group. Each age group has rates of growth, natural mortality, fishing mortality, and reproduction that can vary with time and differ from those of other age groups. Age-structured models incorporate these differences in an attempt to provide realistic predictions of population behavior.

Population processes most frequently included in age-structured models are illustrated in Figure 5.9. This graph shows that the number of fish alive at any age is a function of the number alive in the previous year, less the numbers lost to humans and natural causes. The number alive at age 0, however, is a function of the reproductive rates and age structure of the entire population; this feature is not included in the surplus-production or yield-per-recruit models discussed earlier. Age-structured models of population number can be converted to biomass models by including a predictor of size at age, such as the von Bertalanffy growth model, or by using observed size-at-age data for a given population.

Walters (1969) formulated an age-structured model for fish population studies and illustrated its application for making management decisions in a brook trout fishery. He used an exponential model to predict the number (N) of trout in a year-class that would survive from year to year, given observed age-specific rates of instantaneous natural mortality (M) and hypothetical rates of instantaneous fishing mortality (F):

$$N_{i+1} = N_i e^{-(F_i + M_i)}. \tag{5.38}$$

The model was used to predict total population size and yield under conditions of either no fishing, a constant rate of fishing over time and among ages, or a periodic fishery.

Taylor (1981) claimed that fishery models available prior to the 1980s were either too simplified or too complicated to be useful to inland fisheries managers, and he attempted to solve this problem by developing a computerized population simulator specifically aimed at inland fisheries. His model, termed a generalized inland fishery simulator (GIFSIM), is fundamentally similar to the one proposed by Walters (1969), but it uses a different mathematical approach and can incorporate a number of population processes not considered by Walters. Examples of the application of GIFSIM have been reported by Taylor (1981) for rainbow trout and Zagar and Orth (1986) for largemouth bass.

Greater versatility for addressing management options of inland fisheries is achieved from age-structured models. However, a great deal of information is needed to estimate model parameters, and in many cases, data are inadequate to develop the model for a specific fishery. According to Taylor (1981), the minimum data requirements are (1) the initial age structure or number of fish in each age class, (2) the average length-at-age relationship for the population, (3) weight–length regression coefficients, and (4) natural and fishing mortality rates.

5.5 CONCLUSION

Recommendations for optimal management of a fishery should be based on changes in population number, composition, and biomass that are expected to result from management efforts. Development of this understanding requires assessment of fish population characteristics, including population size, mortality, and growth rates, as well as predictions of future trends by using mathematical models. This chapter provides an introduction to basic methods for studying population dynamics; fisheries managers should consider the suitability of more advanced techniques in specific cases.

The development of models useful for inland fisheries management has been difficult because of complicated processes of fish population regulation and multiple management objectives. Unlike commercial marine fisheries, where maximum sustainable yield of biomass has been the traditional management objective, inland recreational fisheries are managed to optimize recreational values as well as economic yield on a sustained basis. Annual and within-year variation of rates of reproduction, growth, and survival often make it necessary to conduct population assessments over several years to obtain estimates needed for model development. In many cases, long-term studies are not feasible, and population statistics like those described in Chapter 6 are used in lieu of predictive models for making management decisions.

5.6 REFERENCES

Backiel, T., and E. D. Le Cren. 1967. Some density relationships for fish population parameters. Pages 261–293 *in* S. D. Gerking, editor. The biological basis of freshwater fish production. Blackwell Scientific Publications, Oxford, UK.

Bailey, N. J. J. 1951. On estimating the size of mobile populations from recapture data. Biometrika 38:293–306.

Beverton, R. J. H., and S. J. Holt. 1957. On the dynamics of exploited fish populations. Fishery Investigations, Series II, Marine Fisheries, Great Britain Ministry of Agriculture, Fisheries and Food 19.

Brownie, C., D. R. Anderson, K. P. Burnham, and D. S. Robson. 1985. Statistical inference from band recovery data—a handbook, 2nd edition. U.S. Fish and Wildlife Service Resource Publication 156.

Carlander, K. D. 1969. Handbook of freshwater fishery biology, volume 1. Iowa State University Press, Ames.

Chapman, D. G. 1951. Some properties of the hypergeometric distribution with applications to zoological sample censuses. University of California Publications in Statistics 1:131–160.

Clady, M. D., D. E. Campbell, and G. P. Cooper. 1975. Effects of trophy angling on unexploited populations of smallmouth bass. Pages 425–129 *in* R. H. Stroud and H. Clepper, editors. Black bass biology and management. Sport Fishing Institute, Washington, D.C.

Cochran, W. G. 1977. Sampling techniques. Wiley, New York.

Cushing, D. H., and J. G. Harris. 1973. Stock and recruitment and the problem of density-dependence. Rapports et Procès Verbaux des Réunions, Conseil Permanent International pour l'Exploration de la Mer 164:142–155.

Galluci, V. F., and T. J. Quinn. 1979. Reparameterizing, fitting, and testing a simple growth model. Transactions of the American Fisheries Society 108:14–25.

Goedde, L. E., and D. W. Coble. 1981. Effects of angling on a previously fished and an unfished warmwater fish community in two Wisconsin lakes. Transactions of the American Fisheries Society 110:594–603.

Graham, M. 1935. Modern theory of exploiting a fishery, and application to North Sea trawling. Journal du Conseil International pour l'Exploration de la Mer 10:264–274.

Gulland, J. A. 1969. Manual of methods for fish stock assessment, part 1. Fish population analysis. FAO (Food and Agriculture Organization of the United Nations) Manuals in Fisheries Science 4:1–154, Rome.

Gulland, J. A. 1983. Fish stock assessment. A manual of basic methods. Wiley, New York.

Malvestuto, S. P. 1983. Sampling the recreational fishery. Pages 397–419 *in* L. A. Nielsen and D. L. Johnson, editors. Fisheries techniques. American Fisheries Society, Bethesda, Maryland.

Otis, D. L., K. P. Burnham, G. C. White, and D. R. Anderson. 1978. Statistical inference from capture data on closed animal population. Wildlife Monographs 62:1–135.

Redmond, L. C. 1974. Prevention of overharvest of largemouth bass in Missouri impoundments. Pages 54–68 *in* J. L. Funk, editor. Symposium on overharvest and management of largemouth bass in small impoundments. American Fisheries Society, North Central Division, Special Publication 3, Bethesda, Maryland.

Ricker, W. E. 1954. Stock and recruitment. Journal of the Fisheries Research Board of Canada 11:559–623.

Ricker, W. E. 1975. Computation and interpretation of biological statistics of fish populations. Fisheries Research Board of Canada Bulletin 191.

Robson, D. S., and D. G. Chapman. 1961. Catch curves and mortality rates. Transactions of the American Fisheries Society 90:181–189.

Robson, D. S., and H. A. Regier. 1964. Sample size in Petersen mark–recapture experiments. Transactions of the American Fisheries Society 93:215–226.

Schnabel, Z. E. 1938. The estimation of the total fish population of a lake. American Mathematical Monographs 45:348–368.

Seber, G. A. F. 1982. The estimation of animal abundance and related parameters, 2nd edition. Griffin, London.

Shaefer, M. B. 1954. Some aspects of the dynamics of populations important to the management of the commercial marine fisheries. Inter-American Tropical Tuna Commission Bulletin 1:27–56.

Shaefer, M. B. 1968. Methods of estimating effects of fishing on fish populations. Transactions of the American Fisheries Society 97:231–241.

Shelton, W. L., W. D. Davies, T. A. King, and T. J. Timmons. 1979. Variation in growth of the initial year class of largemouth bass in West Point Reservoir, Alabama–Georgia. Transactions of the American Fisheries Society 108:142–149.

Smith, C. L. 1983. Evaluating human factors. Pages 431–446 *in* L. A. Nielsen and D. L. Johnson, editors. Fisheries techniques. American Fisheries Society, Bethesda, Maryland.

Taylor, M. W. 1981. A generalized inland fisheries simulator for management biologists. North American Journal of Fisheries Management 1:60–72.

von Bertalanffy, L. 1938. A quantitative theory of organic growth. Human Biology 10:181–213.

Walters, C. J. 1969. A generalized computer simulation model for fish population studies. Transactions of the American Fisheries Society 98:505–512.

Weithman, A. S. 1986. Measuring the value and benefits of reservoir fisheries programs. Pages 11–17 *in* G. E. Hall and M. J. Van Den Avyle, editors. Reservoir fisheries management: strategies for the 80's. American Fisheries Society, Southern Division, Reservoir Committee, Bethesda, Maryland.

White, G. C., D. R. Anderson, K. P. Burnham, and D. L. Otis. 1982. Capture–recapture and removal methods for sampling closed populations. Los Alamos National Laboratory, LA-8787-NERP, Los Alamos, New Mexico.

Zagar, A. J., and D. J. Orth. 1986. Evaluation of harvest regulations for largemouth bass populations in reservoirs: a computer simulation. Pages 218–226 *in* G. E. Hall and M. J. Van Den Avyle, editors. Reservoir fisheries management: strategies for the 80's. American Fisheries Society, Southern Division, Reservoir Committee, Bethesda, Maryland.

Zippin, C. 1956. An evaluation of the removal method of estimating animal populations. Biometrics 12:163–169.

Zippin, C. 1958. The removal method of population estimation. Journal of Wildlife Management 22:82–90.

Chapter 6

Practical Use of Biological Statistics

JOHN J. NEY

6.1 INTRODUCTION

Managers are often responsible for conducting routine surveys of fish communities. Fish are captured, counted, measured for length and weight, and scales are removed for age analysis. The resulting data are conscientiously recorded, but too often they receive only superficial analysis; a small payoff for a large effort. If properly designed and consistently executed, surveys can yield information required for management of fish populations. Samples of fish can provide statistics concerning recruitment, growth, and survival that describe how the fish stock of interest is responding to physical, biological, and human factors that control its health and abundance.

Management decisions should always begin with a thorough assessment of available biological statistics. At the very least, this evaluation will identify aspects of population and community dynamics that may need more focused study. When biological statistics are sufficiently descriptive, they can be related to the socioeconomic concerns of the fishing clientele (Chapter 7) to determine options for enhancing angler satisfaction. Management decisions involving habitat modifications (Chapters 8–11), community manipulations (Chapters 12–15), or altered fishing regulations (Chapter 16), will then be dictated by their potential for success and the availability of resources.

6.2 ASSESSING INFORMATION NEEDS

The nature and amount of data required to periodically evaluate a fishery depend on the management intentions for that resource. The first step in designing a program should be to formulate those intentions as a goal (e.g., to develop a trophy muskellunge fishery, to maintain balanced bluegill–largemouth bass populations, to maximize sustainable crappie harvest). From this general goal, specific objectives with precise and measurable characteristics (e.g., to grow muskellunge to 80 cm in 3 years with 40% annual survival) are derived. The objectives in turn direct the search for biological data which accurately measure the desired characteristics.

The final step in identifying data needs is the choice of statistics to estimate population or community parameters. In most instances, several alternative data collection methods and resultant statistics can be used to provide a measure of each parameter. Selection of the best statistic is determined by a balance between the cost (effort) to obtain it and the amount of information that it provides.

The population statistics described in this chapter are simple to obtain, can be incorporated in monitoring studies, and provide information on the dynamics of target stocks. They are intended to provide explanations of how species are responding to their total environment in terms of the dynamic factors which dictate abundance–recruitment, growth, and survival. Understanding why the population performs as it does often requires more in-depth study. Declining growth, for example, results from an inadequate amount of food available per consumer, which could either be due to too many consumers (including competing species) or poor food production. Sometimes, satisfactory explanations can be developed from the manager's knowledge of the system and the environmental influences on it. Community, rather than population-level statistics, particularly those describing predator–prey relations, also frequently provide explanations that guide management decisions.

6.3 SOURCES OF BIOLOGICAL DATA

The manager should make a careful review of the recent literature before implementing a sampling design to assure that the statistics to be gathered can be compared with existing information and that they be as accurate as possible by current standards. Detailed discussions of sampling designs to monitor fish stocks, as well as techniques to evaluate angler harvest, are provided by Johnson and Nielsen (1983) and Malvestuto (1983). The sampling design assures two attributes for the data—consistency and representativeness.

6.3.1 Consistency

Most uses of fish statistics are comparative. The health of a population or community may be evaluated by relating current statistics to those previously reported for fishes from the same system (temporal or trend comparisons). Statistics from several bodies of water may also be compared to assess relative performance (spatial comparisons). When adequate statistics have been obtained on the performance (abundance, recruitment, growth, survival) of key species in a regional array of systems, standards may be developed to gauge how well fish are doing in a particular system. Valid comparison, whether over time or among different waters, requires that the data always be obtained in the same manner. It may be invalid, for example, to compare the relative abundance of crappies caught in gill nets and trap nets because crappies are much better at avoiding gill nets. Consistency in sampling design (timing, site, gear, and effort) does not assure that the statistics being compared are true estimates of populations, because the sampling designs can be biased.

6.3.2 Representativeness

Bias occurs when the portion of the population sampled is not typical of the population as a whole. As a consequence, the statistics developed from sampling will not accurately estimate population parameters. The assumption that what can be collected is representative of the population is one of the most troubling to fisheries scientists. Several measures can be employed to evaluate the accuracy of biological samples and provide a degree of confidence in statistics (Chapter 5). However, very precise samples, where the difference from one to the next is slight, might also not be accurate if an atypical segment of the population is consistently collected. If trap nets always catch a disproportionate share of young crappies, the age structures depicted by repeated trap-net collections could be very similar, but still fail to reflect the real age distribution in the crappie population. A degree of precision (repeatability) is essential to have confidence in statistical estimates, but the only way to assure true representativeness is to compare the sample statistic directly with the population parameter. Because all the individual fish that compose a population can rarely be examined (as in draining a pond), accuracy of sample statistics is usually evaluated indirectly by comparing estimates developed from two or more sampling methods.

6.3.3 Assessment Survey Data

Fish captured in the course of surveys provide simple, easily measured data, including length, weight, and appearance. Relationships between length and weight are indicative of the fish's growth and overall health or condition. Occurrence of sores, tumors, ectoparasites, and scars can also be noted as indicators of health. The frequency of evidence of disease in a sample can indicate the health of a population and has been used as a measure of water quality in U.S. streams (Leonard and Orth 1986).

Enumeration and measurement of the fish captured (or a representative subset) can lead to several statistics which estimate population parameters. The species composition (by number and biomass) of the sample can be useful to evaluate predation, competition, and resource use within the fish community, as well as water quality (Leonard and Orth 1986).

For a target species, the number of individuals captured with a given amount of fishing effort provides a measure of its absolute abundance or an estimate of relative abundance for temporal or spatial comparisons. Relative abundance is best expressed as the number caught per unit of effort. Relative abundance of young fish is widely interpreted to assess recruitment.

Plotting the number of fish captured by length interval provides a length-frequency distribution. Where overlap of lengths between successive age groups are minimal or can be discounted (as is particularly likely for younger fish), the length-frequency distribution will describe the age structure of the population. Under steady state conditions, the length-frequency distribution also describes the patterns of growth with age and can be used to estimate annual mortality rates by the catch-curve method (Chapter 5).

When survey designs encompass several different habitats (e.g., depths or substrates), a record of the location of each catch can provide information on fish distribution by size, sex, and species. Analysis of spatial distribution data can help define habitat use.

6.3.4 Data from Anglers

The single most important measure of a fishery is the catch. Harvest (or yield) is that portion of the catch that is removed from the system. The number of anglers and the time they spend fishing vary greatly over time and among locations. Consequently, total fishing effort is routinely estimated and applied as the divisor to total catch to provide the mean catch per unit effort statistic (e.g., kilograms/day; number/hour) to describe average angler success for comparative purposes.

If a representative fraction of angler harvest is examined, population data can be obtained which are similar to data obtained by sampling procedures used by biologists. Length, weight, growth, size and age distributions, relative abundance, recruitment, mortality, and species composition can all be measured or estimated from hook-and-line samples. Fisheries managers increasingly recognize the recreational catch as an inexpensive and rapid alternative to obtaining samples. Although conventional creel surveys are often labor intensive, large catches can be quickly examined at fishing tournaments. An alternative and complementary approach is to provide anglers with diaries in which to record details of their catch. The time required of fisheries managers varies among the different approaches. For example, in a study of largemouth bass in Lake Minnetonka, Minnesota, one person could examine 1 fish per hour during a creel survey, 5 per hour by electrofishing, 15 per hour through angler diaries, and 40 per hour at tournaments (Ebbers 1987).

However, there is a strong probability that fish taken by hook and line will not be representative of the population, and statistics derived from such samples will inaccurately estimate population parameters. Anglers may use techniques that are not only species-selective but size-selective as well. Comparison of biological and angler-generated sample statistics for largemouth bass indicate that bias will vary with the body of water and the type of angler. Although Ebbers (1987) reported strong similarity in length-frequency distributions, growth and mortality rates, and population estimates obtained by electroshocking and angler surveys, Gabelhouse and Willis (1986) found that tournament anglers caught a disproportionate number of intermediate-size bass. In general, the angler catch should not be substituted for conventionally obtained samples to provide population statistics unless comparison to biological data from each sample shows no real differences.

6.4   STATISTICS FOR STOCK ASSESSMENT

Samples of fish can quickly be described by measuring lengths and weights of individual specimens. These basic data can be transformed into simple statistics to evaluate the two fundamental characteristics of fish stocks—well-being and abundance.

6.4.1   Weight–Length Relationships

The ratio of the weight of a fish to its length varies with the species, size, and ecological conditions under which it feeds and expends energy to live. When species and size effects can be discounted, the ratio provides a measure of the fish's health or well-being. Comparison of weight–length ratios of fish in a body of water over time or between different bodies of water helps a manager to assess how well fish are able to feed and grow. Weight–length ratios, when obtained from precise and consistent measurements, have proven to be accurate indicators of relative growth and so can be substituted for the more laborious procedure of constructing growth histories by ageing fish (Wege and Anderson 1978).

Because weight–length ratios are derived from individual specimens in a sample, the data show both averages and variance, essential properties for statistical tests comparing different groups. When comparing populations, weight–

length ratios should always be taken in the same season when tissue accumulation is neither extremely high nor low (e.g., avoid pre- and post-spawning samples); the mid- to late growing season is preferred (Wege and Anderson 1978).

Weight–length ratios are usually expressed in whole numbers, called condition factors. Weight (W) of fish tends to increase as a cubic function of length (L), and the weight–length relationship can be expressed as a power curve:

$$W = aL^b \tag{6.1}$$

where a and b are population-specific constants. For many populations, b will be close to 3. Condition (K) of an individual fish is sometimes expressed as:

$$K = \frac{W \cdot X}{L^3} \tag{6.2}$$

where X is a scaling constant, dependent upon metric units used, to achieve integer status (see Anderson and Gutreuter 1983 for example). In this form, known as the Fulton type, the condition factor has achieved its widest usage. Comparisons to other populations can be made from published values, such as those in the *Handbook of Freshwater Fishery Biology* (Carlander 1977). However, comparisons are not valid between species or even between length groups within species because b is not truly 3 and fish have different or changing body shapes.

In an effort to eliminate the size bias, Le Cren (1951) developed the relative condition factor (K_n) which uses the weight–length relation developed over all size groups in a particular population:

$$K_n = \frac{W}{\hat{W}} \tag{6.3}$$

where W is weight and $\hat{W}$ is equal to aL^b. The relative condition factor expresses the deviation of an individual's weight from the average for fish of its length in that population. As such, its use is limited to within-population comparisons, as for seasonal effects or sexual differences in growth.

Recently, the concept of relative condition has been expanded for inter-population comparisons by replacing the population-specific weight–length relation with a standard for the species (Wege and Anderson 1978). The resulting condition factor is termed relative weight (W_r) and is determined as follows:

$$W_r = \frac{W}{W_s} \times 100 \tag{6.4}$$

where W_s is the standard weight for a specimen of the measured length. The standard weight–length relations are developed from available weight–length relations for the species (Wege and Anderson 1978; Murphy et al. 1990). Relative weight may become the most commonly used index of condition because it enables direct comparison of different sizes and species of fish and provides an instant benchmark for evaluating the well-being of a population without a literature search. Standard weight equations have been developed for a number of prominent freshwater species; some examples are presented in Table 6.1.

Condition factors in their various forms facilitate a rapid assessment of the food availability- consumption-growth phenomenon. Within populations, they can be used to monitor the influence of environmental change over time, as Colle and

Table 6.1 Values of the intercept (*a*) and slope (*b*) in the standard weight equations ($\log_{10}$ $W_s = a + b \log_{10} L$) for North American freshwater fishes. Equations for metric units (g and mm).

Species	*a*	*b*	Source
Largemouth bass (northern)	−5.316	3.191	Anderson and Gutreuter (1983)
Bluegill	−5.374	3.316	Anderson and Gutreuter (1983)
Gizzard shad	−5.376	3.170	Anderson and Gutreuter (1983)
Walleye	−5.453	3.180	Murphy et al. (1990)
Northern pike	−5.369	3.059	Willis (1989)

Shireman (1980) did to evaluate the response of centrarchids to hydrilla *Hydrilla verticillata* infestations in Florida lakes. Analysis of condition factor dynamics can also identify life stages or seasons where available food is inadequate. In inter-population comparisons, poor condition factor may signal the need for more focused study of the amount of food available per consumer (a function of both food production and intensity of competition) or the environmental factors (such as cover and water quality) that affect its efficient use.

6.4.2 Abundance

Knowledge of the numerical abundance of a fish stock is a component of the information required for its management. When numbers are matched to weight at length data, total stock biomass (usually referred to as standing crop or standing stock) can be calculated. Methods to estimate absolute abundance (Chapter 5) frequently require more effort and expense than can be allocated in assessment surveys. Instead, the catch of fish in a survey sample is related to the effort expended to collect it in a statistic known as relative abundance. Division of the catch (usually number but sometimes weight) by effort yields catch per unit effort (C/f = CPUE) and, in theory, removes the effect of variable effort in the measurement of abundance. Relative abundance is used to make temporal or spatial comparisons. If a high correlation can be demonstrated between relative and absolute abundance, CPUE may also be used to estimate actual stock size.

Sampling effort must be precisely measured and consistently expended for developing relative abundance statistics. A wide variety of effort units are in common use; most involve measures of the amount of gear deployed, duration of sampling, or area sampled, alone or in combination. Consistency of sampling effort is a requisite for valid comparisons; capture efficiency can not be affected by how the sample was collected. Gear should be standardized in type, technical specifications (e.g., gill net mesh sizes, trawl throat dimensions), and manner of operation. Similar habitats should always be used as sampling locations. Sampling to compare relative abundances should also be conducted at the same time of year, but avoiding unusual weather or water conditions (e.g., storms, turbidity) likely to affect capture success. Prolonged sampling effort could change fish behavior or otherwise cause the gear to be less effective. Catching efficiency can be altered by even slight changes in the sampling regime.

Application of the relative abundance statistic is predicated on the assumption that catch per unit effort is directly proportional to population size (*N*). For this to be true, catchability (*q*, the probability of catching an individual fish in one unit of effort) has to be constant (from Chapter 5, $C/f = qN$). In reality, catchability is never constant. Fish activity patterns, weather, and water quality change rapidly and influence capture success. Gear saturation (excess effort) also reduces catchability. These influences can usually be minimized by sampling under typical environmental conditions with a moderate amount of well-distributed effort. Sampling gear should be chosen to neither attract nor repel the target species, but rather to catch them by random encounter. In several commercial and sport fisheries catchability has declined as abundance increased (Bannerot and Austin 1983), causing catch per unit effort to stabilize. The opposite relationship appears to result from intense fishing or from changes in angler behavior, but it does not seem likely to occur in typical assessment surveys. The only method to check the validity of the proportionality assumption is to directly compare relative and absolute abundance, and this is usually not possible. Sampling programs to obtain relative abundance statistics have been developed for most prominent freshwater species in different types of systems.

If a paired series of CPUE statistics and absolute abundance indices are highly correlated, a single measure of relative abundance can be used to predict actual population size. The technique is illustrated by Hall (1986), who compared shoreline electrofishing CPUE for largemouth bass with concurrent mark–recapture population estimates in 12 Ohio impoundments (Figure 6.1). The paired data were analyzed by linear regression with CPUE as the independent variable (X) and population density as the dependent variable (Y).

Relative abundance statistics obtained for a single stock over time can be used to assess whether stock size is changing. Analysis of temporal patterns of relative abundance is valuable both to identify trends and to evaluate phenomena (management actions, fishing pressure, environmental alterations) with the potential to cause change in population sizes. Although it is often possible to visually spot trends from graphs of CPUE versus time, short-term variability might mask them. Linear regression, in this instance with CPUE as the dependent variable and time as the independent variable, should be applied to determine whether a trend actually exists (slope of the regression line is not zero), the direction and rate of change (the slope itself), and the strength of the CPUE-time relationship (Figure 6.2). Measures of the suspected agent of change in fish abundance (e.g., fishing pressure, nutrient concentration) may be available over the time period of assessment; these can then substituted for time as the independent variable in the regression.

Relative abundance of juvenile fish is frequently used to assess reproductive success and to project the future abundance of the adult stock. To be effective as a predictor of recruitment, relative abundance of juvenile fish must be positively related to their later abundance in the fishable stock. Larval and juvenile survival rates are highly variable within many fish populations because these life stages are particularly vulnerable to starvation and predation (Ploskey and Jenkins 1982). It is imperative that the validity of a recruitment index be established prior to its use. Comparison of the relative abundances of a cohort as juveniles and following recruitment by regression analysis can provide that assessment (Willis 1987).

Management agencies commonly employ standardized sampling schemes in assessment surveys (Johnson and Nielsen 1983). Relative abundances of species and life stages obtained by standard procedures can then be compared among similar systems to give an overview of density patterns, provide insight into influencing factors, develop management strategies, and adjust angler expectations. Although sampling effort is easily controlled, lakes and rivers are all unique and have inherent differences in the ease with which fish may be taken from them

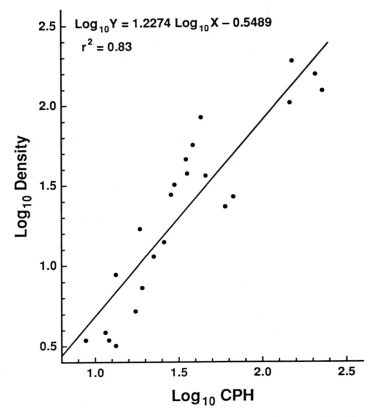

Figure 6.1 Linear regression of estimated population density as a function of relative abundance (electroshocking catch per hour, CPH) of largemouth bass in Ohio impoundments. Both variables are transformed to log-base 10 to increase linearity of the relationship. (From Hall 1986.)

by a particular method. For example, CPUE for gill-net samples from clear and turbid lakes will likely be affected by differential net avoidance, regardless of true population densities. Care must therefore be taken to limit inter-system comparisons to waters in which catchability should be similar.

6.4.3 Population Structure

When all ages or sizes of a species are taken in proportion to their true abundances, the sample is representative of the structure of the population. Analysis of the relative abundance of the various age groups in the sample provides measures of the dynamic processes that govern population number and biomass. Size-group comparisons can be formulated in indices for rapid assessment of the harvest potential of the population. Coupling of size-structure indices (usually length-based) with size-at-age statistics permits understanding of how the size composition of the population is achieved.

The age-frequency distribution of a representative population sample describes the status of the successive cohorts. Changes in age-frequency distributions over time will identify instances of excessive mortality. Age-frequency distributions can also be used to describe the pattern of growth in length within populations. If

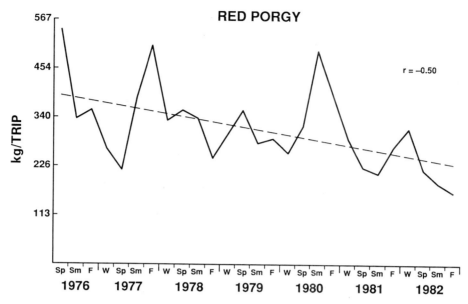

RED PORGY

r = −0.50

kg/TRIP

567
454
340
226
113

Sp Sm F | W Sp Sm F | W Sp Sm F | W Sp Sm F | W Sp Sm F | W Sp Sm F | W Sp Sm F
1976   1977   1978   1979   1980   1981   1982

Figure 6.2 Trend analysis of CPUE (kilogram/trip) over time in the South Carolina commercial fishery for red porgy. Dashed line is linear regression. (From Low et al. 1985.)

growth conditions are assumed to be static, plotting means and standard deviations of lengths at successive ages will be sufficient; otherwise, particular cohorts must be tracked over their lifespans from annual samples.

The age composition of a sample is best determined by ageing individual fish from hard body parts that deposit annual rings (Jearld 1983). Ageing fish from scales or other structures is a laborious process and often not possible where growth is year-round, such as in tropical waters. Consequently, length-frequency analysis, in which length modes (most frequent lengths) are assigned ages, has become a popular alternative to direct ageing procedures.

Assignment of ages from length-frequency distributions is widely practiced but of limited utility. A length-frequency histogram is likely to show distinct modes for the youngest age groups but less distinct or indistinct modes for older fish (see Figure 16.12 in Jearld 1983). Short spawning periods and low variability in early growth rates cause these modes to be sharp and well separated. However, a multi-modal length distribution can develop if spawning is prolonged or intermittent or if differential growth occurs (Shelton et al. 1979). The distance between true modes and the variability within each associated distribution will determine their degree of resolution in the overall length-frequency distribution. For normal distributions of length at age, only a single mode will be apparent for two adjacent age groups if the difference between modes is less than twice the minimum standard deviation (Figure 6.3). Growth slows and becomes more variable as fish age, causing high overlap in distributions and smaller distances between modes. Both damping of modes and spurious modes confound the assignment of ages to length-frequency distributions. Resolution for early age groups can be improved by proper choice of measurement interval for graphical display (Anderson and Gutreuter 1983), but assignment of ages by visual inspection will still likely be limited to the first few age groups. Microcomputer programs that define overlap-

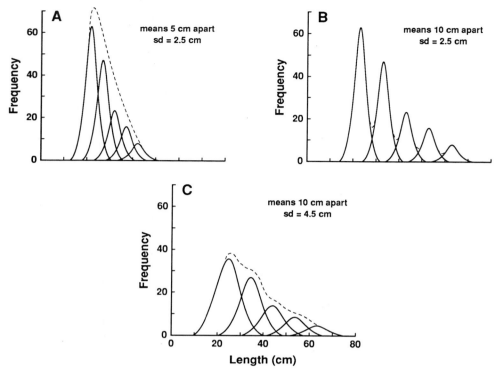

Figure 6.3 Effects of distribution about means on the distinctness of successive modes in length frequency plots. Dashed curves are frequency plots for large samples. In panel B, dashed curves follow solid curves very closely. Proportions of the population represented by each successive age group are the same in all three panels. (From Macdonald 1987.)

ping distributions are available, but they require informed decisions on input parameters to minimize error and have not as yet received much use for inland, temperate-latitude fishes. Jearld (1983) reviewed both graphical and computer-assisted techniques for determining age distributions from length frequencies.

For comparative purposes, the length at age of only a single age group may be adequate to characterize growth in the population. Kruse (1988) used total length at age 3 to assess growth rates of largemouth bass in Missouri impoundments and to clarify interpretation of the length-frequency distributions of these populations. It generally will be best to choose an intermediate age to indicate both cumulated growth performance over several years and how rapidly fish are reaching harvestable size. However, the effort required to estimate mean total length at any single age should be a small fraction of that needed to determine length at each successive age in the population.

Length-frequency distributions are better suited to describe the status of a fishable population than its dynamics. The size distribution of fish in a representative sample can be quickly analyzed without reference to age composition to assess the balance of the population or community. As defined by Swingle (1950), a population is balanced when it can sustain a satisfactory (for management objectives) harvest of good-size fish in proportion to the productivity of the habitat. Balance therefore depends on the density of fish of various sizes in the population, both adequate numbers of catchables and sufficient smaller fish to provide replacement.

Indices of population balance derived from length-frequency distributions have been most widely used in small (<200 hectares) midwestern impoundments with largemouth bass–bluegill fisheries. The basic index is proportional stock density (PSD) developed by Anderson (1976) as:

$$PSD\ (\%) = \frac{\text{number} \geq \text{quality size}}{\text{number} \geq \text{stocksize}} \times 100. \tag{6.5}$$

Minimum stock and quality lengths are defined as some length within 20–26% and 36–41%, respectively, of angling world record length (Anderson and Weithman 1978). These correspond roughly to the minimum sizes at which anglers will first catch the species (stock length) and consider the specimen desirable (quality length). Gustafson (1988) has developed a confidence interval estimator for PSD when sample sizes are large and PSD values are not close to 0 or 100.

Two approaches can be used to determine the range of PSD values indicative of a population structure that meets management objectives for a particular species/ system situation and is therefore "in balance." In the empiric method, PSDs can be calculated where management objectives (e.g., harvest, predation intensity) are being met and the range of these is then used as the standard. Alternatively, a simple model can be employed to generate PSDs using average or optimal annual growth and mortality rates and constant recruitment (Anderson and Weithman 1978). Desirable PSD values will vary with management objectives. For example, control of stunting in bluegill populations could require a much higher proportion of stock-size largemouth bass than is optimal to support a trophy bass fishery.

Generally, PSDs indicative of balance for target species are based on sustainable harvest of sizes preferred by anglers (Table 6.2). The relative stock density (RSD) index has been developed to better assess the size distribution within the quality-length portion of the population. Although RSD can be the proportion of any size group within the total sample of stock-size and larger fish, it is usually used to track subsets of the quality- size group (RSD_{30} for example, would refer to all fish over 30 cm). A standard categorization of quality-size fish linked to the angling experience has been proposed by Gabelhouse (1984; see also Anderson and Gutreuter 1983). These additional length groups are also percentages of world-record lengths: preferred (minimum 45–55%); memorable (minimum 59–64%); and trophy (minimum 74–80%). Determination of RSDs of gamefish in

Table 6.2 Ranges of proportional stock density (PSD) values suggested as indicative of balance when the population supports a substantial fishery.

Species	PSD	Source
Largemouth bass	40–70	Anderson (1980)
Bluegill	20–40	Anderson (1980)
Yellow perch	30–50	Anderson and Weithman (1978)
Walleye	30–60	Anderson and Weithman (1978)
Northern pike	30–60	Anderson and Weithman (1978)
Muskellunge	30–60	Anderson and Weithman (1978)
Smallmouth bass	30–60	Anderson and Weithman (1978)

these categories provides a readily understandable description of the fishing opportunity provided by the population. Precision can be increased by calculating specific RSDs as the proportion of total stock within each category length range rather than as the proportion of all fish longer than the minimum category length.

Although the PSD and RSD indices assess population structure in terms of angling potential, management actions should not be undertaken without additional information. Balance requires not only size but also number; density of the population should also always be estimated (most often this will be as CPUE). Management responses to adjust population size structure may also require knowledge of growth, mortality, and recruitment rates. A PSD value judged to be too high could be due to rapid growth, low mortality, or recruitment of an extremely strong year-class; the inverse is true for low PSD. The appropriate management response to rectify an extreme PSD might be quite different if the source of the perceived imbalance is recruitment rather than growth or survival.

Size selectivity in sampling can be a major impediment to obtaining an accurate length-frequency index, particularly in large systems with diverse habitats. Sampling bias associated with gear selectivity or differential seasonal distributions can cause PSD values to vary severalfold. Comparison of the length distributions in different gears or in successive samples can be used to identify and control bias (Carline et al. 1984).

A potentially greater obstacle to the general adoption of length-frequency indices in inland fisheries management is system instability. Most management alternatives to correct imbalance in fish populations assume that recruitment, growth, and mortality are density dependent. Although density dependence does appear to control these processes in small central-latitude impoundments, abiotic factors may play a greater role in larger waters or at higher latitudes. In particular, reproductive success and eventual recruitment may fluctuate widely from year to year in response to climatic factors. Extremes in year-class strength will be reflected in length-frequency indices. The PSDs of largemouth bass in an Ohio impoundment and of walleye in a natural Wisconsin lake varied threefold in one year, largely as a function of recruitment (Carline et al. 1984; Serns 1985). In these situations, the usefulness of structural indices is likely to be limited to tracking population changes beyond management control.

6.4.4 Angler Catch Statistics

The three fundamental descriptors of a fishery are the catch (C, and its subcomponent, harvest), fishing effort expended (f) and catch per unit effort (C/f or CPUE). Any two of these statistics can be used to calculate the third. All are typically estimated by the creel survey technique which consists of interviews of anglers, inspection of the catch, and tabulation of hours spent fishing. For most bodies of water, the creel survey is a sample requiring statistical expansion over time and area to achieve total estimates of catch and effort in a particular period. Choice of sampling design and the method to contact anglers can strongly influence the resulting estimates; options are discussed by Malvestuto (1983).

Quantitative description of a catch usually requires interviewing anglers in addition to inspecting their fish to characterize the fraction discarded. Size and bag limits have long restricted angler harvest, and the rising popularity of catch-and-release fishing for many species is widening the discrepancy between what is caught and what is retained. However, the number and weight of fish harvested by fishing remains the most critical piece of information to the manager responsible for sustaining a balanced yield.

Fishing effort is best measured as the number of angler hours because trip length is highly variable. Total effort in time fished is frequently related to the size of the body of water (e.g., hours per surface hectare or stream kilometer) to describe the distribution of fishing pressure over time or among systems. Profiles of fishing pressure can prompt management actions to redistribute fishing effort to alleviate crowding, promote catch rates, and enhance angler satisfaction. In multispecies fisheries, effort should be partitioned by hours spent pursuing target species to avoid underestimating angler success.

The ratio of the catch of target species to the effort expended in their pursuit may be the single best indicator of fishing quality. However, fishing success as measured by CPUE should not be equated with angler satisfaction. The objectives for "going fishing" are varied, reflecting not only the angler's personal philosophy but also his expectations for the total experience. Definition of angler satisfaction remains elusive, but a better approximation will be achieved by including social as well as fishery metrics in the assessment (Chapter 7).

Catch per unit effort can be estimated from a representative sample of the angling population and used to derive total catch from total effort ($C = f[C/f]$) or total effort from total catch ($f = C/[C/f]$) if either of these statistics is not available. The CPUE statistic has also sometimes been used as an index of abundance for intensively fished species. Use of angler CPUE as an abundance estimator should be avoided because hook-and-line capture rate is influenced by many factors other than abundance, especially the interactive behavior of the fishers and their prey.

Creel surveys on large waters are laborious and expensive, but they are usually the only way to obtain accurate total catch and effort statistics. However, anglers and their catch can be monitored in a variety of less intensive ways to obtain useful statistics. Citation and tag return programs have both proved useful in this regard.

Rewarding successful anglers with certificates or other forms of recognition is an inexpensive way to monitor species catch on a state or regional basis. Citation programs usually recognize trophy fish with minimum species-specific weights required for entry. This information must be supplemented by biological and creel statistics to appraise the status of the whole population.

Tags recovered by anglers from previously marked fish (M = number of fish marked) are commonly used in estimating fishing mortality. When the number of tag returns (R) is totalled for the year following marking, the ratio R/M is a direct estimate of the annual exploitation rate (u) (i.e., the percent of the initial population that has been harvested). The exploitation rate can be used alone or in conjunction with the total mortality rate (A) to assess the impact of fishing on the population (Chapter 5).

The validity of mortality and abundance estimates based on angler tag returns depends on, among other things, the ability of all anglers to recognize tags and their willingness to report them. The tag reporting rate can best be determined by comparing the number of tags observed in a creel survey with the number subsequently reported by interviewed anglers (Green et al. 1983).

Table 6.3 Single-variable predictors of fish standing stock or yield for North American waters showing good predictive power.

Predictor	Prediction	Waters tested	Source
Surface area (km^2)	Total yield (kg/yr)	Northern natural lakes	Youngs and Heimbuch (1982)
Total phosphorus (μ/L)	Total yield (kg/ha/yr)	Northern natural lakes	Hanson and Leggett (1982)
Chlorophyll-a (mg/m^2)	Sportfish yield (kg/ha/yr)	Midwestern U.S. lakes and reservoirs	Jones and Hoyer (1982)
Benthos standing stock (kg/ha)	Total yield (kg/ha/yr)	Northern natural lakes	Matuszek (1978)
Total phosphorus (μ/L)	Total standing stock (kg/ha)	Southern Appalachian reservoirs	Yurk and Ney (1989)
Morphoedaphic index	Total yield (kg/ha/yr)	Northern natural lakes	Ryder (1965)
Morphoedaphic index	Total standing stock (kg/ha)	U.S. hydropower storage reservoirs	Jenkins (1982)

6.4.5 Fishery Productivity

The capacity of a body of water to support fish biomass (carrying capacity) and to provide a sustainable fish harvest is extremely useful information to the fisheries manager. Knowledge of system capacity allows the manager to adjust angler expectations and manipulate the harvest to make best use of the resource. However, direct determination of biomass carrying capacity and sustainable yield for individual waters generally requires detailed long-term data.

Much research has been devoted to empirical prediction of fish standing stock and yield. A measure of standing stock or yield as the dependent variable is paired with an independent variable or variables without substantiation of a cause-and-effect relationship. Paired data are obtained for a set of waters where they have been accurately measured and then analyzed by regression analysis (Ryder 1965).

The search for suitable predictors of fishery productivity has been long and intense because of the obvious benefits. Early efforts to predict fishery productivity in lakes focused on morphometric features such as surface area and mean depth. Predictive power was improved by using nutrient or other chemical (edaphic) variables as predictors (Hanson and Leggett 1982). The most well-known predictor of fishery productivity in natural lakes is the morphoedaphic index (MEI) which is the concentration of total dissolved solids divided by mean depth (Ryder et al. 1974). The MEI remains among the best predictors, but it must be applied judiciously to lakes that meet particular climatic and chemical criteria. Biological variables such as phytoplankton or benthic invertebrate standing stock or chlorophyll-a activity have also demonstrated good predictive power for fishery productivity in lakes, but may be difficult to measure accurately. Regression models that predict fishery productivity in North American streams are also numerous, but their usefulness to date has been limited by small sample size, narrow focus, complex (multivariate) structure and data requirements, and low predictive power.

The number of simple regression models that can predict fish standing stock or harvest potential in standing waters is substantial (Table 6.3) and can be expected to increase for both lakes and streams. Selection of a model to project fishery productivity in a particular system should be based on meeting three criteria: (1) feasibility of consistent and representative measurement of the predictor, (2) predictive power, and (3) similarity of the system to those in the regression data set.

6.5 STATISTICS FOR COMMUNITY ASSESSMENT

An aquatic community includes all plants and animals living in or closely associated with a body of water. In practice, community assessment refers to evaluation of any multispecies combination within the ecosystem. Community assessment statistics have been used for two general purposes by fisheries managers. Status of the relationship between prey (forage) and predatory fishes is described to determine community balance—the potential to provide a satisfactory angling harvest. Composition of the fish species assemblage is analyzed to assess the ability of the system to support a healthy aquatic biota. The statistics employed for each purpose are varied and still evolving. No single estimator of either prey–predator balance or water quality has been proven suitable for general application.

6.5.1 Prey–Predator Relations

The PSD index developed to assess balance within the population can be extended to simultaneously visualize the status of both the prey and predator components of the community. Prey PSD is plotted versus predator PSD in a tic-tac-toe grid (Figure 6.4). Parallel lines bound the PSD percentages indicative of balance (as determined by management objective; Gabelhouse 1984) for each trophic group. The central rectangle formed by the intersection of horizontal and vertical lines defines the desired state of mutual balance for prey and predators. The PSD grid offers a rapid visual status assessment. Originally developed to describe largemouth bass–bluegill relationships in ponds, it can be expanded to encompass all predators and prey by weighing for the relative abundance of species in each category (Anderson and Weithman 1978). Drawbacks of the PSD graph are the same as for the single-species PSD index. Additional information on density, growth, recruitment and mortality are required to explain the status graphically depicted; management actions should not be undertaken on the basis of graph results alone. Uncontrollable variability in population dynamics, particularly recruitment, will be reflected in extreme and potentially misleading annual changes in both prey and predator PSDs. Plots of PSD have been used primarily to track response to management actions in small, relatively simple ponds.

Because biomass is the product of numbers and weight, biomass ratios incorporate densities (absolute or relative abundance) of prey and predators directly. The ratios also serve as indicators of the adequacy of the food supply to meet predator demand (Ney 1990). Biomass ratios were first developed for small southern U.S. ponds to indicate balance between largemouth bass and their forage (Swingle 1950). The F/C ratio is the total weight of forage species (usually predominately bluegill) divided by the total weight of predator species; the range 3–6:1 is desirable. The Y/C ratio is a refinement that recognizes size limitations on predation in which Y is the total weight of forage fish small enough to be eaten by the average-size adult predator, and C is as before. Balance is indicated by Y/C ratios of 1.0 to 3.0. Use of both ratios has been largely limited to southern pond management (see Chapter 20). Desirable pond values do not apply in southern reservoirs, but appropriate values for these and other systems could conceivably be generated through empiric analysis of extensive abundance and harvest data sets.

Jenkins and Morais (1978) developed the available prey:predator ratio (AP:P) as

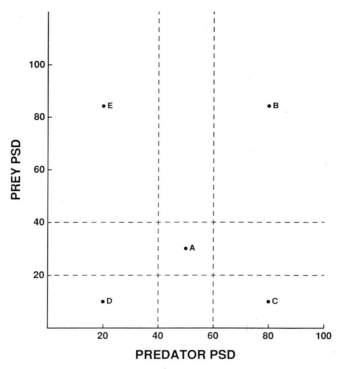

Figure 6.4 Tic-tac-toe graph comparing proportional stock densities of predators and prey. Parallel lines bound the desired PSD ranges. Potential interpretations for the different combinations are: A = mutual balance for satisfactory fishing; B = community comprised of large, old specimens, indicative of an unfished population; C = large predators excessively cropping large prey; D = overfishing of predators and stunting of prey; and E = high population of small predators excessively cropping young prey.

an elaboration of Y/C to account for size limitations on prey consumption for predators of all lengths. The basis of the AP:P determination is a series of curvilinear equations which predict the maximum total length of a forage fish species which can be ingested by a largemouth bass of a given total length. Availability equations are derived from comparisons of mouth diameter of bass and maximum body depth of prey; all predators are standardized in terms of largemouth bass length equivalents on the basis of their relative throat diameters. Using standing stock estimates and length distributions for all species, available prey biomass is plotted as a function of cumulative predator biomass for successively larger predators (Figure 6.5). Jenkins and Morais (1978) considered that a 1:1 AP:P ratio was indicative of prey sufficiency for southern reservoirs sampled in August. The AP:P ratio has been used to identify periods and sizes at which predators encounter prey deficiencies in southern reservoirs (Timmons et al. 1980). The AP:P is most easily applied in these systems because standing stocks and size distributions of all species are routinely estimated by cove-rotenone sampling. The AP:P approach can be adapted to consider behavioral and distributional effects in addition to size influences on prey availability (Ney 1990). The ingestibility limit equations can be used independent of abundance data to

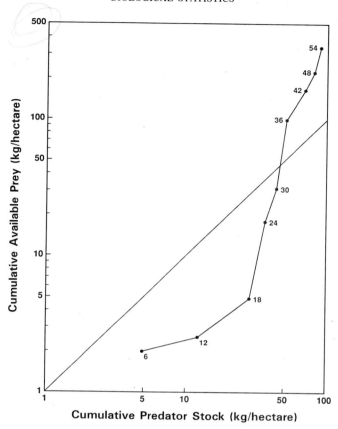

Figure 6.5 Logarithmic plot of prey standing stock morphologically available to the cumulative standing stock of predators at successive 6–cm total length intervals. Points below the 1:1 AP:P line indicate clear deficiencies in available prey for predators of these sizes. (After Jenkins 1979.)

compare relative availability of different prey species or suggest predator stocking strategies to maximize use of prey (Ney and Orth 1986).

It is now recognized that top carnivores in lakes can influence the abundance and composition of lower trophic levels (Carpenter et al. 1985). By consuming planktivorous forage fishes, piscivores relax the predation pressure on zooplankton. The abundance of large zooplankters (e.g., *Daphnia* spp.), which are positively selected by forage fishes, should therefore be directly related to the intensity of predation by piscivores. Mills et al. (1987) confirmed that mean zooplankton size increases with the ratio of predator to prey fishes in New York lakes and advocated zooplankton size as an index of predator–prey balance. They suggested monitoring zooplankton by vertical-haul sampling in spring and again in summer, the latter to gauge the abundance of age-0 fishes. Annual changes in mean zooplankton size might be used to quickly assess the success of game and forage fish introductions or the impact of major perturbations, such as winterkill.

6.5.2 Water Quality Indicators

Characterization of the fish assemblage to indicate the well-being of an aquatic ecosystem is attractive because fish represent aquatic life to the general public.

Table 6.4 Diversity indices frequently used to characterize fish communities and similarity indices with potential for inter-site comparisons.

Name	Index	Source
Diversity Indices		
Margalef's	$D = \dfrac{S - 1}{\log_e N}$	Margalef 1958
Shannon–Weaver	$H' = -\displaystyle\sum_{i=1}^{s} \dfrac{n_i}{N} \log_e \dfrac{n_i}{N}$	Shannon and Weaver 1949
Similarity Indices		
Sorenson's	$SI = \dfrac{2C}{A + B}$	Sorenson 1948
Percentage similarity	$PSC = 100 - 0.5 \displaystyle\sum_{i=1}^{s} |a_i - b_i|$	Whittaker 1952

A = number of species in sample A
B = number of species in sample B
C = number of species common to both samples
a = percentage of sample A represented by the species
b = percentage of sample B represented by the species
N = number of individuals in the sample
S = number of species in the sample
n_i = number of individuals of species i in the sample

Fish, as the dominant vertebrates in most systems, also integrate the multiple impacts on the water resource. Most attempts to use fish as ecosystem indicators have been directed at flowing waters, where impacts are more potentially severe and identifiable than in lakes.

The most commonly used single measurement to describe a taxonomic grouping (such as fish) within a community is diversity. Although ecologists disagree on the exact definition of diversity, the consensus is that it is a function of both the number of species present (richness) and the equitability of the distribution of individuals within these species (evenness). Diversity indices that simultaneously consider both richness and evenness in mathematical formulations are included in Table 6.4. High values of diversity indices have generally been interpreted to indicate relatively unspoiled systems, with diversity inverse to the degree of degradation. However, it is now recognized that diversity can actually increase with disturbance of the system, and that diversity is not inevitably linked with system stability (Washington 1984).

Similarity indices (Table 6.4) compare diversity between two areas. Although their use in aquatic ecosystems has been limited, they have particular potential for contrasting sites which differ only in exposure to degradation, as for upstream and downstream locations with a point-source discharge between them. Of course, the assumption of uniformity of sites is difficult to verify because a stream is a physical and biological continuum, progressively changing from headwaters to its mouth. The merits of the many diversity and similarity indices are critically reviewed by Washington (1984).

Table 6.5 Fish community metrics, scoring criteria, and class boundaries proposed for the index of biotic integrity. (From Leonard and Orth 1986.)

Category or class	Metric	Scoring criteria		
		5 (best)	3 (fair)	1 (worst)
Species composition	Total number of fish species			
	Species richness and composition of darter species			
	Species richness and composition of sunfish species	Varies with stream size and region		
	Species richness and composition of sucker species			
	Presence of intolerant species			
	Proportion of individuals as green sunfish	<5%	5–20%	>20%
Trophic composition	Proportion of individuals as omnivores	<20%	20–45%	>45%
	Proportion of individuals as insectivorous cyprinids	>45%	20–45%	<20%
	Proportion of individuals as top carnivores	>5%	1–5%	<1%
Fish abundance and health	Number of individuals in sample	Varies with stream size		
	Proportion of individuals as hybrids	0	0–1%	>1%
	Proportion of fish with disease or anomalies	0	0–1%	>1%

The "indicator organism" approach to assessing ecosystem degradation has only recently been applied with fish in a formal manner, although subjective appraisal based on the presence or absence of tolerant (e.g., common carp) and intolerant (e.g., rainbow trout) species have been made for decades. Karr (1981) proposed the index of biotic integrity (IBI) to assess stream degradation from measurable attributes of the fish assemblage which can be easily derived from a representative sample. As applied to midwestern streams in agricultural areas, the IBI consists of 12 attributes in three categories: species composition, trophic composition and health, and abundance of fish (Table 6.5). Species composition attributes focus on overall richness and richness within major taxonomic groups as well as the occurrences of notably tolerant and intolerant species. Food habits of the fish assemblage as categorized by trophic composition are products of the diversity and productivity of the lower tropic levels in the community. Fish abundance and fish health reflect system productivity and habitat stability. A fish sample is assigned one, three, or five points for each attribute by comparison to expectations for a pristine stream of similar size in the same region. Total scores define stream health in six classes ranging from excellent (pristine) to no fish (extremely degraded).

Effective use of the IBI requires a knowledge of the structure and function of regional stream fish communities and of species' tolerances. Species composition attributes will vary as functions of stream size and region; Fausch et al. (1984) have developed methods for adjusting expectations relative to these variables. The number and nature of the indicator attributes will clearly require modification when applied to different fish assemblages in other geographic and ecological provinces. Initial tests of the IBI have been promising. Scores agreed well with expert opinion for midwestern streams (Fausch et al. 1984). The simplicity and practical relevance of the IBI make it an attractive potential tool for assessing stream health.

6.6 CONCLUSION

The population and community statistics described in this chapter are not inclusive, and many approaches to estimating key parameters are still evolving. It is critical, therefore, that the manager give careful consideration to the choice of statistics in the design of assessment and creel surveys. Rarely will a single biological statistic provide adequate information for decision making. Rather, the sampling program should be designed to provide an array of statistics which will describe the response of the population and community to the multiple factors that influence its performance. Indices of growth, abundance, recruitment, survival, harvest, and population–community structure may all be needed to develop an understanding of that response.

Choice of parameters to be estimated should be dictated by the potential uses of that information. Selection of the appropriate statistic to estimate each parameter should then be based on the criteria of: (1) effort required, (2) consistency with which the statistic can be measured and thus confidently compared, and (3) the probable representativeness of the sample and consequent accuracy of the estimator. Evaluation of how well statistics meet these criteria will require knowledge of their performance for the target organisms in similar systems. Although biological descriptors of fishery resources are essential to effective management, fishery statistics are not so routine or universal that a standard recipe formula can be expected to provide them.

6.7 REFERENCES

Anderson, R. O. 1976. Management of small warmwater impoundments. Fisheries (Bethesda) 1(6):5–7, 26–28.

Anderson, R. O. 1980. Proportional Stock Density (*PSD*) and Relative Weight (*Wr*): interpretive indices for fish populations and communities. Pages 27–33 *in* S. Gloss and B. Shupp, editors. *Practical fisheries management: more with less in the 1980's.* Proceedings of the American Fisheries Society, New York Chapter, Ithaca, New York.

Anderson, R. O., and S. J. Gutreuter. 1983. Length, weight and associated structural indices. Pages 283–300 *in* Nielsen and Johnson (1983).

Anderson, R. O., and A. S. Weithman. 1978. The concept of balance for coolwater fish populations. American Fisheries Society Special Publication 11:371–381.

Bannerot, S. P., and C. B. Austin. 1983. Using frequency distribution of catch per unit effort to measure fish stock abundance. Transactions of the American Fisheries Society 112:608–617.

Carlander, K. D. 1977. Handbook of freshwater fishery biology, volume 2. Iowa State University Press, Ames.

Carline, R. F., B. L. Johnson, and T. J. Hall. 1984. Estimation and interpretation of proportional stock density for fish populations in Ohio impoundments. North American Journal of Fisheries Management 4:139–154.

Carpenter, S. R., J. F. Kitchell, and J. R. Hodgson. 1985. Cascading trophic interaction and lake productivity. Bioscience 35:635–639.

Colle, D. E., and J. V. Shireman. 1980. Coefficients of condition for largemouth bass, bluegill, and redear sunfish in hydrilla-infested lakes. Transactions of the American Fisheries Society 109:521–531.

Ebbers, M. A. 1987. Vital statistics of a largemouth bass population in Minnesota from electrofishing and angler-supplied data. North American Journal of Fisheries Management 7:252–259.

Fausch, K. D., J. R. Karr, and P. R. Yant. 1984. Regional application of an index of biotic integrity based on stream fish communities. Transactions of the American Fisheries Society 113:39–55.

Gabelhouse, D. W., Jr. 1984. A length-categorization system to assess fish stocks. North American Journal of Fisheries Management 4:273–283.

Gabelhouse, D. W., Jr., and D. W. Willis. 1986. Biases and utility of angler catch data for assessing size structure and density of largemouth bass. North American Journal of Fisheries Management 6:481–489.

Green, A. W., G. C. Matlock, and J. E. Weaver. 1983. A method for directly estimating the tag-reporting rate of anglers. Transactions of the American Fisheries Society 112:412–415.

Gustafson, K. A. 1988. Approximating confidence intervals for indices of fish population size structure. North American Journal of Fisheries Management 8:139–141.

Hall, T. J. 1986. Electrofishing catch per hour as an indicator of largemouth bass density in Ohio impoundments. North American Journal of Fisheries Management 6:397–400.

Hanson, J. M., and W. C. Leggett. 1982. Empirical prediction of fish biomass and yield. Canadian Journal of Fisheries and Aquatic Sciences 39:257–263.

Jearld, A., Jr. 1983. Age determination. Pages 301–324 in Nielsen and Johnson (1983).

Jenkins, R. M. 1979. Predator–prey relations in reservoirs. Pages 123–134 in H. Clepper, editor. Predator–prey systems in fisheries management. Sport Fishing Institute, Washington, D.C.

Jenkins, R. M. 1982. The morphoedaphic index and reservoir fish production. Transactions of the American Fisheries Society 111:133–140.

Jenkins, R. M., and D. I. Morais. 1978. Prey–predator relations in the predator- stocking-evaluation reservoirs. Proceedings of the Annual Conference Southeastern Association of Fish and Wildlife Agencies 30 (1976):141–157.

Johnson, D. L., and L. A. Nielsen. 1983. Sampling considerations. Pages 1–21 in Nielsen and Johnson (1983).

Jones, J. R., and M. V. Hoyer. 1982. Sportfish harvest predicted by summer chlorophyll-a concentrations in midwestern lakes and reservoirs. Transactions of the American Fisheries Society 111:176–179.

Karr, J. R. 1981. Assessment of biotic integrity using fish communities. Fisheries (Bethesda) 6(6):21–27.

Kruse, J. S. 1988. Guidelines for assessing largemouth bass fisheries in large impoundments. Missouri Department of Conservation, Final Report. Jefferson City.

Le Cren, E. D. 1951. The length-weight relationship and seasonal cycles in gonad weight and condition in the perch Perca fluviatilis. Journal of Animal Ecology 20:201–219.

Leonard, P. M., and D. J. Orth. 1986. Application and testing of an index of biotic integrity in small, coolwater streams. Transactions of the American Fisheries Society 115:401–414.

Low, R. A., G. F. Ulrich, C. A. Barans, and D. A. Oakley. 1985. Analysis of catch per unit effort and length composition in the South Carolina commercial handline fishery, 1976–82. North American Journal of Fisheries Management 5:340–363.

Macdonald, P. D. M. 1987. Analysis of length-frequency distributions. Pages 371–384 in R. C. Summerfelt and G. R. Hall, editors. Age and growth of fish. Iowa State University Press, Ames.

Malvestuto, S. P. 1983. Sampling the recreational fishery. Pages 397–419 in Nielsen and Johnson (1983).

Margalef, R. 1958. Information theory in ecology. General Systems Bulletin 3:36–71. University of Louisville, Systems Science Institute, Louisville, Kentucky.

Matuszek, J. M. 1978. Empirical predictions of fish yields of large North American lakes. Transactions of the American Fisheries Society 107:385–394.

Mills, E. L., D. M. Green, and A. Schiavone. 1987. Use of zooplankton size to assess the community structure of fish populations in freshwater lakes. North American Journal of Fisheries Management 7:369–378.

Murphy, B. R., M. L. Brown, and T. A. Springer. 1990. Evaluation of the relative weight (W_r) index, with application to walleye. North American Journal of Fisheries Management 10:85–97.

Nielsen, L. A., and D. L. Johnson, editors. 1983. Fisheries techniques. American Fisheries Society, Bethesda, Maryland.

Ney, J. J. 1990. Trophic economics in fisheries: assessment of demand/supply relationships between predators and prey. Reviews in Aquatic Sciences 2:55–81.

Ney, J. J., and D. J. Orth. 1986. Coping with future shock: matching predator stocking programs to prey abundance. Pages 81–92 in R. H. Stroud, editor. Fish culture in fisheries management. American Fisheries Society, Fish Culture Section and Fisheries Management Section, Bethesda, Maryland.

Ploskey, G. R., and R. M. Jenkins. 1982. Biomass model of reservoir fish and fish-food interactions with implications for management. North American Journal of Fisheries Management 2:105–121.

Ryder, R. A. 1965. A method for estimating the potential fish production of north-temperate lakes. Transactions of the American Fisheries Society 94:214–218.

Ryder, R. A., S. R. Kerr, K. H. Loftus, and H. A. Regier. 1974. The morphoedaphic index, a fish yield estimator—review and evaluation. Journal of the Fisheries Research Board of Canada 31:663–688.

Serns, S. L. 1985. Proportional stock density index—is it a useful tool for assessing fish populations in northern latitudes? Wisconsin Department of Natural Resources, Research Report 132, Madison.

Shannon, C. E., and W. Weaver. 1949. The mathematical theory of communication. University of Illinois Press, Urbana.

Shelton, W. L., W. D. Davies, T. A. King, and T. J. Timmons. 1979. Variation in the growth of the initial year class of largemouth bass in West Point Reservoir, Alabama and Georgia. Transactions of the American Fisheries Society 108:142–149.

Sorensen, T. 1948. A method establishing groups of equal amplitude in plant society based on similarity of species content. Koneglige Danske Videnskabernes Selskab 5:1–34.

Swingle, H. S. 1950. Relationships and dynamics of balanced and unbalanced fish populations. Agricultural Experiment Station, Auburn University, Alabama, Bulletin 274.

Timmons, T. J., W. L. Shelton, and W. D. Davies. 1980. Differential growth of largemouth bass in West Point Reservoir, Alabama–Georgia. Transactions of the American Fisheries Society 109:176–186.

Washington, H. G. 1984. Diversity, biotic and similarity indices. A review with special relevance to aquatic ecosystems. Water Research 18:653–694.

Wege, G. J., and R. O. Anderson. 1978. Relative weight (W_r): a new index of condition for largemouth bass. Pages 79–91 in G. D. Novinger and J. G. Dillard, editors. New approaches to the management of small impoundments. American Fisheries Society, North Central Division, Special Publication 5, Bethesda, Maryland.

Whittaker, R. H. 1952. A study of summer foliage insect communities in the Great Smokey Mountains. Ecological Monographs 22:1–47.

Willis, D. W. 1987. Use of gill-net data to provide a recruitment index for walleyes. North American Journal of Fisheries Management 7:591–592.

Willis, D. W. 1989. Proposed standard length-weight equation for northern pike. North American Journal of Fisheries Management 9:203–208.

Youngs, W. D., and D. G. Heimbuch. 1982. Another consideration of the morphoedaphic index. Transactions of the American Fisheries Society 111:151–153.

Yurk, J. J., and J. J. Ney. 1989. Analysis of phosphorus-fishery productivity relationships in southern Appalachian reservoirs: can lakes be too clean for fish? Lake and Reservoir Management 5:83–90.

Chapter 7

Socioeconomic Benefits of Fisheries

A. STEPHEN WEITHMAN

7.1 INTRODUCTION

Careful consideration of the full range of fisheries benefits began when the fisheries profession evolved from a guiding philosophy of maximum sustainable yield (MSY) to a philosophy of optimum sustainable yield (OSY). In MSY, the term yield can be defined as weight of fish harvested. In OSY, however, yield refers to all socioeconomic as well as biological benefits associated with fisheries, including the joy of the sport. This chapter identifies socioeconomic benefits of fisheries, outlines methods for measuring benefits, and discusses management implications of socioeconomic data.

Modern fisheries management requires a mastery of more than the technical problems associated with fish and water. Habitat and fish populations have been studied extensively, but today this traditional approach is too limited. Since fisheries are managed for people, an effort must also be made to understand angler attitudes, preferences, characteristics, and needs. If fisheries managers accept this challenge of incorporating social sciences into the evaluation of recreational and commercial fisheries benefits, we will improve our ability to justify important management programs.

7.2 IDENTIFICATION OF FISHERIES BENEFITS

Fisheries provide a myriad of benefits to society. Steinhoff et al. (1987) presented the benefits of wildlife through the use of an excellent historical perspective and classification system; similar benefits are applicable to fisheries. This section identifies groups of people that benefit from fisheries, and develops a logical framework for categorizing the benefits they receive.

First, consider the angler's perspective. When anglers plan a fishing trip, they often envision the location, fishing gear, company of friends or family, and fish they will catch. Fishing is just one type of recreation that individuals can choose. Reasons for fishing are personal; however, angler expectations are related to realizing certain tangible or intangible benefits in addition to harvesting fish (Figure 7.1). Most people, with the exception of tournament anglers and people who depend on anglers for sales revenue, do not think of recreational fishing as a business, nor are they concerned about what happens to the money spent on

Figure 7.1   The benefits of fishing include more than just the catch. (Photo courtesy of Missouri Department of Conservation.)

recreational fishing. Individual anglers repeat their fishing trips if their expectations have been reasonably fulfilled. Thus, the experience has value.

The perspective of charter operators or fishing guides is different. Recreational fishing is their business (Figure 7.2). Their benefits include employment, income, and the satisfaction of helping anglers realize their expectations by showing them where and how to catch fish. These people are a part of the community and the local economy. They continue to provide their services as long as they (1) make money, (2) are successful guiding anglers to fish, and (3) enjoy what they are doing.

Some businesses, including marinas, boat and bait dealerships, and fishing tackle wholesalers and retailers, depend heavily upon fishing-related expenditures. Their benefits include employment, income, and the satisfaction of providing needed goods and services to anglers. These people are also a part of the community and local and regional economies. They continue to provide goods and services as long as they are profitable and there is a demand. Other businesses, such as restaurants, service stations, and motels, that derive a portion of their business directly or indirectly from anglers, could also be included in this group.

Figure 7.2 To the charter boat operator, fishing is a way of life. (Photo courtesy of Missouri Department of Conservation.)

Commercial fishers have yet another perspective. For them, fishing is a means of support and a way of life. Their benefits include employment, income, and the satisfaction of providing a source of nutritious food. As with the charter operators and other local businesses, they are a part of the community and local economy and they continue to fish as long as it can be done profitably and they enjoy what they are doing.

Native American fishers must also be considered. Their fishing rights have been protected by treaties for cultural reasons and permit use of the resources as subsistence fisheries and as a source of income. Their benefits include a way of life, a tradition that is maintained, and a source of food and cultural pride. They continue to fish as long as their rights are upheld and the rewards are worth the effort.

Fisheries benefits depend upon the perspective and circumstances of the participant, given that the recreational, commercial, and Native American fishers have access to a place to fish. The quality of fishing, which is based on the species, size, number, and diversity of fish caught by an angler with a distinct set of attitudes about those fish, dictates the magnitude of benefits available for anglers and businesses.

All fisheries-related benefits can be considered social or economic, but an important distinction must be made between two types of benefits: values and impacts (Table 7.1). Socioeconomic values represent the importance that people place on the resource (Rockland 1985)—that is, both the satisfaction that people

Table 7.1 Socioeconomic benefits of fisheries.

Effect	Benefits	
	Values	Impacts
Social	Cultural	Quality of life
	Societal	Social well-being
	Psychological	
	Physiological	
Economic	User	Direct
	Consumptive	Indirect
	Nonconsumptive	Induced
	Indirect	
	Nonuser	
	Option	
	Existence	

derive from a resource (intangible) and the worth they place on it (tangible). Values describe what people receive related to their expenses—for recreational anglers, the satisfaction of the trip (which can be converted to dollars as described in section 7.2.1); for commercial fishers, profit. As an example, values include the opportunity to fish. For recreational anglers, tangible values could include money spent for bait, tackle, other equipment, services, gasoline, lodging, and meals. For commercial fishers, money is invested in equipment and wages. Socioeconomic impacts, in contrast to values, represent the social and economic effects that are generated by the use of the resource (Rockland 1985). In other words, impacts represent the effects on the community and local, regional, and national economies in terms of jobs, income, and tax receipts.

7.2.1 Socioeconomic Values

Socioeconomic values can be subdivided into social values that are held and economic values that are assigned. Held values are usually associated more with ideas, behaviors, outcomes, and experiences, whereas assigned values are usually associated more with goods, services, and opportunities.

Brown and Manfredo (1987) identified four categories of social values that are important considerations for fisheries managers: cultural, societal, psychological, and physiological. Cultural and societal values are more generic, pertaining to nations and communities, whereas psychological and physiological values relate to individual anglers. Cultural values represent a collective feeling towards fish and wildlife. Customs such as fear or worship of animals in primitive cultures are examples. Today, subcultures have evolved around the use of specific methods for harvesting particular species such as snagging for paddlefish, flyfishing for rainbow trout, or bait-casting for largemouth bass. Societal values are based on relationships between people as part of a family or community. The togetherness of a family fishing trip would be representative of societal values. Psychological values are those that relate to the satisfaction, motives, or attitudes associated with the use or knowledge of the existence of a fishery. An example would be an angler's perception of the quality of fishing. Physiological values relate to improvements in human health (better conditioning and reduction of stress) related to the sport of fishing.

Two types of economic values associated with fisheries have been identified: user and nonuser (Bishop 1987; Rockland 1985). User values result from direct or indirect participation. Nonuser values are potential or intrinsic values of the resource.

User values can be further subdivided into consumptive, nonconsumptive, and indirect use values (Table 7.1; Bishop 1987). An example of consumptive use is harvest (removal) of fish by an angler. The fact that people go fishing despite the time and money required indicates there are values associated with participation. If expenses exceed perceived values, people might change location, species, or method, or could stop fishing altogether and substitute other activities. Willingness to pay, which includes actual expenditures and excess value to users, is an appropriate measure of the economic value of a recreational fishery.

Nonconsumptive uses of fishery resources include sight-seeing, fish watching, snorkeling, scuba diving, visiting aquaria, studying nature, and photographing fishes and their habitat. These unlicensed uses do not deplete fishery resources, but are important. Just as with consumptive use, money is spent to participate, and values are realized or participation in the activities ceases. Measurement of values is a little more difficult because many nonconsumptive uses occur in conjunction with recreational activities such as camping, swimming, or boating.

Indirect users do not actually come into contact with the resource or habitats. Their uses are limited to such things as reading about fish, seeing pictures of fish, and watching television programs about fish. Values are derived from these activities because people still invest time and money in them.

Nonuser values can be subdivided into option and existence values. An option value means that individuals are leaving the door open to future participation although they currently choose not to participate. A type of option value, the quasi-option value, allows postponement of the decision to participate until more information is gathered. Existence values occur when people value a fishery despite the fact that they do not currently use the resource or ever plan to in the future. Bequest value is an existence value in which a person wants a particular resource to exist for future generations. Altruistic value is an existence value where the motivation is to allow for the survival of a species.

7.2.2 Socioeconomic Impacts

Socioeconomic impacts are the effects of money spent by recreationists (primarily by anglers, but also by nonconsumptive and indirect users) on local, regional, and national economies. The initial or obvious impacts are economic, but social changes are also likely to result (Table 7.1).

The total economic impact within a defined region exceeds actual expenditures by recreationists, because some of the money spent in a local economy continues to circulate in that area. A portion of money leaves the area to pay for goods, services, and taxes, while money that remains is used to pay wages or represents profit. Wages and profits are then spent by residents in the local economy, resulting in another round of economic impact. Money spent by recreationists is said to have a "direct" economic impact. "Indirect" economic impacts result from businesses that procure materials locally to produce goods and provide services to meet the needs of recreationists. Money spent by people who earn wages or profits in the region is said to have an "induced" economic impact (Table 7.1).

An example illustrating the different economic impacts follows. Assume anglers enter a bait-and-tackle shop and buy fishing rods and reels, lures, and night crawlers. The direct economic impact occurs when anglers exchange money for merchandise. An indirect economic impact is the purchase of materials by businesses that manufacture the fishing rods and lures. Wages paid by those businesses and the wholesaler who provides night crawlers are also indirect economic impacts. All products and services have an indirect economic impact as long as they are produced and provided within the region of interest. If the reel is imported, however, the money spent to bring it to the region is lost to the regional economy and no longer represents a local indirect economic impact. Induced economic impacts occur when clerks in the bait-and-tackle shop spend their wages locally. A similar set of direct, indirect, and induced economic impacts can be envisioned for a commercial fishing operation.

Economic impacts can be quantified by determining the number of jobs created, wages paid, sales, or profits. These figures depend upon a definition of the region of interest, which could range from an individual community to a county, state, multi-state, or national area. Most economic impacts can be converted to dollars for comparison (measurement of the impacts will be discussed in section 7.3.2).

Social impacts are more elusive and are not as readily quantifiable. They relate to quality of life and social well-being. Social impacts are reflected by changes in social relationships and the social organization of fisheries production, distribution, and marketing systems (Vanderpool 1987). Social impacts due to changes in fisheries management are most likely the result of changes in business conditions or employment. Improvements or changes in fisheries can cause jobs to be created, transferred, or lost. Changes in employment and business can result in population shifts and the need for more or fewer housing and public services.

Creation of jobs, especially in environmentally compatible industries, is definitely a benefit to people and their communities. Employment provides people with a sense of worth and identity. Adequate employment strengthens community cohesion and allows for development of social institutions such as schools and churches. Loss of jobs has the exact opposite social effect. Changes in types of employment can be either positive or negative. For instance, if satisfying, career-oriented positions are replaced with minimum-wage jobs, the effect on individuals and the community will be negative. Besides career considerations related to employment, another factor is income. More income generally creates greater social well-being.

7.3 MEASUREMENT OF FISHERIES BENEFITS

In this section two general categories of benefit assessment techniques—nonmonetary and monetary—are discussed with a variety of approaches listed as part of each group (Table 7.2). Total value assessment, a third approach, is a synthesis of nonmonetary and monetary evaluations. The focus is on assessment techniques for fisheries that have broad acceptance by sociologists and economists as well as biologists. Some methods target a specific set of benefits, while

Table 7.2 Methods of assessing fishery benefits.

Benefit measurement	Approach
Nonmonetary	
Social well-being measurement	Angler survey on changes in a fishery.
Psychophysical measures	Angler survey on aesthetic appeal.
Multiattribute choice approach	Angler survey on a variety of fishery characteristics.
Attitude measures	Angler survey on factors that affect the quality of fishing.
Social impact assessment	Projection of changes that will likely result from a new policy or program.
Monetary	
Economic impact assessment	Angler expenditures used as input for regional economic models.
Economic value assessment	Angler expenditures used as input for travel cost or contingent valuation.
Total economic valuation	Angler expenditures used as input for contingent valuation, plus nonconsumptive user values.
Combination	
Total value assessment	Comprehensive evaluation that combines nonmonetary and monetary evaluations to determine social and economic values.

other methods provide a comprehensive assessment. Techniques discussed hold the greatest promise for measuring socioeconomic benefits associated with recreational and commercial fisheries.

Almost every method of assessing fishery benefits requires survey data from people who fish or who in some other way might be affected by a change in a fishery (Table 7.2). The three most common approaches for collecting information are on-site, mail, and telephone surveys. Each method has advantages and disadvantages that must be considered for individual projects. Thorough reviews are available to help decide how to collect survey data (Malvestuto 1983).

7.3.1 Nonmonetary Methods for Assessing Social Benefits

Nonmonetary assessments are being conducted because of the need for improved information about social and environmental effects resulting from changes in commercial and recreational fisheries. Use of social sciences in solving fishery problems began in earnest in the 1970s after economic valuations gained acceptance. Gregory (1987) provided additional details and references on social benefit measurement.

The social well-being measurement approach assesses the effects of a fishery project or policy on the well-being of individuals in order to determine the overall effects on society. Theoretically, an index of well-being should reflect changes observed by people in their quality of life, and hence changes in the community and in social relationships. A typical social assessment would describe current conditions and identify groups of people who would be affected by a change. Interviews can be as basic as asking people how they feel about a possible or actual change in a fishery and how seriously it would affect them. Indicators of well-being include employment of women, crime, and divorce rates (Gregory 1987).

Psychophysical measures determine the sensitivity of people to physical

Figure 7.3 Point sources of pollution are not only deleterious to the integrity of the offended aquatic ecosystem, but severely detract from the aesthetic appeal and thus diminish the angling experience. (Photo courtesy of Missouri Department of Conservation.)

features of a fishery environment. Sites can be ranked based on aesthetic appeal, or a single site can be judged after modification to reflect a change in the environment. Rating scales are used to differentiate changes in value. This approach has been applied in studies on the importance of landscape features to participants and for assessing the scenic beauty of forest environments. This technique could be used to evaluate the sensory impression of the fishery setting to anglers (Figure 7.3).

The multi-attribute choice approach analyzes preferences to aid in making decisions on multifaceted fishery problems. The procedure involves selecting relevant attributes or characteristics of a fishery, specifying how the attributes will be measured, identifying all possible outcomes, weighing the attributes based on relative importance, and determining the effect a management policy will have on each attribute so a single, overall score can be computed.

Walker et al. (1983) used the multi-attribute approach to analyze trade-offs with respect to allocation and production of wild and hatchery coho salmon in Oregon. As a result of the analysis, they selected the most effective management policy of the 12 proposed, determined that harvest rate was the most important decision variable, and found that the number of smolts released should be increased only up to the point that survival is affected.

Attitude measurements focus on the response of anglers to any of a number of factors that can influence the quality of a fishing trip. Attitude measures address indirect behavioral associations with environmental conditions, such as the setting or quality of the experience (Gregory 1987). Numerical scales list conditions used to rate angler response. Many researchers have recently focused on angler attitudes as being important to recreational fisheries management.

A specific example of attitude measurement was the development of indices by Weithman and Anderson (1978a) and Weithman and Katti (1979) to measure the quality of fishing. An investigation of Missouri angler attitudes revealed the importance of getting outdoors, the setting, companionship, and catching fish (Weithman and Anderson 1978b). An analysis of memorable fishing trips emphasized the importance of species, size, number, and diversity of the catch. When a person chooses to fish rather than participate in another type of outdoor recreation, obviously they want to catch fish or at least have that opportunity.

Weithman and Anderson (1978a) developed equations to rate fishing quality based on angler attitudes about the importance of species, size, number, and diversity of fish caught. The accuracy and precision of these indices were tested by Weithman and Katti (1979) by calculating index values and comparing the results to actual angler preferences between two fish or groups of fish. Index values were able to discriminate between relatively small differences in fishing quality. Further work in this area has been directed at asking anglers to rate fishing quality on a 10-point scale for a particular day compared to their "normal" success. Mean values of this numerical rating also appear to be quite sensitive to differences in fishing quality.

Social impact assessments determine economic and social costs, and benefits associated with development of fisheries policies and programs. This type of assessment is more comprehensive than other nonmonetary techniques previously discussed. The emphasis is on understanding and projecting likely changes in social relationships, social structures and institutions, and normative systems and world views (Vanderpool 1987). A social impact assessment should occur prior to any changes in a fishery and include an analysis of historical, cultural, economic, ecological, and demographic dimensions. This approach is especially applicable to evaluation of commercial fisheries, although not exclusively so, compared to the other nonmonetary measures. Social impact assessments in fisheries have been extremely limited, but based on applications in other natural resource areas they hold great potential.

7.3.2 Monetary Methods for Assessing Economic Benefits

Monetary methods have been used for a number of years to evaluate the economic impacts and values of recreational and commercial fisheries. This section discusses accepted valuation techniques and their use in conducting three types of economic benefit assessment: (1) economic impact, (2) economic value, and (3) total economic valuation (Table 7.2). Several researchers have studied valuation techniques in recent years and have presented specific information that would allow the reader to develop similar models and analyses for assessing economic benefits. In general, impacts are best determined by using economic models to track angler expenditures through the economy, and values are best estimated by using travel cost or contingent valuation models to determine to what extent values derived by anglers exceed their expenses.

Economic impact assessments are conducted to determine the effect of fishing-related expenditures on the local, regional, or national economy. This approach is used to quantify socioeconomic impacts identified in section 7.2.2, including direct, indirect, and induced impacts. An economic impact assessment is the process of summing all three types of impacts with respect to the effects of angler expenditures on income, employment, and taxes in the region of interest.

Rockland (1986) discussed four models which can be used in the analysis of the economic impacts due to recreational fisheries: (1) economic base, (2) econometric, (3) input-output, and (4) modified input-output. Given relevant employment data and angler expenditures, it is possible to estimate employment, earnings, business output, and the resulting socioeconomic impacts. Rockland (1986) discussed strengths, weaknesses, and recommendations for application of the four models.

Gross expenditures by anglers in the study area are needed as an input for an economic impact assessment. Multipliers (for output, employment, or income) can be applied to expenditures to determine the full impact on the economy if, and only if, they are properly developed and specified (Hushak 1987). These multipliers express the total amount of income generated in an economy due to anglers spending money by measuring the circulation of these expenditures through individual components of the economy. Multipliers reflect general relationships. Labor-intensive industries, such as amusement and recreation, have high multiplier values. Much of the revenue from these industries is converted to employee wages—money that remains, in part, in the local economy to be spent again. Industries that market products, such as grocery stores, have low multiplier values. Much of their revenue pays the cost of goods supplied by wholesalers outside the community, and is lost to the local economy.

Economic impact assessments have been conducted for a number of recreational activities. Two fisheries management examples include a determination of the value of a salmonid fishery in New York (Brown 1976) and a rainbow trout fishery in Missouri (Weithman and Haas 1982).

Economic value assessments determine the value derived by participants from recreational fishing. This approach is used to quantify economic values (identified in section 7.2.1) for consumptive users of the resource. Although many valuation techniques have been proposed, two in particular—travel cost and contingent valuation (Dwyer et al. 1977)—deserve further discussion because of their applicability to fisheries evaluations. Both methods have been designed to estimate "consumer surplus," which is the value that anglers receive in excess of their expenditures.

The travel cost and contingent valuation methods allow a more complete and accurate analysis of value provided by fisheries. They take into account shifts in demand for recreation due to changes in the quantity and quality of recreation provided. This type of evaluation is the best measure of the value of management programs. At the same time, an economic evaluation allows comparisons with other fisheries, or other activities that compete with fisheries for limited resources.

Both methods are based on the relationship between angler preference for, and costs associated with, particular fisheries. The travel cost method depends on observed days of fishing and actual angler expenses; contingent valuation is based on an estimate of days fished and angler response to hypothetical questions about the value of the fishery.

Two important assumptions are involved in applying these methods. The travel cost method depends on the assumption that anglers would respond the same to a fee for fishing as they would to an increase in their travel expenses. The basic assumption for the contingent valuation method is that an angler's response to a hypothetical question about the value of a fishery would actually fit his behavior in a real-market situation. Information is available which outlines the methods of applying the travel cost (Hansen 1986) and contingent valuation (Hoehn 1987) techniques.

Both methods have been used repeatedly to evaluate benefits from outdoor recreation. Specific examples related to recreational fisheries management follow. A travel cost model was used by Burt and Brewer (1974) to predict recreation benefits at a proposed reservoir in Missouri, and Knetsch et al. (1976) to estimate benefits for a number of reservoirs in California. The contingent valuation method was used by Oster (1977) to determine willingness to pay to clean up the Merrimack River and Walsh et al. (1978) to determine recreation benefits from high-mountain reservoirs in Colorado.

Use of the travel cost and contingent valuation methods should be emphasized in future valuation of recreational fisheries. Studies already completed provide good examples for future applications, but there is room for improvement. Some areas currently under investigation include incorporating travel time and grouping participants in the travel cost model, determining the importance of site quality and substitute sites, and refining questioning techniques for contingent valuation.

Incorporation of travel time into travel cost models is an important consideration because this time is considered an expense of the trip (Wilman 1980). However, since travel time is usually correlated with distance traveled or total expense, inclusion in the model has been difficult. Travel time is taken into account most often by assuming a trade-off between money and time based on some function of the average wage rate (Cesario 1976). McConnell and Strand (1981) developed a method to set the cost of travel time using an independent estimate of travel time and round-trip distance to the site. Two problems are apparent: (1) travel time might not be a liability in every instance, because some people derive benefits during the trip by sight-seeing; and (2) if trips for which the site of interest is not the only destination are included in the valuation, site benefits will be overestimated. Few attempts have been made to separate benefits provided by primary and secondary destinations (Dwyer et al. 1977).

The travel cost model uses data on distance traveled from an angler's residence to the site under evaluation to produce a demand curve. Grouping anglers into categories for consideration and comparison based upon origin has become an art. Binkley and Hanemann (1976) recommended categorizing people according to areas within concentric rings surrounding the site being evaluated. Others, however, have separated participants by county of origin or geographic areas with relatively homogeneous populations to make their evaluation. Another approach has been to consider individual responses and ignore aggregation over a particular area (Brown and Nawas 1973). How anglers are grouped is a matter of choice for individual researchers, but each type of grouping can provide valid estimates.

Site quality, including factors such as congestion, aesthetics, and fishing quality, can be an important variable for consideration in predicting visitation by anglers (Figure 7.4). Much research has been conducted on the effects of congestion as they relate to the management of natural resources. Overcrowding,

Figure 7.4 For many, the setting for fishing is as or more important than the expectation to catch fish. (Photo courtesy of Missouri Department of Conservation.)

while not a problem at most fisheries, can affect the quality of the experience at sites near some metropolitan areas and on opening day for certain species. Cesario (1969) approached the question of site quality by observing actual behavior of recreationists. Fishing quality might be the most important site-quality variable, yet little consideration has been given to this topic. Stevens (1966) included catch rates as an indication of fishing quality. The overall quality of fishing index proposed by Weithman and Katti (1979) could be used as one indication of site quality. Further investigation of these variables is needed.

Visitation to one reservoir or stream is affected by availability of other nearby fishing sites. Substitute sites can play a major role in the development of realistic models to predict fishery visitation and benefits. Several investigators have developed multiple-site models to evaluate benefits associated with aquatic resources in a state or region of the country. If the site being evaluated is unique, substitute sites are not a factor. Additional research should be directed at the comparability of a variety of sites for recreational fishing.

Many potential problems have been identified in regard to the questioning procedure or design of contingent valuation surveys (Hoehn 1987). If respondents cannot relate the questions to real-world situations, the survey could be biased. Strategic bias could be another problem. Survey results could be affected if people think they might benefit by giving a particular answer, but evidence of strategic behavior usually has not been found. Another concern is that a change in fees in the survey from a hypothetical standpoint might be objectionable to a segment of the respondents. Care must also be taken not to infer that an actual fee for entry could change. In questions to determine willingness to pay, a bias can be introduced by selecting a given starting point for dollars an angler is willing to spend to participate. Thayer (1981) developed a test and adjustment to eliminate this problem (starting-point bias).

Total economic valuation expands on economic value assessment by incorporating indirect, option, and existence values (described in 7.2.1), in addition to consumptive use values. Option prices or existence values of a variety of natural resources or sites for recreation have been considered. These values also are accrued by nonparticipants solely from knowing a particular opportunity exists. For example, Greenley et al. (1981) measured the option value of maintaining the water quality of the South Platte River basin in Colorado, while Brookshire and Randall (1978) estimated activity, option, and existence values for hunting and preserving wildlife resources in Wyoming.

The traditional contingent valuation and travel cost approaches can each measure portions of total economic value. Contingent valuation is the only technique, however, that currently can be adapted to measure the entire total economic value including indirect-use values and intrinsic values (Randall 1987). The approach of estimating total economic value (all assigned values) is particularly useful for placing a value on endangered species or other nonexploited resources (Bishop et al. 1987).

7.3.3   Total Value Assessment

Total value assessment is a new, comprehensive concept of valuation in which all values—social, anthropological, political, philosophical, and economic—are considered concurrently (Table 7.2; Talhelm and Libby 1987). The problem is that no single valuation method addresses all areas of interest, so they must be addressed individually. Total value assessment considers all different kinds of value, including held and assigned values, in resolving fisheries conflicts. In particular, the use of social sciences must be emphasized to aid in management of fisheries resources (Talhelm 1987). Multidisciplinary research should expedite the integration of social sciences with traditional approaches to evaluating fisheries.

7.4   MANAGEMENT IMPLICATIONS OF SOCIOECONOMIC DATA

Socioeconomic data are essential for effective fisheries management because fisheries are managed for people. People includes the person who goes fishing with the family on vacation, the dedicated angler who fishes 50 days a year, commercial and Native American fishers who make a living from the resource, and the person who owns the bait-and-tackle shop. Fisheries biologists state that their intent is to manage the fish to provide high-quality fishing. Criteria for making fishing better need to be defined to lay the groundwork for these management efforts. Fisheries managers will be successful if they (1) carefully consider the entire array of benefits fisheries offer, (2) make the effort to determine what angler and other groups want, and (3) develop a process to explain to people the differences between their expectations and a manager's ability to effect change in a particular ecosystem.

Fisheries management is the process of working with a given aquatic habitat and assemblage of organisms for the benefit of people in a recreational or commercial setting. Goals and objectives are generally set to approach OSY (see Chapter 2). Important benefits have been identified (section 7.2) and methods have been recommended to assess those benefits (section 7.3). This section documents the importance of socioeconomic data and explains use of the data in a decision-making process. Four general fisheries-related examples are discussed to demonstrate the utility of socioeconomic data.

7.4.1 Agency Planning

Planning is essential for the efficient, long-term operation of a natural resource agency. Typically, such an agency has the constitutional responsibility to preserve, protect, restore, and manage fishery resources. The agency manages the resources for the people (users and nonusers alike), but administrators must be cognizant of the potential for socioeconomic impacts on regional economies as well. Socioeconomic data can provide clues to allow identification of the clientele and their expectations from the fisheries under management. Entire programs can be evaluated on the basis of projected needs and desires.

One of the most important resource agency functions is asset allocation—selecting purposes and priorities for spending money. Budgets are prepared based on a program review and prioritization. From an agency standpoint, a benefit:cost analysis is appealing. Costs can be calculated accurately, but socioeconomic data are needed to estimate benefits. The choices of building a fish hatchery, buying access sites on a river, and building community lakes could then be evaluated rationally. Often the choice is not either-or, but rather identifying the best mix of expenditures. Another example would be justification for an endangered species program. An analysis of nonuser economic values would allow direct comparisons with user-oriented programs designed to enhance recreational fishing.

7.4.2 Recreational Fisheries Management

One objective that frequently appears in recreational fisheries management plans is to improve the quality of fishing (i.e., increase catch or harvest rates and the size of fish caught, or increase angler use or satisfaction). Monitoring fishing quality is an important part of fisheries management. Examination of angler attitudes reveals that almost all anglers rate their fishing trips as excellent or good. If those same anglers are asked about their fishing success, however, a majority often rates it as fair or poor. Even without any management efforts, anglers can apparently have a good time given the opportunity to fish. Fisheries managers have the opportunity, though, to improve fishing success and increase the number of memorable fishing trips by considering angler preferences in management plans.

We must delve deeper into anglers' minds to maximize benefits they derive from recreational fisheries, specifically concentrating on fishing success. Opportunities exist at most recreational fisheries to make improvements through new species introductions (see Chapter 12), supplemental stocking (see Chapter 13), and fish habitat manipulation (see Chapters 8–11). Selection of the best approaches from a list of those that are biologically satisfactory should be based on calculations of socioeconomic values. As an example, assume an additional predator is needed to balance a fish community. If the choice is between two of equal biological value, which one should be selected? In another instance, is it desirable to double the catch rate of a species if harvest is reduced by 90%? Or, should a supplemental stocking program be dropped because the overall agency budget is reduced by 10%? These questions cannot be answered without more

Figure 7.5 The urgency for environmental recovery efforts (oil spill shown here) is predicated on socioeconomic concerns. (Photo courtesy of Missouri Department of Conservation.)

information, preferably socioeconomic data related to people who use the resource or who could be affected by changes in policy.

7.4.3 Resolution of Environmental Problems

Knowing the complete array of fishery benefits, including socioeconomic values and impacts, is essential for protecting aquatic resources and giving decision makers comparable data to evaluate competing projects. Destruction of aquatic resources jeopardizes every value discussed in section 7.2. A catastrophic event could eliminate current recreational fishing (consumptive use), degrade water quality (nonconsumptive and indirect uses), and even eliminate nonuser values if the damage is irreparable (Figure 7.5). Socioeconomic impacts would also undoubtedly be significant as the reduction in recreational use would affect the local and regional economies.

Consider three examples of environmental problems for analysis: (1) an extensive oil spill into a river, (2) a dam failing to meet state water quality standards for tailwater releases, and (3) withdrawal of water from the Missouri River for nonfishery uses. If an oil spill results in an extensive fish kill, recovery of the replacement cost for the fish would be insufficient to cover the benefits foregone. Anglers would be affected because fishing quality would be diminished, and eventually businesses would be affected when anglers stopped returning to the area. Extirpation or threats to rare or endangered species would be another concern. Failure of Table Rock Dam discharges to meet state standards of 6 mg/L dissolved oxygen affects the rainbow trout fishery in downstream Lake Taneycomo, Missouri. The results are reduced angler success and visitation to the area in the fall and a variety of socioeconomic impacts (Weithman and Haas 1982). Proposals have been made to divert water from the Missouri River for irrigation and industrial purposes. This water could yield measurable benefits to interests on both sides of the issue. The best way to resolve the problem would be to conduct a complete study of the socioeconomic values and impacts associated with both proposals.

7.4.4 Conflict Resolution

Conflicts between groups of fishers who simultaneously exploit a fishery resource are becoming more common. The problem is that certain freshwater fish stocks, especially those that are harvested commercially, run the risk of overfishing and excessive harvest to the point of depletion of stocks well below MSY. Native American fishing rights have also caused concern. Tensions mount, especially when the species of interest periodically have weak year-classes.

In general, from a value standpoint, recreational use is more efficient than commercial use of aquatic resources. The socioeconomic impacts resulting from changes in fisheries could be significant, though, and therefore must be considered. What effects will occur in terms of employment and income with the closure of a commercial fishery? What reaction will people have when they lose their livelihood? A recent trend has been for states to buy out commercial fishing interests and thereby eliminate commercial harvest when conflicts with recreational anglers are perceived.

An example of a potential problem is the striped bass fishery in the Hudson River. An important recreational fishery has developed there even during closure of commercial fishing due to PCB contamination. Should commercial fishing be permitted after the water quality problems are solved? A total value assessment would be in order to make this determination.

Native American fishing rights are another issue (see Chapter 4). Socioeconomic data could be used to compare benefits provided by recreational fishing versus a Native American fishery. The economic values and impacts of a recreational fishery might be very important. Likewise, the social values and impacts of a Native American fishery might be substantial. In this instance, politics might take precedence, but a socioeconomic valuation is still important in making all interests aware of the tradeoffs involved with each potential decision.

7.5 SUMMARY

Fishery benefits are real and important (Figure 7.6), but measuring them is an art, not a science. No single approach to estimating socioeconomic benefits is perfect for all situations. Selection of an appropriate method depends on knowledge of the fishery in question, a list of the benefits of interest, importance of the decision being made, and fiscal constraints.

People are the ultimate beneficiaries of recreational and commercial fisheries management. Fisheries professionals need to know the social and economic benefits of their management efforts. Knowledge of socioeconomic benefits is useful in program development, program evaluation, and crisis management. Fisheries professionals are learning that the essence of management is establish-

Figure 7.6 Fishery benefits are real and important! (Photo courtesy of Missouri Department of Conservation.)

ing priorities, and this process involves not only biological data, but fiscal constraints, legislative influence, management goals of an agency or its commission, public opinion, and a thorough analysis of socioeconomic benefits provided by fisheries.

7.6 REFERENCES

Binkley, C. S., and W. M. Hanemann. 1976. The recreation benefits of water quality improvements: analysis of day trips in an urban setting. United States Environmental Protection Agency, NTIS (National Technical Information Service) PB-257719, Springfield, Virginia.

Bishop, R. C. 1987. Economic values defined. Pages 24–33 *in* Decker and Goff (1987).

Bishop, R. C., K. J. Boyle, and M. P. Welsh. 1987. Toward total economic valuation of Great Lakes fishery resources. Transactions of the American Fisheries Society 116:339–345.

Brookshire, D. S., and A. Randall. 1978. Public policy alternatives, public goods, and contingent valuation mechanisms. Proceedings of the Western Economic Association, Honolulu, Hawaii, 53:20–26.

Brown, P. J., and M. J. Manfredo. 1987. Social values defined. Pages 12–23 *in* Decker and Goff (1987).

Brown, T. L. 1976. The 1973–75 salmon runs: New York's Salmon River sport fishery, angler activity, and economic impact. Cornell University, New York Sea Grant Institute NYSSGP-RS-76-025, Ithaca.

Brown, W. G., and F. Nawas. 1973. Impact of aggregation on the estimation of outdoor recreation demand functions. American Journal of Agricultural Economics 55:246–249.

Burt, O. R., and D. Brewer. 1974. Evaluation of recreational benefits associated with the Pattonsburg Dam Reservoir. University of Missouri, Office of Industrial Development Studies, Columbia.

Cesario, F. J. 1969. Operations research in outdoor recreation. Journal of Leisure Research 1:33–52.

Cesario, F. J. 1976. Value of time in recreation benefit studies. Land Economics 52:32–41.

Decker, D. J., and G. R. Goff, editors. 1987. Valuing wildlife: economic and social perspectives. Westview Press, Boulder, Colorado.

Dwyer, J. F., J. R. Kelly, and M. D. Bowes. 1977. Improved procedures for valuation of the contribution of recreation to national economic development. University of Illinois, Water Resources Center, Research Report 128, Champaign–Urbana.

Greenley, D. A., R. B. Walsh, and R. A. Young. 1981. Option value: empirical evidence from a case study of recreation and water quality. Quarterly Journal of Economics 96:657–674 (Wiley, New York).

Gregory, R. 1987. Nonmonetary measures of nonmarket fishery resource benefits. Transactions of the American Fisheries Society 116:374–380.

Hansen, W. J. 1986. Valuating the recreational use of fishery resource sites with the travel cost method. Sport Fishing Institute, Technical Report V, Washington, D.C.

Hoehn, J. P. 1987. Contingent valuation in fisheries management: the design of satisfactory contingent valuation formats. Transactions of the American Fisheries Society 116:412–419.

Hushak, L. J. 1987. Use of input-output analysis in fisheries assessment. Transactions of the American Fisheries Society 116:441–449.

Knetsch, J. L., R. E. Brown, and W. J. Hansen. 1976. Estimating expected use and value of recreation sites. Pages 103–115 in C. E. Gearing, W. W. Swart, and T. Var, editors. Planning for tourism development: quantitative approaches. Praeger, New York.

Malvestuto, S. P. 1983. Sampling the recreational fishery. Pages 397–419 in L. A. Nielsen and D. L. Johnson, editors. Fisheries techniques. American Fisheries Society, Bethesda, Maryland.

McConnell, K. E., and I. Strand. 1981. Measuring the cost of time in recreation demand analysis: an application to sport fishing. American Journal of Agricultural Economics 63:153–156.

Oster, S. 1977. Survey results on the benefit of water pollution abatement in the Merrimack River basin. Water Resources Research 13:882–884.

Randall, A. 1987. Total economic value as a basis for policy. Transactions of the American Fisheries Society 116:325–335.

Rockland, D. B. 1985. The economic benefits of a fishery resource: a practical guide. Sport Fishing Institute, Technical Report I, Washington, D.C.

Rockland, D. B. 1986. Economic models: black boxes or helpful tools? A reference guide. Sport Fishing Institute, Technical Report II, Washington, D.C.

Steinhoff, H. W., R. G. Walsh, T. J. Peterle, and J. M. Petulla. 1987. Evolution of the valuation of wildlife. Pages 34–48 in D. J. Decker and G. R. Goff (1987).

Stevens, J. B. 1966. Recreation benefits from water pollution control. Water Resources Research 2:167–181.

Talhelm, D. R. 1987. Recommendations from the SAFR symposium. Transactions of the American Fisheries Society 116:537–540.

Talhelm, D. R., and L. W. Libby. 1987. In search of a total value assessment framework: SAFR symposium overview and synthesis. Transactions of the American Fisheries Society 116:293–301.

Thayer, M. A. 1981. Contingent valuation techniques for assessing environmental impacts: further evidence. Journal of Environmental Economics and Management 8:27–44.

Vanderpool, C. K. 1987. Social impact assessment and fisheries. Transactions of the American Fisheries Society 116:479–485.

Walker, K., B. Rettig, and R. Hilborn. 1983. Analysis of multiple objectives in Oregon coho salmon policy. Canadian Journal of Fisheries and Aquatic Sciences 40:580–587.

Walsh, G., D. A. Greenley, R. A. Young, J. R. McKean, and A. A. Prato. 1978. Option values, preservation values, and recreational benefits of improved water quality. U.S. Environmental Protection Agency EPA-600/5-78-001, Washington, D.C.

Weithman, A. S., and R. O. Anderson. 1978a. An analysis of memorable fishing trips by Missouri anglers. Fisheries (Bethesda) 3(1):19–20.

Weithman, A. S., and R. O. Anderson. 1978b. A method of evaluating fishing quality. Fisheries (Bethesda) 3(3):6–10.

Weithman, A. S., and M. A. Haas. 1982. Socioeconomic value of the trout fishery in Lake Taneycomo, Missouri. Transactions of the American Fisheries Society 111:223–230.

Weithman, A. S., and S. K. Katti. 1979. Testing of fishing quality indices. Transactions of the American Fisheries Society 108:320–325.

Wilman, E. A. 1980. The value of time in recreation benefit studies. Journal of Environmental Economics and Management 7:272–286.

HABITAT MANIPULATIONS

Chapter 8

Watershed Management and Land-Use Practices

THOMAS A. WESCHE

8.1 INTRODUCTION

A hydrologist may be better able than a biologist to evaluate the dynamic fisheries production potential of a drainage basin. Certainly the two disciplines working together will come much nearer to accurately assessing the potential than will either working alone. (G. L. Ziemer 1971.)

The fisheries management profession has found itself faced with the challenge of evolving from just fish managers to water managers, land managers, and habitat managers. Today, successful fisheries managers must approach their work not only from a classical fisheries perspective, but also from a watershed perspective. The objective of this chapter is to present an overview of watershed structure, process, and function, stressing the interdependence of fishery resources within a watershed upon not only species biology, but also on the geomorphology, hydrology, and land use of the system.

8.1.1 Watershed Definition and Characterization

A watershed is any sloping land surface that sheds water. A more functional definition is, "all land enclosed by a continuous hydrologic drainage divide and lying upslope from a specified point on a stream." Typically, watershed is synonymous with drainage basin or catchment. A drainage basin is a watershed that collects and discharges its surface streamflow through one outlet or mouth, while a catchment is generally considered to be a small drainage basin (Hewlett and Nutter 1969). For most purposes, these terms can be applied interchangeably.

What is implied when we use the term watershed management? The Society of American Foresters defined watershed management in 1944 as:

The management of the natural resources of a drainage basin primarily for the production and protection of water supplies and water-based resources, including the control of erosion and floods, and the protection of esthetic values associated with water.

The key words here are "...production and protection of water supplies *and water-based resources*..." which includes fisheries.

181

The most important function of a watershed is to produce water. From a water supply standpoint, we must be concerned not only with the total quantity of water yielded, but also with the timing of that yield (flow regimen) and its quality. These factors vary between watersheds depending upon the characteristics of the drainage basin.

Numerous variables interact within a watershed to control streamflow and the nature of the stream channels through which water flows. In general, these variables can be categorized as climatic, topographic, geologic, and vegetative. One must be aware of the strong interrelations between these controlling factors, their dominant influence on the character of watersheds, and ultimately, the role they play in regard to the zoogeography of fish species, population size, and fishery quality.

Climate can be defined as the long-term aggregate atmospheric condition produced by daily weather. The climate of an area has direct bearing on both the quantity of water yield and the flow regimen. Those measurable factors we typically use to describe a climate include:

- air temperature (mean, maximum, and minimum),
- freeze dates (frost-free period) soil temperature,
- precipitation (quantity, type, temporal, and spatial distribution),
- humidity (quantity of water vapor in the air),
- evaporation (process by which water changes phase from liquid to vapor),
- wind (speed and direction),
- solar radiation,
- atmospheric pressure.

The climate of an area is governed by latitude, elevation, proximity to oceans, and local topographic features. Together, these factors determine or modify airflow patterns and weather systems which dictate the ranges of precipitation, temperature, humidity, wind, and evaporation that a watershed experiences.

Geomorphology, the science that deals with relief features of the earth's surface, has been the discipline most responsible for developing quantitative techniques to measure topographic characteristics of drainange basins and attempting to explain their influence in terms of watershed functions and processes (Brown 1970). Leopold and Langbein (1963) summarized the relation of these two objectives when they wrote, "In geomorphologic systems the ability to measure may always exceed ability to forecast or explain."

Over the century that geomorphology has been recognized as a science, numerous topographic measures have been devised to describe watersheds. These measures characterize the basin as a whole, the total channel system or network, reaches within the total channel network, and cross sections within a channel reach. For each of these components, measures of area, length, shape, and relief have been developed. In general, the area of the watershed dictates the quantity of water yield, while length, shape, and relief control the streamflow regimen and the rate of sediment yield.

Commonly measured topographic attributes of watersheds are listed in Table 8.1 along with their definition or derivation and a literature source for locating additional information. Figure 8.1 depicts the stream ordering method developed

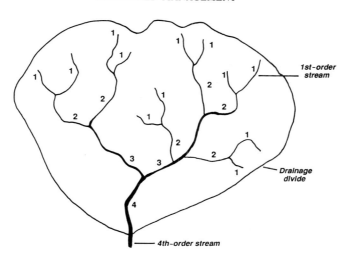

Figure 8.1 Stream ordering within a 4th-order watershed using the Strahler method. (After Hewlett and Nutter 1969.)

by Strahler (1952) and Figure 8.2 illustrates the typical patterns of river channel reaches. River cross sections can be described as shown in Figure 8.3.

Watershed descriptions should include reference to the rocks and sediments that underlie the basin, because geologic characteristics determine the nature and extent of groundwater storage, the type of material available for erosion and transport, and the chemical quality of the water. Geologists categorize rocks as igneous, sedimentary, and metamorphic based upon their mineral composition, texture, and formation process. Groundwater-bearing formations sufficiently permeable to transmit and yield water are termed aquifers. The most common aquifer materials are unconsolidated sands and gravels that occur in river valleys, old stream beds, coastal plains, dunes, and glacial deposits. Sandstone also serves as aquifer material; other sedimentary rocks such as shale and solid limestone do

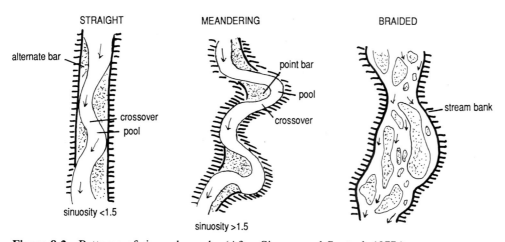

Figure 8.2 Patterns of river channels. (After Simons and Senturk 1977.)

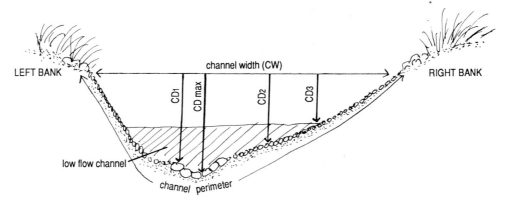

Figure 8.3   Cross section of a stream channel (looking downstream). Channel depth = CD; channel width = CW.

not. Where these rocks and others, such as granite (igneous) and gneiss (metamorphic), are highly fractured, small water yields may be possible. Volcanic materials such as basalt and lavas can form aquifers if they are highly fractured or porous (Bouwer 1978).

Weathering is the physicochemical process whereby the minerals in rocks at or near the earth's surface are altered under normal conditions of temperature and pressure. Weathering proceeds at widely varying rates depending on the type of rocks, climate, and vegetation. Topography in large measure determines the rate at which the products of weathering—sediment—are removed. The weathered products that remain in place are the basic materials on which soil is developed. Those rock fragments, which range in size from large boulders to fine particles, are termed fluvial sediments when they are moved by water. Movement may occur as sheet erosion from land surfaces (particularly for the fine particles) or as channel erosion once runoff accumulates in streams. Within streams, fluvial sediments are transported as bedload, where the particles move within a few particle diameters of the streambed, or as suspended sediment in the turbulent flow.

Weathering of rock material can also contribute to the overall nutrient budget of a watershed. Studies at Hubbard Brook, a northeastern hardwood experimental forest, revealed that while precipitation and atmospheric inputs were major sources of sulfur, nitrogen, chloride, and phosphorous, weathering was the primary source of calcium, magnesium, potassium, and sodium in the ecosystem (Likens et al. 1977).

The vegetative character of a watershed is a function of climate, soils, topography, and land use. The vegetation that covers the land surface plays a dominant role in controlling the timing and amount of water yield from a basin, as well as the quantity of soil available for transport to the stream channel as fluvial sediment. Hence, when soils have been denuded by land-use activities, erosion control is focused primarily on rapid establishment of the plant community. Vegetation also contributes to the aesthetic qualities of a basin. Watersheds are commonly characterized by the composition of their vegetation (e.g., coniferous forest, deciduous woodland, grassland, shrubland).

Table 8.1 Commonly measured topographic attributes of watersheds.

Symbol	Attribute	Derivation or definition	Units	Source
BA	Basin area	For a specified stream location, that area, measured in a horizontal plane enclosed by a drainage divide	km²	Horton (1945)
BP	Basin perimeter	Length of the boundary line along a topographic ridge which separates two adjacent drainage basins	km	Smith (1950)
BL	Basin length	Length of the line, parallel to the main drainage line, from the headwater divide to a specified stream location	km	Potter (1961)
BR	Basin relief	Highest elevation on the headwater divide minus the elevation at a specified stream location	m	Schumm (1956)
TSL	Total stream length	Sum of the length of all streams within a drainage basin	km	Horton (1945)
MBE	Mean basin elevation	Sum of the products of the areas between contour lines and the average elevation between contour lines divided by the total area of the basin	m	Burton and Wesche (1974)
SO	Stream order	Method of classifying streams as part of a drainage basin network (See Figure 8.1)		Strahler (1952)
DD	Drainage density	TSL/BA	km/km²	Horton (1945)
CS	Channel slope	Elevation at 85% of stream length minus elevation at 10% of stream length divided by stream length between these two points		Craig and Rankl (1978)
S	Sinuosity	Ratio of stream length to valley length		Knighton (1984)

8.1.2 Hydrologic Processes on Watersheds

The circulation of water from the oceans to the atmosphere, to the land, and back to the oceans is referred to as the hydrologic cycle. Anderson et al. (1976) considered the hydrologic cycle as a system of water-storage compartments and the flow of water within and between them, as shown in Figure 8.4. Water is stored in the atmosphere and the oceans as well as in the soil, stream channels, lakes, ponds, the groundwater, and vegetation. We call these individual storage compartments and the flows between them hydrologic processes. Water moves between storage areas in either the solid, liquid, or gaseous state. Definitions of the various hydrologic processes of concern to watershed managers are presented in Table 8.2.

Another way to consider the hydrologic cycle is in terms of the disposition of precipitation. This relationship can be expressed as the water balance equation,

$$P = R + E_T + S, \tag{8.1}$$

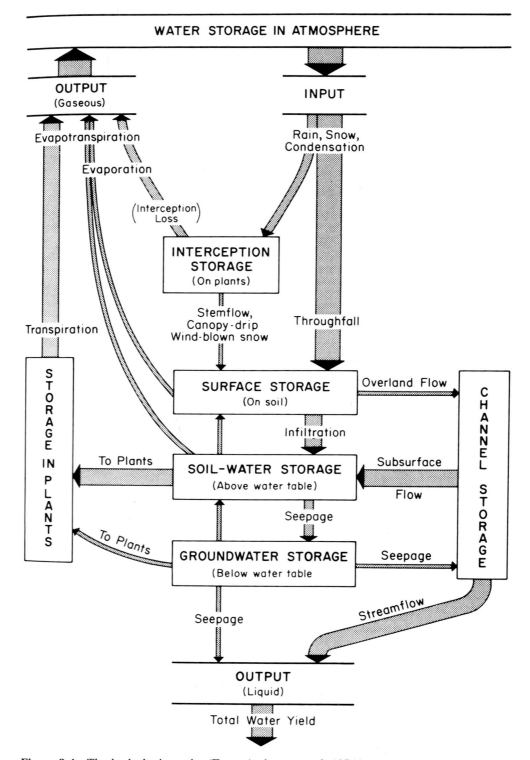

Figure 8.4   The hydrologic cycle. (From Anderson et al. 1976.)

Table 8.2 Definition of hydrologic processes.

Hydrologic process	Definition
Precipitation	The discharge of water, in liquid or solid state, from the atmosphere onto a land or water surface.
Throughfall	In a vegetated area, the precipitation that falls directly to the ground, including streamflow, canopy drip, and windblown snow.
Interception loss	That portion of the precipitation caught and held by vegetation and lost by evaporation, never reaching the ground.
Overland flow	The flow of rainwater or snowmelt over the land surface toward stream channels. Upon entering a stream, it becomes runoff.
Infiltration	The flow of liquid water into the soil above the water table.
Subsurface flow	Water which infiltrates the soil surface and moves laterally through the upper soil layers until it enters a channel.
Streamflow	The discharge that occurs in a natural channel.
Seepage	The slow movement of water from soil-water storage to groundwater storage.
Transpiration	The process by which water vapor escapes from a living plant, principally through the leaves, and enters the atmosphere.
Evaporation	The process by which water is transformed from the liquid or solid state into the vapor state.
Evapotranspiration	The volume of water evaporated and transpired from soil and plant surfaces per unit land area.
Water yield	The total runoff from a drainage basin through surface channels and aquifers.

where P = Precipitation during a given time interval t,
R = Streamflow or total water yield during a time interval t,
E_T = Evapotranspiration during a time interval t,
S = Change in storage or that portion of the precipitation which is retained or lost from storage in the earth's mantle during the time interval t. As t increases, S approaches zero.

Rearranging this equation, the controlling variables with which a watershed manager has to work to increase water yields can be identified,

$$R = P - E_T - S. \tag{8.2}$$

These variables are typically expressed in terms of inches or centimeters of water. This contrasts somewhat with the measures of water normally applied by water users (acre-feet or cubic meters) and stream habitat biologists (cubic feet per second, cfs; cubic meters per second, cms). Perhaps an example is the simplest way to relate these units. If 48 in of precipitation falls on a 1,000-acre watershed during a year and 50% of this total is realized as streamflow, the water yield is 24 in (48 in × 0.50). The total volume of water yielded during the year would be equal to 2,000 acre-ft (24 in × 1,000 acres × 1 ft/12 in). The average discharge for the

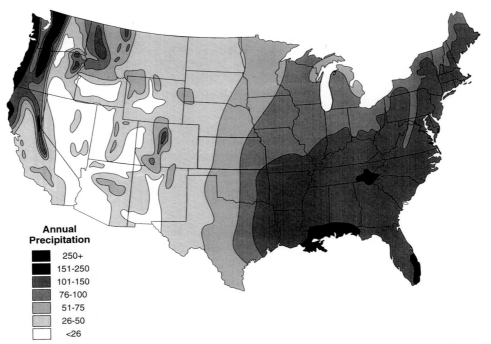

Figure 8.5 Average annual precipitation (in centimeters) in the contiguous United States. (From Satterlund 1972.)

stream draining this watershed over the course of the year would be 2.76 cfs (2,000 acre-ft/year $\times$ 43,560 ft^2/acre-ft $\times$ 1 year/365 d $\times$ 1 d/86,400 s).

Average annual precipitation over the contiguous United States is 30 in, the equivalent of 4.75 billion acre-ft of water. Of this total, 3.4 billion acre-ft (70%) are lost due to evapotranspiration from nonirrigated lands and an additional 3% are consumptively used by irrigators, industry, and municipalities (Hewlett and Nutter 1969). The remaining 27% is accounted for by streamflow.

Precipitation is highly variable across the United States (Figure 8.5). Extremes range from more than 500 cm annually on the Olympic Peninsula of Washington to less than 13 cm across much of the desert southwest (Satterlund 1972). Such variation is due to storm paths, the distance and direction from large bodies of water, location with respect to barriers to atmospheric flow, and elevation. For the most part, alpine and forested zones of the Unites States receive at least 50 cm of precipitation annually, while brush and grassland receive 25 to 50 cm of precipitation in semiarid zones and less than 25 cm of precipitation in arid zones.

The proportion of precipitation that results in streamflow is variable between geographic regions and vegetation zones. Based upon Figure 8.5, it is apparent why the impetus for maintaining instream flows for fisheries has come primarily from the western United States. Table 8.3 summarizes precipitation–water yield relations from several locations and vegetation zones around the country. In general, the percent of precipitation resulting in streamflow increases with increasing precipitation. This is because evapotranspiration is the most constant variable in the water balance equation. Once this requirement is met, remaining

Table 8.3 Precipitation–water yield relations for different types of watersheds. Wyoming data are from Wesche (1982); all other data are from Anderson (1976).

Location	Vegetation type	Mid-area elevation (m)	Precipitation (cm)	Streamflow (cm)	% Water yield
New Hampshire	Northern hardwoods	625	122	69	56
West Virginia	Central Appalachian hardwoods	763	147	61	41
North Carolina	Southern Appalachian hardwoods	824	183	91	50
Minnesota	Bog black spruce	422	79	18	23
Colorado	Lodgepole pine, spruce-fir	3,172	56	30	54
Arizona	Ponderosa pine, white fir	2,181	81	8	9
Arizona	Chaparral	1,296	69	5	7
Oregon	Douglas fir	763	239	155	65
Wyoming	Alpine	3,355	127	97	76

precipitation is available for runoff. Within a given watershed, the percent yielded as runoff will typically increase with elevation.

Often, water supply difficulties are not problems of water yield quantity, but rather are related to the timing of runoff. As anyone who has ever watched a river for any length of time will attest, streamflow can be highly variable, not only from one year to the next, but in some cases, from one minute to the next. The hydrograph is a basic tool used by watershed managers to describe the streamflow regimen of a river as well as the volume of runoff from a basin. A hydrograph is a plot showing some measure of discharge with respect to time for a given stream location. The components of a simple hydrograph are shown in Figure 8.6.

Hydrographs, which vary within and among watersheds, are dependent upon the watershed factors already discussed as well as weather factors which include amount, intensity, duration and distribution of the precipitation, storm direction, and temperature. Within a watershed, storms of differing intensities and durations yield hydrographs of differing magnitudes and shapes.

Figure 8.7 compares annual hydrographs from two Midwest streams and a mountain stream from the central Rocky Mountains. River A shows flashy responses to both precipitation and snowmelt runoff, which is indicative of poor storage due to impermeable or shallow soil layers. The River B hydrograph shows little variation throughout the year, although the river had a precipitation regime similar to River A. The River B watershed, however, is characterized by deep, permeable, sandy soils which have good infiltration and storage capability. The mountain stream hydrograph indicates the dominance of snowmelt runoff on the hydrology of the region. Approximately 75 to 80% of the annual water yield results from snowpack which accumulates during the winter months. Baseflow during the remainder of the year is generally stable. Human use of the land surface within a watershed can also exert a strong influence on both hydrograph size and shape.

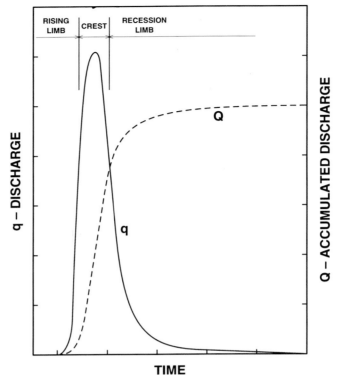

Figure 8.6 Component parts of a simple hydrograph. Q is the volume of runoff as related to time, and q is the discharge rate at any point in time.

Sediment is a major product from watersheds and is produced in part by the hydrologic cycle. Sediment yield can be defined as the sediment transported out of the drainage basin from all fluvial erosion sources. It is typically measured in tons per unit land area per unit time. From a watershed management perspective, the sediment yield from a basin is important because (1) an estimated 80% of water quality degradation results from erosion, (2) sediment interacts strongly with other water quality components, and (3) sediment yield is directly affected by land-use activities (Anderson et al. 1976).

Erosion and sedimentation are the two phases of the process of detaching particulate material from one location and depositing it in another location. Erosion refers to the removal of the material while sedimentation refers to its deposition. Once transport has begun, the material is referred to as sediment and can be either organic or inorganic in nature (Satterlund 1972). Within a watershed, there are three types of erosion: surface, mass movement, and channel cutting, all of which can contribute significantly to the sediment yield of the basin.

Surface erosion is the detachment and removal of individual soil particles from the land surface and includes sheet erosion, and rill and gully formation. Two hydrologic processes are primarily involved: raindrop splash and overland flow. Raindrops striking bare soil cause minor explosions that dislodge soil particles. These particles can be forced upward into suspension and moved downslope in overland flow.

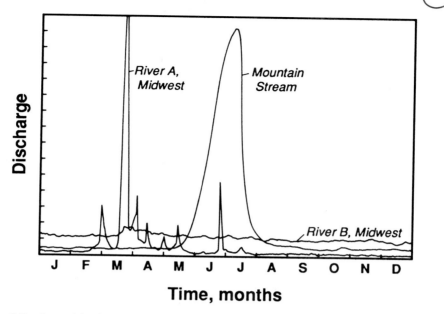

Figure 8.7 Annual hydrographs for the streams in the midwestern United States compared to a typical mountain stream in the central Rocky Mountains. (Adapted from Hewlett and Nutter 1969.)

Mass movement of slopes is caused by gravity. It occurs when the cohesive and frictional strength of an earth mass fails across an area of weakness. When the movement occurs at a rate perceptible to the eye, it is called a landslide. Creep occurs at a rate slower than the eye can perceive. Mass movement can result from either an increase in shear stress, a decrease in resistance, or both. Shear stress is the sum of all forces acting to displace the earth mass on the slope. Although the weight of the soil itself is generally the primary stress component, water, vegetation, wind, vibrations (e.g., earthquakes, explosions), downcutting of streams, and soil creep on steep slopes may all contribute. Land-use activities such as road building and timber harvest can also significantly influence stress.

Channel cutting refers to the detachment and movement of material from stream channels by flowing water. While much of the suspended sediment load transported by a stream is the product of erosion processes on the land surface of the watershed, the bulk of the sediment transported as bedload is the result of channel erosion. Channel erosion, like runoff, is highly variable over space and time.

Sediment transport is a function of the availability of material to be eroded and the sediment transporting ability of the streamflow. Each stream has a fixed sediment carrying capacity, based upon its channel dimensions and water velocity (Leopold and Maddock 1953). A concept fundamental to sediment transport is that of stream power, the rate at which a stream does work. This expression of sediment transport capability is usually defined in English units by the relationship

$$\omega = \gamma \bar{D} \bar{V} \bar{S},\tag{8.3}$$

where ω = stream power (pounds per foot of channel width per second),
  γ = specific weight of water (62.4 lbs per ft³),
  $\bar{D}$ = mean depth of the flow (ft),
  $\bar{V}$ = mean velocity of the flow (ft per second),
  $\bar{S}$ = stream slope (ft per ft).

As discharge changes, both stream power and sediment carrying capacity change at a particular stream location. Natural stream channels are seldom uniform from one location to another. Even though discharge may be constant at two locations on a given stream, depth, velocity, and slope will differ, resulting in variations of stream power and transport capacity.

As a result of this temporal and spatial variation, streams—even though they may be considered stable—are characterized by a continuously shifting equilibrium. Sediment is eroded and transported primarily on the rising limb of the hydrograph and deposited on the receding limb. At any given time, sediment is moved from one channel location and deposited in another. Pools may be scoured of deposited sediment at high flows while riffles may be filled. The opposite may occur at lower discharge levels. The net result is that most sediment moves downstream in a sort of "leapfrog" pattern, responding to the changing hydrograph and the hydraulic characteristics of the channel.

8.2   WATERSHED–FISHERIES RELATIONS

8.2.1   A Watershed Comparison

Broad-scale watershed differences can be equated to differences in fishery habitats and species composition. For example, consider two small watersheds, one located in the spruce-fir forest of the central Rocky Mountains, the other in the shortgrass prairie of the northern Great Plains. While drainage basin size and stream order may be similar, differences occur in the aquatic habitat provided.

In the mountain watershed, basin elevation is high, precipitation is approximately 76 cm annually, and the hydrograph is dominated by spring snowmelt. Soils are thin with high infiltration rates. Overland flow seldom occurs even though drainage density is low. Water yield as a percent of precipitation is in the 40–50% range and streamflow is perennial. The valley bottom is quite narrow; the adjacent conifer-covered slopes rise steeply. Basin relief is high resulting in a relatively steep channel slope. Within the channel itself, the thin soil mantle has been eroded exposing the underlying geology of the basin. The streambed is dominated by bedrock, boulders, and cobbles in the straight steeper reaches; in the flatter sections, gravels and sands are deposited and a meandering pattern may develop. Streambanks are poorly developed in the steeper reaches due to the shortage of building material and sediment, and reflect the composition of the substrate. Where the valley floor widens and gradient lessens, sediments transported from the steep reaches can settle out along the channel boundaries, vegetation can become established, and bank development occurs. Sediment yield from the watershed is limited, which results in quite low suspended sediment transport and turbidity. Water temperatures remain cool and nutrient loadings low. Bedload transport of cobbles and gravels occurs primarily during the rising limb of the spring runoff hydrograph when available stream power is at or near its maximum. At this time, pocket pools are scoured behind the boulders which form much of the habitat in the steeps and meander pools, and mid-channel pools and undercut banks are scoured in the flats. On the receding limb of the hydrograph, as stream power lessens, these larger sediments can no longer be transported and are deposited along the alternate, point, and mid-channel bars. Through these integrated watershed processes, salmonid rearing and spawning habitats are created and maintained on an annual basis.

Consider now the same processes but under the different climatic hydrologic, topographic, geologic, and vegetative conditions of the shortgrass prairie watershed. Here, basin elevation is relatively low and the climate is semiarid with annual precipitation of approximately 36 cm, over 40% of which occurs as rainfall during the summer months, primarily the product of thunderstorm activity. Soils are heavy and deep, having a high clay content and a low infiltration rate. The annual hydrograph is flashy. Streamflow is ephemeral; surface flow in the channel occurs only in response to precipitation, during which overland flow may occur even though the topography is gently rolling. Surface flow constitutes only a small percentage of the annual precipitation due primarily to the high potential evapotranspiration rate and the limited contribution of groundwater. Surface runoff is highly turbid due to the sparse vegetative groundcover on the uplands. In a small, low-relief watershed such as this, the low-gradient channel may show little development, often more closely resembling a grass-covered waterway. The only aquatic habitat present may be a series of unconnected "potholes" which support small populations of hardy species such as black bullheads and green sunfish. These potholes may be formed at low spots in the channel where ephemeral standing water is present long enough to kill the grass cover. Subsequent high flows can then scour these areas to a depth sufficient to intersect the shallow groundwater aquifer and create a perennial pond environment.

Such a macroscopic comparison allows managers to make gross estimates of the fishery potential of diverse watersheds. Certainly even the most casual observer could discern the obvious differences between these streams. The important point to be made, however, is not that they are different, but why they are different.

8.2.2 Watershed–Fisheries Models

In recent years, researchers and managers have shown increasing interest in discerning more detailed differences between watersheds and relating them to fisheries potential. If a functional link can be defined between drainage basin characteristics and stream habitat quality or fish carrying capacity, then mathematical models can be developed to (1) prioritize streams for management purposes based upon their potential, (2) estimate the influence various types and intensities of land use have had on fisheries, and (3) better integrate fisheries into existing land-classification systems.

There has been a great deal of interest in watershed–fisheries modeling for salmonid streams in western North America. Fausch et al. (1988) reviewed 99 models that predicted standing stock of salmonids in stream systems. While the majority were based upon field-measured instream habitat variables, several models used only measures of drainage basin geomorphology as independent variables. Each model was developed for a specific geographic area, within which

Table 8.4 Two models that relate watershed characteristics to a measure of fishery quality.

Source	Independent variables	Dependent variables	Geographic area
Burton and Wesche (1974)	Mean basin elevation Basin Area Forested area Total stream length	Resident trout standing stock	Small mountain streams in SE Wyoming
Platts (1979)	Stream order	Stream channel characteristics Fish numbers Number of fish species	South Fork of Salmon River, Idaho

variables such as climate, geology, and water quality were relatively constant. Table 8.4 summarizes two of these geomorphic models.

The findings of Lanka et al. (1987) tended to reinforce the results of earlier investigators. Lanka et al. developed physical regression models using geomorphic variables, instream habitat variables, and combinations of the two. In all cases, the combined models yielded the best test results. In general, smaller, more gently sloping watersheds produced the best salmonid habitats. A strong link existed between geomorphic variables and measured physical habitat variables. Highest standing stocks were often associated with the transition zone between the high-gradient, boulder-strewn forest stream type and the lower gradient, gravel habitat characteristic of rangeland streams. Such zones offer maximum habitat diversity. Increasing stream size, reflected by such geomorphic parameters as basin area and stream order, was in general inversely related to salmonid density. This relation could in part be the result of a decrease in the contribution of the riparian zone to the aquatic habitat resulting from more intensive land use in lower stream reaches (Wesche et al. 1987).

8.2.3 Riparian Zones

Riparian, or streamside, areas serve as the transition zone between the terrestrial and aquatic environments. During the settlement of the western United States in the 1800s, these areas provided much of the food, shelter, and fuel necessary for pioneer survival. Today, riparian areas are recognized as being transportation corridors, high producers of timber and forage, key habitats for a diversity of wildlife, major components of quality fisheries habitat, prime recreation areas, and critical to the overall management of any watershed. Given this array of uses, it is not surprising that riparian areas have been the focus of much debate in recent years between user groups and management agencies.

The availability of water makes riparian areas different from the uplands. Riparian zones exist because water is available to plants during their entire growing season and it is the availability of this water that promotes the dominance of species that need a water table continually near their root zone. If the water table is lowered in relation to the root zone, water-loving species are replaced by plants better adapted to xeric conditions. Along mountain streams where moisture is abundant, little difference may be observed in plant speciation and density between streamside and upland areas. Observation of the "river of green" during late summer alongside a basin stream in a semiarid region clearly indicates the water dependence of riparian vegetation.

From a watershed management and a fisheries perspective, riparian areas serve several important functions. Streamside vegetation plays a role in controlling channel morphology. Not only does the available root biomass serve to stabilize otherwise erosive streambank soils, but when overbank flooding occurs, the above-ground portions of the plants serve to increase the roughness of the floodplain, thereby slowing the water and promoting infiltration and recharge of the alluvial aquifer. In this manner, instantaneous water yield from a particular hydrologic event may be reduced. At least a portion of this water can then be released from shallow groundwater storage later in the season to improve baseflow conditions and water regimen. Sediment being transported by the overbank flow is deposited due to the stilling effect of the vegetation, thereby reducing sediment yield from the watershed. As many plant nutrients (especially phosphorus and to some extent nitrogen) in surface runoff are attached to the sediment particles, nutrient loadings to the aquatic system can also be reduced by storage in the riparian zone (Karr and Schlosser 1978). In this way, streambanks are built and stabilized. Also, riparian vegetation acts to control nonpoint-source pollution by filtering out sediments delivered from upland slopes by overland flow.

As a result of modified flood peaks, reduced sediment yields, and stabilized channel morphology, the quality of the aquatic habitat is maintained. Spawning beds for fish and microhabitats for macroinvertebrates remain relatively free of damaging fine sediment deposits. Filling of pools is reduced thereby maintaining water depths and structural diversity. Escape cover provided by well-sodded undercut streambanks is maintained. Primary production is enhanced by reduced turbidity and critical low-flow habitat is improved by augmented baseflow levels. In cold regions, narrower, deeper channels promote snow bridging which can play an important role in the food web and large woody debris which can provide cover and help to control channel form. Finally, shading can help to maintain cooler water temperatures, especially in small streams, thereby also increasing the capacity of the water to hold dissolved oxygen.

8.3 LAND-USE EFFECTS

The term land use encompasses many activities, including timber harvest, mining, agriculture, urbanization, and recreation. These activities can affect our stream resources directly through the destruction of habitat by trampling, construction, and channelization, as well as by their influence on watershed processes which govern water yield, water regimen, and sediment production.

8.3.1 Effects on Water Yield and Regimen

A review of the hydrologic cycle (Figure 8.4) and the water balance equation (equation 8.1) indicates the processes that can be influenced by land use to effect a water yield change. In general, those land-use activities that produce augmented water yields do so by reducing storage losses through evaporation and transpiration or by decreasing infiltration rates, thereby increasing overland flow to the channel.

The removal of vegetation or the conversion of one vegetative type to another are traditional means for increasing water yields. Land uses, such as timber harvest, urbanization, and agriculture directly influence the density and composition of vegetation, thereby effecting water loss by evapotranspiration.

Possibly more research has been conducted in the area of timber harvest–water yield relations than in other land-use areas. Much of the work has focused on the western United States where periodic shortages of useable water seriously impact the growth and economy of the region. Because approximately 75% of the western United States' water supply originates on forested lands administered by the Forest Service, this agency has had a long involvement with such research.

U.S. Forest Service research conducted at the Fraser Experimental Forest, located in the forest snowpack zone in the central Rocky Mountains of Colorado, has provided insight regarding timber harvest–water yield relations. Here, snowfall is the key to water yield, as it is throughout much of western North America and the northern latitudes. Research results from the Fool Creek watershed, where timber was harvested in the mid-1950s by clear-cutting alternate strips of varying width, indicated that over the past 30 years water yield was increased by almost 40%, the result of an extended spring runoff. Peak flows and late summer streamflow appear to be little affected. Analysis indicates that the effect of the timber harvest treatment is diminished by about 0.04 in per year. At this rate, 70 to 80 years will be required for water yield to return to pretreatment levels (Alexander et al. 1985).

The water yield increase on Fool Creek can be attributed to several factors. Evapotranspiration losses were reduced by the clearing of timber from the alternating strips. Redistribution of the snowpack also contributed to the water yield gains. Overall, the watershed realized only a small net increase in total snow accumulation. However, in the cleared strips, 30% more snow, in terms of water equivalent, was deposited. Because more snow was deposited in the openings where soil moisture deficits were lowest and higher melt rates caused meltwater to become available earlier in the season before evapotranspiration could deplete it, overall efficiency of the treatment was enhanced and water yields were augmented (Alexander et al. 1985). Forested watersheds outside the snowpack zone tend to produce even greater water yield increases in response to timber harvest (Satterlund 1972). In areas such as these where precipitation is predominantly in the form of rain, late summer streamflow can be augmented, although timing is largely dependent upon distribution of precipitation.

Increasing spring snowmelt runoff may result in damage to stream channels. Studies have established relationships between various flood flow levels and the hydraulic geometry of stream channels (Leopold and Miller 1956). Should flood peaks be augmented, channel adjustments may be necessary in terms of channel width, depth, cross section area and slope to accommodate the additional flow. Data from the Fool Creek experiment and a similar earlier study at Wagon Wheel Gap, Colorado, indicate that while the duration of flood peaks was increased, the magnitude of the peaks changed very little and stream channels were not effected (Bates and Henry 1928; Leaf 1970).

Removal of woody vegetation, such as chaparral and juniper, from semiarid watersheds has resulted in less consistent water yield changes than observed under wetter, forested conditions (Satterlund 1972). Most increases have been observed on moist lowland sites having deep soils during years of above-normal precipitation. Failures have generally occurred during normal or dry years at upland sites having shallow soils.

Much interest has also been shown in the eradication of phreatophytes (plants which obtain their supply of moisture primarily from the groundwater or capillary fringe just above the groundwater) from streamside areas to increase water yields. As water seldom limits evapotranspiration losses from phreatophytes, and because such communities often exhibit an "oasis effect" due to the dry, hot uplands that surround them, water losses per unit area are probably greater than in other vegetation types. Therefore, removing riparian vegetation should result in large water yield gains.

Interest in this management strategy has been especially high in the desert Southwest. Heindl (1965) estimated that if all riparian vegetation was removed along major streams in Arizona, New Mexico, and western Texas, 37 km³ of water could be salvaged annually. However, Campbell (1970) pointed out that predictable water yield increases following total removal of riparian vegetation have not been realized. Based upon the limited research conducted in the southwestern United States, he concluded (1) removal of riparian vegetation increases surface flow if sufficient plants existed on the site, (2) enhanced water yields are modest, most likely because of increased surface evaporation, and (3) periodic treatment is necessary, the cost of which must be included in the total cost of the water harvested. Certainly, given the immense value of riparian areas for multiple use, such a management strategy should proceed cautiously.

Agriculture is a major land use which has had a significant influence on water yield. Over much of the semiarid western United States, water withdrawal for irrigation of crop and meadow lands is a major water use. In Wyoming, for example, over 90% of appropriated water is for irrigation. The remaining 10% meets the needs of industry, municipalities, and other uses within the state.

The storage and withdrawal of water for irrigation has played a major role in shaping the stream channels and riparian areas that we see today in the western United States. Depletion of streamflow during the spring runoff period reduces stream power available for transporting deposited sediments and seeds. In the absence of these "flushing flows," riparian vegetation encroaches, reducing the width and conveyance capacity of the active channel. For example, the North Platte River in western Nebraska has shrunk from a wide, shallow, braided channel 763 m in width down to only 91.5 m over the past 100 years (Williams 1978).

Often, agriculture and water development are blamed for degraded aquatic and riparian habitats. Although in some cases such criticism may be justified, the positive aspects should be rationally examined from a watershed perspective. Water removed from the channel and applied to the land for crop production may (1) encourage development of new riparian vegetation in portions of the former active channel; (2) reduce evaporative loss by storing water in the shallow, unconfined aquifer connected to the stream; (3) augment late season baseflows by means of return flows to the channel; (4) reduce the power of peak runoff flows thereby preventing channel damage; and (5) encourage development of a narrower, deeper channel where aquatic habitat quantity may be reduced but quality is improved. In some cases, impacts may be negative. For example, return flows may increase water temperatures and salinity dependent on soil type and conditions. In other cases, river regulation may lead to the development of premier aquatic habitats, such as the blue-ribbon trout waters of the North Platte and Green rivers of Wyoming. Each project must be considered individually on its own merits. The labelling of all water development as harmful to aquatic–riparian systems is poor management policy. Without healthy riparian zones, quality fisheries habitat for many species cannot exist.

Livestock grazing can also have an effect on water yield and regimen. Lusby (1970) compared runoff from ungrazed and grazed semiarid rangeland watersheds in western Colorado. Results indicated a 30% reduction in runoff from the ungrazed watershed. As pointed out by Holechek (1980), soils compacted by overgrazing with little vegetative cover have greatly reduced infiltration rates. Water moves over the soil rather than into it causing rapid delivery of heavy runoff to stream channels. The resultant hydrograph is flashy and groundwater storage may not be adequate to maintain late season base flows.

The need to provide rapid drainage of tillable agricultural lands and prevent the flooding of crops has led to the channelization of thousands of river miles in North America. Channelization is the process whereby channel bed, bank, and form roughness (e.g., boulders, snags, meander patterns), which impede flow, are reduced so that flood waters pass more quickly and the channel conveys greater flood peaks without overtopping the streambanks. Channelization, in association with improved field drainage, may result in higher peak flows of shorter duration than would be realized under more natural watershed conditions.

Urbanization, including the development of roads, buildings, other municipal-industrial structures, parking lots, and such recreational developments as ski areas and second homes, can have significant effects on the hydrology of a watershed. Around metropolitan areas, asphalt and concrete have replaced soil, buildings have replaced trees and other native vegetation, and sewer systems have replaced streambeds. The result has been and continues to be rapidly changing watershed conditions. Four interrelated, but separable, effects can occur to the hydrology of an urbanized area: (1) changes in peak flow characteristics, (2) changes in total runoff, (3) changes in water quality, and (4) changes in hydrologic amenities such as aesthetic characteristics (Leopold 1968).

Several hydrologic processes are affected as a small forested watershed becomes urbanized. These include infiltration and overland flow, soil-moisture storage, evapotranspiration, runoff, and peak flows. Overall, the total water yield from an urbanized area can be expected to exceed that from a forested watershed having a similar climate. Urban runoff is enhanced by reduced infiltration, soil-moisture storage and evapotranspiration, and increased overland flow. Peak flows are generally increased as a result of the augmented overland flows while the lag time between precipitation and runoff is reduced.

Since about 1960, the growing American population has found increasing time for recreational pursuits. Much of this time is spent outdoors on our forest and rangeland watersheds enjoying a variety of activities such as fishing, hunting, camping, sight-seeing, hiking, motorbiking, four-wheel driving, and snowmobiling. From a hydrologic perspective, such relatively dispersed activities have minimal impact when use remains low. As use increases, and is concentrated, so do associated trampling, compaction, and trailing effects, especially within small watersheds. Trail and road building which may accompany such uses often contribute more to soil compaction, enhancement of overland flow, and hydrograph alteration than the uses for which they were built.

Table 8.5 Representative rates of erosion for various land uses. (From USEPA 1973.)

Land use	Tons per square mile per year	Relative to forest = 1
Forest	24	1
Grassland	240	10
Abandoned surface mines	2,400	100
Cropland	4,800	200
Harvested forest	12,000	500
Active surface mines	48,000	2,000
Construction	48,000	2,000

8.3.2 Effects on Sediment

Sediment is widely regarded as the greatest source of water pollution in the United States (USEPA 1976). Not only can excess sediment degrade the quality of aquatic habitat, it can also have numerous other effects, including:

- reducing water storage in reservoirs, lakes and ponds;
- reducing conveyance capacity of stream channels thereby encouraging over-bank flooding;
- creating turbidity which reduces primary productivity and detracts from recreational use of water;
- degrading water quality for consumptive uses;
- increasing water treatment costs;
- damaging water distribution systems;
- serving as a carrier of other pollutants such as heavy metals, herbicides, insecticides, and plant nutrients;
- acting as a carrier of bacteria and viruses.

In 1966, the annual damage caused by sediment in streams was placed at $262 million. Sediment pollution in 1988 will exceed the $1 billion mark.

Any land-use activity that removes vegetative cover, disturbs the soil mantle, reduces infiltration rates, decreases soil-moisture storage, and increases overland flow has the potential to increase sediment yield from a watershed. Mining, construction, agriculture, and silviculture are major sources of sediment pollution. Of these four, agriculture accounts for approximately 50% of the total contribution, although mining and construction have the highest erosion rates per unit land area. Table 8.5 compares these rates for various land uses.

There are two general approaches for preventing soil erosion: runoff control and soil stabilization. Runoff control practices are designed to reduce the ability of runoff to cause erosion, while soil stabilization practices are intended to protect soil from the erosive action of precipitation, runoff, and wind. Specific sediment control practices can be broken down into four categories: biological control, structural control, mechanical control, and procedural control. For any particular land-use activity, the manager may apply a variety of techniques from any or all of these categories.

The reestablishment of vegetation on disturbed soil is critical for both runoff control and soil stabilization. Different plant species will have different erosion control values for particular applications, dependent upon climate, soil type, moisture conditions, land surface slopes, degree of cultivation, and specific biological and ecological characteristics of the plants themselves. Many reports have been written regarding the selection and application of plant species for different reclamation uses in different areas of the country (USEPA 1976). Another example of biological control would be the introduction of beaver into degraded streams. Their dam building can serve to trap sediments, store water, provide grade control for the channel, and encourage recovery of the riparian zone.

Numerous structural sediment control methods have been developed. These may include, but certainly are not limited to, the proper installation of culverts, ponds, drainage pipes, and diversion channels to control runoff, and the placement and construction of rock riprap, gabions, revetments, and grade-control structures (small dams) to stabilize eroding stream channels and watercourses (see Chapters 9, 17, and 18).

Mechanical control measures deal primarily with the grading, shaping, and conditioning of the soil surface, often in association with revegetation efforts. These measures may include such treatments as manipulation of slope gradient and length, scarification of the soil surface by chiseling and gouging, and development of contour terraces or furrows to disrupt and slow surface runoff.

The fourth category, procedural control, is equally important. By procedural, we refer to the careful planning, conduct, and timing of activities associated with a land use to prevent or reduce soil erosion. Such measures range in complexity from closing a steep, unpaved forest road to vehicular traffic during wet seasons, to the design of a four-lane highway through a region having unstable soils, or the development of an operations plan for a major surface mine. Also included as procedural would be the selection of methods to remove harvested timber from a watershed. In summary, if soil erosion is to be minimized, climate, topography, geology, hydrology, and vegetation need to be carefully considered before the land-use activity begins, as well as during and after the activity is completed.

8.3.3 An Example

A stream and the fisheries habitat it provides is a reflection of its watershed. Historic and present land-use activities such as timber harvest, mining, grazing, water development, urbanization, and recreation all can leave their mark, however subtle, on the stream environs.

An important question a fisheries manager must address is, Why is the stream habitat I am trying to manage in the condition that it is? Often, the answer is not found by looking just at the wetted habitat between the streambanks. Examine the photographs presented in Figure 8.8. From these, what can be perceived about historic land use in the Douglas Creek basin and the long-term trend of the aquatic habitat?

Douglas Creek is a moderate-size stream located high in the forest snowpack zone of the central Rocky Mountains. Average discharge is 30–40 cfs (0.85–1.13 cms). Several aspects of the channel evident from the photographs are its width in relation to its depth, the lack of well-developed streambanks even in lower-gradient sections, the absence of woody debris, the establishment of vegetation on bars located on the inside of meanders, the lack of erosion on the banks opposite these bars, and the establishment of young lodgepole pine on the floodplain. As a result, habitat quality is marginal for the brown trout which dominate the population.

Figure 8.8   Douglas Creek, Medicine Bow National Forest, Wyoming.

The Douglas Creek channel is the product of a series of different land uses which date back to before the turn of the century. Timber was first harvested throughout the watershed to provide ties for the railroad. In the absence of roads, massive "tie drives" down Douglas Creek were held each year during the spring snowmelt runoff to transport logs. To facilitate these drives, the channel was cleared of woody debris and boulders, thereby reducing channel roughness and increasing water velocities. As a result, channel widening was encouraged. Gold mining became popular on Douglas Creek during the 1920s and 1930s. Evidence of the activity still remains in the form of long, partially vegetated dredge spoil piles which line the channelized streambanks in many areas. During the early 1960s Rob Roy Dam was built on the upper Douglas Creek to provide municipal water. The vegetative encroachment present today, both in the active channel and on the floodplain, is a result of flow regulation. Rob Roy Dam has stabilized the natural flow regime, especially through the reduction of peak spring runoff flows, and as a result vegetation has become established, streambanks have been rebuilt, and the oversized channel has gradually narrowed. While this process is a slow one due to the short growing season and the limited sediment supply (at 2,745 m in elevation), the trend for the stream and its habitat is in a positive direction.

As with Douglas Creek, each stream bears the indelible mark of its watershed and human use of the watershed. A challenge for fisheries managers is to work with scientists, engineers, and resource managers from other disciplines to minimize impacts to our aquatic habitats in the face of increasingly intensive land use.

8.4 REFERENCES

Alexander, R. R., C. A. Troendle, M. R. Kaufmann, W. D. Shepperd, G. L. Crouch, and R. K. Watkins. 1985. The Fraser Experimental Forest, Colorado: research program and published research 1937–1985. U.S. Forest Service General Technical Report RM-118.

Anderson, H. W., M. D. Hoover, and K. G. Reinhart. 1976. Forests and water: effects of forest management on floods, sedimentation, and water supply. U.S. Forest Service, General Technical Report PSW-18/1976.

Bates, C. G., and A. V. Henry. 1928. Forest and streamflow experiment at Wagon Wheel Gap, Colorado. U.S. Department of Agriculture, Weather Bureau Monthly Weather Review Supplement 30, Washington, D.C.

Bouwer, H. 1978. Groundwater hydrology. McGraw-Hill, New York.

Brown, E. H. 1970. Man shapes the earth. Journal of Geography 136:74–84.

Burton, R. A., and T. A. Wesche. 1974. Relationship of duration of flows and selected watershed parameters to the standing crop estimates of trout populations. University of Wyoming, Water Resources Research Institute Series Publication 52, Laramie.

Campbell, C. J. 1970. Ecological implications of riparian vegetation management. Journal of Soil and Water Conservation 2:49–52.

Craig, G. S., Jr., and J. G. Rankl. 1978. Analysis of runoff from small drainage basins in Wyoming. U.S. Geological Survey, Water-Supply Paper 2056.

Fausch, K. D., C. L. Hawkes, and M. G. Parsons. 1988. Models that predict standing crop of stream fish from habitat variables: 1950–1985. U.S. Forest Service, General Technical Report PNW-213.

Heindl, L. A. 1965. Groundwater in the Southwest—a perspective. In J. E. Fletcher and G. L. Bender, editors. Ecology of groundwater in the southwestern United States. Arizona State University, Tempe.

Hewlett, J. D., and W. L. Nutter. 1969. An outline of forest hydrology. University of Georgia Press, Athens.

Holechek, J. 1980. Livestock grazing impacts on rangeland ecosystems. Journal of Soil and Water Conservation 4:162–164.

Horton, R. E. 1945. Erosional development of streams and their drainage basins, hydrophysical approach to quantitative morphology. Geological Society of America Bulletin 56:275–370.

Karr, J. R., and I. J. Schlosser. 1978. Water resources and the land-water interface. Science (Washington, D.C.) 201:229–234.

Knighton, D. 1984. Fluvial forms and processes. Edward Arnold, London.

Lanka, R. P., W. A. Hubert, and T. A. Wesche. 1987. Relations of geomorphology to instream habitat and trout standing stock in small Wyoming streams. Transactions of the American Fisheries Society 116:21–28.

Leaf, C. F. 1970. Sediment yields from central Colorado snow zone. American Society of Civil Engineers, Hydraulics Division Journal 96:87–93.

Leopold, L. B. 1968. Hydrology for urban land planning. U.S. Geological Survey, Circular 554.

Leopold, L. B., and W. B. Langbein. 1963. Association and indeterminacy in geomorphology. In C. C. Albrilton, editor. The fabric of geology. New York.

Leopold, L. B., and T. Maddock. 1953. The hydraulic geometry of stream channels and some physiographic implications. U.S. Geological Survey, Professional Papers 252.

Leopold, L. B., and J. P. Miller. 1956. Ephemeral streams—hydraulic factors and their relations to the drainage net. U.S. Geological Survey, Professional Papers 282-A.

Likens, G. E., F. H. Bormann, N. M. Johnson, D. W. Fisher, and R. S. Pierce. 1970. Effects of forest cutting and herbicide treatment on nutrient budgets in the Hubbard Brook watershed-ecosystem. Ecological Monographs 40:23–47.

Lusby, G. C. 1970. Hydrologic and biotic effects of grazing versus nongrazing near Grand Junction, Colorado. U.S. Geological Survey, Professional Papers 700B.

Platts, W. S. 1979. Relationships among stream order, fish populations, and aquatic geomorphology in an Idaho river drainage. Fisheries (Bethesda) 4(2):5–9.

Potter, W. D. 1961. Peak rates of runoff from small watersheds. U.S. Department of Commerce, Bureau of Public Roads, Hydraulic Design Series 2, Washington, D.C.

Satterlund, D. R. 1972. Wildland watershed management. Wiley, New York.

Schumm, S. A. 1956. The evolution of drainage systems and slopes in badlands at Perth Amboy, New Jersey. Geological Society of America Bulletin 67:597–646.

Simons, D. B., and F. Senturk. 1977. Sediment transport technology. Water Resources Publications, Fort Collins, Colorado.

Smith, K. G. 1950. Standards for grading texture of erosional topography. American Journal of Science 248:655–668.

Strahler, A. N. 1952. Hypsometric (area-altitude) analysis of erosional topography. Geological Society of America Bulletin 63:1117–1142.

USEPA (U.S. Environmental Protection Agency). 1973. Methods for identifying and evaluating the nature and extent of nonpoint sources of pollutants. EPA-4030/9-73-014, Washington, D.C.

USEPA (U.S. Environmental Protection Agency). 1976. Erosion and sediment control, surface mining in the eastern U.S., planning, volume 1. EPA-625/3-76-006, Washington, D.C.

Wesche, T. A. 1982. The Snowy Range Observatory: an update. University of Wyoming, Water Resources Research Institute Series Publication 81, Laramie.

Wesche, T. A., C. M. Goertler, and C. B. Frye. 1987. Contribution of riparian vegetation to trout cover in small streams. North American Journal of Fisheries Management 7:151–153.

Williams, G. P. 1978. The use of shrinking river channels—the North Platte and Platte rivers in Nebraska. U.S. Geological Survey, Circular 781.

Chapter 9

Stream Habitat Management

DONALD J. ORTH AND RAY J. WHITE

9.1 INTRODUCTION

Habitat for fish is a place—or for migratory fishes, a set of places—in which a fish, a fish population, or a fish assemblage can find the physical and chemical features needed for life, such as suitable water quality, migration routes, spawning grounds, feeding sites, resting sites, and shelter from enemies and adverse weather. Although food, predators, and competitors *are not* habitat, proper places in which to seek food, escape predators, and contend with competitors *are* part of habitat, and a suitable ecosystem for fish includes habitat for these other organisms, as well. The characteristics of the habitat play a large role in determining the numbers, sizes, and species of fish that can be sustained. Habitat management is a vital aspect of stream fisheries management because it affects fish populations. Habitat management should, therefore, be coordinated with population assessment (Chapters 5 and 6) because characteristics of fish populations and harvest can be used to monitor stream habitat quality (Fausch et al. 1990). Stream habitat management must be coordinated with regulation of land use (Chapter 8) because streambanks, their vegetation, and watershed conditions ultimately affect habitat quality for fish. There is a close interaction between riparian (streamside) vegetation and stream habitat in small as well as large streams. Consequently, an effective stream habitat management program should treat the riparian–stream ecosystem as a single unit. Modern stream habitat managers have a thorough knowledge of the influence of stream habitat on the many benefits that streams can provide to society (cf. Meehan 1991; Boon et al. 1992).

9.1.1 Objectives of Stream Habitat Management

The objectives of stream habitat management are to protect or improve stream habitat such that fish populations or assemblages can be sustained or increased. Habitat protection involves preventing changes that would adversely affect habitats. Habitat improvement can have either of two sub-objectives: restoration or enhancement. Habitat restoration means returning degraded habitat to predisturbance condition. Habitat enhancement means improving habitat beyond some naturally less-than-optimal condition. Often the same methods are appropriate for restoration and enhancement.

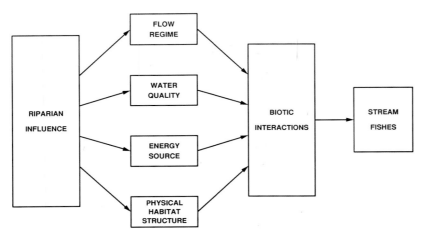

Figure 9.1 Stream habitat factors affecting distribution and abundance of stream fishes. (Modified from Karr et al. 1986.)

9.2 STREAM HABITAT FACTORS

Stream fisheries management requires an understanding of the interaction of habitat factors and the fish populations of interest. Five classes of factors affect the distribution and abundance of stream fish (Figure 9.1): streamflow, water quality, energy source, physical habitat structure, and biotic interactions (Karr et al. 1986). These interact to determine the characteristics of the fish community in a particular stream; characteristics of the riparian zone and the watershed also influence these five factors.

9.2.1 Riparian and Watershed Influence

The riparian zone and watershed characteristics intimately affect stream habitat. Riparian plants filter sediments and nutrients, provide shade, stabilize streambanks, provide cover in the form of large and small woody debris, and produce leaf litter energy inputs. Altered riparian zones or watershed land use can detrimentally affect stream habitat (Meehan 1991). It is essential that riparian zones be protected and managed with awareness of their influence on stream habitats.

9.2.2 Streamflow Discharge

Streamflow discharge, the volume of water flowing past a cross section in a channel per unit time, can have great short-term variability, but in most streams the seasonal pattern (flow regime) is often predictable within a region. The stream biota is adapted to the seasonal cycle of flooding and low flows; however, streamflow can limit the biomass of fishes. For example, the late summer streamflow and the magnitude of seasonal variation in streamflow significantly affect trout biomass in Wyoming streams (Binns and Eiserman 1979). Streamflow also affects fish populations indirectly because it affects the other four classes of habitat factors. For example, water temperatures will generally change more rapidly and become more extreme as streamflows decline.

Streamflow regime refers to the permanence and seasonal patterns of stream-flow. Some streams have stable flow due to ground or glacial water sources, whereas others fluctuate erratically or are intermittent because they are fed primarily by overland runoff. Although stable streamflow is advantageous to some kinds of fish, it is not always necessary or even desirable for all kinds of fish. High flows form channels by scouring pools and sorting and cleansing streambed materials. High flows also serve as cues for timing of migration and spawning. Low flows often correspond to time of recruitment and growth of young fish and, therefore, extreme low flows may limit production of young fish. Alteration of a stream's natural flow regime may be undesirable if it adversely modifies normal seasonal high and low flows to which the biota is adapted.

9.2.3 Water Quality

Factors that determine water quality, such as temperature, pH, turbidity, dissolved oxygen, alkalinity, dissolved nutrients, and presence or absence of anthropogenic toxicants, can directly affect stream fish or can indirectly affect them through effects on food production. Protection of water quality is encouraged by maintaining regional water quality standards for point-source dischargers (Alabaster and Lloyd 1984) and by regulating nonpoint sources of pollutants.

9.2.4 Energy Source

Energy to support organisms in streams is divided into two sources: allochthonous refers to energy sources produced outside of the aquatic system; autochthonous refers to those produced within it. The composition of the benthic invertebrate and fish communities reflects the dominant energy sources available. The amount of energy available from allochthonous and autochthonous sources can be directly altered by streamflow, water quality, and changes in the riparian vegetation. Physical habitat structure can modify the efficiency with which energy sources are consumed and incorporated into animal tissue.

9.2.5 Physical Habitat Structure

Physical habitat structure refers to inchannel and immediately adjacent riparian characteristics of bed materials, water depth, current velocity, bank slope, and cover that determine the amount of suitable living space. Physical habitat requirements vary among species and life history stages. Many stream fishes spawn over or in stoney material or appropriately sized rock (Shirvell and Dungey 1983). Fry and juvenile fish may occupy special rearing or nursery habitat affording protection from predators. Nursery habitats include shallow areas protected by rocks, live vegetation, or woody debris. Adult fish select microhabitats close to foraging locations and escape cover. During winter, when they face extremely low water temperatures (and may be lethargic), some fishes seek deep crevices in streambeds as protection from current. A diversity of habitats is often correlated with a diversity of fish species. Also, in streams in which water quantity or quality are not limiting, cover is often positively correlated with fish density or biomass (Fausch et al. 1988), and management techniques that increase the amount of cover in a stream can substantially increase fish abundance.

9.2.6 Biotic Interactions

Biological interactions, particularly predation and competition, can affect a fish's habitat selection and use as well as its survival in a given habitat. The abundance of fish that can be sustained in a given habitat may depend on the outcome of competition and predation. Among sites of comparable habitat quality, for example, those with piscivorous fishes may have smaller biomasses of nonpiscivorous fishes than sites lacking piscivores (Bowlby and Roff 1986). Therefore, species introductions (Chapter 12) can result in changes in abundance of fishes even if habitat quality or exploitation does not change.

9.3 EXAMINING AND ANALYZING STREAM HABITAT

A stream habitat project is conducted as a sequence of steps, a process typically lasting several years and requiring an interdisciplinary team to help the fisheries biologist. Stream hydrologists and morphologists are usually essential (cf. Gordon et al. 1992). Services from such fields as geology, engineering, entomology, plant ecology, and livestock grazing are often needed. Some government agencies have the requisite professionals on staff and have well-established procedures for following the appropriate steps. In other situations, members of the management team may have to be contracted.

At minimum, the following 12 steps should be included. The first seven form a planning phase. Manipulating habitat without proper planning is a common mistake. In complex projects it can be impractical to plan the whole project at one time. It may be better to refine project methods with the actual experience gained from pilot studies or from executing a small part of the project initially. Some projects are repeatedly cycled through such adaptive refinement. At all stages, all pertinent parties should be given progress reports and their approval should be obtained at each stage.

1. Broad objective-setting. Determine the overall objective. For many streams, the objective will be to maximize resident game fish populations. For others, such as in anadromous or adfluvial fisheries, it may be to supply more young to an ocean, lake, or river for commercial or sportfishing. For still others, it may be to preserve or restore endangered fishes. Specific objectives (White, 1991a) are set later.

2. Examination. Measure and analyze the stream's biological, physical, and chemical characteristics. It is fundamental to analyze the stream's hydrology and morphology—sources and flow regimes of water and sediment, gross channel form and mechanics, and underlying processes of the drainage basin. The objective is to determine if basin processes—erosion, sedimentation, water transport, and channel development among them—are in dynamic equilibrium and, if they are not, to estimate how and when equilibrium might be reached (Kondolf and Sale 1985; Heede and Rinne 1990). Fish habitat is inventoried by various procedures (Wesche 1980; Binns 1982; Oswood and Barber 1982; Platts et al. 1987). Inspecting the drainage basin's vegetation, soils, geology, and human uses, and analyzing water chemistry data can help managers assess productive potential and pollution hazards. Stream temper-

ature should be surveyed during extreme hot and cold weather, and the progression of ice conditions—a major influence on fish and habitat—should be observed in winter and during spring break up. Fish population inventory can reveal habitat deficiencies and perhaps rule out the need for certain other data. For example, an abundance of fast-growing young in a population that has few trophy-size trout, despite being lightly fished, may indicate little likelihood of problems with water quality, food supply, or reproductive habitat, and further investigation might concentrate on structural habitat. It is always advisable to survey aquatic and riparian plants, beaver workings, and major predatory wildlife. In special cases, it can be useful to analyze the benthic fauna.

3. Diagnosis. Identify the stream's fish production potential and habitat problems, based on information from examining the resource. Some of the problems revealed (e.g., fish population deficiencies due to overfishing) may point to needs for management other than—or integrated with—habitat work. Resource examination and diagnosis is often a repeated cyclic process done in several steps appropriate to local conditions (Figure 9.2).

4. Decision on appropriateness. If no significant habitat problem exists, or if the stream potential does not coincide with the broad objective, cancel the project. If potentials reasonably match objectives and there is a significant habitat problem to address, proceed to step 5.

5. Specific objective-setting. Write definite, quantified results to be obtained by certain times (White 1991a). Typically, a hierarchy of increasingly specific objectives is stated. At one level, the objective might be to increase the population of age-1 cutthroat trout by 80% within 5 years; appropriate sub-objectives might include increasing spawning and juvenile-rearing habitat by 80% within 2 years. Even more detailed objectives would state the amounts of riffle (containing specific sizes of gravel) to be created for spawning and the numbers of pools and cover installations of various types to be created for rearing.

6. Design. Methods of habitat treatment are selected and developed to meet the specific objectives. The methods are based on information from the examination and diagnosis steps (often with further mapping and detailed elevational surveying), and on the management team's knowledge about stream mechanics and the ecology of the species involved. The locations of riparian and inchannel treatments are designated on stream maps, detailed drawings of typical structures are made, and specifications and procedures are written.

7. Estimation of materials, equipment, labor, and costs. Based on the design and such factors as travel distances and physical accessibility of the project area, lists of the kinds and amounts of materials, equipment, and labor are drawn up. Prices are obtained from a variety of sources and a cost estimate is compiled. This step may be accompanied by efforts to secure additional funds.

8. Decision on feasibility. If funding is not sufficient to cover costs, alter or cancel the project. If funding is adequate, proceed to step 9.

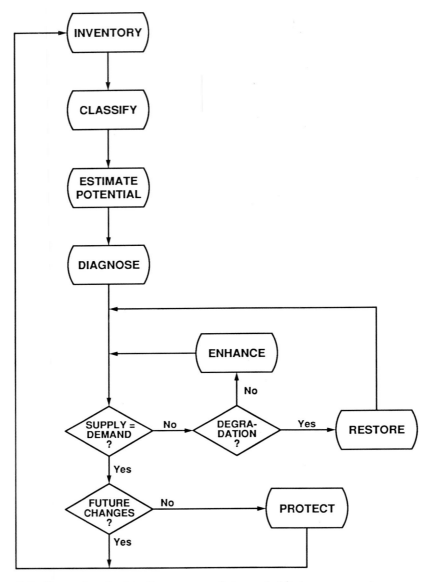

Figure 9.2 Stages involved in the process of stream habitat management.

9. Permitting. Submit the project plan for requisite approval of government agencies. If the needed permits are denied, revise plans or cancel the project. If the permits are granted, proceed to step 10.

10. Organization. Arrange for materials, equipment, and labor. Draw up a work schedule. Train personnel.

11. Management. Conduct the restoration or enhancement. Monitor funding. For quality control and correct interpretation of project design, it is especially important that members of the design team help direct the habitat treatments.

12. Evaluation of results. Periodically remeasure physical attributes of the stream and reinventory fish populations to see if results meet the stated objectives and to refine methods for better results in the future. Because most physical responses to stream management techniques develop for months or years and fish population responses lag behind these, biological evaluation must span five or more years (Hunt 1976).

Special conditions of the stream, species involved, management objectives, and sociopolitical setting may require other steps. Negotiation for trespass to the stream or surrounding property may be needed. Maintenance of the restorations or enhancements involved in many kinds of habitat projects is an ongoing process that must be considered in planning and budgeting.

9.4 HABITAT PROTECTION: PREVENTING AND REMEDYING ADVERSE HABITAT MODIFICATIONS

9.4.1 Channelization

Channelization creates unfavorable stream habitat; the resulting uniform channels lack the pools, riffles, and boulders or log jams that are essential for sustaining fish abundance. Channelization typically involves some combination of straightening, widening, or deepening. In addition, streambank vegetation and instream obstructions (e.g., snags) are often removed to increase channel capacity. Artificial reinforcement of streambanks is often done because stream power has been increased and stabilizing bank vegetation has been removed. Channelization is done to

1. drain wetlands and thereby increase the amount of usable land for agriculture,
2. reduce flooding in localized areas by increasing channel capacity,
3. relocate channels in highway and other construction projects, and
4. permit navigation.

Stream straightening results in a loss of important fish habitat features associated with natural meandering and pool–riffle patterns, in addition to decreasing stream length and increasing stream slope and power. As a consequence, habitat diversity is reduced, and the altered channel has higher and more uniform velocity. Abundance of sport fish is often 8 to 10 times greater in natural channels than in channelized parts of the same stream.

Although altered stream channels eventually recover through natural fluvial and biotic processes, recovery may require decades, perhaps centuries. Artificial riffles and pools, spaced 5–7 channel widths apart along the down-valley line (as in natural streams; Leopold et al. 1964), have been successful in restoring a more natural fish and invertebrate fauna than typically exists in channelized streams without such mitigation (Edwards et al. 1984).

The following design principles can be used to avoid adverse impacts of channel modifications (Keller 1978; Nunnally 1978).

1. Where possible, avoid straightening and steepening channels.
2. Preserve and promote bank stability by leaving trees and shrubs in place, minimizing channel alteration, restoring vegetation in disturbed areas, and judiciously placing riprap or tree revetment.
3. Emulate nature in designing channel form by creating appropriately spaced pools, riffles, and meanders.
4. Conduct snagging and debris clearing by hand rather than with heavy equipment.

Instead of making severe channel modifications, it will be advantageous to undertake alternatives such as building levees far from the channel, constructing floodways, removing buildings and other structures from the floodplain, zoning to prohibit building in the flood-prone areas, and converting land uses in the floodplain to those that flooding will not harm. Many hard lessons have been learned from our repeated attempts to channelize streams rather than understand their dynamic nature; therefore, many costly restoration projects are needed (cf. Brookes 1988).

9.4.2   Removal of Woody Debris

Woody debris accumulates naturally in streams and plays important roles in stream mechanics and fish habitat; its clearance has often reduced carrying capacity for fish. Woody material creates pools, increases structural complexity, provides fish cover, forms substrate for invertebrates, traps gravel for spawning and invertebrate production, holds other organic matter, and increases channel stability. Logs and smaller woody matter enter channels by windthrow, the toppling effects of snow and ice, bank erosion, avalanches, and beaver activity. Riparian logging may increase woody debris in streams immediately, but reduces long-term input. Recent studies (Chamberlin et al. 1991; Hicks et al. 1991) indicate log jams and other woody debris accumulations have commonly been removed to restore channel capacity and reduce impoundment, eliminate barriers to fish migration and human navigation, and return streams to pre-logging condition.

9.4.3   Barriers to Fish Migrations

Barriers to fish migrations have had detrimental effects on valuable anadromous and adfluvial fish stocks (Chapter 23). Constructed dams are common barriers. Fishways are often required to allow fish passage to upstream spawning grounds. Downstream passage of fish through turbines, spillways, and conduits may cause excessive mortality and guidance to downstream passage structures may be required. Roadway culverts often block or hinder fish passage, unless installed to meet special criteria for the fish involved.

9.5   HABITAT RESTORATION: REMEDYING STREAM ABUSES

9.5.1   Streambank Stabilization

Economical and effective techniques for stabilizing eroding streambanks often include revegetation, installation of revetments made of trees and brush, and

creation of riprap. In designing streambank protection consider the basic erosional processes and conditions in streams (Henderson 1986) as well as the needs
of fish. Riprap is often used along the lower parts of banks. Large, dense, angular
rock that is resistant to freeze-shattering provides the best habitat for fish. The
rock should be loosely laid to create refugia and turbulence. The minimum size
rock to be used depends on the density of the rock and the stream power it must
withstand. Tree and brush revetments can be installed along with riprap, to
provide even more cover for fish. Trees should be anchored parallel to banks and
overlapped one-third to one-half (Binns 1986). Grasses, shrubs, and trees can be
planted on the upper sides and tops of the banks.

9.5.2 Fishway and Culvert Design

To design structures that will not impede migrating fish, biologists must
understand the behavior and swimming capacity of fish and work with hydraulic
engineers who know how to apply principles of fluid mechanics. Swimming
speeds of fish vary according to species, body size, and environmental factors,
especially temperature (Beamish 1978). The critical swimming speeds and leaping
abilities of the species and the sizes of fish in a particular stream must be known
in order to design effective fishways, culverts, and guiding devices.

The basic principles of design are that fishways should be readily passable by all
migratory species in the stream; operate at all water levels in the forebay and
tailrace of the barrier; be navigable by fish without injury or undue stress; and
have entrances that fish can find, enter, and pass without delay. Fishways can be
categorized based on whether the design incorporates (1) a series of steps
(ladders), (2) baffles to reduce velocities (chutes or vertical-slot fishways), or (3)
lift systems. The earliest fishway designs had a series of steps created by
alternating weirs and pools (Orsborn 1987), hence the name fish ladder (Figure
9.3). Weir-and-pool fishways were modified with submerged orifices or vertical
slots to permit passage at a wider range of flow through the fishway. A chute-type
fishway consists of a narrow channel with many closely spaced baffles to reduce
water velocity. Designs that mechanically lift fish over barriers include various
elevators and locks, the effectiveness of which may be limited by intermittent
operation and need for frequent maintenance. Designs of the step-and-chute
fishways must accommodate swimming capabilities of fish. For weir-and-pool
type fishways, a maximum drop of 0.3 m is often recommended to permit rapid
and easy migration, but this criterion is based on gross underestimates of leaping
abilities of some species. New designs that attempt to more closely match fishway
design with leaping and swimming capabilities will likely lead to more efficient and
cost-effective fishways (Orsborn 1987).

The most common problems associated with culverts are (1) excessive drop at
the downstream end, (2) velocities exceeding critical swimming speeds, (3)
shallow flow, (4) lack of resting pools at inlets and outlets, and (5) upstream
blockage due to debris. Each of these can be avoided by proper design.

A culvert should be designed so the mean velocity of water passing through it
is significantly less than the critical prolonged swimming speed for the species and
size of fish involved. The critical prolonged swimming speed is the maximum
velocity a fish can maintain for a given time. Designs should permit passage by fish
in 5 minutes or less (Bovee 1982). Time required for passage is obtained by

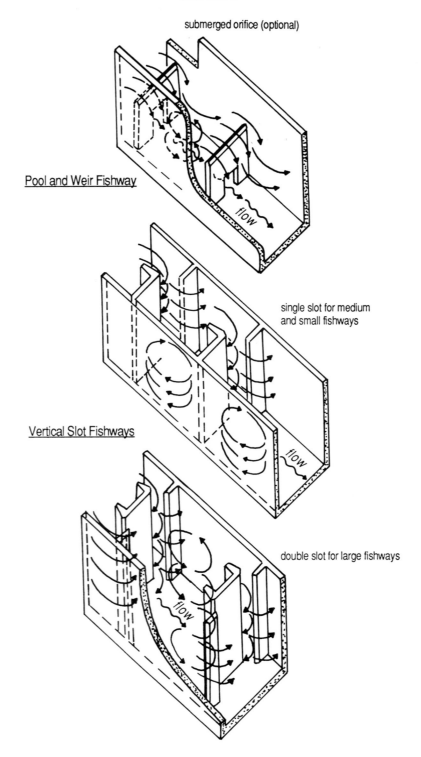

submerged orifice (optional)

Pool and Weir Fishway

flow

single slot for medium
and small fishways

Vertical Slot Fishways

flow

double slot for large fishways

flow

Figure 9.3 Diagram of common fishway types.

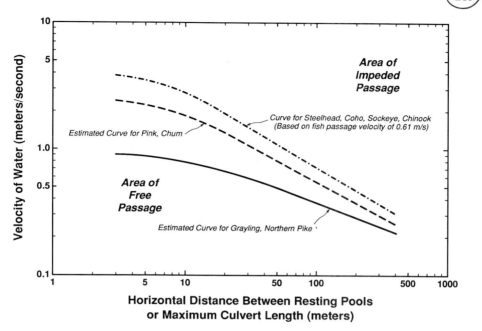

Figure 9.4 Maximum swimming capabilities of several species as they relate to velocity and culvert length. These figures can be used to determine suitable velocities, culvert lengths, and distances between resting pools for fishways and culverts. (Modified from Slaney et al. 1980.)

dividing the length of culvert by net velocity of the fish (swimming speed minus mean current velocity in culvert). Therefore, longer culverts should be designed with lower average velocities (perhaps including resting areas such as baffles) to avoid fatiguing fish (Figure 9.4). In existing culverts, where velocity or shallowness limits passage, baffles or other devices that impart roughness can be installed to create more suitable water velocities and depths, if capacity of the culvert is sufficient. Resting areas should be available at the upstream and downstream ends of the culverts.

9.5.3 Screens and Guiding Devices

Alterations of stream habitat caused by dams and water withdrawals require screens or guiding devices to prevent fish from entering certain areas, such as cooling water intakes, turbines, or irrigation systems. Many designs are available depending on the application.

9.6 STREAMFLOW MANAGEMENT

Streamflow discharge in many streams is artificially regulated by dams and diversions. Consequently, flow regimens may be adversely disrupted, but sometimes the type of regulation allows for manipulation of flow to benefit fisheries. In arid and semiarid regions, water withdrawals for out-of-stream uses have often left large segments of stream channels partially or completely dewatered. In these regions, management agencies and concerned public groups have worked to reserve minimum instream flows for fish habitat. In water-rich regions, dewatering of stream channels is less common, but other aspects of flow regulation have significant effects on stream habitat and biota. To avoid or reduce adverse effects, fisheries managers should understand effects of various flow modifications and be able to use a variety of analyses in determining impacts and recommending suitable flow regimens.

Responses of fish and invertebrates to flow regulation vary (Gore and Petts 1989). In a review of case histories of regulated streams in the Pacific Northwest, Burt and Mundie (1986) found that salmonid populations declined in 76% of cases and increased or showed no change in 24% of cases following flow regulation. Most of the positive or benign effects occurred where flows had been little changed, had been increased, or had been decreased in critical months. Reduced stocks were commonly associated with flow decreases that reduced habitat. Other reasons for reduced stocks included barriers, sedimentation, fluctuating flows, impoundment, and altered temperature or water quality. The effects of flow modification on fish stocks are complex and difficult to predict.

Modifying a streamflow regimen often requires a permit or environmental impact statement to comply with state, provincial, or federal statutes. Site-specific evaluations should be conducted, where possible. The hierarchy developed by Petts (1984) serves to organize the many possible impacts according to time scales (Figure 9.5). First-order impacts are those that occur shortly after dam construction, such as migration barriers and altered flow, sediment load, water quality, and plankton. Second-order impacts occur later, typically as a result of first-order impacts. For example, major reduction in sediment loads and peak flows can result in gradual change in channel shape and bed materials. Third-order impacts are the cumulative effects on fish and invertebrates stemming from first- or second-order impacts.

A major first-order impact of flood control impoundments is that maximum seasonal flow is so reduced that significant channel changes occur. In addition, downstream water quality and temperature patterns can change, depending on reservoir stratification, outlet depth, and flow releases. Releases of hypolimnetic water are often cold and anoxic and have elevated nutrient, iron, manganese, and hydrogen sulphide concentrations. The downstream result is often conversion from warmwater to coldwater fishes and to a benthos composed of only a few tolerant taxa.

Water diversions without large impoundments usually significantly modify flow during the low-flow season only, and impacts tend to be most severe during drought years. Where low flows are made more extreme, the reduced living space, reduced cover availability, and elevated temperature can significantly reduce fish populations.

Although "run-of-river" (low-head) hydropower facilities may have minimal effects on flows, operations causing large daily fluctuations in flow can have dramatic effects. In hydroelectric "peaking" operations, water is stored in a reservoir, often with no minimum release, and run through turbines to satisfy peak demands, typically with sudden starts and stops. Rapidly fluctuating flow changes habitat faster than some fish and invertebrates can endure. Periodic dewatering of the stream margins reduces insect abundance and diversity. Stranding of stream organisms may occur, depending on stream channel shape and the rate of change

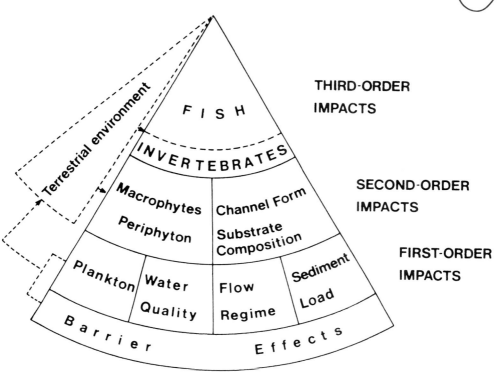

Figure 9.5 Hierarchical framework for examining the impacts of river impoundment. (Modified from Petts 1984.)

in discharge or depth. Fishes adapted to the shallow, slower-current areas along stream margins are less abundant in sites subjected to daily flow fluctuations (Bain et al. 1988).

Many methods exist to determine acceptable flow regimens. These analytic methods, reviewed by Trihey and Stalnaker (1985), Morhardt and Altouney (1985), and Estes and Orsborn (1986), can be categorized as (1) discharge methods, (2) hydraulic rating, (3) habitat rating, and (4) biological response (ordered according to increasing resolution, data needs, and costs). The appropriate type of analysis depends on the stage of planning and amount of controversy involved. Presently, no methods are adequate for predicting biological responses. The first three methods are described in the following paragraphs.

Discharge methods require no field work and rely on streamflow statistics or drainage basin variables. The Tennant (1976) method is most commonly used. Because most streams show a similar pattern in the relation between depth, velocity, top width, and discharge, expressed as a percentage of the mean annual flow (Figure 9.6), the Tennant method prescribes a fixed percentage of the unregulated mean annual flow. For the streams Tennant (1976) used, flows below 10% of the mean annual flow result in degraded habitat conditions due to dramatic reductions in depth, velocity, and width (Figure 9.6) and are considered minimum acceptable flows; flows of 30 and 60% result in more satisfactory conditions for fish, wildlife, and related recreational values.

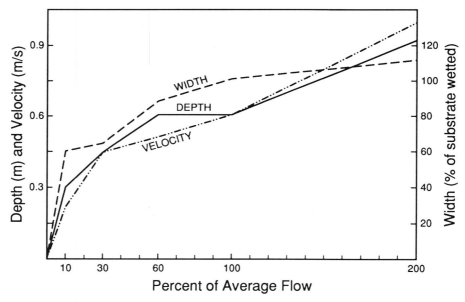

Figure 9.6 Relations between average width, depth, velocity, and percent of flow from ten tests using the Montana Method (Tennant 1976).

Hydraulic rating methods require field measurements at different flows in order to relate hydraulic geometry variables to discharge. The most common hydraulic rating method uses the relation between wetted perimeter and discharge to recommend flows to preserve some habitat. Wetted perimeter, the length of wetted streambed along a transect perpendicular to the direction of flow, is related to wetted surface area of a streambed. The rating of habitats by hydraulic methods cannot be determined without additional measurements, such as current velocity, bed materials, or cover.

Habitat rating methods relate quantity and quality of habitat to streamflow. These methods usually address the specific requirements of selected fish species. The instream flow incremental methodology (IFIM), the most common habitat rating method, involves a variety of models and information sources to describe a time series of usable habitat for various life stages of the fish under natural conditions and proposed project conditions (Bovee 1982). Figure 9.7 describes the major components involved in determining total usable habitat. Microhabitat suitability criteria, channel structure data, and hydraulic simulations are used to calculate amount of usable microhabitat. Weighted usable area is the measure of microhabitat area and is calculated by weighting the surface area relative to optimum habitat. The physical habitat simulation system consists of computer software to determine relations between weighted usable area and streamflow, which can be used to negotiate acceptable flows (Cavendish and Duncan 1986). Other components of IFIM consist of water quality and temperature models and species suitability criteria for these factors.

Parts of IFIM have been widely applied in the western United States. However, controversy exists over its general applicability. Stream habitat managers should be cautious in developing streamflow regimes to protect stream fisheries because

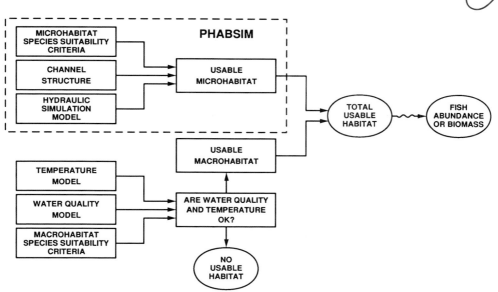

Figure 9.7 Components of the instream flow incremental methodology. (Modified from Shirvell 1986.)

our ability to predict biological responses to streamflow change is not well developed.

9.7 STREAM HABITAT IMPROVEMENT

Various technical guides to stream habitat improvement—primarily for salmonids—have appeared in recent decades (White and Brynildson 1967; Slaney et al. 1980; Hall and Baker 1982; Reeves and Roelofs 1982; Wesche 1985; Hunter 1991; Hunt 1993). In streams containing nonmigratory trout, most habitat work centers on cover and otherwise creating proper spaces for survival and growth of large fish that anglers seek (Hunt [1988] reviewed 45 evaluations of such work in Wisconsin); reproductive and juvenile rearing habitat often are not major objectives because they are not in short supply or are achieved concomitantly. For migratory fish that use streams primarily as reproductive areas, the principle objective is to create conditions for producing more out-migrating young so work is commonly focused on enhancing spawning grounds and rearing habitat for juveniles.

9.7.1 Cautions in Habitat Enhancement and Restoration

It is paramount to have sound fish- and fishing-oriented purposes for habitat enhancement and to let the design of projects follow from these. A common mistake is to regard stream habitat improvement as a collection of techniques, a matter of simply finding ideas in a how-to book and installing structures. Cookbook habitat work characterized by infatuation with techniques fails again and again to achieve proper effects. How to build a habitat structure is important, but not as important as why—and, deriving from the why, selecting the right structure for a valid ecological purpose and placing it properly in relation to channel form.

The trend in stream habitat management is toward considering the whole ecosystem of the stream and its drainage basin, and operating on a landscape scale, not just in the stream itself. Managers should avoid radically altering streams in ways that, while enhancing habitat for the target organism(s), destroy it for other desirable organisms. Fish and wildlife managers have been criticized for narrowly focusing on habitat for game and food species and disregarding broader ecosystem concerns such as habitat diversity and species diversity. The danger of this may be reduced if stream habitat enhancement simulates natural features of channel form and flow.

In improving habitat for sport fishes, the manager should avoid structures that impart an unnatural appearance. Most anglers value naturalness of the setting in which they conduct their sport (Chapter 7). The physical aspects valued by anglers who have long fished the project stream should be preserved if possible. In structural work, even the use of artificial materials (e.g., steel stakes, pins and wire, plastic fabrics and meshes) for concealed components should be held to a minimum. Despite maintenance, everything built eventually disintegrates. When structures made of natural materials such as logs and "native" rock break up due to rot or extreme high water, the result will be debris that looks natural, not a clutter of metal and plastic objects.

The effective life of many types of habitat work will be greatly extended if periodic postconstruction maintenance is done. Just about anything that is built (a house, highway, or motor vehicle) must be inspected and maintained from time to time if it is to last. Maintenance is a frequently neglected aspect of habitat projects. Sometimes it is not thought of in the overall project plan. Those who carry out a project tend to become so tired of it by the time it is done that they drop the matter completely to turn to other things. Also, maintenance is a low-visibility function that is easy to drop from budgets during financial crunches.

A widespread tendency to overstabilize channels stems from undiscriminating fear of channel erosion. Moderate channel erosion is natural and creates fish habitat. People often wish to have streams locked into unchanging courses to facilitate waterway navigation; to protect roads, buildings, cropland and other property in flood plains; or just to keep things as they are. The result has been an emphasis on "hard stabilization" in channel engineering, and this tends to carry over into fish habitat work, sometimes overwhelming the fish habitat aspects.

Fish and other organisms may benefit by gradual shifts of stream course, but these may destroy installed habitat structures. A solution to this dilemma may often lie in relying less on "permanent" structures, and placing more emphasis on streamside vegetation and its "flexible stabilization." The thatch and binding of streamside vegetation can, through flexibility and regrowth, change interactively as erosion and deposition take place in the channel. Use of living plants in waterway engineering has been developed in Europe, perhaps most elaborately in the German "Lebendbau" (live construction) methods (Kirwald 1964).

Streams exist in a state of dynamic equilibrium (Leopold et al. 1964). Habitat managers should be cautious about unfavorably reducing the natural dynamism of streams. Toward this end, it will be helpful to let floodplains serve their natural function—as areas to be flooded.

9.7.2 Direct Improvement of Channel Form and Current Pattern

Stream-dwelling fishes are adapted to the channel forms that occur most commonly in nature. Therefore, simulating those forms is likely a sound principle to follow when improving existing channels or when relocating channels. Channel patterns are broadly classified as (1) steep, erosive channels, in which sediments erode away faster than they are replaced, and in which water follows an irregular mix of rather straight and jagged courses over bedrock or over and through boulders; (2) meandering channels of moderate to low slope with pronounced pools at the outsides of bends and shallower riffles or silt bars between bends (these have sediment entering at about the same rate as it is carried away downstream and are for this and other reasons often considered the most stable of the forms); and (3) braided channels, choked with sediment that is being fed in faster than it can be washed away. Other, less common patterns exist, all or most of which may be intergrades or combinations of the three main ones. The meandering type is probably the most common. We infer from Leopold and Langbein (1966) and Hasfurther (1985) that meandering tends to be well developed in alluvial streams of less than about 0.5% slope, but that it can occur also in other situations.

Two major large-scale aspects of channel structure prevail: the meandering form itself and, in steeper streams, the stairstepped form. Features typically associated with meandering and stepping provide essential elements of habitat for fish. The remarkably regular, graceful winding pattern of meanders is the tendency of streams in traversing gentle slopes in various media, notably in soils (including gravels and cobble) that are easily eroded and transported but cohesive enough to form firm banks (Leopold and Langbein 1966). Therefore, the features of meandering predominate in many sections of most alluvial streams.

Intergrades and combinations of meandering and stepping may exist in the same stream. Natural stairstepping is caused by log jams, accumulations of other woody debris, and beaver dams, as well as by bedrock ledges and groups of boulders. Some of these obstacles also occur in low-gradient meandering streams and may tend to obscure the meander pattern and interrupt its regularity but do not destroy the tendency; we can think of them as resetting the meander cycle.

Within-channel forms are broadly classified as pools, riffles, glides (or runs), and falls. Bisson et al. (1982) identified types of riffles and pools in small, steep streams of western Washington as:

1. low-gradient riffles, having slope less than 4% and current of 20–50 cm/s;
2. rapids, having rather even slope greater than 4% over large boulders and current >50 cm/s;
3. cascades, also steeper than 4% but with uneven slope consisting of series of small steps and pocket pools.

In streams of lower gradient (less than 0.5%), other types of relatively shallow, sloping beds called bars are composed of fine gravel, sand, silt, and mixtures of these.

The six pool types recognized by Bisson et al. were:

1. lateral-scour pools, by far the most prevalent type, occurring where for various reasons flow veers against a channel bank;

2. trench pools, longitudinal grooves having bedrock sides;
3. plunge pools, formed by water dropping vertically from a complete or nearly complete channel obstruction;
4. dammed pools, upstream from channel obstructions;
5. backwater pools, formed along channel margins by major highwater eddies behind or beside large partial obstructions such as root wads or boulders; and
6. secondary channel pools, which become isolated or dry up during normal low flow.

The habitat features of meandering channels are especially important because the meandering form is so prevalent. Pool-and-riffle undulation of the streambed results from the same process that causes meandering and is often a feature to enhance. Long lateral-scour pools develop along the current-bearing banks of meander bends, and, as described above, offer advantageous habitat, particularly for large fish. Riffles, which provide spawning and nursery habitat for many fish, tend to form in the crossover areas between bends. The distance from pool to pool or riffle to riffle along the down-valley line (not along the channel unless it is straight) is the same as the bend-to-bend spacing of the meander system—about five to seven times channel width. This interval is a guideline for spacing of habitat improvement structures.

The lateral-scour pools of meander bends can be enhanced by using vegetation (often restoring it) to increase cohesiveness of the current-bearing banks. This leads to decreased lateral erosion of the channel and greater downcutting and undercutting. Curved artificial reinforcements of banks, such as certain forms of riprap (rock), log, or whole-tree revetment can also be effective in achieving this (Figures 9.8 and 9.9)—often in combination with promoting healthier bank vegetation. Properly using current deflectors in conjunction with the bank revetments (Figure 9.10) further develops the lateral-scour effect. This accentuates the meander form by making bends slightly tighter, i.e., shortens the radius of meander curvature. Another way to guide current into creating or enhancing lateral scour is, as described shortly below, by installing low oblique sills. To fit in with the meandering form, these should be placed as shown in Figure 9.11.

In steep streams, especially the smaller ones, the steps formed by log jams, beaver dams, and accumulations of large rock are important in stabilizing the channel, creating areas of slow current, trapping sediments such as gravel, and causing plunge pools and lateral-scour pools. Particularly in old-growth forest streams, such elements may be abundant. In contrast to the rather regular spacing of habitat features associated with meandering channels, the spacing of stairsteps and associated features tend to be irregular. Steep streams may sometimes meander, but they are often highly irregular in course pattern, having relatively straight reaches interrupted by sharp bends. If the obstacles that create steps in steep streams are removed, the relatively straight parts of the channel develop very swift current, which, during high water, erodes the channel forcefully until large rough elements are exposed or until the streambed downcuts to low gradient. Improving fish habitat in steep channels will largely involve enhancing or restoring the stepped form. This can be done by installing sills (check dams, drop structures, or overpours) made of boulders or logs. Better effects for fish habitat

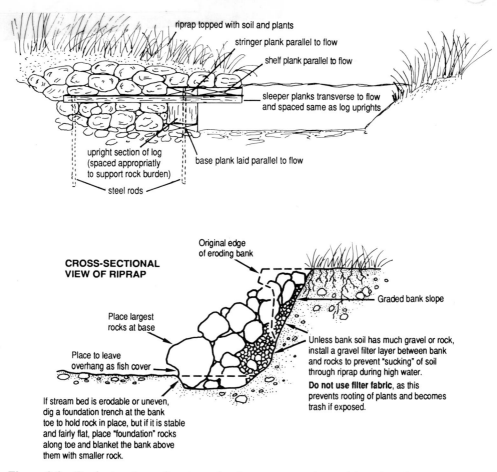

riprap topped with soil and plants

stringer plank parallel to flow

shelf plank parallel to flow

sleeper planks transverse to flow
and spaced same as log uprights

upright section of log
(spaced appropriatly
to support rock burden)

base plank laid parallel to flow

steel rods

**CROSS-SECTIONAL
VIEW OF RIPRAP**

Original edge
of eroding bank

Graded bank slope

Place largest
rocks at base

Place to leave
overhang as fish cover

Unless bank soil has much gravel or rock,
install a gravel filter layer between bank
and rocks to prevent "sucking" of soil
through riprap during high water.
Do not use filter fabric, as this
prevents rooting of plants and becomes
trash if exposed.

If stream bed is erodable or uneven,
dig a foundation trench at the bank
toe to hold rock in place, but if it is stable
and fairly flat, place "foundation" rocks
along toe and blanket the bank above
them with smaller rock.

Figure 9.8 Rock structures for streambank revetment: riprap (above) and a cantilevered ledge to provide for fish along current-bearing banks.

are often more possible with logs than with boulders. Promoting colonization by dam-building beavers can also create useful stepping (but in low-gradient streams can be disadvantageous).

A suitable holding area for a stream fish is a space having proper water depth, volume, and currents for the activities of a fish of a particular size and species, as well as having ample cover. Particularly in shallow streams, cover for hiding and security is crucial. It consists of objects or channel formations that offer fish concealment from predators or visual isolation from competitors. General categories of security cover are (1) overhead bank cover (overhead referring to the fish's head, not ours), such as undercuts and objects associated with the streambank that are submerged coverts, under which fish can hide; (2) water depth, in small streams especially that afforded by deep pools; (3) midstream objects, such as boulders and aquatic plants; and (4) hydraulic features, such as areas of broken or turbulent water surface and masses of bubbles or foam. Cover that is part of, or closely associated with, streambanks is usually the most

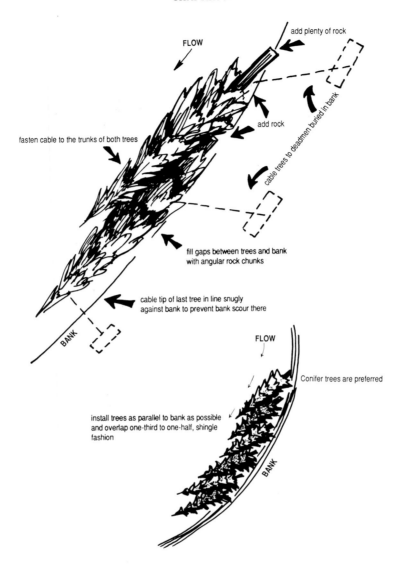

Figure 9.9 Using trees for streambank revetment.

prevalent kind of cover in natural streams. Deep water is discussed above under the category of pools. Pools are made much more beneficial for fish if they adjoin bank cover or have midstream cover objects within them. Various methods for making cover also create spaces with more favorable shape and current.

Generally, the most effective way to enhance cover is to create physical niches in, or objects jutting out from, the streambank, under which fish can position themselves when fleeing from attack or when resting. Trees, bushes, and bank ledges that overhang the water surface cast shadows in which trout lie during sunlit times (Fausch and White 1981) and such objects may, to varying degrees, directly obstruct the view of aerial predators. The closer the overhanging object is to the streambed, the darker and more protective its shadow will be. For

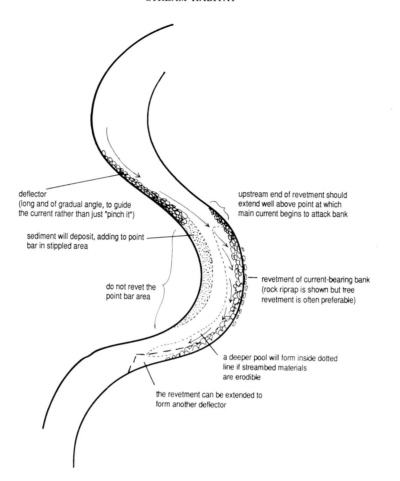

deflector
(long and of gradual angle, to guide
the current rather than just "pinch it")

upstream end of revetment should
extend well above point at which
main current begins to attack bank

sediment will deposit, adding to point
bar in stippled area

do not revet the
point bar area

revetment of current-bearing bank
(rock riprap is shown but tree
revetment is often preferable)

a deeper pool will form inside dotted
line if streambed materials
are erodible

the revetment can be extended to
form another deflector

Figure 9.10   Current deflector in conjunction with revetment.

daytime hiding, trout prefer submerged overhead covers, particularly those close
to the bed. Abundance of trout, particularly large trout, is often closely correlated
with the amount of overhead bank cover in streams.

Installing a rock revetment, also called a riprap or riprap blanket (Figure 9.8),
is the most common method for artificially protecting streambanks and, if
properly placed along current-bearing banks (e.g., concave banks of meander
bends) can benefit fish. Riprapping inner (convex or point-bar) banks of meanders
may waste money and harm the riparian ecosystem. Guidelines for installing
riprap in ways that benefit fish are found in Binns (1986).

Tree revetment (Figure 9.9), although less permanent, may be relatively
inexpensive and much more beneficial to fish than riprap. Series of whole trees
can be fastened in place with long steel stakes and cables. When installed as a
dense, overlapping "thatch," such revetment can simulate beneficial aspects of
trees that naturally lodge along current-bearing banks. In some situations, as in
restoring habitat along steep, forested rivers that would naturally have large
masses of haphazard log jams, it may be appropriate to install trees in a jumbled
fashion, with some trunks and root masses jutting out in various directions.

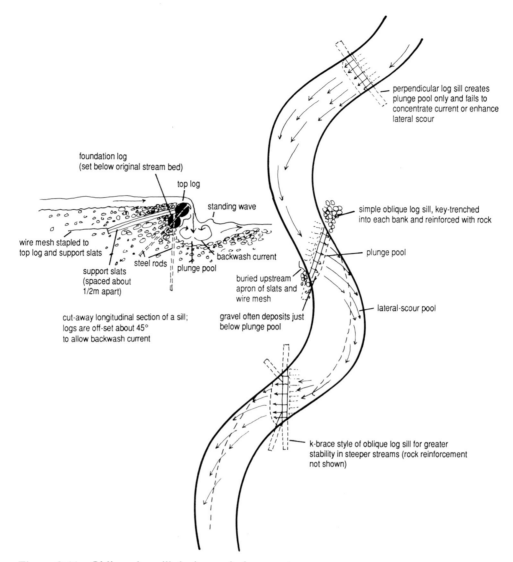

Figure 9.11   Oblique log sill design and placement.

Tree revetment immediately creates cover and—by trapping fine sediment—speeds the development of riparian vegetation, making it a particularly suitable technique for restoring habitat in streams damaged by overgrazing and other destabilizing activities. The damaging activity should, of course, be halted before revetment is undertaken.

Stream pools and glides (runs) can be formed either by slightly raising the water level (damming, but not enough to reduce water velocity to pond-like conditions), by excavating, or by causing current to scour down the bed. Scour is by far the more important process in stream fish habitat improvement, and there are several general methods for achieving it.

The roughening of current-bearing surfaces, such as the leading edges of deflectors or banks lined with riprap, tree revetment, or natural vegetation, can create a "current-pulling" or velocity-concentrating effect that can cause lateral-scour pools to form. Often restoring or enhancing streambank vegetation induces considerable lateral-scour pool formation. A lateral-scour pool can also be formed by installing a sill oblique to channel alignment (White, 1991b), so as to redirect current against one bank (Figure 9.11). The oblique sill can be built very low—a mere linear bump on the streambed—and still have the current-bending effect. Alternatively, a higher oblique drop structure can be built (bed slope permitting), to create a plunge pool, as well as a lateral-scour pool (Figure 9.11). Such structures simulate some of the important effects of large woody debris in streams.

In low-gradient streams (perhaps <0.5% slope), relatively long, gradually protruding wing deflectors can guide the current to flow more strongly against an area of the opposite bank that is downstream from the deflector. There is no standard angle for deflectors to jut from the bank. The receiving bank will, if cohesive and rough, resist lateral erosion, and scouring will occur in the bed along it. If the current-bearing edge of the deflector is also rough and tough, then it will have the current-pulling effect. Thus, a lateral-scour pool (or glide) will form along the structure and downstream from it on the opposite side of the stream (Figure 9.10).

Improving fish reproductive success can also be an objective in habitat management. This may involve creating better quantity or quality of spawning or nursery (juvenile rearing) areas, or both. For fish such as salmonids that spawn in gravel, certain channel modifications can increase the amount of gravel streambed available to them. General approaches are (1) removal of fine streambed sediments that overlie gravel beds, (2) removal of interstitial deposits of fine sediments within gravel beds, and (3) causing increased deposition of gravel at key locations in streambeds. The first and second approaches may involve reducing input of fine sediment from surrounding land, from streambanks, or from upstream, or installing deflectors to scour sand off gravels. In some situations, sediment can be removed by digging streambed pits (e.g., 1.5 m deep and 30–50 m long), then periodically re-excavating the sand and silt that drift into such traps (Hansen et al, 1983).

In temperate and colder climates, winter is generally a time of severe hardship and high mortality for stream-dwelling fish. Not only does the water become unfavorably cold—near 0°C in many streams—but various types of ice, particularly anchor (slush) ice and frazil (needle) ice form, and sudden movements of thick ice slabs can mechanically injure fish or cause other threatening conditions. Other hazards are the extreme winter low flows that occur in some regions and near-0°C snowmelt floods, which are much more dangerous than warmer floods. Stream fish move to special winter habitats that ameliorate the hardships, and it is important in habitat management to provide such refuges. The manager should know the special winter habitat requirements of the target species and its life stages. Various smaller fishes take refuge from predators and water current in the interstices of wood debris or under streambed rocks. Those too large to fit in such spaces may swim to lakes or to the quiet, protective depths of deep pools in larger streams. The deep habitat must be available and the routes to it passable.

9.8 CONCLUSION

Successfully managing habitat for fish (and fishing) in streams requires thorough knowledge of and concern for streams and stream processes, and the kinds of fish and other organisms involved. It requires knowing and caring about the uses and values that people have in mind for the resource, such as angling, commercial fisheries, species protection, and aesthetics. It requires knowing when not to manage—to leave streams as they are. Increased abundance of fish, better fishing and ecosystem health are basic objectives. The modern trend is toward more professional management, toward more attention to design and planning, and toward managing in ways that derive from and are increasingly attuned to natural processes in streams, the processes to which the fishes are adapted. This trend—in contrast to the artificiality and concern for tidiness that characterized some past work—involves increased focus on the drainage basin; on riparian grazing and logging practices; on the roles of streambank vegetation, woody debris, and beaver; and on structural complexity within the channel.

9.9 REFERENCES

Alabaster, J. S., and R. Lloyd, editors. 1984. Water quality criteria for freshwater fish, 2nd edition. Butterworth, London.

Bain, M. B., J. T. Finn, and H. E. Booke. 1988. Streamflow regulation and fish community structure. Ecology 69:382–392.

Beamish, F. W. H. 1978. Swimming capacity. Pages 101–187 in W. S. Hoar and D. J. Randall, editors. Fish physiology, volume 7. Academic Press, New York.

Binns, N. A. 1982. Habitat quality index procedures manual. Wyoming Game and Fish Department, Cheyenne.

Binns, N. A. 1986. Stabilizing eroding stream banks in Wyoming. A guide to controlling bank erosion in streams. Wyoming Game and Fish Department, Cheyenne.

Binns, N. A., and F. M. Eiserman. 1979. Quantification of fluvial trout habitat in Wyoming. Transactions of the American Fisheries Society 108:215–228.

Bisson, P. A., J. L. Nielsen, R. A. Palmason, and L. E. Grove. 1982. A system of naming habitat types in small streams, with examples of habitat utilization by salmonids during low streamflow. Pages 62–73 in N. B. Armantrout, editor. Acquisition and utilization of aquatic habitat inventory information. American Fisheries Society, Western Division, Bethesda, Maryland.

Boon, P. J., P. Calow, and G. E. Petts, editors. 1992. The conservation and management of rivers. Wiley, New York.

Bovee, K. D. 1982. A guide to stream habitat analysis using the instream flow incremental methodology. U.S. Fish and Wildlife Service FWS/OBS-82/26.

Bowlby, J. N., and J. C. Roff. 1986. Trophic structure in southern Ontario streams. Ecology 67:1670–1679.

Brookes, A. 1988. Channelized rivers: perspectives for environmental management. Wiley, New York.

Burt, D. W., and J. H. Mundie. 1986. Case histories of regulated stream flow and its effects on salmonid populations. Canadian Technical Report of Fisheries and Aquatic Sciences 1477.

Cavendish, M. G., and M. I. Duncan. 1986. Use of the instream flow incremental methodology: a tool for negotiation. Environmental Impact Assessment Review 6:347–363.

Chamberlin, T. W., R. D. Harr, and F. H. Everest. 1991. Timber harvesting, silviculture, and watershed processes. American Fisheries Society Special Publication 19:181–205.

Edwards, C. J., B. L. Griswold, R. A. Tubb, E. C. Weber, and L. C. Wood. 1984.

Mitigating effects of artificial riffles and pools on the fauna of a channelized warmwater stream. North American Journal of Fisheries Management 4:194–203.

Estes, C. C., and J. F. Orsborn. 1986. Review and analysis of methods for quantifying instream flow requirements. Water Resources Bulletin 22:389–398.

Fausch, K. D., C. L. Hawkes, and M. G. Parsons. 1988. A review of models that predict standing crop of stream fish from habitat variables: 1950–1985. U.S. Forest Service General Technical Report PNW-GTR-213.

Fausch, K. D., J. Lyons, J. R. Karr, and P. L. Angermeier. 1990. Fish communities as indicators of environmental degradation. American Fisheries Society Symposium 8:123–144.

Fausch, K. D., and R. J. White. 1981. Competition between brook trout (*Salvelinus fontinalis*) and brown trout (*Salmo trutta*) for positions in a Michigan stream. Canadian Journal of Fisheries and Aquatic Sciences 38:1220–1227.

Gordon, N. D., T. A. McMahon, and B. L. Finlayson. 1992. Stream hydrology. An introduction for ecologists. Wiley, New York.

Gore, J. A., and G. E. Petts, editors. 1989. Alternatives in regulated river management. CRC Press, Boca Raton, Florida.

Hall, J. D., and C. O. Baker. 1982. Rehabilitating and enhancing stream habitat: 1. Review and evaluation. U.S. Forest Service General Technical Report PNW-138.

Hansen, E. A., G. R. Alexander, and W. H. Dunn. 1983. Sand sediment in a Michigan trout stream, part I. A technique for removing sand bedload from streams. North American Journal of Fisheries Management 3:355–364.

Hasfurther, V. R. 1985. The use of meander parameters in restoring hydrologic balance to reclaimed stream bed. Pages 21–40 in J. A. Gore, editor. The restoration of rivers and streams: theories and experience. Butterworth, Stoneham, Massachusetts.

Heede, B. H., and J. N. Rinne. 1990. Hydrodynamic and fluvial morphologic processes: implications for fisheries management and research. North American Journal of Fisheries Management 10:249–268.

Henderson, J. E. 1986. Environmental designs for streambank protection projects. Water Resources Bulletin 22:549–558.

Hicks, B. J., J. D. Hall, P. A. Bisson, and J. R. Sedell. 1991. Responses of salmonids to habitat changes. American Fisheries Society Special Publication 19:483–518.

Hunt, R. L. 1976. A long-term evaluation of trout habitat development and its relation to improving management-related research. Transactions of the American Fisheries Society 105:361–364.

Hunt, R. L. 1988. A compendium of 45 trout stream habitat development evaluations in Wisconsin during 1953–1985. Wisconsin Department of Natural Resources, Technical Bulletin 162, Madison.

Hunt, R. L. 1993. Trout stream therapy. The University of Wisconsin Press, Madison.

Hunter, C. J. 1991. Better trout habitat: a guide to stream restoration and management. Island Press, Washington, D.C.

Karr, J. R., K. O. Fausch, P. L. Angermeier, P. R. Yant, and I. J. Schlosser. 1986. Assessing biological integrity in running waters: a method and its rationale. Illinois Natural History Survey, Special Publication 5, Champaign.

Keller, E. A. 1978. Pools, riffles, and channelization. Environmental Geology and Water Sciences 2:119–127.

Kirwald, E. 1964. Bewässerpflege. GLV Verlagsgesellschaft, Munich, Germany.

Kondolf, G. M., and M. J. Sale. 1985. Application of historical channel stability analysis to instream flow studies. Pages 184–194 in F. W. Olson, R. G. White, and R. H. Hamre, editors. Symposium on small hydropower and fisheries. American Fisheries Society, Western Division and Bioengineering Section, Bethesda, Maryland.

Leopold, L. B., and W. B. Langbein. 1966. River meanders. Scientific American 214(6): 60–70.

Leopold, L. B., M. G. Wolman, and J. P. Miller. 1964. Fluvial processes in fluvial geomorphology. Freeman, San Francisco.

Meehan, W. R., editor. 1991. Influences of forest and rangeland management on salmonid fishes and their habitats. American Fisheries Society Special Publication 19.

Morhardt, J. E., and E. G. Altouney. 1985. Instream flow requirements: what is the state of the art? Hydro Review 4:66–69.

Nunnally, N. R. 1978. Stream renovation: an alternative to channelization. Environmental Management 2:403–411.

Orsborn, J. F. 1987. Fishway design practices. American Fisheries Society Symposium 1:122–130.

Oswood, M. E., and W. E. Barber. 1982. Assessment of fish habitat in streams: goals, constraints, and a new technique. Fisheries (Bethesda) 7(4):8–11.

Petts, G. E. 1984. Impounded rivers. Wiley, New York.

Platts, W. S., and twelve coauthors. 1987. Methods for evaluating riparian habitats with applications to management. U.S. Forest Service General Technical Report INT-221.

Reeves, G. H., and T. D. Roelofs. 1982. Rehabilitating and enhancing stream habitat: 2. Field application. U.S. Forest Service General Technical Report PNW-140.

Shirvell, C. S. 1986. Pitfalls of physical habitat simulation in the instream flow incremental methodology. Canadian Technical Report of Fisheries and Aquatic Sciences 1460.

Shirvell, C. S., and R. G. Dungey. 1983. Microhabitats chosen by brown trout for feeding and spawning in rivers. Transactions of the American Fisheries Society 112:355–367.

Slaney, P. A., R. J. Finnigan, D. E. Marshall, J. H. Mundie, and G. D. Taylor. 1980. Stream enhancement guide. Canada Department of Fisheries and Oceans and British Columbia Ministry of Environment, Stream Enhancement Research Committee, Vancouver.

Tennant, D. L. 1976. Instream flow regimens for fish, wildlife, recreation and related environmental resources. Fisheries (Bethesda) 1(4):6–10.

Trihey, E. W., and C. B. Stalnaker. 1985. Evolution and application of instream flow methodologies to small hydropower developments: an overview of the issues. Pages 176–183 in F. W. Olson, R. G. White, and R. H. Hamre, editors. Symposium on small hydropower and fisheries. American Fisheries Society, Western Division and Bioengineering Section, Bethesda, Maryland.

Wesche, T. A. 1980. The WRRI trout cover rating method: development and application. University of Wyoming, Water Resources Research Institute, Laramie.

Wesche, T. A. 1985. Stream channel modifications and reclamation structures to enhance fish habitat. Pages 103–163 in J. A. Gore, editor. The restoration of rivers and streams. Butterworth, Stoneham, Massachusetts.

White, R. J. 1991a. Objectives should dictate methods in managing stream habitat for fish. American Fisheries Society Symposium 10:44–52.

White, R. J. 1991b. Resisted lateral scour in streams—its special importance to salmonid habitat and management. American Fisheries Society Symposium 10:200–203.

White, R. J., and O. M. Brynildson. 1967. Guidelines for management of trout stream habitat in Wisconsin. Wisconsin Department of Natural Resources Technical Bulletin 39.

Chapter 10

Lake and Reservoir Habitat Management

ROBERT C. SUMMERFELT

10.1 INTRODUCTION

The program theme of the 121st Meeting of the American Fisheries Society, September 1991, was "Habitat: A Place for Fish, A Place for Fishing, A Place for Fisheries." The theme highlights the importance of habitat and identifies it as one of three major components of a fishery: the aquatic habitat, the fish, and the people who use the resource. This chapter focuses on habitat management in standing bodies of water (lacustrine habitats), natural or artificial, which are larger than ponds, but smaller than the Great Lakes.

10.1.1 Lacustrine Habitats

Before characterizing lake management, it is useful to classify the kinds of surface waters. Small natural or artificial bodies of water, less than 4 hectares in surface area, are called ponds. Larger water bodies are natural or artificial lakes. Intermediate-size lakes have areas of 4 to 200 hectares. Large lakes have areas exceeding 200 hectares. Artificial lakes in either size category usually are called reservoirs (the term used in this chapter) or impoundments.

Ponds (see Chapter 20) are dispersed management units. Ponds on public lands often are difficult to reach, and public use of privately owned ponds typically is restricted or prohibited. Thus, ponds generally have low fishing pressure and most public fisheries agencies cannot justify providing comprehensive management services for them. However, some agencies provide management advice to individual pond owners, and they may provide fingerling fish for new or renovated ponds, sometimes for a fee.

Intermediate-size lakes often have heavy fishing pressure, but they are of manageable size and a modest investment of money and effort will generally produce results obvious to the public. In 1985, 14.8% of the 38.4 million freshwater anglers 10 years old and older who fished fresh water—excluding the Great Lakes—fished in intermediate-size lakes of 4 to 16 hectares (USFWS 1988). Many lake management strategies have evolved from studies of artificial lakes less than 200 hectares; especially those smaller than 20 hectares.

Artificial lakes also are called reservoirs or impoundments. Recreational uses are only part of many functions of large artificial lakes. Other uses include hydropower, flood control, navigation, irrigation, power plant cooling, municipal and industrial water supply, and streamflow augmentation. There are few large, single-purpose reservoirs, and multiple use often results in competitive and conflicting demands on the reservoir management plan. These diverse purposes place limitations on fisheries management options because most reservoirs are storage elements of a local or regional water resource system. The operation plan, which is a statement of the principal social aim that dominates reservoir function, involves basinwide considerations (Peters 1986).

Reservoirs are an intermediate habitat type between natural lakes and streams. Flooding riverine habitats to make a reservoir has provided lacustrine fishery habitat in many localities where few or no natural lakes previously existed. Large reservoirs have inundated substantial portions of major rivers in North America, thereby limiting the stretches of free-flowing river, and resulting in trade-offs in fishery resources. Permanently flooding the gravel bars of rivers in the Midwest (Pfleiger 1975) and the Columbia River system (Collins 1976) has eliminated natural spawning sites for paddlefish and has adversely effected Pacific salmon, respectively.

In some regions, the term flowage (literally, the state of being flooded) is used for a reservoir or enlarged natural lake. Some natural lakes, however, are classified as reservoirs after their surface elevation is artificially raised and the original area or volume more than doubled (i.e., flowage).

The majority of natural lakes in North America are of glacial origin (8,000–10,000 years old). These lakes dot most of Canada and the northern United States from the Dakotas to New England, areas that were overrun by the last continental glaciation (Frey 1963); other lakes in western mountains were created by alpine glaciers. Ecologically, reservoirs and natural lakes are quite different (see (Chapter 21). Reservoirs often are built on rivers draining large basins with diverse land uses, and the quantity and quality of their water can change greatly during the year. Their water levels can fluctuate widely as they are filled and drawn down for flood control, irrigation, or power production. Because of their variable ecological conditions, reservoirs pose challenging fisheries management problems. Many biotic characteristics of lakes and reservoirs have been evaluated to find indices of fish yield and standing crop in natural and artificial lakes (Carline 1986).

Pump-storage and cooling impoundments are special categories of reservoirs. Pump-storage reservoirs are used for hydropower generation; water is pumped to a higher-level impoundment during periods of low power demand so it can drive turbines by gravity during periods of peak demand. Cooling impoundments are constructed primarily for the dissipation of waste heat from steam electric generating plants (Olmsted and Clugston 1986).

Modified riverine systems, called run-of-the-river navigation pools, formed by locks and dams, such as those on the upper Mississippi River, are not classified as reservoirs because they have storage ratios (the ratio of the reservoir water volume at the listed surface area to the annual discharge volume) of less than 0.01, which means that the annual flow-through is 100 times greater than the average annual pool volume (Jenkins and Morais 1971). The short turnover time (also called residence time) of these environments makes them more riverine than lacustrine. In comparison, Dale Hollow Lake, Tennessee, has a storage ratio of 1.16; Fort Peck, Montana, 2.36; and Grandby Reservoir, Colorado, 3.00 (Jenkins and Morais 1971).

Reservoirs have a tailwater, the river area downstream from a large reservoir that is strongly influenced by the fluctuations in reservoir discharge. Fish abundance in reservoir tailwaters is often substantial (Moser and Hicks 1970), and angler harvest in the tailwater may rival that in the reservoir. Obviously, tailwater fisheries are affected by temperature, dissolved oxygen and gas pressure in the discharge, the quantity and timing of reservoir releases (Peters 1986), and quality of angler access.

Why are fish so abundant in reservoir tailwaters? The abundance of fish in tailwaters is related to tailwater ecology and the effects of dams on upstream fish migration. Tailwater fishes originate from both the river and the reservoir. The dam blocks upstream movement of fish, thereby concentrating them in the tailwater, but it also provides an abundance of food for predacious fishes when the current and water temperature are favorable for feeding and spawning activity. Although fishing may be better in the tailwater than in the reservoir, fluctuations in abundance of fish stocks is similar to the fluctuations in the reservoir (Walburg 1971). For the most part, management of tailwater fish habitat must be directed at modifications of the quality and quantities of reservoir releases. This includes using multiple penstock discharges to control water temperature or structural modifications to avoid deep plunge pools that cause gas supersaturation.

10.1.2 Lake Surveys

Lake and reservoir habitat management is a process of problem solving that requires knowledge of limnology, fisheries, economics, hydrology, and civil engineering. The manager's basic tool for identifying problems is surveys of (1) the fish community, (2) the environment (limnological), and (3) the people (creel) who use the resource. A survey is a formal inspection according to a predefined plan using standardized sampling methods with statistically acceptable rationale. A fishery survey is an essential first step in developing a lake management plan. The lake survey provides perspective and insight to real or imaginary problems of the three components of a fishery:

Fish	Environment	People
missing year-class	lack of spawning habitat	reduced angling effort
poor growth	pollution	increased angling pressure
excessive exploitation	shoreline erosion	declining catch rate
presence of problem animals	hypolimnetic anoxia	poor catch quality
size structure	aquatic weed problems	declining socioeconomic value
fish kills	sedimentation	angler dissatisfaction

A precise statement of the management problem is needed to focus on remedial action. Corrective action, as with other aspects of fisheries management, is usually complex and must be undertaken with an understanding of each lake's environment (Weithman and Haas 1982).

10.2 MANAGEMENT TO IMPROVE ENVIRONMENTAL QUALITY

10.2.1 Watershed Considerations

Land–water interactions modify and control the physicochemical environment of lakes and reservoirs. It is folly to consider lake habitat management without first understanding the linkage between runoff and point- and nonpoint-source pollution and the quality of the lake environment because this relationship will ultimately limit management options and recreational use. In the long term, soil erosion and nutrient problems must be controlled at their source rather than focusing on inlake treatments of their symptoms. A quality fishery cannot be developed or sustained when there are substantial pollution problems and periodic fish kills.

The relationship of lakes to their watershed affects reservoirs more than natural lakes because reservoirs have higher watershed area to surface area ratios. Reservoirs receive higher sediment and nutrient loading than do most natural lakes (Thornton et al. 1990). As a consequence, the aging process of reservoirs is faster than natural lakes because of more rapid basin filling and higher nutrient loading (Kimmel and Groeger 1986).

Nonpoint-source pollutants in contrast to a point source, such as the effluent pipe of a papermill, originate from diffuse sites throughout the drainage basin (i.e., watershed or catchment). The major kinds of nonpoint-source pollutants are silt and clay, inorganic nutrients, particulate and dissolved organic matter, and pesticides. Lake sediments originate from croplands, eroded pastures, forests, and construction sites; organic matter and nutrients (nitrogen and phosphorus) originate from barnyards, feed lots, manured fields, and row crops; and pesticides originate from orchards, croplands, and forests.

Eutrophication and sedimentation from human activities have severely deteriorated fisheries and limited the recreational opportunities of many older impoundments (Likens 1972; Born et al. 1973; Henderson-Sellers and Markland 1987). It is difficult to reduce inputs of nutrients and sediments from diffuse agricultural and residential lands in large watersheds. Nevertheless, interagency efforts through the U.S. Environmental Protection Agency Clean Lakes Program combined with effective public hearing and other public relations efforts have demonstrated watershed improvements can protect and restore lakes, at least in modest-size basins (Born et al. 1973). Basically, sediment and excessive nutrients have been kept out of lakes by improved tillage practices and proper waste management. If nutrient inputs cannot be reduced, the only course of action is to manage around the problem (see section 10.2.2).

10.2.1.1 Sediment

Reservoirs are effective at trapping suspended solids, therefore, the filling of reservoirs is the dominant aging process (Kimmel and Groeger 1986). The deposition of solids decreases lake storage capacity and smothers fish spawning sites (e.g., lake trout and walleye) and affects the diversity and abundance of many kinds of aquatic life. Reduction in depth increases the possibility for winterkill, and increases the area suitable for invasion of vascular aquatic plants. Small lakes can be easily overwhelmed by sediment deposition.

Sediment originates from erosion processes within the drainage area, the river channel and the lake shoreline, the interface between the water and the land. In terms of absolute volume, sediment is the major nonpoint-source pollutant in reservoirs. Basin filling and reduced water depth changes the entire ecology of lotic systems (Brugam 1978; Thornton et al. 1990). Throughout the corn belt and other areas with watersheds dominated by row crops, the greatest threat to water quality and lake aging is soil erosion exceeding tolerable soil loss guidelines (''T'' value) established by the U.S. Soil Conservation Service. Lake volume may be totally lost to siltation in the worst cases (with sediment at the lip of the spillway) as happened at Lake Ballenger, Texas, and Mono Reservoir, California (Owen 1980). Even much less basin filling may result in major ecological changes.

In addition to reducing lake volume, suspended solids make the water turbid, and lakes located where colloidal clays are abundant may have persistent muddy water conditions long after a given runoff event. Suspended solids, except for colloidal clay, settle (i.e. they are deposited) when they reach a lake, the larger particles first, forming deltas at the headwaters of the reservoir. The lighter clays are dispersed throughout the lake. Turbidity reduces the depth to which light can penetrate, thereby limiting primary production.

The U.S. Department of Agriculture's Rural Clean Water Program was initiated in 1980 to evaluate best management practices to reduce the loads of sediment and nutrients entering lakes and streams from agricultural sources. Best management practices include tillage, crop rotation, vegetative cover, crop residue, nutrient management, and use of structural devices such as grassed waterways, sediment retention basins, and erosion control weirs (Johengen et al. 1989).

The U.S. Soil Conservation Service Watershed Management Program has provided cost-share assistance for terrace and pond construction and best soil management practices to reduce soil erosion on watersheds above publicly owned lakes. Activities may include sediment dikes at the upper reaches of major arms and riprapping the shoreline (McGhee 1990). The sediment dikes are low-head, rock-covered dams that sufficiently slow the sediment ladened inflowing water to deposit much of the sediment above the dam.

Sediment is usually removed by hydraulic dredging. In theory, a purposeful reduction in lake level over many years may allow erosional processes within the lake basin to wash out sediment through the discharge, operating much like thalweg current in rivers is used for dredge disposal—the current carries the soil downstream. However, this strategy for removal of reservoir sediment has not been adequately evaluated. On a limited scale a pronounced overwinter draw-down may be somewhat effective in small lakes to consolidate a flocculent sediment and deepen the lake (Beard 1973).

10.2.1.2 Organic Matter

Inputs of organic matter into lakes can be harmful or beneficial, depending on the relative amounts and the inherent fertility of the lake. Detritus, the particulate, unrecognizable organic matter of plant and animal origin, is the basis for food chains of large, turbid reservoirs, because primary productivity is often light limited (i.e., photosynthesis is limited by poor light penetration) rather than nutrient limited (Marzolf and Osborne 1971). Because the major source of organic matter for fish production in many turbid mainstream impoundments is derived from the watershed (allochthonous), not from primary productivity (autochthonous), food chains of most large reservoirs are heterotrophic rather than autotrophic.

Even though organic matter inputs in large reservoirs may form the basis for the food chain, small- and intermediate-size lakes can be overwhelmed by excessive inputs from feedlots and runoff from frozen fields fertilized with manure from herds of dairy cattle. The biochemical oxygen demand (BOD) from decomposition of organic matter may remove oxygen faster than it can be produced by algae or entrained at the surface. High BOD levels result in hypolimnetic oxygen depletion and increased incidence of fish kills.

10.2.1.3   Nutrient Enrichment

Phosphorus additions from external sources and regeneration of phosphorus from lake sediments are the major forces driving biological productivity of lakes. Lakes with quantities of total phosphorus in excess of 30 μg/L are eutrophic, and >100 μg/L are hypereutrophic (Wetzel 1975). Nutrient enrichment from human sources, called cultural eutrophication, is the major cause of lake eutrophication in North America (Edmondson 1969). Eutrophication is the major and most persistent water quality problem throughout the United States. For example, based on total phosphorus concentrations, 65% of Minnesota's 12,034 lakes are classified as eutrophic or hypereutrophic (Heiskary 1985). Phosphorus inputs usually come from agricultural sources and municipal effluents. The consequence of eutrophication is manifested by frequent occurrence of algal blooms, decreased water transparency, scums and mats of blue-green algae, and dense littoral zone beds of submergent and emergent vegetation.

Dense blooms of algae are deleterious to almost all uses of water, making a lake less desirable as a place for fishing, swimming, and boating, and causing taste and odor problems in drinking water which are costly to treat. Algae may be so dense that they shade-out, or limit the light needed for development of aquatic macrophytes. When abundant, however, rooted macrophytes may pump nutrients from the sediment into plant biomass which is put back into circulation when the plants decompose. Excessive macrophytes can inhibit fishing and boating, and adversely affect predator–prey balance by affording too much cover for the prey. The survival of the entire fish community may be jeopardized when a dense plant growth decomposes. A huge biochemical oxygen demand from decomposition results in depletion of oxygen concentrations.

Aside from situations where light necessary for algae abundance is limited by the presence of inorganic materials, particularly suspended colloidal clay, transparency varies in relationship to the mass of algae which varies in proportion to the amounts of phosphorus in the water (Edmondson 1969; Jones and Bachman 1974). Quantification of nutrient loading rates can be used to predict lake trophic state (Carlson 1977). A trophic status index, which refers to the nutritional status of a lake, can be used as a relative assessment of the degree to which nutrients and algal biomass are present in a lake (Carlson 1977).

When the immediate job is to manage a eutrophic lake, and to effectively channel the productive potential to beneficial means, a variety of management practices are available: dilution and flushing, phosphorus precipitation and inactivation, sediment removal, water-level drawdown, whole lake aeration (destratification), hypolimnetic aeration, and mechanical or biological harvesting of macrophytes (Dunst et al. 1974; Cooke et al. 1986; Henderson-Sellers and Markland 1987).

10.2.1.4   Acidification

Acidification of aquatic ecosystems is one of the most serious types of environmental pollution. The acidification of surface waters is evident in North America (Figure 10.1) and Europe (van Breeman et al. 1984). Acidification of

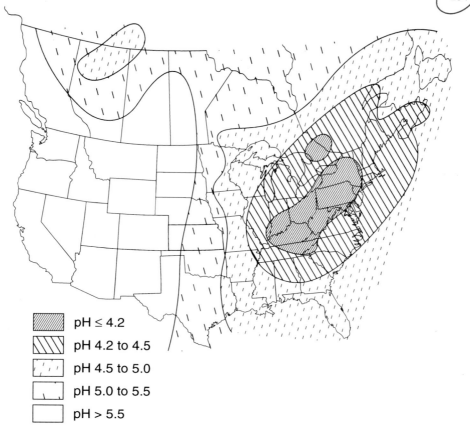

Figure 10.1 Acidity (pH) of precipitation across Canada and the United States. Data compiled from North American monitoring networks.

aquatic ecosystems requires the combination of two factors: (1) poor buffering capacity (i.e., low acid-neutralizing capacity), and (2) acid deposition (i.e., low average pH of precipitation). Acid deposition includes various forms of precipitation (rain, sleet, snow, dew, and fog) containing sulfuric acid (H_2SO_4) and nitric acid (HNO_3) and dry deposition of particles containing sulfate and nitrate salts. Sulfuric and nitric acids are secondary pollutants from acid precursors (SO_2 and NO_x), the primary pollutants released from tall smokestacks of coal-burning electrical utilities, smelters, and steel mills and emitted from trucks and automobiles. In the United States in 1980, 27 million tons of SO_2 and 21 million tons of NO_x were emitted, of which 80% of the SO_2 and 65% of the NO_x came from 31 states bordering or east of the Mississippi River (Office of Technology Assessment 1985).

Acid precipitation is a major threat to fisheries and all other forms of aquatic life in acid-sensitive lakes in the northeastern United States and eastern Canada (Beamish 1976; Schofield 1976; Johnson 1982). The sulfur in the emission originates from the sulfur in the coal; the nitrogen, however, originates from the air; NO_x is formed from combustion at high temperatures. Direct effects of acidification on fish follow from low pH and solubilization of an aluminum ion which is toxic to fish embryos and larvae (Baker 1982). The no-effect level of pH depression on reproduction in fishes is around 6.5. Low pH upsets calcium metabolism of fishes and protein deposition in developing eggs. Acid stress may coagulate gill mucous, damage gill epithelia, and upset electrolyte homeostasis (Fromm 1980). Acidification is especially serious for glacial lakes of Scandinavia, nearly all of Canada east of Lake Winnipeg, the Adirondacks, the upper midwestern United States, high elevation lakes of mountain regions of the West, and also the Appalachians.

Tall stacks, which disperse the plume of the acid precursors, literally ''air mail'' the pollution over large geographical areas. The dispersion effectively converts a point-source pollution problem into a nonpoint one. In the United States, the 1990 Clean Air Act mandates substantial reductions in sulfur dioxide emission by the year 2000. Lakes already acidified by acidic deposition may be limed to counteract acidification. Liming neutralizes acidity directly, and it increases the acid-neutralizing capacity. It is not economically feasible to lime large numbers of remote lakes. Without control over acid deposition, liming must be repeated to prevent reacidification (Driscoll et al. 1987).

10.2.1.5   Contaminants

Fish kills are caused by a diversity of factors, not just toxic substances (Meyer and Barclay 1990). Contaminants have both direct and indirect effects on aquatic life (Cairns et al. 1984; Nriagu and Simmons 1984); they may kill selectively or they may kill all of the biota. Contaminants emanating from point sources (mines, petrochemical, municipal and industrial sites) may be chemical or physical in nature. Physical pollutants include solids (mine tailings, fiber from paper mills) and waste heat (thermal effluents). Chemical contaminants include a large variety of toxic substances, particularly those that have been associated with elevated tumor incidence in fishes (Black 1983). Some substances from paper mills and other sources produce taste and odor problems which render the fish unsuitable for human consumption. Environmental contaminants are a major problem in many aquatic environments (Cairns et al. 1984). Insecticides, are often toxic to fish in concentrations of parts per billion (micrograms per liter). Moreover, accumulation of contaminants in fish flesh may require restrictions on marketing the fish, hazard warnings to the public about eating the fish, and closure of the fishery.

10.2.1.6   Heated Effluents

Thermal pollution is a physical form of point-source pollution that occurs where lakes and rivers are used as a water source for cooling condensers of electric power production facilities. Heated water effluents of steam electric plants may affect fish spawning and growth, cause gas bubble disease, and sudden shutdowns may result in fish kills from temperature shock (Olmsted and Clugston 1986). Additionally, pumping large volumes of water for condenser cooling causes fish entrainment and impingement problems at water intake areas (Figure 10.2). Eggs, larval, and fingerling fish and invertebrates (zooplankton) that are pulled along with the current through the mesh of barrier screens are said to be entrained; larger fishes that are held against the barrier screens are said to be impinged (Fletcher 1990). In the United States, once-through cooling is used by 60% of the

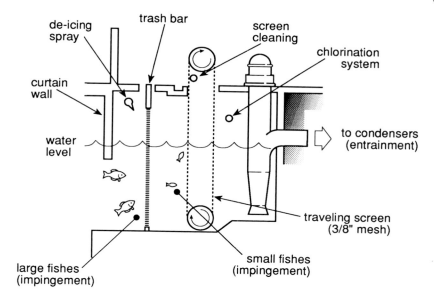

Figure 10.2 Impingement and entrainment problems at a steam electric generating plant. Large fishes are killed by impingement (entrapment) on the trash bars, and small fish and zooplankton that pass through the screens (entrainment with the current) are killed by high temperatures, pressure changes, and mechanical damage from passage through the condensers and plumbing of the power plant. (Redrawn from Zweiacker and Bowles 1976.)

fossil-fuel steam electric plants and 74% of the operating nuclear-fueled plants to condense exhaust steam from the turbines (Schubel and Marcy 1978). An entrained organism going through a once-through cooling system will experience a variety of lethal physical stressors including shear force, impact, abrasion, and a temperature rise of 11°C when passing through the condensers.

Generally, enforcement of pollution laws and even the basic monitoring of pollutant levels in the environment and in fish are not the legal responsibility of most fisheries agencies. The fisheries biologist, however, plays a key role in observing problems and violations and brings these to the attention of appropriate agency officials. Close cooperation between fisheries and environmental (pollution) control agencies is essential.

10.2.2 Lake Management

Habitat management considerations include lake treatments to protect and enhance the environment for the benefit of the aquatic community and the quality of the angling experience.

10.2.2.1 Erosion Control

Waves generated by wind or the wake of power boats may cause damage to earthen dams and exposed shorelines. The suspended solids and sediment derived from shoreline erosion may adversely impact spawning areas, increase lake turbidity, and have other effects similar to nonpoint source sediment inputs. Barren, windswept shorelines are poor food producers, unsuitable habitat for spawning of nest-building fish, and poor nursery habitat for young game and prey fishes.

Management options to protect shorelines from erosion include (1) grading steep slopes and terracing above the water line to prevent erosion and slumping; (2) construction of a retaining wall (bulkhead) along the waterfront, or riprapping the land–water interface with rocks, gabbions, or beach stabilization mats; (3) reducing the energy of incoming waves by use of breakwaters (armored earthen structures of floating-tire breakwaters); and (4) wake limits for power boats in nearshore areas with heavy boat traffic.

Steep slopes may need to be reshaped by terracing above the water line and riprapping the foot of the slope. Riprapping the face of earthen dams is usually done at the time of lake construction to protect the dam, but is also valuable for spawning or nursery habitat. Given the lay of the lake in relation to prevailing winds, selected segments of shoreline can be protected by wave dissipation revetments, erosion control dikes, and riprapping. The most sensitive sites are those exposed to a long fetch (distance over water from which the wind will blow) or where wake is produced by boat traffic. Riprapping the face of a dam or selected shorelines with rock can provide protection from erosion and provide spawning sites for fishes such as walleye because they scatter their eggs over shoals of gravel or boulders. Structures made of used automobile tires (erosion control mats) have also aided in controlling shore erosion (Candle 1985).

A naturally vegetated littoral zone diminishes wave action against the shoreline. Likewise, a stable and well-vegetated land–water interface is a desirable alternative to riprapping as a means to protect and stabilize shorelines. Aquatic macrophytes or flooded terrestrial vegetation provide habitat for vegetation spawners (e.g., northern pike and yellow perch) and nursery areas for young-of-the-year fish such as largemouth bass. Unfortunately, the fluctuating zone of large reservoirs may have prolonged intervals when the shorelines are dewatered, and as a consequence, aquatic macrophytes are generally absent. It is also difficult to find mesophytic or terrestrial plants that can tolerate intermittent flooding or prolonged dewatering. In flood control reservoirs, plants above the normal pool are sometimes inundated for days as the reservoir meets the obligation for storage of floodwaters; in other lakes (e.g., irrigation reservoirs) water levels may sometimes remain below the permanent pool for several years.

The edge of a fluctuating reservoir is an unstable ecotone, which is unfavorable for permanent colonization of most aquatic or terrestrial plants. Maiden cane *Panicum hemitomon* has shown promise for control of wave action at the waterline because it can actually grow in water under a variety of site and soil conditions in southern areas of the United States (Young 1973). Weeping love grass *Eragrostis* sp. and *Sericea lepedeza* are usually beneficial in producing a quick cover of bare soil surfaces after construction or following drawdown.

10.2.2.2 Fertilization

Lake fertility is indicated by nutrient concentration and algal density; therefore, concentrations of total phosphorus or chlorophyll-*a*, the principal photosynthetically active plant pigment, are useful indices of lake trophic state (see Chapter 21). Treatment of infertile small ponds with organic or inorganic fertilizers increases fish production. Artificial fertilization has also been used to increase biological productivity of natural lakes and reservoirs. There is a positive relationship between fish standing crop and harvest and chlorophyll-*a* and total phosphorus concentrations in U.S. reservoirs. Experimental fertilization of the coves of large reservoirs has been done to enhance primary productivity and zooplankton populations which will in turn enhance survival of larval or fingerling fish. Success has been limited to oligotrophic lakes. Generally, rapid exchange rates between water in the coves and the main reservoir will dilute the application. It is unrealistic to obtain a favorable cost–benefit ratio for fertilization of a large reservoir on a sustained basis given the rapid flushing rate.

Although fertilization of ponds has been successful when carefully undertaken, excessive fertility is more often a problem in reservoirs and natural lakes. Shortly after Swingle's early studies on pond fertilization in Alabama (see Chapter 20), warnings of problems related to fertilization quickly surfaced (Hasler 1947; Ball 1952). It is now obvious that enhancement of biological production of lakes by application of inorganic fertilizers (phosphorus compounds) is undertaken with great risks. Nutrient enrichment increases the hypolimnetic oxygen deficit, encourages algae blooms and macrophyte problems, and the incidence of summer and winter fish kills. Moreover, substantial reduction in water clarity (transparency) by suspended inorganic solids or blooms of algae may be regarded as a degradation in water quality.

An inverse relationship often occurs between macrophyte abundance and chlorophyll-*a* concentration in natural lakes, i.e., a large influx of nutrients from surface runoff often promotes an algal bloom which reduces light penetration and abundance of aquatic macrophytes (Colle et al. 1987). However, the application of fertilizers to ponds with the objective of stimulating phytoplankton blooms to shade-out extensive macrophytes may create further macrophyte problems. Given some control of lake levels, the "trophic upsurge" phenomena of a new reservoir can be recreated (Ploskey 1986).

10.2.2.3 Vegetation Control

More attention has been given to controlling vegetation than to enhancement. Infestations of certain nonnative aquatic plants such as hydrilla *Hydrilla verticillata* and Eurasian watermilfoil *Myriophyllum spicatum* have certainly received attention because they interfere with fishing and boating and have replaced native species of aquatic plants. Reduction in angling in Orange Lake, Florida, caused a 90% loss in revenue from the sport fishery (Colle et al. 1987). The economic value of an unimpaired sport fishery will typically far exceeded the cost of aquatic plant control.

Aquatic weed control may be undertaken by mechanical, chemical, and biological methods. Mechanical methods include cutting, dredging to deepen the water, covering bottoms with plastic film or other synthetic materials, and drawdown (Dunst and Nichols 1979). Mechanical harvesting (cutting) has improved in cost-effectiveness with innovations in equipment, but it is still expensive, labor intensive, and a never-ending process akin to lawn mowing. However, mechanical harvesting of heavy growths of aquatic macrophytes in high public-use areas avoids the hazards of chemicals and use of exotic organisms. Cutting and removal of aquatic vegetation, along with the nutrient content in the vegetation, is regarded as an inlake tool for water quality management.

Chemical methods usually involve use of aquatic herbicides. Only certain herbicides and algicides should be used in the aquatic environment; see Schnick et al. (1986) for a list of registered herbicides. Many herbicides have withdrawal intervals specified for different uses—the period of time that must pass after the last chemical treatment before the water can be used for drinking, swimming, irrigation, or livestock. The label on many herbicides contains precautionary statements regarding hazards to humans and domestic animals. There are restrictions such as "Do not use water for any purpose for 7 days after treatments"; "Water may not be used for irrigation of agricultural crops, or for watering cattle, goats, hogs, horses, poultry, or sheep, or for human consumption until 12 months following treatment"; or "Do not use fish from treated water for 90 days after application." Because most lakes have multiple uses, such regulatory restrictions greatly reduce the choice of registered herbicides. Moreover, herbicides are costly, and their effectiveness is usually limited to the year of application.

Biological control of aquatic macrophytes with crayfish or herbivorous fishes (e.g., grass carp or tilapia) has been effective in some lakes but biological control is controversial. There is a widespread fear that grass carp may become a problem species similar to the common carp. These concerns have resulted in restrictions and prohibitions on using any kind of nonnative fish (see Chapter 12). If a herbivorous fish escapes into a river system, it may invade a waterfowl refuge where the fish would compete directly with waterfowl for rhizomes or tubers of aquatic plants such as *Valisneria americana* and *Potamogeton*. However, some state agencies allow use of sterile (triploid) grass carp.

10.2.2.4 Landscaping

Vegetation management can serve to enhance environmental quality for fish and anglers, as well as improve aesthetics. A windbreak around the periphery of the lake can reduce fetch and wave height. A long windbreak perpendicular to the prevailing winds can reduce low-level wind velocities even in low-relief topography (Young 1973). A windbreak can provide protection for shallow-water fish spawning as well as create quiet-water areas for anglers on windy days. In many areas, a mowed grassy area is preferred around the periphery of ponds or intermediate-size lakes for aesthetic reasons and to eliminate habitat for ticks, chiggers, and poisonous snakes. Downed trees on the shoreline add natural brushy spots for spawning and concentrate fish for anglers.

10.2.2.5 Waterscaping

Management of vegetation for production of desirable food and cover is a major theme of terrestrial wildlife management. In aquatic habitats, however, much more attention has been given to controlling vegetation than to enhancement. Flowering aquatic macrophytes may enhance the aesthetic appearance of a lake; provide food for waterfowl, muskrats, and other wildlife; serve as cover for fish; and function as attachment surfaces for periphyton and many kinds of macroinvertebrates (phytomacrofauna) that are food for fish and live attached to submerged plants (Figure 10.3).

Crowder and Cooper (1979) found that fish growth was greater at a macrophyte density sufficient to provide cover for invertebrate food so the fish do not overexploit their prey, yet not too great to prevent effective foraging. The idea of an optimal macrophyte density implies a vegetation enhancement or control program that considers effects on fish–invertebrate relationships and piscivory.

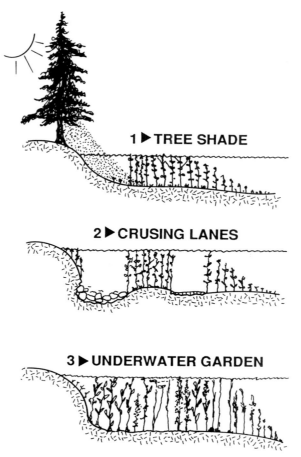

Figure 10.3 (1) A row of trees planted on a south shoreline will shade the water and retard rooted aquatic plants growing near shore; (2) in small, shallow lakes with heavy fishing pressure, cruising lanes through heavy vegetation can be provided by covering the bottom with fiberglass screens topped with rock; (3) waterscaping may be done with broad-leaved ponds weeds and water lilies to provide invertebrate food for fish and waterfowl, shelter for prey species, and spawning habitat. (Redrawn from Engle 1989.)

Loucks et al. (1979) recommended a patterned macrophyte harvesting procedure to obtain edge effects.

In Texas, survey data from 30 reservoirs showed that a strong positive linear relationship existed between submerged vegetation (up to 20% coverage) and largemouth bass recruitment and standing crop (Durocher et al. 1984). In small ponds, the relationship between plant standing crops during the growing season and largemouth bass production is parabolic with an optimum at mid-densities, (i.e., largemouth bass production is reduced by very low or very high plant biomass).

Enhancement of aquatic vegetation may include direct planting of tubers and fertilization. In reservoirs with a barren fluctuating zone where aquatic macrophytes are unable to colonize, the shoreline may be planted by aerial seeding with an annual rye grass or some other cover crop. The cover crop provides benefits similar to aquatic macrophytes, but it decomposes quickly after inundation.

10.2.2.6   Stratification and Destratification

Many North American lakes develop stratification, summer and winter, because the density of water changes with temperature, reaching maximum density at 4°C. In the summer, stratification results in three layers: (1) a warm well-mixed upper layer with oxygen levels near saturation for the temperature and atmospheric pressure (epilimnion), (2) a middle transition layer (thermocline), and (3) a lower area of cold water (hypolimnion). In winter, the coldest water (0°C) is adjacent to the ice and the warmest, heaviest water (4°C) is near the bottom.

By midsummer, many stratified lakes have an oxygen-depleted hypolimnion, often with oxygen concentrations near zero. Because oxygen concentrations need to be more than 2 mg/L for a water stratum to be continuously occupied by even the most low-oxygen-tolerant fishes, the hypolimnion of eutrophic lakes is generally devoid of fish. The anoxic hypolimnion is also impoverished of many desirable kinds of benthic invertebrates that are important fish foods. Species of fish that cannot tolerate the warm temperatures of the epilimnion may be "sandwiched" between thermally lethal epilimnetic water above, and cooler but anoxic water below. The environmental deterioration that accompanies stratified conditions has caused summerkill of some coldwater species.

Even the inhabitants of the epilimnion are adversely affected by pronounced stratification. Warmwater fishes may cease growing because of high epilimnetic temperatures. Summer stratification may also be an important contributing factor to the occurrence of winterkill because accumulated hypolimnetic biological oxygen demand is not metabolized during the relatively short fall turnover and carryover into the winter exacerbates wintertime oxygen demand.

Lake destratification strategies have been used since the early 1960s to improve water quality of municipal water supply lakes. Also, destratification of a reservoir can be used to improve a tailwater fishery. Considerable research effort has gone into means to destratify reservoirs to improve the quality of water in and released from reservoirs (Toetz et al. 1972; American Society of Civil Engineers 1979; Burns and Powling 1981). In lakes where it is not necessary to maintain a cold hypolimnion for trout or other fishes, the lake may be destratified by various means or hypolimnetic water may be discharged. Multilevel outlets of some multipurpose artificial lakes makes selection of the depth of water discharge possible, i.e., the discharge may be from epilimnial, intermediate, or hypolimnial levels (Moen and Dewey 1978). Of course, the effect of a hypolimnetic discharge on the tailrace fishery always needs to be considered.

Potential benefits of artificial destratification include expanded fish habitat, amelioration of signs of eutrophication (changes in algae composition dominance from blue-green to green algae or diatoms), an increase in diversity, and average size of the benthos, and avoidance of summerkill and winterkill (Figure 10.4). For example, destratification of a 9.7-hectare lake, where as much as 82% of the lake bottom was below the thermocline when the lake was stratified, increased channel catfish harvest by 227% (Mosher 1983). The cause of the increase in angler harvest of catfish is not clear, but it is known that anglers often fish in the hypolimnion of stratified lakes where there are no fish. Increased temperature, decreased vege-

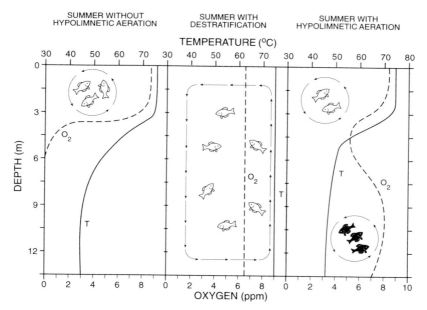

Figure 10.4 Influence of summer hypolimnetic aeration on stratification and vertical distributions of fish in small reservoirs. (Redrawn from Fast 1981.)

tation, and increased diversity of benthic macroinvertebrates may also be reasons for the increased harvest.

The two major kinds of destratification systems are air bubble and mechanical pumps. Air bubble systems are compressors or air blowers that force air through diffusers or Helixors. The rising air bubbles, by entraining an envelope of cold water, create an upflow of cold water. Destratification by air injection is a common reservoir management technique to expand fish habitat (Fast and Hulguist 1989). The upward movement of air bubbles develops a mixing cell and pulls in surrounding water. Mechanical pumping is most efficiently done by use of a high-volume axial flow pump. A mechanical pump may be used to move large volumes of warm surface water to the bottom of shallow lakes (Garton et al. 1978).

10.2.2.7 Hypolimnetic Aeration

Methods are available to increase the oxygen content of the hypolimnion without disrupting thermal stratification. By aeration without warming of the hypolimnion, coldwater habitat can be preserved for trout. Hypolimnetic water may be pumped within a pipe to an aeration box on the lake surface then the aerated water returned to the hypolimnion through a separate pipe without appreciable warming (Wirth 1981).

10.2.2.8 Dewatering

Deep drawdowns of lake levels may occur when water is drawn off for irrigation, navigation, and from prolonged drought. Short-term water level fluctuations occur as a result of electric power generation and flood control. Both short-term and chronic reductions in lake level are detrimental to development of littoral zone plant communities. A decline in lake level during spawning may cause nest abandonment and stranded fish eggs. However, purposeful drawdowns may be used to manipulate year-class strength and predator–prey relations. Common carp spawning has been affected by exposure of spawning areas (vegetated shallow water) immediately following spawning.

Prey may be too large (gizzard shad) or too well protected (vegetation), for the largemouth bass to easily capture. In most established largemouth bass populations, fish typically have an empty stomach, or the stomach contains only a single prey item. Consequently, largemouth bass growth rates are much less than maximum.

Largemouth bass predation on small bluegill has been enhanced by lowering water levels to force the bluegill out of the shelter of littoral vegetation. Bennett et al. (1969) reduced the bluegill population 60% from a fall drawdown at Ridge Lake, Illinois, which decreased surface area 35%. On the other hand, the vulnerability of a pelagic (open water) prey species such as a gizzard shad may be unaffected by drawdown (Wood and Pfitzer 1960; Lantz et al. 1967). A summer drawdown that reduced surface area by 42% and volume by 58% on Little Dixie Lake in central Missouri resulted in increased feeding activity by largemouth bass, a reduced percentage of fish with an empty stomach, and accelerated growth (Heman et al. 1969). The density of fry and intermediate-size bluegill was reduced by the process of stranding small sunfish in vegetation and through increased predation by largemouth bass on the small sunfish. Reduction in abundance of bluegill may result in improved survival of largemouth bass fry the following spawning season.

Purposeful water level manipulation has been used for ameliorating the effects of eutrophication in older, intermediate-size reservoirs (Born et al. 1973). During drawdown, managers have the opportunity to excavate nutrient-rich sediments and deepen shallow areas prone to excessive macrophytes.

10.2.2.9 Water Level Fluctuation

Reservoir littoral zones are highly unstable because fluctuations in reservoirs are more extreme on an annual basis than those observed in natural lakes. Water level fluctuations (daily, seasonal, and annual) are often found to be critically important environmental variables. It has been inferred that reservoirs with large annual changes in water level would likely have less stable fish assemblages (Carline 1986). Because recruitment and growth of many fishes are responsive to the hydrological regime of reservoirs, water level management plans may be used to meet management objectives for different species of fish (Willis 1986). Spring surface area, annual change in area, summer area, and spring flooding are important factors affecting August fish standing crop (Ploskey 1986). An increase in year-class strength occurs in many fishes after spring flooding, but this is nullified when summer drawdown is extensive. Maintaining above-average water levels for as long as possible after successful spawning is necessary to protect young-of-the-year of some fishes (Mitzner 1981).

Given some control of lake levels, the trophic upsurge phenomena of a new reservoir can be recreated (Ploskey 1986). Trophic upsurge refers to the 5 to 10 years of highly successful sport fish production that follows initial reservoir filling. A period of 3 to 4 years of low water levels, when growth of terrestrial vegetation takes place in exposed areas, followed by flooding of substantial areas of vegetation can simulate the trophic upsurge, increase productivity, and bring on strong year-classes of certain fishes (Ploskey 1986). Water level management has been an effective fisheries management strategy for walleye, white bass, and white crappie in Kansas reservoirs (Willis 1986).

Daily fluctuations in lake levels are pronounced in pump-storage reservoirs used for peaking power and these fluctuations are basically detrimental to fish and fishing. At this time, it is unlikely that a quality fishery will be found in a reservoir managed for pump storage hydropower generation that produces more than a 3-m diurnal flux in the lake level.

10.2.2.10 Winterkill Lakes

It has long been evident that winterkill of a fish population in ice-covered lakes is the result of an oxygen reduction below a concentration at which fish can survive (Greenbank 1945). The oxygen decline begins when ice cover seals the water surface from direct contact with the air, and snow-covered ice reduces light penetration needed for algal photosynthesis. A partial or total winterkill of a lake's population is a serious lake management problem in northern latitudes (Schneberger 1971). Shallow and fertile lakes with a thick organic substrate have a high metabolic demand for oxygen. Shallow lakes have a small storage capacity for oxygen, and the close association of the water mass with the substrate may quickly exhaust the oxygen when snow cover is heavy or persists for long periods.

Fish vary in the minimum values of oxygen they require for survival. As a consequence, winterkills rarely kill the entire fish population. Usually the game fish are more sensitive to low oxygen levels than are common carp, buffalofish, and bullheads. Following a winterkill, the surviving adult carp and bullheads spawn successfully and develop a strong year-class because of the absence of predatory fish, then the carp and bullheads become ecologically dominant. They strongly influence environmental quality by making the water extremely turbid. The turbidity limits or prevents development of aquatic algae and macrophytes. Consequently, a winterkill often requires whole lake renovation (poisoning) prior to restocking game species.

Many kinds of aeration devices are available (bubbles or mechanical pumps) to melt an opening in the ice or to keep a section of lake from freezing (Johnson and Skrypek 1975; Smith et al. 1975). Pumping water melts the ice by expending the heat stored in the lake. The opening in the ice allows direct contact between water and air and also allows light penetration. Actual oxygen input from the aeration apparatus itself is a small proportion of the total oxygen input that occurs through the agitated lake surface in ice-free areas. An opening in the ice, or a thinning of the ice near the opening is a hazard. The affected area must be cordoned off with snow fencing, signs must be posted, and educational programs must be carried out to make the public aware of the danger. Ice-cold water is life threatening.

10.3 MANAGEMENT OF FISH POPULATIONS: CONTROLS AND ENHANCEMENT

10.3.1 Managing Habitat

In natural lakes with inlet waterways (i.e., drainage lakes in contrast to seepage lakes) and nearly all reservoirs, common carp make upstream spawning migrations to natural marshes and floodplains inundated by spring runoff. Constructing obstacles (weirs) across streams or the mouth of marshes can prevent common carp access to spawning sites. Weirs to block upstream migration of common carp have been used in Nebraska and Wisconsin. Weirs are used in Iowa lakes to prevent common carp from migrating into connecting marshes. Unfortunately, weirs are not discriminating and they may block migration of desirable species and block boat traffic; weirs are also vulnerable to destruction by floods.

Sharp reductions in lake level may reduce sites for spawning or expose eggs to drying. In Fort Randall Reservoir, South Dakota, a drawdown of 0.5–0.6 m immediately after major episodes of common carp spawning in shallow (0.3 m) flats killed the exposed eggs; the drawdowns were considered responsible for poor year-class strength of carp in 3 years when drawdowns were conducted (Shields 1958). Drawdowns, however, may adversely impact the reproductive success of other fish that spawn in the littoral zone.

Spawning habitat management should be considered when recruitment of desirable species is inconsistent, inadequate, or nonexistent. Spawning of fish such as rainbow trout, striped bass, and white bass may be assisted by removal of barriers to upstream passage. When spawning habitat is suitable but larval fish survival is poor, other approaches are needed. Larval fish survival, and thus year-class strength, is influenced by many density-independent abiotic (wind, temperature, water level fluctuations) and density-dependent biotic factors (abundance of prey and predators of larval fish). Maintenance stocking may be a more practical and effective alternative to habitat manipulation because the spawning requirements of some fish (e.g., striped bass or rainbow trout) are difficult to accommodate. In many lakes, walleye populations are routinely maintained by stocking fry or fingerling fish produced in hatcheries (Conover 1986). Of course, hatchery production is necessary to provide hybrid muskellunge (tiger muskellunge), hybrid striped bass, and hybrid walleye (see Chapter 13).

Management manipulations to enhance spawning habitat include flooding of terrestrial vegetation and use of artificially constructed spawning sites. Fish that spawn in lakes include gravel spawners (lake trout and walleye), vegetation spawners (northern pike and yellow perch), nest builders (largemouth bass, crappie, and other fishes), and cavity spawners (channel and flathead catfish). Artificial spawning reefs of gravel, connecting marshes, and shoreline protection are useful techniques to enhance spawning of these fishes.

In lakes, walleye generally spawn at depths of 0.3–3.0 m on gravel and rubble shoals. In reservoirs, walleye generally spawn in water less than 1.5 m deep over gravel and rubble substrates, commonly on the rock riprap of the dam face. Egg survival from walleye spawning on other bottom types is poor (Prentice and Clark 1978). Newberg (1975) has described criteria for construction and placement of walleye spawning shoals.

Although it has been suggested that spawning devices, such as drain tiles, milk cans, and grease cans, be added to lakes to provide spawning sites for channel catfish (Nelson et al. 1978), this will probably be of little value in large reservoirs because of the access to riverine spawning sites. In small and intermediate-size lakes that also contain largemouth bass, it is not the lack of spawning habitat that causes poor recruitment of channel catfish, rather, it is the exceptional vulnerability of channel catfish fry to predation by largemouth bass. Maintenance stocking of channel catfish in ponds and small lakes requires a nonvulnerable-size fish (>17.5 cm).

Recruitment of largemouth bass is inadequate in many lakes and reservoirs because of density-independent environmental factors. A sudden decline in water temperature may cause male largemouth bass to abandon the nest and stop spawning (Kramer and Smith 1962). Sometimes the fish will resume spawning when the water rewarms. On the other hand, shoreline areas exposed to a long fetch will receive continuous exposure to wind-driven waves, and such environments are generally inhospitable habitats for spawning of nest-building fishes such as largemouth bass (Summerfelt 1975) and crappie (Mitzner 1991), or vegetation spawners such as yellow perch (Clady 1976). A reservoir dominated by great stretches of barren, windswept shoreline will probably have poor recruitment of largemouth bass and crappie. A reservoir with an irregular shape will have a greater proportion of protected spawning habitat (Figure 10.5).

After young largemouth bass have dispersed from the nest, predation becomes a factor in survival. At that time, survival is enhanced by the presence of aquatic vegetation, or flooded terrestrial vegetation present when a reservoir first fills or develops around the periphery following one or more years of low water levels. A high correlation between the amount and duration of flooded terrestrial vegetation and abundance of young-of-the-year largemouth bass in a given year indicates that flooded terrestrial vegetation is essential for providing food and protection (Aggus and Elliott 1975; Shirley and Andrews 1977). It has been recommended that fisheries management effort be directed to produce a high water level through most of a growing season every 3 or 4 years to flood terrestrial vegetation. An annual inundation will not produce a strong year-class each year because the terrestrial vegetation will not be available unless the shoreline is exposed for two or three successive years.

There are pros and cons to clearing trees from reservoir basins (Ploskey 1986). However, inundated timber increase food abundance, provide cover, and increase spawning habitat (Figure 10.6). Studies on distribution of larval gizzard and threadfin shad showed they were more abundant in offshore inundated timber than open-water areas reflecting both extensive use of the flooded trees as spawning sites or increased availability of food or shelter (Van Den Avyle and Petering 1988).

Submerged tire reefs may be used for spawning sites by some fish, but given the small size of the reefs it is unlikely that an artificial reef will substantially influence the year-class strength of fishes in large reservoirs. Research has shown that submerged structures of either natural or artificial materials are soon colonized by an abundance of aquatic invertebrates and a surface film of periphyton. Artificial structures, however, contain such small surface areas relative to other objects in the lake such as rock riprap or standing timber that they cannot be expected to impact primary production or invertebrate biomass of an entire reservoir. Large areas of standing timber in the lake basin will more likely provide meaningful habitat than post-construction additions of artificial reefs.

10.3.2 Subimpoundments as Nursery Areas

Arms of coves, or a segment of a reservoir, partitioned from the larger body, may be stocked with fry and used for rearing northern pike, largemouth bass,

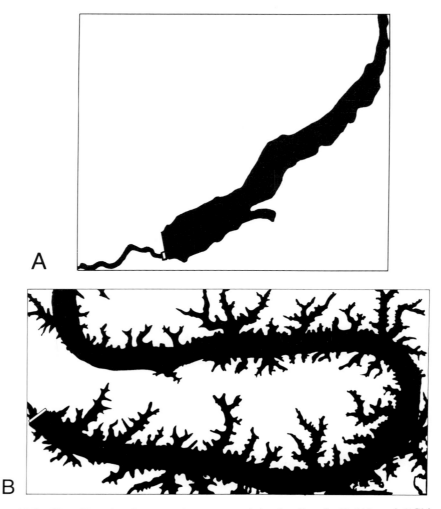

Figure 10.5 Shoreline development is contrasted in the ''wedge'' (A) and ''Chinese dragon'' shapes (B). The ''wedge''-shaped reservoir typically has only a few major arms, minor coves and shoreline irregularities; it is often dominated by windswept beach zones and it lacks suitable habitat for nest spawning fishes. The ''Chinese dragon''-shaped reservoir has a large shoreline development due to flooded tributaries that provide an abundance of wind-sheltered coves with submerged standing timber.

walleye, striped bass, and other species to a nonvulnerable size for stocking. Essentially, a drainable watershed pond located to facilitate draining directly into the adjacent reservoir is best. In Arkansas, a nursery pond was used to rear fry of walleye, channel catfish, largemouth bass, and striped bass (Keith 1970). It was particularly successful in developing a walleye fishery in Norfork and Bull Shoals reservoirs from fingerlings released from the nursery ponds. Once again, efficient hatchery production may provide a lower cost–benefit ratio than the costs of constructing and managing subimpoundments.

Figure 10.6 Retaining standing timber at the time of reservoir construction provides excellent habitat for many kinds of fishes. (Reproduced from Linder 1990.)

10.3.3 Fluctuating Zone Vegetation as Nursery Areas

Provision for protected spawning and nursery areas during reservoir planning can prevent negative effects of the exposed fluctuation zone. As an operational procedure, however, it is difficult to maintain vegetation in the fluctuation zone of large reservoirs. Nelson and Walburg (1977) found that strong yellow perch year-classes were produced in upper Missouri River impoundments when spring water flooded shorelines that had developed terrestrial (mesophytic) plants as a result of being out of the water for one or more years.

Fluctuation zone seeding has been used on Tennessee Valley Authority and Kansas reservoirs (Groen and Schroeder 1978) with generally favorable results on the fish populations. Strange et al. (1982) found that fluctuation zone seeding (fertilized rye grass) had a positive effect on the abundance and growth of young-of-the-year largemouth bass in early summer. Numbers of aquatic insects and small sunfish also increased.

10.3.4 Managing Water Levels to Enhance Recruitment

Lakes with seasonally stable water levels usually develop a variety of macrophyte species in the littoral zone. However, most large multipurpose reservoirs are characterized by extreme seasonal fluctuations in water level which eliminates macrophytes from the fluctuating zone. The lack of vegetative cover in the littoral zone and the coincidental harsh physical environment results in poor spawning habitat for nest-building fishes or excessive cannibalism of the year-class. The absence of cover (i.e., suitable nursery habitat) eliminates protection needed by young-of-the-year fishes. The impacts of this type of habitat problem for largemouth bass results in poor or inconsistent year-class strength in many large reservoirs. Variation in young of the year for abundance in fall samples are related to early season conditions that affect spawning and survival to the swim-up state, or factors affecting growth and survival of the fingerlings from early to late summer. Predation, including cannibalism, and starvation are the most important factors affecting survival in the post-spawning period.

Given the priority of flood control, power generation, and navigation when government management agencies (Army Corps of Engineers, Tennessee Valley Authority, Bureau of Reclamation) and public utilities establish water level schedules, adjustments to a reservoir plan for fisheries are usually limited, but they have been negotiated. In Kansas, Groen and Schroeder (1978) developed a water level management plan for walleye, and Beam (1983) modified the plan for white crappie. A careful analysis of the relationship between water storage and year-class strength can be a useful predictor of potential crappie harvest (Mitzner 1981). In natural lakes, lake levels have substantial impacts on the quality and quantity of spawning habitat and year-class strength (Kallemeyn 1987).

10.3.5 Loss of Fish from Lake Discharges

Large discharges of water following the spawning season may result in loss of large numbers of planktonic fish larvae (Walburg 1971). Water discharge rates may be too swift for many young-of-the-year fish to avoid entrainment in the discharge. Fish (larvae to adults) may be lost from an artificial lake through subsurface and surface discharges. The loss of walleye fry in subsurface discharges substantially affected year-class strength in Lewis and Clark Lake, a mainstream impoundment on the Missouri River, South Dakota (Walburg 1971). Loss from natural lakes will be limited to surface discharges. Walleye fry survival in Kansas reservoirs has been related to the storage ratio (reservoirs with low storage ratios have high discharges); reservoirs with low storage ratios had lower densities of walleye (Willis and Stephen 1987).

Loss of adult fish from a reservoir may occur through subsurface penstock discharges but more often fish are lost from surface discharges over the spillway, and through roller- and taintor-gates. Epilimnetic discharges from reservoirs may also result in substantial losses of fish. For example, Moen and Dewey (1978) found that sunfish, shad, and crappie made up 97% of an estimated 83.3 million fish larvae loss in one year and 95% of 122.4 million larvae lost in the second year. Clark (1942) estimated a loss of 47 fish per hectare in 68 days in a surface discharge from Lake Loramie, Ohio (607 hectare). Louder (1958) reported fish losses over the spillway from two reservoirs in southern Illinois were as high as 116 fish per hectare from Lake Murphysboro (64.8 hectares) during 1,152 hours of overflow (duration of actual spill over the dam) and 1.5 fish per hectare from Little Grassy Lake (486 hectares) in 885 hours. Fish loss is more common over modern concrete spillways with laminar flow at the tip than over a drop-board spillway with turbulent noisy flow (the noise may repel the fish); fish loss also varies by species.

There is no practical way to prevent loss of larvae or adult fish through roller- or taintor-gates of a large reservoir or from the emergency spillway. However, loss of pelagic fish larvae, such as walleye, from a large reservoir, can be reduced if a reservoir water level management plan is in effect. Such a plan anticipates the average springtime rise in water level and the reservoir is prepared to store as much water as possible at the time the fish are hatched (Kallemeyn 1987).

On smaller lakes that have a small spillway overflow for a short spring interval, it is possible to construct simple fish barriers that are effective. Powell and Spencer (1979) devised a parallel-bar barrier to prevent fish loss over spillways.

The horizontal bar barrier allowed small debris to slip through the bars, but retained the fish. This reduces maintenance and risk of dam failure. Bar spacing can be designed to retain fish of different size, but they cannot prevent fry and fingerlings from passing through. Spillway barriers used on ponds that are made of welded wire, hardware cloth, and chicken wire should not be used as they entrap debris which may endanger the dam.

10.4 MANAGEMENT TO ENHANCE FISHING AND QUALITY OF THE EXPERIENCE

10.4.1 Access Roads, Ramps, and Jetties

Earthen dikes (riprapped as needed to reduce erosion) protruding from the shoreline provide bank anglers with access to deeper water sites. Most anglers will not walk far from a parking lot to bank fish, and boat access is generally essential.

10.4.2 Artificial Reefs and Fish Attractors

Popular angling magazines emphasize the location and use of structure as a guide to effective sport fishing. In this sense, structure refers to any type of unique feature of underwater topography (drop-off, presence of a large boulder, a change from rocky to soft substrate, hard bottom, rock or rubble bottom, the edge of macrophyte beds, a submerged tree, or other features that will concentrate fish). This type of structure provides a surface for attachment of invertebrate foods of fish, spawning habitat, and protective cover for small fish.

Artificial structures placed in lakes to attract and concentrate adult and juvenile fishes for the purpose of improving angling catch rates, may be called artificial reefs or fish attractors. Brush piles, structures made with automobile tires, and stakebeds are artificial structures used to concentrate fish to increase angler catch rates. Artificial structures are valuable additions to habitat when lake basins are barren and the average angler has difficulty locating fish. When properly located and marked with buoys or identified on lake maps, artificial structures can help anglers locate fish in lakes without flooded timber or natural features to concentrate the fish. A useful pocket guide to the construction of freshwater artificial reefs has been prepared by Phillips (1990). It illustrates a variety of traditional and new types of artificial reefs.

Artificial reef performance and longevity are determined by materials, design (configuration), and construction (Brown 1986; Seaman and Sprague 1991). There are also legal, economic, and regulatory issues related to siting and construction that must not be overlooked (D'Itri 1986). Submerged trees left standing in coves, brush piles, and artificial structures concentrate fish and improve angler catch rates (e.g., see Prince and Brouha 1974; Wilbur and Crumpton 1974). They are attractive to sunfish, crappie, largemouth bass, and catfish. Crappie are more likely to be the species caught in greatest numbers near reefs. Several studies have shown that catch rate of sport fishes is greater near reefs than in otherwise comparable areas, but there is evidence that artificial structures have increased the potential for overharvest of largemouth bass, defined as the number of days for anglers to catch 40% of the largemouth bass. However, overharvest should not be a problem where appropriate size and creel limits are promulgated and enforced.

Artificial structures concentrate fish by providing cover, food, and spawning habitat (Prince et al. 1975). Artificial structures or standing timber will be colonized by periphyton (aufwuchs) and they provide invertebrate food for fishes (Pardue 1973; Prince et al. 1977). However, it is unlikely that even a hectare of artificial structure in a large, multipurpose reservoir will be sufficient to significantly affect population rate functions (growth, recruitment, and mortality) of even a single species over the entire reservoir. It is more accurate to consider that the function of artificial structures is to attract fish.

10.5 DESIGNING LAKES FOR FISHERIES MANAGEMENT

Creation of new fishing lakes provides the opportunity to create habitat diversity and to add conveniences for anglers. It is far better to include access, boat launching facilities, jetties, and submerged fish attractors in the planning stages and in the overall project budget than to try to include them in the annual operating budget of the fisheries management agency. Also, it costs less to construct structures before flooding a new impoundment than it costs to build them later. Access facilities may include all-weather roads, docks, drinking water, lighting, picnic facilities, restrooms, fish-cleaning stations, and camping facilities.

It is desirable to leave trees in coves to reduce the effect of waves on shorelines, to provide surface for periphyton and food for small fish, to provide spawning habitat, and to attract fish to sites convenient to anglers or out of the way of boat traffic. Because several sport fishes concentrate in coves with standing timber (Willis and Jones 1986), angler harvest is higher from timbered coves than nonwooded areas. If the addition of surface area from artificial structures (brush piles, tire reefs) is important, then it is even more important to leave a substantial area of natural woody vegetation standing in locations (e.g., small coves) where the structure will not be a hazard for boaters and water skiers.

The pre-impoundment construction phase is a good time to build brush piles in proximity to boat access points. Structures consisting of large trees, brush piles, tire reefs, and stake beds are more easily constructed before the lake basin is filled when they can be tied down with cables or anchored with concrete, than after the basin is filled.

Design of the outlet structure is one of the most important considerations in reservoir planning. Multiple penstock discharge points will provide options to regulate water quality of discharges. The discharge level may be above or below the thermocline as determined by water quality and by biological needs within the reservoir and the tailrace (Figure 10.7). In ponds and intermediate-size lakes, a drain should be installed. It can be used to manipulate predator–prey relations (summer drawdown) and for reducing lake volume for renovations or total poisoning.

Depth and slope of the basin are important factors determining environmental quality and lake fish populations (Hill 1986). Extremely deep lakes are generally unproductive, but very shallow lakes are often turbid, or they may have excessive growths of algae or vascular aquatic plants. Sediment basins, essentially subimpoundments, can be used to retain settleable solids in small upstream areas. Site

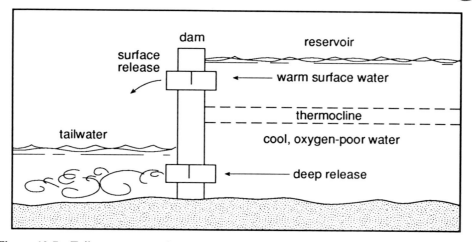

Figure 10.7 Tailwater areas of reservoirs are strongly influenced by level of discharge (surface, multiple, or bottom) from the reservoir. Surface discharges are warm, but deep near-bottom releases are cold and may be anoxic.

selection, dam height, and construction initiatives to maximize shoreline development will increase the land–water interface, enhance spawning habitat and the productive potential of the littoral habitat (Ryder 1978; Olmsted and Clugston 1986).

Gravel shoals may be provided for spawning of walleye and smallmouth bass, shallow vegetated bays are useful for yellow perch and northern pike, and reducing the extent of fetch and barren wind-swept beaches benefit largemouth bass and crappie. Enclosures should be considered to prevent carp from reaching interconnected marshy habitat for spawning. Selected areas of shoreline should also be considered for riprapping when erosion would adversely impact the quality of either the aquatic or surrounding terrestrial environment.

10.6 CONCLUSIONS

Lake and reservoir environments are affected by the watershed. Reservoirs are also dynamic, with seasonal and annual cycles superimposed on long-term trends in weather and ecological succession. Surveys and monitoring are needed to develop data bases, to observe trends, to evaluate management, and to plan new management strategies. In addition, long-term basic research is needed to provide new understanding of the complexity of biological interactions in lake communities (Carpenter 1988). There will also be a continual need for applied research to evaluate new habitat management options. Ecological studies of fish and lacustrine environments have provided a basic understanding of trophic relationships and environmental factors that control fish production and harvest. Fisheries scientists, however, need more information to manage habitat of lakes and reservoirs to achieve specific fisheries management goals.

10.7 REFERENCES

Aggus, L. R., and G. V. Elliott. 1975. Efforts of cover and food on year class strength of largemouth bass. Pages 317–322 *in* H. Clepper, editor. Black bass biology and management. Sport Fishing Institute, Washington, D.C.

American Society of Civil Engineers. 1979. Symposium on re-aeration research. American Society of Civil Engineers, Hydraulics Division, New York.

Baker, J. P. 1982. Effects on fish of metals associated with acidification. Pages 165–167 *in* R. E. Johnson, editor. Acid rain/fisheries. American Fisheries Society, Northeastern Division, Bethesda, Maryland.

Ball, R. C. 1952. Farm pond management in Michigan. Journal of Wildlife Management 16:269.

Beam, J. H. 1983. The effect of annual water level management of population trends of white crappie in Elk City Reservoir, Kansas. North American Journal of Fisheries Management 3:34–40.

Beamish, R. J. 1976. Acidification of lakes in Canada by acid precipitation and the resulting effects on fishes. Water, Air, and Soil Pollution 6:501–514.

Beard, T. D. 1973. Overwinter drawdown: impact on the aquatic vegetation in Murphy Flowage, Wisconsin. Wisconsin Department of Natural Resources Technical Bulletin 61.

Bennett, G. W., H. W. Adkins, and W. F. Childers. 1969. Largemouth bass and other fishes in Ridge Lake, Illinois, 1941–1963. Illinois Natural History Survey Bulletin 30:1–67.

Black, J. J. 1983. Field and laboratory studies of environmental carcinogens in Niagara River fish. Journal of Great Lakes Research 9:326–334.

Born, S. W., T. L. Wirth, J. O. Peterson, J. P. Wall, and D. A. Stephenson. 1973. Dilutional pumping at Snake Lake, Wisconsin—a potential renewal technique for small eutrophic lakes. Wisconsin Department of Natural Resources Technical Bulletin 66.

Breck, J. E., R. T. Prentki, and O. L. Loucks, editors. 1979. Aquatic plants, lake management, and ecosystem consequences of lake harvesting. University of Wisconsin, Institute for Environmental Studies, Madison.

Brouha, P., and E. D. Prince. 1973. How to build a freshwater artificial reef. Virginia Polytechnic Institute and State University, Sea Grant Extension Publication VPI-SG-73-03, Blacksburg.

Brown, A. M. 1986. Modifying reservoir fish habitat with artificial structures. Pages 98–102 *in* Hall and Van Den Avyle (1986).

Brugam, R. B. 1978. Human disturbance and the historical development of Linsely Pond. Ecology 59:19–36.

Burns, F. L., and I. J. Powling, editors. 1981. Destratification of lakes and reservoirs to improve water quality. Australian Government Publishing Service, Canberra.

Cairns, V. W., P. V. Hodson, and J. O. Nriagu. 1984. Contaminant effects on fisheries. Wiley, New York.

Candle, R. D. 1985. Scrap tires as artificial reefs. Pages 293–302 *in* F. M. D'Itri, editor. Artificial reefs: marine and freshwater applications. Lewis Publishers, Chelsea, Michigan.

Carline, R. F. 1986. Assessment of fish populations and measurements of angler harvest. Pages 46–56 *in* Hall and Van Den Avyle (1986).

Carlson, R. E. 1977. A trophic state index for lakes. Limnology and Oceanography 22:361–369.

Carpenter, S. R., editor. 1988. Complex interactions in lake communities. Springer-Verlag, New York.

Clady, M. D. 1976. Influence of temperature and wind on the survival of early stages of yellow perch, *Perca flavescens*. Journal of the Fisheries Research Board of Canada 33:1887–1893.

Clark, C. F. 1942. A study of the loss of fish from an artificial lake over a wasteweir, Lake Loramie, Ohio. Transactions of the North American Wildlife Conference 7:250–256.

Colle, D. E., J. V. Shireman, W. T. Haller, J. C. Joyce, and D. E. Canfield, Jr. 1987. Influence of hydrilla on harvestable sport-fish populations, angler use, and angler expenditures at Orange Lake, Florida. North American Journal of Fisheries Management 7:410–417.

Collins, G. G. 1976. Effects of dams on Pacific salmon and steelhead trout. U.S. National Marine Fisheries Service Marine Fisheries Review 38(1):39–46.

Conover, M. C. 1986. Stocking cool-water species to meet management needs. Pages 31–39 in R. H. Stroud, editor. Fish culture in fisheries management. American Fisheries Society, Fish Culture Section and Fisheries Management Section, Bethesda, Maryland.

Cooke, G. D., E. B. Welch, S. A. Peterson, and P. R. Newroth. 1986. Lake and reservoir restoration. Butterworth, Stoneham, Massachusetts.

Crowder, L. B., and W. E. Cooper. 1979. The effects of macrophyte removal on the feeding efficiency and growth of sunfishes: evidence from pond studies. Pages 251–268 in Breck et al. (1979).

D'Itri, F. M. 1986. Artificial reefs: marine and freshwater applications. Lewis Publishers, Chelsea, Michigan.

Driscoll, C. T., G. F. Fordham, W. A. Ayling, and L. M. Oliver. 1987. The chemical response of acidic lakes to calcium carbonate treatment. Lake and Reservoir Management 3:404–411.

Dunst, R., and S. Nichols. 1979. Macrophyte control in a lake management program. Pages 411–418 in Breck et al. (1979).

Dunst, R. C., and nine coauthors. 1974. Survey of lake rehabilitation techniques and experiences. Wisconsin Department of Natural Resources Technical Bulletin 75.

Durocher, P. P., W. C. Provine, and J. E. Kraai. 1984. Relationship between abundance of largemouth bass and submerged vegetation in Texas reservoirs. North American Journal of Fisheries Management 4:84–88.

Edmondson, W. T. 1969. Eutrophication in North America. Pages 124–149 in Eutrophication: causes, consequences, correctives. National Academy of Sciences, Washington, D.C.

Engle, S. 1989. Lake use planning in local efforts to manage lakes. Pages 101–105 in Proceedings of a national conference on enhancing states' lake management programs. Northeastern Illinois Planning Commission, Chicago.

Fast, A. 1981. Hypolimnetic aeration. Pages 201–218 in F. L. Burns and I. J. Powling, editors. Destratification of lakes and reservoirs to improve water quality. Australian Government Publishing Service, Canberra.

Fast, A. W., and R. G. Hulguist. 1989. Oxygen and temperature relationships in nine artificially aerated California reservoirs. California Fish and Game 75:213–217.

Fletcher, R. I. 1990. Flow dynamics and fish recovery experiments: water intake systems. Transactions of the American Fisheries Society 119:393–415.

Frey, D. G., editor. 1963. Limnology in North America. University of Wisconsin Press, Madison.

Fromm, P. O. 1980. A review of some physiological and toxicological responses of freshwater fish to acid stress. Environmental Biology of Fishes 5:79–93.

Garton, J. E., R. G. Strecker, and R. C. Summerfelt. 1978. Performance of an axial-flow pump for lake destratification. Proceedings of the Annual Conference Southeastern Association of Fish and Wildlife Agencies 30(1976):336–347.

Greenbank, J. 1945. Limnological conditions in ice-covered lakes, especially as related to winterkill of fish. Ecological Monographs 15:343–392.

Groen, C. L., and T. A. Schroeder. 1978. Effects of water level management on walleye and other coolwater fishes in Kansas reservoirs. American Fisheries Society Special Publication 11:278–283.

Hall, G. E., and M. J. Van Den Avyle. 1986. Reservoir fisheries management: strategies for the 80's. American Fisheries Society, Southern Division, Reservoir Committee, Bethesda, Maryland.

Hasler, A. D. 1947. Eutrophication of lakes by domestic drainage. Ecology 28:383–395.

Heiskary, S. A. 1985. Trophic status of Minnesota lakes. Minnesota Pollution Control Agency, Division of Water Quality, Monitoring and Analysis, Minneapolis.

Heman, M. L., R. S. Campbell, and L. C. Redmond. 1969. Manipulation of fish populations through reservoir drawdown. Transactions of the American Fisheries Society 98:293–304.

Henderson-Sellers, B., and H. R. Markland. 1987. Decaying lakes: the origin and control of cultural eutrophication. Wiley, Chichester, UK.

Hill, K. R. 1986. Classification of Iowa lakes and their standing stocks. Lake and Reservoir Management 2:105–109.

Jenkins, R. M., and D. I. Morais. 1971. Reservoir sport fishing effort and harvest in relation to environmental variables. American Fisheries Society Special Publication 8:371–381.

Johengen, T. H., A. M. Beeton, and D. W. Rice. 1989. Evaluating the effectiveness of best management practices to reduce agricultural non-point source pollution. Lake and Reservoir Management 5:63–70.

Johnson, R., and J. Skrypek. 1975. Prevention of winterkill of fish in a southern Minnesota lake through use of a helixor aeration and mixing system. Minnesota Department of Natural Resources, Division of Fish and Wildlife, Investigational Report 336, St. Paul.

Johnson, R. E., editor. 1982. Acid rain/fisheries. American Fisheries Society, Northeastern Division, Bethesda, Maryland.

Jones, J. R., and R. W. Bachmann. 1974. Prediction of phosphorus and chlorophyll levels in lakes. Journal of the Water Pollution Control Federation 48:2176–2182.

Kallemeyn, L. W. 1987. Correlations of regulated lake levels and climatic factors with abundance of young-of-the-year walleye and yellow perch in four lakes in Voyageurs National Park. North American Journal of Fisheries Management 7:513–521.

Keith, W. E. 1970. Preliminary results in the use of a nursery pond as a tool in fishery management. Proceedings of the Annual Conference Southeastern Association of Game and Fish Commissioners 23(1969):501–511.

Kimmel, B. L., and A. W. Groeger. 1986. Limnological and ecological changes associated with reservoir aging. Pages 103–109 in Hall and Van Den Avyle (1986).

Kramer, R. H., and L. L. Smith, Jr. 1962. Formation of year classes in largemouth bass. Transactions of the American Fisheries Society 91:29–41.

Lantz, K. E., J. T. Davis, J. S. Hughes, and H. E. Schafer, Jr. 1967. Water level fluctuation—its effect on vegetation control and fish population management. Proceedings of the Annual Conference Southeastern Association of Game and Fish Commissioners 18(1964):482–492.

Likens, G. E. 1972. Eutrophication and aquatic ecosystems. Pages 14–348 in F. G. Howell, J. B. Gentry, and M. H. Smith, editors. Mineral cycling in southeastern ecosystems. National Technical Information Service, CONF-740513, Springfield, Virginia.

Linder, A., and eight coauthors. 1990. Largemouth bass in the 1990s. In-Fisherman, Brainerd, Minnesota.

Loucks, O. L., and seven coauthors. 1979. Conference findings: an overview. Pages 421–434 in Breck et al. (1979).

Louder, D. 1958. Escape of fish over spillways. Progressive Fish-Culturist 20:38–40.

Marzolf, G. R., and J. A. Osborne. 1971. Primary production in a Great Plains reservoir. Internationale Vereinigung für theoretische und angewandte Limnologie Verhandlungen 18:126–133.

McGhee, M. 1990. Prescription for an aging lake. Iowa Conservationist 49(10):28–31.

Meyer, F. P., and L. A. Barclay. 1990. Field manual for the investigation of fish kills. U.S. Fish and Wildlife Service Resource Publication 177.

Mitzner, L. 1981. Influence of floodwater storage on abundance of juvenile crappie and subsequent harvest at Lake Rathbun, Iowa. North American Journal of Fisheries Management 1:46–50.

Mitzner, L. 1991. Effect of environmental variables upon crappie young, year-class strength, and the sport fishery. North American Journal of Fisheries Management 11:534–542.

Moen, T. E., and M. R. Dewey. 1978. Loss of larval fish by epilimnial discharge from DeGray Lake, Arkansas. Arkansas Academy of Science Proceedings 32:65–67.

Moser, B. B., and D. Hicks. 1970. Fish production of the stilling basin below Canton Reservoir. Proceedings of the Oklahoma Academy of Science 50:69–74.

Mosher, T. D. 1983. Effects of artificial circulation on fish distribution and angling success for channel catfish in a small prairie lake. North American Journal of Fisheries Management 3:403–409.

Nelson, R. W., G. C. Horak, and J. E. Olson. 1978. Western reservoir and stream habitat improvements handbook. U.S. Fish and Wildlife Service FWS/OBS-76/56.

Nelson, R. W., and C. H. Walburg. 1977. Population dynamics of yellow perch (*Perca flavescens*), sauger (*Stizostedion canadense*) and walleye (*S. vitreum vitreum*) in four main stem Missouri River reservoirs. Journal of the Fisheries Research Board of Canada 34:1748–1763.

Newberg, H. J. 1975. Evaluation of an improved walleye (*Stizostedion vitreum*) spawning shoal with criteria for design and placement. Minnesota Department of Natural Resources, Section of Fisheries Investigational Report 340.

Nriagu, J. O., and M. S. Simmons. 1984. Toxic contaminants in the Great Lakes. Wiley, New York.

Olmsted, L. L., and J. P. Clugston. 1986. Fishery management in cooling impoundments. Pages 227–237 *in* Hall and Van Den Avyle (1986).

Owen, O. S. 1980. Natural resource conservation. Macmillan, New York.

Office of Technology Assessment. 1985. Acid rain and transported air pollutants: implications for public policy. U.S. Congress, Office of Technology Assessment, Washington, D.C.

Pardue, G. B. 1973. Production response of the bluegill sunfish, (*Lepomis macrochirus*) Rafinesque, to added attachment surface for fish food organisms. Transactions of the American Fisheries Society 102:622–626.

Peters, J. C. 1986. Enhancing tailwater fisheries. Pages 278–285 *in* Hall and Van Den Avyle (1986).

Pfleiger, W. L. 1975. The fishes of Missouri. Missouri Department of Conservation, Jefferson City.

Phillips, S. H. 1990. A guide to the construction of freshwater artificial reefs. Sport Fishing Institute, Washington, D.C.

Ploskey, G. R. 1986. Effects of water-level changes on reservoir ecosystems with implications for fisheries management. Pages 86–97 *in* Hall and Van Den Avyle (1986).

Powell, D. H., and S. L. Spencer. 1979. Parallel-bar barrier prevents fish loss over spillways. Progressive Fish-Culturist 41:174–175.

Prentice, J. A., and R. C. Clark, Jr. 1978. Walleye fishery management program in Texas—systems approach. American Fisheries Society Special Publication 11:408–416.

Prince, E. D., and P. Brouha. 1974. Progress of the Smith Mountain Reservoir artificial reef project. Texas A&M University Sea Grant College TAMU-SG-74-103:68–72.

Prince, E. D., O. E. Maughan, P. Brouha. 1977. How to build a freshwater artificial reef. Virginia Polytechnic Institute and State University, Sea Grant Extension Publication VPI-SG-77-2, 2nd edition, Blacksburg.

Prince, E. D., R. F. Raleigh, and R. V. Corning. 1975. Artificial reefs and centrarchid basses. Pages 498–505 *in* H. Clepper, editor. Black bass biology and management. Sport Fishing Institute, Washington, D.C.

Ryder, R. H. 1978. Fish yield assessment of large lakes and reservoirs—a prelude to management. Pages 403–423 *in* S. D. Gerking, editor. Ecological freshwater fish production. Blackwell Scientific Publications, Oxford, UK.

Schneberger, E., editor. 1971. A symposium on the management of midwestern winterkill lakes. American Fisheries Society, North Central Division, Special Publication 1, Bethesda, Maryland.

Schnick, R. A., F. P. Meyer, and D. F. Walsh. 1986. Status of fishery chemicals in 1986. Progressive Fish-Culturist 48:1–17.

Schofield, C. L. 1976. Acid precipitation: effects on fish. Ambio 5:228–230.

Schubel, J. R., and B. C. Marcy, Jr. 1978. Power plant entrainment: a biological assessment. Academic Press, New York.

Seaman, W., Jr., and L. M. Sprague. 1991. Artificial habitats for marine and freshwater fisheries. Academic Press, San Diego.

Shields, J. T. 1958. Experimental control of carp reproduction through water drawdowns in Fort Randall Reservoir, South Dakota. Transactions of the American Fisheries Society 87:23–33.

Shirley, K. E., and A. K. Andrews. 1977. Growth, production, and mortality of largemouth bass during the first year of life in Lake Carl Blackwell, Oklahoma. Transactions of the American Fisheries Society 106:590–595.

Smith, S. A., D. R. Knauer, and T. L. Wirth. 1975. Aeration as a lake management technique. Wisconsin Department of Natural Resources Technical Bulletin 87.

Strange, R. J., W. B. Kittrell, and T. D. Broadbent. 1982. Effects of seeding reservoir fluctuations zones on young-of-the-year black bass and associated species. North American Journal of Fisheries Management 2:307–315.

Summerfelt, R. C. 1975. Relationship between weather and year-class strength of largemouth bass. Pages 166–174 *in* H. Clepper, editor. Black bass biology and management. Sport Fishing Institute, Washington, D.C.

Thornton, K. W., B. L. Kimmel, and F. E. Payne. 1990. Reservoir limnology. Wiley, New York.

Toetz, D. W., R. C. Summerfelt, and J. Wilhm. 1972. Biological effects of artificial destratification in lakes and reservoirs—analysis and bibliography. U.S. Department of Interior, Bureau of Reclamation, Report RED-ERC-72-33, Washington, D.C.

USFWS (U.S. Fish and Wildlife Service). 1988. 1985 national survey of fishing, hunting, and wildlife associated recreation. USFWS, Washington, D.C.

van Breeman, N., C. T. Driscoll, and J. Mulder. 1984. Acidic deposition and internal proton sources in acidification of soil and waters. Nature (London) 307:599–604.

Van Den Avyle, J. J., and R. W. Petering. 1988. Inundated timber as nursery habitat for larval gizzard shad and threadfin shad in a new pumped storage reservoir. Transactions of the American Fisheries Society 117:84–89.

Walburg, C. H. 1971. Loss of young fish in reservoir discharge and year-class survival, Lewis and Clark Lake, Missouri River. American Fisheries Society Special Publication 8:441–448.

Weithman, A. S., and M. A. Haas. 1982. Socioeconomic value of the trout fishery in Lake Taneycomo, Missouri. Transactions of the American Fisheries Society 111:223–230.

Wetzel, R. G. 1975. Limnology. Saunders, Philadelphia.

Wilbur, R. L., and J. E. Crumpton. 1974. Florida's fish attractor program. Texas A&M University Sea Grant College TAMU-SG-74-103:39–46.

Willis, D. W. 1986. Review of water level management of Kansas reservoirs. Pages 110–114 *in* Hall and Van Den Avyle (1986).

Willis, D. W., and L. D. Jones. 1986. Fish standing crops in wooded and non-wooded coves of Kansas reservoirs. North American Journal of Fisheries Management 6:105–108.

Willis, D. W., and J. L. Stephen. 1987. Relationship between storage ratio and population density, natural recruitment, and stocking success of walleye in Kansas reservoirs. North American Journal of Fisheries Management 7:279–282.

Wirth, T. L. 1981. Experiences with hypolimnetic aeration in small reservoirs in Wisconsin. Pages 457–467 *in* F. L. Burns and I. J. Powling, editors. Destratification of lakes and reservoirs to improve water quality. Australian Government Publishing Service, Canberra.

Wood, R., and D. W. Pfitzer. 1960. Some effects of water-level fluctuations on the fisheries of large impoundments. Pages 118–138 *in* Soil and water conservation, volume 4. International Union for the Conservation of Nature and Natural Resources, Brussels, Belgium.

Young, W. C. 1973. Plants for shoreline erosion control in southern areas of the United States. Geophysical Monograph 17:798–803.

Zweiacker, P. L., and L. G. Bowles. 1976. Aquatic effects of steam-electric generating plants. Annals of the Oklahoma Academy of Science 5:124–135.

Chapter 11

Maintenance of the Estuarine Environment

WILLIAM H. HERKE AND BARTON D. ROGERS

11.1 INTRODUCTION

Estuarine zones are extremely complex transition areas between freshwater and marine habitats. They are the permanent home for numerous species and the temporary home for many others. Without estuarine zones, our saltwater commercial and recreational fisheries would be drastically reduced. Even so, human actions continue to degrade estuaries, threatening these fisheries.

11.1.1 Habitat Description

The classic definition of an estuary is a semi-enclosed coastal body of water having a free connection with the open sea and within which the sea water is measurably diluted with fresh water deriving from land drainage (Cameron and Pritchard 1963). Pritchard (1967) categorized estuaries into four basic types: drowned river valleys, bar-built, fjord-type, and tectonically produced. Drowned river valleys form the typical estuaries along the eastern coast of North America (e.g., Chesapeake Bay). Bar-built estuaries occur when offshore sand bars build above sea level and create a sound; the Outer Banks of North Carolina form barrier islands to make Albemarle and Pamlico sounds. On the West Coast, glacial gouging, such as along the coast of British Columbia and southeastern Alaska, has created fjord-type estuaries; tectonic processes, such as faulting or subsidence, formed others (e.g., San Francisco Bay). Estuaries provide various gradients from mouth to head, surface to bottom, and side to side, resulting in many diverse habitats and complex natural interactions within a relatively small area.

Lagoons also occur in the transition zone between the land and the sea; they bear strong similarities to bar-built estuaries.

Coastal lagoons are usually differentiated from estuaries on a geomorphological basis; an estuary is commonly considered as the mouth of a river while a coastal lagoon is an embayment separated from the coastal ocean by barrier islands. . . . From an ecological point of view, however, coastal lagoons and estuaries constitute a similar type of ecosystem and we can speak of a lagoon–estuarine environment. . . .

The lagoon–estuarine environment is characterized as a coastal ecotone, connected to the sea in a permanent or ephemeral manner. They are shallow bodies of water, semi-enclosed of variable volumes depending on local climatic and hydrologic conditions. They have variable temperatures and salinities, predominantly muddy bottoms, high turbidity, and irregular topographic and surface characteristics. The flora and fauna have a high level of evolutionary adaptation to stress conditions, and have originated from marine, freshwater and terrestrial sources. The biota of these coastal ecosystems include a varied flora and fauna; this biota is directly important to man for what it yields, and to many marine and fresh water organisms which use estuaries. In this natural conditions [sic], the ecosystem incorporates a balanced network of biotic interrelationships. The natural balance is all too easily upset by human impacts. (From Day and Yáñez-Arancibia 1985.)

The above description of the lagoon–estuarine environment differs from that of Cameron and Pritchard (1963) and Pritchard (1967) by requiring that the water bodies be shallow and by not requiring a free connection to the open sea. Moreover, neither description specifically includes adjoining transition zones, such as coastal marshes, tidally influenced freshwater habitats, or areas offshore of estuaries, particularly where large rivers empty into the ocean. In this chapter, an estuarine zone is defined as an environmental system consisting of the estuary and those transitional areas consistently influenced or affected by water from the estuary (with the understanding that the term estuary shall also include lagoons).

Even in the natural state, sedimentation and sea level variations cause changes on a geologic time scale, whereas climatological (such as El Niño changes, annual sea level changes, droughts, and hurricanes) or tectonic events cause shorter-term effects (Wolfe and Kjerfve 1986). Compared with most ecosystems, many scientists consider the estuarine zone to be a harsh environment because of its rapid and sometimes extreme—short- and long-term—changes in physicochemical factors.

Wind-driven currents and water level fluctuations are the dominant physical forces in certain bar-built estuaries and shallow coastal marshes. However, in most estuaries, tides and their associated tidal currents exert the primary short-term control of many of the physical processes. Generally there are three types of tides: semidiurnal, diurnal, and mixed. Semidiurnal tides, such as those on the Atlantic coast, produce two high-water and two low-water periods during an approximate 24-h period. Diurnal tides occur on the northern Gulf of Mexico coast and generally produce one high and one low tide per day. Mixed tides oscillate between semidiurnal and diurnal and are characteristic of Pacific coast estuaries and some areas of the Gulf of Mexico. Tidal amplitudes vary regionally; for example, they range from less than 0.3 m in the northern Gulf of Mexico, to 2–2.5 m along Georgia and South Carolina, to over 16 m in the Bay of Fundy. They also vary monthly and seasonally in each region. The physicochemical impacts of the tides on the resident biota are extensive and highly variable; the abiotic variability which characterizes coastal environments, compared to more static environments such as inland lakes, increases the complexity of biotic–abiotic interactions. The subject of estuarine physical processes is much more complex than presented here (for more see Kjerfve 1989).

Chemical processes are no less complex than physical processes, and salinity is the dominant factor. Salinities of open oceans generally range from 33 to 37‰ (Sverdrup et al. 1942), while in the estuarine zone they usually vary from 0 to about 35‰. Hypersaline lagoons, such as the Laguna Madre on the southeastern Texas coast, may have considerably higher salinities—as high as 290‰ (Copeland 1967).

Most physicochemical factors vary during the hours of a tidal cycle, seasonally, and from year to year. As the tidal current ebbs and floods, the physicochemical factors at any particular point will vary according to the characteristics of the water passing the point at that moment. Increased riverine input on a seasonal basis (e.g., spring thaw or rainy season), can reduce salinities throughout the estuary for long periods. Shallow estuaries that tend to be dominated by wind forces may show tremendous variability in water levels, temperature, and salinity on an irregular basis. For the shallow marshes of southwestern Louisiana, short-term changes in some physicochemical factors are caused by passages of strong atmospheric cold fronts (Figure 11.1). Climatic variations of longer duration also result in long-term variability.

Water and soil salinities, along with physical factors (such as periodicity of flooding or desiccation, subsurface circulation, drainage, porosity, etc.) that control these salinities, determine the types of vegetation that occur in the estuarine zone. Lower variations in salinity and water levels characterize the landward zones, and if salinities are generally less than 5‰, freshwater vegetation predominates (Mitsch and Gosselink 1986). As salinity and tidal amplitude increase in the seaward zones, the vegetation changes to plants tolerant of brackish and saline water, frequently resulting in less diverse or essentially monotypic stands of marsh plants or mangroves, or one of the submerged seagrasses. Many studies have shown these marsh or mangrove areas, or seagrass beds, to be among the areas of highest primary productivity in the world. Their primary production, and the nutrients from both the land and sea that are periodically or constantly supplied to, or regenerated within, the estuarine zone make it extremely productive for fisheries. The complexity and productivity in the estuarine zone, with many alternate pathways, give the nekton communities some year-to-year stability in a variable physical ecosystem.

11.1.2 Life Cycles of Estuarine Zone Organisms

Biological adaptations are the key factors in the life histories of coastal fisheries species. Yáñez-Arancibia et al. (1985) discussed the hypothesis that the life histories of the dominant species are coupled with the physical processes in the ecosystem as evolutionary adaptations to stress conditions, and that the species have migration patterns from marine, freshwater, and truly estuarine sources. Organisms that inhabit estuarine zones have adapted to an ever-changing environment. Many of these organisms are euryhaline and eurythermal; that is, they can tolerate a wide range of salinities or temperatures, respectively. Some do this at all stages of their life cycle, whereas others accommodate wide ranges of physical factors by varying their location along the numerous estuarine gradients as their tolerance changes during their successive life states.

Although several different categorical schemes have been applied to fisheries organisms in the estuarine zone, two basic categories will be discussed here: residents and transients. Anadromous species are discussed in Chapter 23.

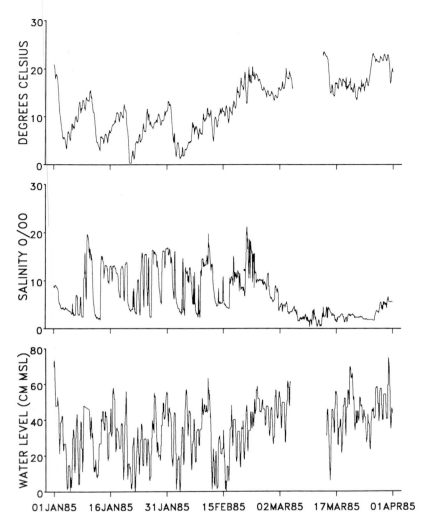

Figure 11.1 Hourly values for temperature, salinity, and water level in Grand Bayou in southwestern Louisiana. The large, abrupt changes in January and early February were caused by the passage of strong cold fronts. Data gaps were due to data logger malfunction.

Residents are organisms that remain in the estuarine zone and accordingly must be able to tolerate varying physicochemical conditions throughout their lives. Except for oysters, crabs, and a few others, most of these species are not of direct commercial or recreational importance. However, some have great ecological significance in that they provide important links in the food web for commercial and recreational species. They may also be important in the process of decomposing the remains of vascular plants to detritus (which forms the base of some food webs) or in recycling or resuspending various nutrients into the water column.

Many of the transient organisms are important recreationally and commercially. Some transient species occur in the estuarine zone only occasionally and do not require its use to complete their life cycle. Others must use the estuarine zone for

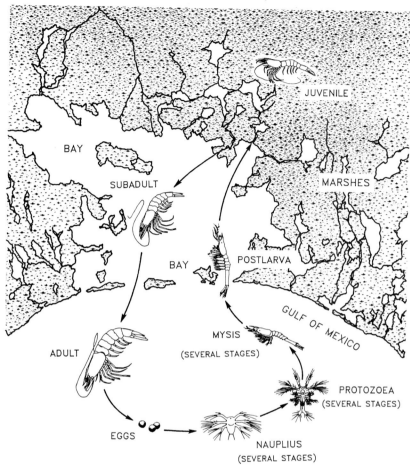

Figure 11.2 Life cycle of penaeid shrimp of the U.S. Atlantic and northern Gulf of Mexico. Although fish undergo fewer morphological changes, these life cycle movements are characteristic of most economically important estuarine-dependent species in this area. (Drawing by Michelle Lagory.)

some portion of their life cycle if their members are to survive in appreciable numbers; these are classified as estuarine-dependent species. (Some scientists prefer the term estuarine-marine.) Excluding anadromous and catadromous species, estuarine-dependent organisms of the Atlantic and northern Gulf coasts typically spawn offshore, nearshore, or in lower bays. The larvae or juveniles, or both, migrate into the estuarine zone, often proceeding into the lower salinity waters. In Louisiana, shallow-water marsh areas tend to have considerably higher nekton densities than the adjacent large open water areas (Herke 1971; Day et al. 1982). The young organisms grow rapidly for a few weeks to a few months in this nursery area and then begin their return to the open sea to complete the cycle. The life cycle of penaeid shrimp (Figure 11.2) serves as a good example of an estuarine-dependent species.

Most estuarine-dependent species have peak migrations in and out of the nursery areas during the year; timing of the peaks varies among species. At any

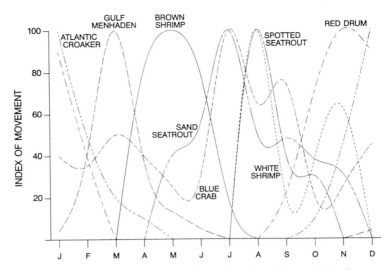

Figure 11.3 Index of movement (immigration and emigration combined), averaged over the period 1980–1982, for eight important species as measured by over 6,000 passive trap samples in southwestern Louisiana. (From Rogers and Herke 1985.) Index = (average catch per trap each month/highest monthly average) × 100; thus, the index for the month of highest catch = 100. At least 120 additional species migrated through the same channels during the year. (Brown shrimp *Penaeus aztecus*; white shrimp *Penaeus setiferus*.)

given time during the year many species are using a nursery area; evolutionary processes have partitioned out the load so that there is efficient use year-round. Figure 11.3 depicts the timing of migrations in the Louisiana coastal marsh of eight commercially or recreationally important species.

There are fewer estuaries on the West Coast, and open coast species predominate. Even so, there are a number of estuarine-dependent species (for example, several species of salmon, and one or more species of flatfishes and crabs) but their life cycles do not necessarily follow the above pattern. (For further details see Chapter 23.)

Tropical estuarine-zone populations are characterized by greater numbers of species, but members of these species occur in less abundance than members of the dominant species of cooler regions. Juvenile fishes and shrimps in temperate estuarine areas appear to respond, by emigration, to the interaction of seemingly minor changes in environmental factors (Herke 1971, 1977). There is less variation in physical factors in the tropics, especially in temperature and salinity. Nevertheless, it has been proposed that even in the tropics small variations in such factors as salinity, temperature, turbidity, sediment type, hydrography, and circulation patterns may be very important elements in controlling species biology and production (see Day and Yáñez-Arancibia 1985).

11.1.3 Value of Estuarine-Dependent Fisheries

Fisheries of the estuarine zone are of great economic importance. In many areas, coastal fisheries dominate the entire economy, from docks and processing plants, to marinas and boat sales, to restaurants, and sometimes even to the tourist industry.

An estimated two-thirds of the commercial fisheries catch in the Atlantic and associated waters is composed of estuarine-dependent species; marine fisheries organisms are more important in the Pacific coast commercial catch, but estuarine-dependent species still account for approximately 46% in landings value (McHugh 1966). In 1987, U.S. commercial landings of fish and shellfish equaled 3.1 million tonnes (6.9 billion pounds) with a $3.1 billion dockside value (NMFS 1988). These amounts are conservative because an unknown proportion of the catch is unreported or underreported. While these values include species that are not considered estuarine-dependent, menhaden made up 39% (1.2 billion kilograms) of the 1987 landings, whereas penaeid shrimp accounted for 17% ($516 million) of the value. These two groups are classically estuarine dependent. Salmon, which are also estuarine dependent, were third in landings and first in value (255 million kilogram valued at $596 million). These three estuarine-dependent groups alone accounted for over 50% by weight, and nearly 40% by value, of the entire U.S. commercial catch. Consider also that the preceding are dockside values only; the total economic value is considerably greater (see Chapter 7). Also, from the commercial viewpoint, the estuarine zone attracts many tourists, and part of its attraction is the prevalence of tasty fresh seafood such as Maryland blue crab *Callinectes sapidus*, Pacific salmon, or Louisiana shrimp Creole.

Determining the recreational and aesthetic value of estuarine fisheries is difficult. The National Marine Fisheries Service (NMFS 1987a) estimated for 1986 that 411.2 million fishes were taken by recreational anglers along the Atlantic and Gulf coasts; over 80% were taken in inland coastal waters (e.g., rivers, sounds, and bays) or within 3 miles of shore. The National Marine Fisheries Service (NMFS 1987b) estimated that 55.3 million fishes were taken on the U.S. Pacific coast during the same year, of which 81% were taken either in inland coastal waters or no more than 3 miles out in the ocean. Estuarine fisheries are obviously of extreme importance both commercially and recreationally.

11.2 HABITAT LOSS AND CHANGE

Habitat loss and change in the estuarine zone is constantly occurring. This loss and change can be categorized as either natural or human induced, although often the effects of the latter compound (or are at least additive to) and are more rapid than natural effects.

11.2.1 Natural Effects

A major natural phenomenon is subsidence, which results when geologic processes cause a general lowering of the land area. Shallow water areas become deeper and areas that were upland become submerged. Many of the plants die and are replaced by other species, or no vascular plants survive and eventually open water occurs. These habitat changes cause a shift in the food web and a subsequent change in the macrofaunal community structure.

Sediment deposition causes essentially the opposite situation. Open water areas fill in, sometimes resulting in the smothering of productive shellfish beds or submerged seagrass flats, and mud flats or emergent marsh appear.

Natural changes in apparent sea level (land subsidence plus change in eustatic sea level) have resulted in estuarine zone changes throughout both geologic and recent history. Stevenson et al. (1986) reviewed 15 areas dated with lead-210 or cesium-137 and found that substrate accretion at four areas was clearly not keeping pace with apparent sea level rise; moreover, these four areas were from regions that compose approximately half the tidal marsh acreage in the United States.

Droughts may cause normally brackish areas to become hypersaline, as happens along the Texas coast. Conversely, abnormally high rainfall sometimes causes extreme drops in salinity, as when much of the Chesapeake Bay became fresh in 1973 following a hurricane. Hurricanes may also increase salinities and turbidities, and cause large-scale physical changes in the estuarine zone.

Tidal passes between barrier islands open and close, causing hydrographic changes, and the barrier islands themselves may migrate to occupy different positions and configurations.

Naturally occurring estuarine habitat losses or changes are usually not too serious, at least in the long term. They may be minor in scope, offset by compensating changes elsewhere; they may be short term; or they may occur slowly enough to allow the flora and fauna to adjust to the changes. Estuarine organisms evolved in a highly variable environment and are generally resilient to natural changes. Populations decimated by natural catastrophes have the ability to rebuild once conditions return to "normal." Unfortunately, the same is not necessarily true of populations diminished by human-induced habitat loss and change because conditions may never return to "normal."

11.2.2 Human-Induced Effects

Construction of ports and marinas produces one of the most obvious human-induced effects since it results in a direct loss of habitat. Construction also causes altered flow patterns and increased pollution. Channelization of existing water-ways, or construction of new ones, also results in observable direct loss of habitat; however, the subsequent alteration of flow patterns frequently causes more serious harm. For example, greater and more rapid tidal exchanges may (1) cause sediment deposition over productive areas, (2) change the faunal community composition, or (3) change the salinity regime sufficiently to kill existing vegetation and prevent establishment of alternate species, thereby reducing the detrital food base and also causing substrate erosion. The myriad canals and associated levees dug through marshes for oil field access and navigation have been indicated as a major cause for the annual coastal erosion of about 13,000 hectares in Louisiana (Turner 1987).

Human population density in coastal areas is increasing rapidly. For example, the U.S. population per square mile in 1950 and 1985, respectively, by coastal region, was: 283.3 and 434.7 (Atlantic); 59.8 and 151.9 (Gulf of Mexico); and 97.6 and 237.6 (Pacific) (U.S. Bureau of the Census 1986). Acceding to the tremendous desire to live near the coast results in a high environmental cost. Beach-front construction destabilizes the natural responses of beaches or sand dunes to storms and erosion. The need for such amenities as housing, roads, bridges, and waste disposal increases as population increases.

Further increases in the rate of sea level rise may result from human activities (i.e., global warming). Combustion of fossil fuels, deforestation, and cement manufacture have caused the atmospheric concentration of carbon dioxide to increase by 20% (Titus 1986). Other trace gases that also trap heat (the greenhouse effect), such as chlorofluorocarbons, have also been increasing as a result of human activities (Ramanathan et al. 1985). The increase in the earth's temperature will cause expansion of seawater and increased seaward drainage from the melting of glacial ice. Estimates of sea level rise range from 10 to 21 cm by 2025 and 57 to 368 cm by 2100 (Titus 1986).

The impacts of sea level rise on the estuarine zone will be wide ranging; the most obvious will be increased inundation and erosion. Saltwater will intrude farther inland, changing the flora and fauna, and even threatening sources of drinking water. Narrow marsh zones on high relief coastlines (e.g., northwestern coast of North America) may move farther inland or may disappear. Wide marsh zones, normally associated with low relief coastlines (e.g., southeastern and Gulf coasts of the United States) may migrate inland, but large areas will be lost if the habitats cannot migrate inland or adjust as quickly as the sea level rises. At present it is difficult to express specific marsh responses to sea level rise.

The impacts of sea level rise on shoreline developments could be tremendous. The choice would be to either retreat, or build levees and install pumps to hold back the water (Titus 1986); either will be expensive and the latter will prevent marshes from migrating landward. Another problem associated with sea level rise is that areas that are now upland, and practically unregulated by governing agencies, will eventually become wetlands. Will these areas then become regulated? Legal battles will probably arise from this issue. We have barely touched on the complex, and still largely conjectural, subject of the effects of global climate change on estuarine-dependent fisheries (see Kennedy 1990).

Changes in river courses cause a decrease in sediment input to some areas and a subsequent increase in other areas; changes in the Santee and Cooper river deltas in South Carolina furnish an example. Some years ago, dams were constructed that created two large reservoirs and diverted some of the Santee River flow down the Cooper River. Siltation in the Cooper River delta increased and brackish marsh crept up the Santee River. Now, water is being rediverted to increase flow in the Santee River and reduce siltation in Charleston harbor at the mouth of the Cooper River. Increased flow down the Santee River will move the interface of fresh and brackish water seaward and increase siltation in the Santee delta.

Water-control structures, impoundments, ditches, and levees or dikes are placed in estuaries and associated wetlands for various reasons. Usually, the main purpose of these structures is to maintain a certain hydrological regime that favors agriculture or particular animal groups such as ducks (or in the case of mosquito control, to disfavor certain species). Ditching or diking has been widely used in the past for mosquito control. Impoundments are constructed and managed in Florida to reduce production of saltmarsh mosquitos. Initially these impoundments were managed strictly for mosquito control; however, with recent scientific information, these management procedures are incorporating measures to mitigate their damages to fish and wildlife habitat. Elsewhere, some levees are placed somewhat incidentally, as in the construction of causeways; other alterations may be caused by the piling of spoil along the side of a dredged channel. These levees and spoil piles can result in unintentional areas of impoundment and soil waterlogging, and eventual dieback of the marsh vegetation (Mendelssohn et al.

1981). In some areas, notably coastal Louisiana, levees and water-control structures are also employed in an attempt to reduce saltwater intrusion and to provide the capability to achieve seasonal drawdowns for reestablishment of emergent vegetation. Since many estuarine-dependent organisms migrate into the estuarine zone as larvae or juveniles (or both), they generally migrate with tidal flow. If tidal flow is substantially altered or prevented, migration pathways to nursery areas are altered or blocked.

Until recently scientific documentation of the effects of impoundment on fisheries has been lacking. A multifaceted study was conducted in South Carolina in which the objectives were to evaluate the effects of coastal impoundments on sedimentation; hydraulic cycles; nutrient exchange; primary productivity; and the plankton, benthos, nekton, and wildlife communities (DeVoe and Baughman 1987). Total annual tidal exchange of dissolved organic carbon, particulate organic carbon, and phytoplankton biomass was similar between the impoundments and an adjoining natural area, but the timing of this exchange varied because exchange from the impoundments could only occur when their water-control structures were open. Timing of water exchange between impoundments and the natural system interfered with the normal life cycles of many nektonic species. On the other hand, management of impoundments for certain wildlife food plants created improved habitat for ducks. Thus, managed coastal impoundments may provide good habitat for target species (ducks in this case), but create barriers to migrations of many estuarine-dependent organisms. The impact of these barriers can be at least partially mitigated by operation of the water-control structures to allow passage of estuarine-dependent fisheries organisms during their periods of peak movement.

Solid, low-level dams are used in Louisiana to create a similar, but much larger, form of semi-impoundment. The dams (called weirs in Louisiana) are placed in tidal bayous or canals with their crests set about 15 cm below average marsh soil level. Water flows into the marsh when the tide rises above the level of the crest and the water inside; as tide level falls, water flows out only until the water level reaches crest level. Semi-impoundments often provide, at least temporarily, improved duck habitat by increasing growth of submergent vegetation, such as widgeongrass *Ruppia maritima*. However, studies indicate semi-impoundments may result in the loss of emergent marsh plants, thus increasing marsh erosion. Moreover, these weirs substantially reduce the export of estuarine-dependent organisms from the semi-impounded areas. When the total export of fisheries organisms from two matched 35-hectare ponds was compared over two complete annual cycles, 107 species were taken from the pond with no weir, whereas only 83 were taken from the pond with the weir (Herke et al. 1992). For both years, export of numbers and biomass—in terms of nearly all estuarine-dependent species—was substantially reduced from the weired pond. Also, a large-scale comprehensive study (Cahoon and Groat 1990) similar to that reported by DeVoe and Baughman (1987), but also involving photo interpretation of 16 existing semi-impoundments, reported results resembling those of DeVoe and Baughman (1987) and cast strong doubt on the ability of semi-impoundments to reduce coastal land loss in Louisiana.

There is an abundance of literature describing the actual, or potential, benefits of mariculture—the farming of saltwater aquatic organisms. However, it must be recognized that such endeavors may adversely affect the estuarine environment. Loss of valuable natural nursery grounds, accelerated eutrophication caused by mariculture effluents, and interference with natural water movements, as well as the host of problems associated with introduced species and stocks (see Chapter 12), are just a few of the potential impacts culture ventures can pose to the estuary. Such risks need to be considered in the benefit–cost ratio of existing and potential culture operations that are in proximity to estuaries.

Pollution is possibly the most serious of all threats to the estuarine zone. Pollution can come from such point sources as sewage and industrial outfalls, and from nonpoint sources such as farm and urban runoff, far up the watershed. Pollution can even come from outside the watershed via intentional transport (barge, truck, rail, etc.) and atmospheric deposition. Acid rain may be a less serious problem in estuarine than in fresh waters because of the buffering effect of sea salts. However, Fisher et al. (1988) estimated that 39% of the nitrogen input to Chesapeake Bay was from atmospheric deposition in 1984.

The effects of pollution on estuarine organisms can be (1) subtle, hard to measure, and difficult to understand, such as reduced growth rates and increased susceptibility to predators, disease, and environmental stresses; or (2) somewhat more apparent, as with increased internal lesions and cancerous growths in fishes correlated with increased levels of toxic materials (e.g., Malins et al. 1982), or statistically associated with externally visible anomalies and disease conditions (e.g., Ziskowski et al. 1987); or (3) obvious and catastrophic, as with the occurrence of massive fish kills. The types of pollution vary from location to location, and the means of reducing it must be tailored to the individual sites.

Human-induced changes in freshwater inflow to estuaries may have poorly understood, subtle, and yet far-reaching effects. Inflow changes may result from nearby stream diversion, or from actions taken hundreds of kilometers up watersheds such as irrigation withdrawal or dam construction and subsequent changes in the discharge cycle. (Dams also block migrations and trap sediment that otherwise would have been transported to the estuarine zone to form substrate for marsh building.)

Changes in freshwater inflow affect the composition of phytoplankton species and the factors responsible for phytoplankton growth such as light, nutrients, and the vertical stability of the water column. Freshwater inflow effects are less well defined for zooplankton, and the elucidation of effects on larval fish and the resulting adult populations is hampered by the lack of understanding of factors governing larval survival and year-class strength. Much work remains before a comprehensive understanding of the role of freshwater inflow to estuaries is achieved (Drinkwater 1986). However, some direct effects of inflow changes can be deduced. For example, in some estuaries, larvae are retained in the estuary by the counterflowing tidal and riverine currents (e.g., Graham 1972), and changes in freshwater inflow undoubtedly affect larval retention and year-class strength.

11.3 THE MANAGEMENT PROCESS

Up to this point we have discussed some natural processes, fisheries values, and human-induced perturbations in the estuarine zone. Responsibility for maintenance of the physical and biological integrity of the estuarine zone lies with individual, private, local, state, regional, national, and international interests. How some of these interests attempt to maintain the estuarine zone will now be summarized.

11.3.1 Federal Agency Responsibilities

There are many federal laws and regulations in North America that are intended to help maintain the physical and biological integrity of the estuarine zone (many of these are discussed in Chapter 4). Sometimes there are overlaps and gaps of responsibility. Box 11.1 presents a partial summary of the regulatory and management functions for some of the major federal agencies that attempt to protect the estuarine zone. The following section elaborates on only a few points.

In the United States, the Water Quality Act of 1987 authorizes the Environmental Protection Agency to work with states and other federal agencies to conduct a National Estuary Program, a comprehensive study of the management

---

Box 11.1   Partial summary of regulatory and management functions for some of the federal agencies of the United States, Canada, and Mexico. (Changes were being proposed when this was written; check with agency headquarters or regional offices for latest developments.)

UNITED STATES[1]

Army Corps of Engineers (ACE)—Administers Section 10 of the River and Harbor Act of 1899 and the permit program of Section 404 of the Clean Water Act. Processes permit applications by soliciting comments from the public and concerned agencies for a 30-day period via a comment period, conducts a full public interest review of probable impacts including cumulative impacts of the proposed activity.

Environmental Protection Agency (EPA)—Comments to the ACE on Section 10 and 404 permit applications. Section 404(c) of the Clean Water Act provides the EPA with authority to prevent discharge of dredged or fill material into a defined area in waters whenever such discharge will have an adverse effect on municipal water supplies, shellfish beds and fishery areas, wildlife, or recreational areas.

Fish and Wildlife Service (FWS)—Under authority of the Fish and Wildlife Coordination Act, reviews and comments on Section 10 and 404 permit applications, federal water resource project proposals, and provides technical assistance to protect fish and wildlife resources and mitigate project impacts. Other FWS activities in estuarine habitats include research on estuarine systems, management of anadromous fisheries, endangered species protection and management, and wetland acquisition and habitat management on wildlife refuges.

National Oceanic and Atmospheric Administration (NOAA)

  National Marine Fisheries Service—Evaluates and provides recommendations on federal regulatory and construction activities to protect living marine resources (LMR) and mitigate adverse impacts. Conducts research on life history of LMR and on factors affecting LMR and its habitat.

  Estuarine Programs Office—Responsible for coordinating estuarine research policy for NOAA under the mandate of Public Law 99-659.

Office of Ocean and Coastal Resource Management—Administers the Coastal Zone Management Act of 1972 through which the federal government offers coastal states financial help to develop state coastal zone management plans that meet federal standards.

CANADA[2]

Environmental Protection—Conducts sanitary and bacteriological surveys on the quality of shellfish growing waters and determines if sale is safe for human consumption; monitors effluent quality and compliance with regulations; monitors nonpoint source pollution; handles all ocean dredging and dumping activities.

Inland Waters-Lands—Develops and implements federal provincial water management agreements; maintains data on water and land use; conducts studies relative to water and land management issues; monitors streamflow and collects chemical, biological and bacteriological data on water, sediment, and biota.

Canadian Wildlife Service—Migratory birds management and habitat conservation; wetland acquisition and management; rare and endangered species management; and effects of pollutants on migratory birds and their habitat.

Atmospheric Environment Service—Provides climatological information on weather and conducts atmospheric research.

Department of Fisheries and Oceans—Responsible for the management of fisheries and fish habitat; reviews all projects that may impact fisheries or fish habitat; provides information on fish stocks, locations, and catch regulations; responsible for regulation of commercial fisheries and enforces the Fisheries Act.

Energy, Mines and Resources—Involved in monitoring of temporal changes in coastal and estuarine morphology and sedimentation rates.

MEXICO[3]

Mexico has almost 150 well-defined lagoon ecosystems. These represent more than 1.5 million hectares of lagoon-estuarine surface, including 12,500 sq km of coastal lagoons. One-third of the Mexican littoral region is composed of lagoon-estuarine ecosystems. The main institutions involved with the regulation of the coastal environment are: Secretaria de Pesca (Instituto Nacional de la Pesca), Secretaria de Marina, Secretaria de Desarrollo Urbano y Ecología, Secretaria de Agricultura y Recursos Hidráulicos, Petróleos Mexicanos, Secretaria de Industria y Comercio, and the State governments. See Chapter 4 for further details.

---

[1] Portions of the United States section were excerpted from a draft manuscript by Steve Gilbert and Roger Banks, U.S. Fish and Wildlife Service.

[2] Information for Canada was taken from a report by Environment Canada entitled "A Profile of Important Estuaries in Atlantic Canada."

[3] Information for Mexico was taken from a letter from Dr. Alejandro Yáñez-Arancibia, Laboratorio de Ictiología y Ecología Estuarina, Instituto de Ciencias del Mar y Limnología, Universidad Nacional Autónoma de México.

needs of selected estuaries of national significance (see Chapter 4 for further information). As of January 1991, the National Estuary Program was addressing 17 estuaries in the contiguous 48 states—10 on the East Coast, 4 on the Gulf Coast, and 3 on the West Coast. Studies of other estuaries may be added in the future. The product of each study will be a Comprehensive Conservation and Management Plan recommending specific pollution abatement and living resource management actions that should be undertaken to reach specific environmental goals for each estuary. Each study involves environmental assessment projects to characterize status and trends in the estuary's health and resources, pollution control feasibility studies, public involvement and information activities, and the development of data management systems to support both the initial study and follow-up monitoring to assess success of any clean-up actions taken. A major objective of the National Estuary Program will be to transfer information gained from its studies to a wide audience of environmental scientists, government agencies, and other interested groups concerned with estuarine environmental protection.

Under the U.S. Coastal Zone Act of 1972 (see Office of Ocean and Coastal Resource Management, Box 11.1) activities of federal agencies directly affecting the coastal zone must be consistent with programs of the adjoining states to the maximum extent possible (Archer and Knecht 1987). As of January 1988, 29 of 35 eligible coastal states and territories had federally approved coastal management programs. In addition, the Act established a national system of estuarine research reserves to be managed jointly by the states and federal government. These reserves represent various biogeographical regions and estuarine types in the United States. They are established to provide opportunities for long-term research, education, and interpretation. As of January 1988, 17 reserves had been established.

In Canada, the federal Department of Fisheries and Oceans' long-term policy objective is the achievement of an overall net gain of the productive capacity of fisheries habitats. This is to be achieved through restoration of degraded fisheries habitat, development of new habitat, and application of the no net loss principle in regard to habitat conservation. Under this principle, the Department will strive to balance unavoidable habitat losses with habitat replacement on a project-by-project basis. For most estuaries where a number of major projects are planned, an estuary management plan is put together. Conflicting goals which could affect fish habitat are then sorted out. Although the results may be mixed, this is usually a better process than examining each development in isolation, which leads to insidious, cumulative harm from numerous projects, each causing at least a small amount of damage to the fisheries habitat.

The Ecological Equilibrium Statute (Chapter 4) of Mexico includes the ecological policy concerning the regulation, protection, management, exploitation, and education about the rational use of water, aquatic ecosystems, and living resources. Estuaries, along with all other water bodies, are within the purview of this all-inclusive environmental legislation.

11.3.2 State and Local Responsibilities

All states in the United States have some means of regulating activities within their boundaries. Most coastal states have met the federal Coastal Zone Management Act requirements and thus receive federal dollars to support a coastal zone management agency. However, names, powers, and administrative procedures of this agency vary from state to state. If a state has a procedure for issuing permits before certain actions may take place in the estuarine zone, it is generally administered by this agency. Many local political jurisdictions have similar permitting requirements. If the actions contemplated are sufficiently local in nature, a state may give automatic approval to permits issued by local authorities.

Besides any coastal zone management agency that may exist, all states have at least one fish and wildlife agency, and sometimes other agencies, that may review and comment on Section 10 and 404 permit requests (see Chapter 4 and Box 11.1) advertised by the Army Corps of Engineers. Some local political jurisdictions also review and comment on permit requests.

In Canada and Mexico, maintaining the biological integrity of the estuarine zone is primarily a responsibility of the federal government; provincial, state, and local governments play relatively minor roles.

11.3.3 Private Organization Responsibilities

Many private organizations work to protect the estuarine zone, of which we have space to name only a few. For example, in the United States two of the larger organizations that buy title to land or obtain conservation easements on estuarine property are The Nature Conservancy and the National Audubon Society. The Environmental Defense Fund and The Natural Resources Defense Council initiate legal action to force compliance with environmental laws, including those pertaining to the estuarine zone; depending on circumstances, they sue lawbreakers or governmental agencies failing to enforce the law, or both. The National Wildlife Federation carries on an extensive public education program on the need to preserve wildlife habitat, with considerable emphasis on wetlands. There are also many other smaller private conservation groups that seek to protect the estuarine zone. For example, the San Juan Preservation Trust provides a low-cost, effective program to encourage coastline conservation on the San Juan Islands in Puget Sound, Washington (Myhr 1987). (Also see Box 11.2.)

In Canada, the Pacific Estuary Conservation Program, headed by The Nature Trust of British Columbia, coordinates two other private groups: Wildlife Habitat Canada and Ducks Unlimited Canada. The main goal of this program is to promote conservation and enhancement of coastal wetland habitats in British Columbia with high fish and wildlife values. This program has secured lands in the Cowichan, Nanaimo, and Fraser estuaries, as well as the Baynes Sound. The Provincial Government Ministries of Environment and Parks, and Forests and Lands, assist in managing acquired lands and rehabilitation projects and in providing technical assistance. The Canadian Wildlife Service and the Department of Fisheries and Oceans have served as advisors to this program.

Organizations similar to those in the United States and Canada also exist in Mexico, but they are fewer in number.

11.4 MANAGEMENT NEEDS

With the multiplicity of laws and agencies dedicated to maintenance of the estuarine zone, one might think its maintenance would be assured. Unfortunately,

---

Box 11.2   Some Information Sources

1. Information about land conservation through conservation easement programs can be obtained from the Land Trust Alliance, 900 17th Street, N.W., Suite 410, Washington, D.C. 20006-2596. The Alliance is the national organization serving hundreds of local and regional land conservation groups in the United States.

2. An excellent 72-page citizen's guide on how to become effectively involved in protecting the environment can be purchased for about $10 from Save the Bay, Inc., 434 Smith Street, Providence, Rhode Island 02908-3732. Entitled "Where the Land Meets the Water," it is written for use in protecting Narragansett Bay, but its basic information is universal. There are chapters on getting started—initiating citizen action; agencies and laws; environmental review of development proposals; planning and zoning; septic systems; runoff pollution; wetlands; barrier beaches; flood zone management; oil spills and hazardous discharges; and source reduction and recycling.

3. Some consistent sources of information on wetlands restoration are the publications of Hillsborough Community College, Institute of Florida Studies, Plant City, Florida; Society of Wetland Scientists, Wilmington, North Carolina; and the Office of Conference Services, Colorado State University.

---

such is not the case. Chapter 4 pointed out many of the legal complexities and problems involved, but there are numerous others. One is loopholes in existing laws. For instance, although Section 404 of the Clean Water Act forbids a developer to discharge dredge or fill material into a wetland without a Corps of Engineers permit, the developer needs no permit to dredge a wetland so long as the dredged material is not deposited in the wetland. Furthermore, it is nearly impossible to draft laws or agency regulations that cover all situations. Therefore, additional means for maintaining the estuarine zone are needed.

Although preservation of the estuarine zone in its pristine condition would be the best means of maintaining it, this is impossible because of actions that occur outside the estuarine zone. Moreover, preservation usually has to be balanced with economic development. In many areas, a passive preservationist approach will not address prior human alteration of the estuarine zone that will continue to cause deterioration. Thus, management is required.

Humans are continually proposing projects that will affect estuaries. Understanding how ecosystems function is a necessity before a project can be evaluated for its impacts; thus, a basic need of management is the availability of sound scientific information. In many cases, this information is readily available but not known by those who need it. (Improved transfer of scientific knowledge would provide for more effective fisheries management [Loftus 1987].) When this is so, besides making a personal literature search, there are many computerized literature searches that can be done to locate useful data; most major university libraries can perform this service. Also, many general answers can be obtained from university scientists or from state fish and wildlife agencies. More specific answers may require a new scientific study; if the study can be justified economically, the key then is planning. It may take several years to authorize, fund, and conduct the study. It should start early enough that the results will be available when a management decision is needed.

11.4.1 Informed Use

Informed use is probably the most realistic approach to the maintenance of the estuarine zone, and the most important aspect of informed use is public awareness and involvement. The public should be told of potential habitat losses (and what these would mean in terms of fisheries losses) that will likely occur with a specific project. Most agencies have a mechanism to hear public comment (Chapters 2 and 3), and sincere efforts should be made to obtain timely public comment and involvement. Sometimes the public comment is all too late in the form of, Where are all the fish? The public should be made aware of just how important a marsh or estuary is to recreational and commercial fishing, and to the enjoyment of seafood and the economic support of restaurants, fish houses, and the economy of the area. If the public is aware of potential losses, and becomes actively involved, it can demand a revised plan that will at least reduce, if not eliminate, impacts that would be detrimental to fisheries. (See Chapter 3 and Box 11.2.)

Individual governmental bodies are also enacting laws to maintain the integrity of the estuarine zone. For instance, Oregon recently passed a law designed to maintain the functional characteristics and processes of an estuary—such as its natural biological productivity, habitats and species diversity, unique features, and water quality—when intertidal or tidal marsh resources are destroyed by removal or fill activities. The procedures for enforcing the law, and the numerical habitat rating system involved, are described in Hamilton (1984).

This chapter has discussed project-specific management, but a preferable management regime is to set a long-term policy for development in an area. The stewards of a community should determine the amount and type of development that will be allowed in the coastal zone. This will involve a much higher degree of governmental land-use control than the public historically has been willing to accept. However, as the human population continues to expand, the public must eventually abandon its frontier ethic. To continue to resist land-use controls will result in intolerable damages to all environments, including the estuarine zone.

One way to a sound management decision is to get all concerned interests informed and working toward a common goal. For example, in Washington State, the Hood Canal Coordinating Council was established in 1985. It is charged with preparing and implementing a comprehensive management plan for environmental resource protection and enhancement in the region, an arm of Puget Sound. Membership on the Council is composed of one county commissioner from each of the three bordering counties, one representative from each of the two affected Indian tribes, and one representative from the state Ecological Commission (Simmons 1987). Such cooperative approaches to environmental protection are potentially much more effective than the uncoordinated, piecemeal approach so often used elsewhere. On the other hand, special management groups are sometimes so large and entrenched in local political issues that meaningful management decisions are grossly impeded.

Recognizing the long-range need for land-use controls, in 1984 the Maryland legislature passed the Chesapeake Bay Critical Area Law. This far-sighted action resulted in land-use restrictions much more stringent and widespread than generally exist elsewhere in North America. The law designated a Critical Area consisting of all waters of the Chesapeake Bay and extending 305 m beyond landward boundaries of state or private wetlands and heads of tides. The law also affects the use of the streamside of all the Bay's perennial tributaries. Local jurisdictions are required to classify and delineate lands within their portion of the Critical Area into one of three categories as defined by criteria set out in the law: intense development, limited development, and resource conservation areas. Limits are set on how much of a local jurisdiction can be in each category. The law specifies many things that must, or must not, be done in the Critical Area. For example: no commercial timber harvesting is allowed within 15.25 m of mean high water of the Bay or its perennial tributaries; all farms must have soil and water management plans within 5 years; and a 30.50-m-minimum vegetated buffer along tidal waters and streams is required for all new development. Further details can be found in Salin (1987).

11.4.2   Restoration

It is possible to restore estuarine habitat, but usually it is very expensive. The Chesapeake Bay again furnishes a good example. Because of the increased intensity of human activities within the Chesapeake Bay watershed, the health of the Bay has declined rapidly since 1950 (Salin 1987). To restore the Bay's health will require more than enforcement of Maryland's Critical Area Law. Throughout the watershed it will require increased sewage treatment plant capacity, urban storm water projects, and construction of industrial pretreatment facilities to remove toxic materials. The total cost may run into billions of dollars. However, failure to undertake these actions will result in even greater losses, in terms of the economy, aesthetics, and health.

The most economical means of restoring marsh in the estuarine zone may be wise use of dredge spoil. Most ports and waterways maintain some level of dredging to keep ship channels open. With proper planning, spoil may be placed or directed such that it results in marsh building. However, this too may be expensive if the spoil must be transported very far.

Sometimes diversion of waterflow may be the most cost-effective means of restoring estuarine habitat. In Louisiana, diversions from the Mississippi River are being used to reestablish more favorable salinity regimes, and build marsh through the additional input of sediment. Of course, this is sometimes a matter of "robbing Peter to pay Paul" as previously mentioned in the case of the rediversion of the Santee-Cooper river system in South Carolina.

Where estuarine habitat is scarce and highly desired, small areas are sometimes restored or created by more labor or energy-intensive methods. This can be a very expensive process. A 12-hectare parcel of lumber and plywood mills, occupying original estuary and salt marsh wetlands in northern California, was redeveloped into estuarine and freshwater wetlands at a cost of $280,000 (Allen and Hull 1987). In southern California there are ongoing plans to restore a 243-hectare estuarine lagoon at an estimated cost of $15–25 million (Marcus 1987). It is readily apparent that it is much more economical to preserve already existent estuarine habitat than to create it.

Box 11.3   Sage Advice

The best and often the only way to deal with global problems is, paradoxically, to look for solutions peculiar to each locality.... If we really want to do something for the health of our planet, the best place to start is in the streets, fields, roads, rivers, marshes, and coastlines of our communities.

Rene Dubos

Examine each question in terms of what is ethically and aesthetically right, as well as what is economically expedient. A thing is right when it tends to preserve the integrity, stability, and beauty of the biotic community. It is wrong when it tends otherwise.

Aldo Leopold

Unfortunately, although restored areas may become vegetated and appear the same as marsh formed naturally, there is question as to whether such human-induced marshes function as well ecologically, or are as productive of fisheries organisms, as naturally occurring marshes (Josselyn et al. 1987; Minello et al. 1987). The techniques have not been used long enough for good evaluation. (See further, Box 11.2.)

11.5   CONCLUSION

Laws, agencies, and scientific information are the foundation for maintenance of the estuarine environment. However, as this chapter has described, numerous human-induced impacts can disrupt these "not quite fresh and not quite marine" ecosystems. In the final analysis, maintenance of the biological integrity of the estuarine zone depends on an interested, informed, and actively participating public that knows what is right for its own locality (Box 11.3).

11.6   REFERENCES

Allen, G. H., and D. Hull. 1987. Restoring of Butcher's Slough estuary—a case history. Pages 3674–3687 *in* Magoon, et al. (1987).

Archer, J. H., and R. W. Knecht. 1987. The U.S. National Coastal Zone Management Program—problems and opportunities in the next phase. Coastal Zone Management Journal 15:103–120.

Cahoon, D. R., and C. G. Groat, editors. 1990. A study of marsh management practice in coastal Louisiana. Final Report (Contract 14-12-0001-30410) to Minerals Management Service, New Orleans, Louisiana.

Cameron, W. M., and D. W. Pritchard. 1963. Estuaries, volume 2. Pages 306–324 *in* M. N. Hill, editor. The sea. Ideas and observations on the progress in the study of the seas. Wiley-Interscience, New York.

Copeland, B. J. 1967. Environmental characteristics of hypersaline lagoons. Contributions in Marine Science 12:207–218.

Day, J. W., Jr., C. S. Hopkinson, and W. H. Conner. 1982. An analysis of environmental factors regulating community metabolism and fisheries production in a Louisiana estuary. Pages 121–136 *in* V. S. Kennedy, editor. Estuarine comparisons. Academic Press, New York.

Day, J. W., Jr., and A. Yáñez-Arancibia. 1985. Coastal lagoons and estuaries as an environment for nekton. Pages 17–34 *in* A. Yáñez-Arancibia, editor. Fish community ecology in estuaries and coastal lagoons: towards an ecosystem integration. Universidad Nacional Autónoma de México Press, Ciudad Universitaria, México, Distrito Federal.

DeVoe, M. R., and D. S. Baughman, editors. 1987. South Carolina coastal wetland impoundments: ecological characterization, management, status, and use, volume 1. Executive summary. South Carolina Sea Grant Consortium, Publication SC-SG-TR-86-1, Charleston.

Drinkwater, K. F. 1986. On the role of freshwater outflow in coastal marine ecosystems: a workshop summary. NATO ASI (Advanced Science Institutes) Series Series G7:429–438.

Fisher, D., J. Ceraso, T. Mathew, and M. Oppenheimer. 1988. Polluted coastal waters: the role of acid rain. Environmental Defense Fund, New York.

Graham, J. J. 1972. Retention of larval herring within the Sheepscot estuary of Maine. National Marine Fisheries Service Fishery Bulletin 70:299–305.

Hamilton, S. F. 1984. Estuarine mitigation: the Oregon process. Oregon Division of State Lands, Salem.

Herke, W. H. 1971. Use of natural, and semi-impounded, Louisiana tidal marshes as nurseries for fishes and crustaceans. Doctoral dissertation. Louisiana State University, Baton Rouge. (Also: University Microfilms, Order 71-29,372, Ann Arbor, Michigan.)

Herke, W. H. 1977. Life history concepts of motile estuarine-dependent species should be re-evaluated. Privately published, available from W. H. Herke, Department of Forestry, Wildlife, and Fisheries, Louisiana State University, Baton Rouge, Louisiana 70803.

Herke, W. H., E. E. Knudsen, P. A. Knudsen, and B. D. Rogers. 1992. Effects of semi-impoundment of Louisiana marsh on fish and crustacean nursery use and export. North American Journal of Fisheries Management 12:151–160.

Josselyn, M. N., J. Duffield, and M. Quammen. 1987. An evaluation of habitat use in natural and restored tidal marshes in San Francisco Bay, California. Pages 3085–3094 *in* Magoon et al. (1987).

Kennedy, V. S. 1990. Anticipated effects of climate change on estuarine and coastal fisheries. Fisheries (Bethesda) 15(6):16–24.

Kjerfve, B. 1989. Estuarine geomorphology and physical oceanography. Pages 47–77 *in* J. W. Day Jr., C. Hall, M. Kemp, and A. Yáñez-Arancibia. Estuarine ecology. Wiley-Interscience, New York.

Loftus, K. H. 1987. Inadequate science transfer: an issue basic to effective fisheries management. Transactions of the American Fisheries Society 116:314–319.

Magoon, O. T., H. Converse, D. Miner, L. T. Tobin, D. Clark, and G. Domurat, editors. 1987. Coastal zone '87. American Society of Civil Engineers, New York.

Malins, D. C., B. B. McCain, D. W. Brown, A. K. Sparks, H. O. Hodgins, and S-L. Chan. 1982. Chemical contaminants and abnormalities in fish and invertebrates from Puget Sound. National Technical Information Service, PB83-115188, Springfield, Virginia.

Marcus, L. 1987. Wetland restoration and port development: the Batiquitos Lagoon case. Pages 4152–4166 *in* Magoon et al. (1987).

McHugh, J. L. 1966. Management of estuarine fisheries. American Fisheries Society Special Publication 3:133–154.

Mendelssohn, I. A., K. L. McKee, and W. H. Patrick, Jr. 1981. Oxygen deficiency in *Spartina alterniflora* roots: metabolic adaptation to anoxia. Science (Washington, D.C.) 214:439–441.

Minello, T. J., R. J. Zimmerman, and E. F. Klima. 1987. Creation of fishery habitat in estuaries. Pages 106–117 *in* M. C. Landin and H. K. Smith, editors. Beneficial uses of dredged material. U.S. Army Corps of Engineers, Waterways Experiment Station, Vicksburg, Mississippi.

Mitsch, W. J., and J. G. Gosselink. 1986. Wetlands. Van Nostrand Reinhold, New York.

Myhr, R. O. 1987. Private coastline conservation management: the land trust in the San Juan Islands, Washington. Pages 3266–3273 in Magoon et al. (1987).

NMFS (National Marine Fisheries Service). 1987a. Marine recreational fishery statistics survey, Atlantic and Gulf coasts, 1986. U.S. NMFS Current Fishery Statistics 8392.

NMFS (National Marine Fisheries Service). 1987b. Marine recreational fishery statistics survey, Pacific coast, 1986. U.S. NMFS Current Fishery Statistics 8393.

NMFS (National Marine Fisheries Service). 1988. Fisheries of the United States, 1987. U.S. NMFS Current Fishery Statistics 8700.

Pritchard, D. W. 1967. What is an estuary: physical viewpoint. Pages 3–5 in G. H. Lauff, editor. Estuaries. American Association for the Advancement of Science Publication 83, Washington, D.C.

Ramanathan, V., R. J. Cicerone, H. B. Singh, and J. T. Kiehl. 1985. Trace gas trends and their potential role in climate change. Journal of Geophysical Research 90(D3):5547–5566.

Rogers, B. D., and W. H. Herke. 1985. Estuarine-dependent fish and crustacean movements and weir management. Pages 201–219 in C. F. Bryan, P. J. Zwank, and R. H. Chabreck, editors. Proceedings of the Fourth Coastal Marsh and Estuary Management Symposium. Louisiana Cooperative Fish and Wildlife Research Unit, Louisiana State University Agricultural Center, Baton Rouge.

Salin, S. L. 1987. Maryland's critical area program: saving the bay. Pages 208–221 in Magoon et al. (1987).

Simmons, D. M. 1987. A new approach to watershed planning. Pages 2726–2740 in Magoon et al. (1987).

Stevenson, J. C., L. G. Ward, and M. S. Kearney. 1986. Vertical accretion in marshes with varying rates of sea level rise. Pages 241–259 in D. A. Wolfe, editor. Estuarine variability. Academic Press, San Diego, California.

Sverdrup, H. U., M. W. Johnson, and R. H. Fleming. 1942. The oceans. Prentice-Hall, New York.

Titus, J. G. 1986. Greenhouse effect, sea level rise, and coastal zone management. Coastal Zone Management Journal 14:147–171.

Turner, R. E. 1987. Relationship between canal and levee density and coastal land loss in Louisiana. U.S. Fish and Wildlife Service Biological Report 85(14).

U.S. Bureau of the Census. 1986. Statistical abstract of the United States: 1987, 107th edition. U.S. Bureau of the Census, Washington, D.C.

Wolfe, D. A., and B. Kjerfve. 1986. Estuarine variability: an overview. Pages 3–17 in D. A. Wolfe, editor. Estuarine variability. Academic Press, San Diego, California.

Yáñez-Arancibia, A., and six coauthors. 1985. Ecología de poblaciones de peces dominantes en estuarios tropicales: factores ambientales que regulan las estrategias biológicas y la producción. Pages 311–365 in A. Yáñez-Arancibia, editor. Fish community ecology in estauries and coastal lagoons: towards an ecosystem integration. Universidad Nacional Autónoma de México Press, Ciudad Universitaria, México, Distrito Federal.

Ziskowski, J. J., L. Despres-Patanjo, R. A. Murchelano, A. B. Howe, D. Ralph, and S. Atran. 1987. Disease in commercially valuable fish stocks in the northwest Atlantic. Marine Pollution Bulletin 18:496–504.

COMMUNITY MANIPULATIONS

Chapter 12

Management of Introduced Fishes

HIRAM W. LI AND PETER B. MOYLE

12.1 INTRODUCTION

The introduction of fish species into waters outside their native range has occurred since the common carp was first moved around by the Chinese over 3,000 years ago and by the Romans 2,000 years ago (Balon 1974). During the Middle Ages, common carp were spread throughout Europe by monastic orders, and Scandinavians were busy stocking "barren" alpine lakes with salmonids. In the following centuries, the great expansion of western civilization was accompanied by the worldwide distribution of species of Euro-American flora and fauna, including fish. This was a reflection of western attitudes that the natural environment was hostile and needed to be tamed, and that natural systems could be improved by introducing familiar, and therefore "superior" species (Crosby 1986). Throughout the world there are now important fisheries based on introduced species, e.g., trout and salmon in New Zealand, Mozambique tilapia in Sri Lanka, common carp in Europe, and brown trout in North America. Only recently have such widely known "success" stories been balanced by the realization that many introductions have done more harm than good.

The damage done by introduced plants and animals to natural systems has caused ecologists and resource managers much concern (Kornberg and Williamson 1986; Mooney and Drake 1986; Holcik 1991; Nesler and Bergersen 1991). The common complaint is that sustainable benefits have been sacrificed for short-term gains (Baltz 1991; Ogutu-Ohwayo and Hecky 1991; Philipp 1991; Spencer et al. 1991). The introduction of new species to improve fisheries is still a common management practice; therefore, it is important to understand the effects introduced species can have on native species and ecosystems. Without such understanding, well-intentioned management programs can create problems that actually subvert the original management intent.

This chapter (1) provides an overview of the reasons for introducing aquatic organisms with examples of successes and failures, (2) presents ecological concepts important for understanding the effects of introductions, (3) suggests some management alternatives to introducing new species, and (4) provides guidelines for evaluating proposed introductions.

12.2 REASONS FOR FISH INTRODUCTIONS

Fishes have been introduced for many and often multiple reasons: (1) to increase local food supplies, (2) to enhance sport and commercial fishing, (3) to manipulate aquatic systems, (4) by accident, and (5) for aesthetic reasons.

12.2.1 Food Supply

The earliest introduced species were semi-domesticated animals and plants that were moved about to create more reliable local food supplies. Domesticated plants and animals were keystones in the development of human culture and their spread was a natural outcome of human expansion across the globe. As human populations grew and moved, commensal species such as rats and weedy plants also spread and feral populations of domestic animals developed. Fish were relatively late additions to the ranks of domestic and commensal species; most species were added after about 1850. However, the worldwide spread of common carp and African tilapias began well before this time. Fishes such as common carp and tilapia are hardy so they can be easily transported, they establish populations quickly in a variety of new environments, and they grow rapidly, especially in ponds. These traits are characteristics of most animals used in aquaculture; they are also the characteristics of pest species.

In response to the increased human demand for fish and static or declining wild fish populations, aquaculture today is a rapidly growing industry. Fish farmers tend to use only familiar species which leads to two problems: (1) native fishes well adapted to local conditions are ignored for aquaculture, and (2) nonnative species escape into local waterways. The latter problem can result in the disruption of local wild fish populations, if not through direct interactions with the introduced fishes, then through exposure to new diseases and parasites.

12.2.2 Fisheries Enhancement

Izaak Walton was one of the first anglers to claim that an introduced fish was superior to native forms. In 1653, he pronounced the recently introduced common carp the "queen" of England's rivers because of its superior qualities as a sport fish. Local fish acclimatization societies formed in various countries, many of them predecessors of present fisheries management agencies. By the late 1800s, introducing species to solve management problems in North America had become a major activity of state and federal agencies. As a result, the dominant fishes in many lakes and rivers in North America are introduced sport fishes (see Courtenay and Kohler 1986; Moyle 1986).

This pattern has been repeated throughout the world, but particularly in the British Empire where sport fishing was a favorite upper-class affectation and local fishes were often considered to be unsuitable prey for the sophisticated angler. Favored sport fishes from Europe and North America were consequently brought to distant lands, often with great difficulty; today species such as largemouth bass, rainbow trout, and brown trout enjoy virtual global distribution. Such fishes are still the backbone of sport fisheries in many areas, but there is a growing realization that they are often poor substitutes for native fishes such as cutthroat trout of the interior basins of western North America. Similarly, the reduction in native cyprinodonts in Lake Titicaca (Peru–Bolivia), important in native subsistence fisheries, resulted from the introduction of rainbow trout from North America and a predatory atherinid from elsewhere in South America.

Some of the most successful uses of introduced sport fishes illustrate the dictum of Giles (1978) that "importations are an admission of defeat in managing native populations to meet existing needs." The best example of this is found in Lake Michigan, which had its native fish communities severely disrupted by the inadvertent invasions of a predator (sea lamprey) and an efficient planktivore (alewife), coupled with the introduction of another planktivore (rainbow smelt) and severe overfishing of native fishes. The result was a nearly complete collapse of the native sport and commercial fisheries (see Chapter 22). Previous experience suggested that predatory Pacific salmonids could greatly reduce the numbers of alewives and rainbow smelt. This proved to be the case and a spectacular fishery for salmon soon developed. The reductions in rainbow smelt and alewives resulted in a marked increase in the numbers of some native planktivores, through reduced competition for zooplankton and, perhaps, reduced predation on plank-tonic larvae. In a way, the introduction of Pacific salmonids worked too well. Salmon and steelhead populations exploded as they "mined" the huge biomass of alewives, creating expectations in anglers that the fabulous fishing would continue indefinitely. However, alewife populations are apparently now being kept at low levels by salmonid predation which in turn limits the number of salmon and trout the system can support (Eck and Brown 1985).

Even though the effort in the Great Lakes to develop a new fish community based on introduced fishes was largely successful, most such efforts create as many problems as they solve. This is well illustrated by the introductions of warmwater sport fishes into reservoirs of the intermountain West, in order to mitigate the losses of riverine salmon and trout fisheries. Unfortunately, the introduced species often invade unimpounded reaches of river and prey upon or compete with native fishes. For example, native fishes of the Columbia and Snake rivers are now exposed to a greater intensity of piscivory than experienced during their evolutionary history because of the presence of introduced walleye, channel catfish, and smallmouth bass (Li et al. 1987; Tabor et al., in press). Juvenile salmonids constitute a significant part of the diet of these fishes and the establishment of these predators therefore counters the official policy of doubling salmonid escapement by the turn of the century.

There have been relatively few successful introductions of commercial fishes, although attempts have been numerous, especially in marine environments. Some of the more successful introductions have been those made to benefit both sport and commercial fisheries, such as the introduction of striped bass and American shad to California in the 1870s, and the spread of lake trout to coldwater lakes around the world. When angler demand for such fishes becomes high, the commercial fisheries are often banned. In North America, the main commercial fisheries on introduced fishes focus on species not favored by anglers, such as buffalofishes in Arizona reservoirs, Sacramento blackfish in a Nevada reservoir, and common carp in rivers and reservoirs of the Midwest.

Because commercial fisheries depend on large quantities of fish, a successful introduction for this purpose is bound to have a major impact on the receiving aquatic ecosystem. For example, in the former Soviet Union the introduction of European perch into Lake Kenon resulted in the extinction of five native species and reduction in the populations of three others (Karasev 1974). In Lake Victoria,

Africa, the introduction of the predatory Nile perch had catastrophic effects on the 200–300 species of endemic haplochromine cichlids that were once important to the native fishers, and many of those species now seem to be extinct (see 12.3.5).

Often the poor performance of a sport fishery is attributed to the absence of an adequate prey base for predatory sport fishes. A common management practice is to introduce a suitable prey organism on the assumption that a new prey base will enhance growth and survival of sport fishes (see Chapter 13). Thus, species such as threadfin shad, gizzard shad, and golden shiner have been widely introduced into warmwater lakes and reservoirs as forage, while rainbow smelt and opossum shrimp *Mysis relicta* have been widely introduced into coldwater lakes and reservoirs. Such introductions have had mixed success. Often growth rates of adult game fishes accelerate following the introduction and an initial period of outstanding fishing develops. However, it is common for the growth and survival rates of juveniles of the same game fishes to decrease because the forage species compete with them for food. Thus, the growth rates of juvenile white and black crappie in a California lake decreased considerably following the introduction of inland silverside, while the growth rates of adult crappie, which preyed on the silverside, increased. The net result was that the large crappie preferred by anglers were the same size at a given age as they were before the introduction (Li et al. 1976). Not surprisingly, some of the most successful forage fish introductions have occurred in reservoirs where the game fish, usually trout, are planted at a size large enough to prey immediately on the forage fishes. A better understanding of predator–prey relationships would allow more efficient use of forage introductions as a management tool.

12.2.3 Manipulation of Aquatic Systems

The use of fish as biological control agents for aquatic pests such as mosquitoes, disease-bearing snails, aquatic weeds, or stunted fishes is a very appealing concept. If effective, a successful introduction for biological control can obviate the need for pesticides, be inexpensive, and have a long-lasting effect. The increased use of fishes for biological control in recent years has been spurred by successes in the use of insects as control agents in agriculture, and the development of biological control theory by applied entomologists. The cost of pesticides is also rapidly increasing, as is public concern about the effects of pesticides on nontarget organisms and human health. As a result, species with good track records for biological control are being spread worldwide, most prominently mosquitofish and grass carp. The mosquitofish is perhaps the most widely distributed fish in the world today, found virtually everywhere in which the climate is suitable. It is successful at controlling mosquitoes because it can live in stagnant water, reproduces and grows rapidly, is a voracious insectivore, and is easy to raise in large numbers. The distribution of the grass carp may someday rival that of the mosquitofish, as it is being widely introduced as a fish that is not only effective at controlling aquatic weeds (often introduced species themselves), but one that converts these weeds to edible fish flesh. The grass carp is hardy, easy to culture, a voracious grazer, and relatively nonselective in its choice of plants to eat.

Even though there are many success stories in the use of mosquitofish, grass carp, and other fishes for biological control, the use of such fish also entails considerable risk. This is because fish typically select a wide variety of prey species and will forage on pest species only when they are abundant and easily available. Fish are also longer lived than most of their prey, so have relatively slow population responses to increases or decreases in prey abundance. As a result, nontarget organisms can be adversely affected by control fishes. For example, if densities of mosquitofish are too low, their presence may actually increase mosquito populations because mosquitofish preferentially prey on larger insects that are natural predators of mosquito larvae; they switch to mosquitoes only after the large insects are depleted. Mosquitofish also can displace native fishes that may actually be better at mosquito control in some types of habitat.

Piscivores have been introduced to control prey, but aside from the Great Lakes experience, evidence of the success of this strategy in improving fisheries is largely equivocal (Noble 1981; Wydowski and Bennett 1981). The main reason for equivocal results is that predator–prey interactions are more complex than generally realized. For example, the tiger muskellunge, which is a sterile hybrid of the muskellunge and northern pike, has been proposed as a safe management tool to reduce populations of stunted sunfish in lakes and ponds. However, tiger muskellunge are only able to prey effectively on sunfish when prey densities are high, when vegetation is sparse, when no alternative prey are available, and when the tiger muskellunge are too large to be eaten by largemouth bass (Tomcko et al. 1984), a set of conditions not often met.

Limnologists have noted that water quality may be improved by manipulating the food chain (Shapiro and Wright 1984). One method is to introduce piscivores in order to reduce populations of small fishes that prey on zooplankton. Zooplankton populations then increase and this results in increased grazing by the zooplankton on phytoplankton, causing an increase in water clarity. Such shifts have been noted by Scavia et al. (1986) in Lake Michigan, but it is unusual for predation to be so effective. In addition, the goals of water quality management and fisheries management may not be compatible; high harvest rates of predatory fishes may result in increases of planktivores and decreased water clarity.

12.2.4 Accidental and Unauthorized Introductions

The idea that an adequate forage base is necessary to produce large populations of game fish is so obvious that the introductions of minnows and other small fishes is a popular pastime of anglers. Often these introductions are simply the result of anglers collecting bait fish from one stream, fishing in a stream in a different drainage, and then releasing their unused bait. Such cryptic introductions are more common than generally realized and can explain many anomalies in fish distribution. Bait-bucket transfers are only one type of accidental introduction. Fishes and invertebrates have also become established in new waters after being transported through canals, carried as contaminants in truckloads of hatchery-reared game fish, or allowed to escape from fish farms. Increasingly, aquatic organisms are making their way into new waters through ships that carry cargo in only one direction and carry enormous quantities of water back as ballast (Carlton 1985). For example, the ruff (a small predatory perch), a predatory cladoceran, and the zebra mussel *Dreissena polymorpha* have recently been added to the fauna of the Great Lakes by this method and they have considerable potential for altering the entire aquatic ecosystem (Scavia et al. 1988; Roberts 1990). In the Sacramento–San Joaquin estuary, recent ballast-water introductions include two species of gobies, two species of copepods, an amphipod, and a euryhaline clam. There is growing evidence that these species have altered food chains that lead to striped bass and other fishes (Nichols et al. 1990, Meng and Orsi 1991).

12.2.5 Aesthetic Considerations

A contributing factor to most deliberate introductions is aesthetics. People who make introductions have a cultural preference for the taste, appearance, or sporting style of the introduced species over native species. A few introductions have been made solely for aesthetic reasons, such as the planting of goldfish in urban waterways or of tropical fish in hot springs.

12.3 INTRODUCTIONS AND ECOLOGICAL THEORY

A species introduction is a type of ecological perturbation that, if successful, will alter the biotic community into which the species has become part. Therefore, the purpose of this section is to discuss the use of ecological theory to predict how a species proposed for introduction is likely to alter the receiving community, as well as the degree of community disturbance it is likely to cause. Relevant theory falls into two general headings: the niche concept and the concept of limiting similarity. These two ideas form the basis of the theory of island biogeography, which can be used as a general framework to explain how colonization and extinction processes shape communities of organisms. Following the presentation of these concepts, the mechanisms that enable introduced species to invade and alter biotic communities will be discussed: (1) competition, (2) parasite–host interactions, (3) predation, (4) habitat modification, (5) indirect interactions, and (6) hybridization. More detailed discussions of these and other concepts in relation to fish introductions can be found in Courtenay and Robins (1989).

12.3.1 Niche Concept

The most widely accepted description of the niche is that of Hutchinson (1958) who developed the idea as a multidimensional attribute of a species or population that contains many axes describing where and how the species lives. There are axes for such dimensions as prey size, prey type, depth of water, and velocity of water. The niche is therefore a characteristic of the organism, not the environment. Lack of understanding of this idea has led to the introduction of species to fill "vacant," "unoccupied," or "empty" niches. The vacant niche concept confuses available resources of the aquatic habitat with the ecological function of the organism (Herbold and Moyle 1986). Usually the vacant niche is identified as an abundant food resource located in a particular habitat. A species is introduced to use that resource based upon observations of its life history in other aquatic systems on the assumption that it will perform in the new system just as it did in the old. This assumption is often wrong because the niche expression of an organism in any particular ecological community is limited by local physicochemical and biological constraints.

The ecological niche as defined by Hutchinson (1958) has two aspects: the fundamental niche and the realized niche (Figure 12.1). The fundamental niche is the total capacity of the organism to perform activities over a wide range of environmental conditions; it is circumscribed by genetically determined physiological limits. The realized niche is the fraction of the fundamental niche that is expressed as an adaptation to local conditions. For example, the presence of a competitor or predator of a species may prevent that species from using food or space it would use in their absence, resulting in a rather narrow realized niche (niche compression). If the constraints imposed by the other organisms are removed, the realized niche becomes larger (niche expansion). Therefore, at any given time and place a population of a species is using only a small part of its fundamental niche. A classic example of how the realized niche of an introduced organism can change is illustrated by the problems caused by the introductions of opossum shrimp into lakes throughout the intermountain West. It was presumed that this small shrimp would function elsewhere as it does in Kootenay Lake, British Columbia, Canada, where it subsists largely on phytoplankton and detritus. Instead, in new environments, the opossum shrimp has proven to preferentially feed on zooplankton. As a result, it has eliminated large zooplankton species that were often important foods for the very game fishes whose populations they were supposed to enhance (Lasenby et al. 1986; Nesler and Bergersen 1991).

The niche of a species can change not only with its environment but also through ontogeny. Most fishes feed on quite different and smaller prey while young and switch to larger prey as they mature. These shifts in diets are often accompanied by shifts in habitat use. Most introductions are made without regard to the distinct niches of the early life history stages of the fishes in question, despite the fact that interspecific interactions for species change with life stage (Werner 1986). For example, introduced forage fishes can compete with juvenile piscivores for invertebrate prey or they can prey upon the young of their predators (Crowder 1980; Kohler and Ney 1980).

Niche expression may differ within a population because of individual differences. Specialization in feeding habits by individuals in populations of rainbow trout has been documented (Bryan and Larkin 1972). Nordeng (1983) found specialization in habitat use and degree of anadromy by individual Arctic char. If the number of genetically determined specialists within a population changes due to shifts in selection pressure, the fundamental niche of the population may change as well. Thus, intense selection pressure through periodic winter kills may select for cold-hardy strains of mosquitofish and swordtail, enabling these fishes to invade areas where they are not wanted. Similarly, rainbow trout in western Australia now live at higher temperatures than other stocks (Arthington and Mitchell 1986). In short, the niche of a species is flexible, its limits can be stretched genetically, and it can change rapidly through natural selection.

12.3.2 Limiting Similarity and Island Biogeography

As a new species invades an ecosystem, five types of outcomes are possible: (1) species addition with species extinction (species replacement), (2) species addition without niche compression of similar species, (3) species addition with niche compression, (4) multiple species extinctions because of alterations of food webs or the environment, and (5) failure of the invading species to become established. These are predictions based upon the theory of limiting similarity (MacArthur and Levins 1967) and the theory of island biogeography (MacArthur and Wilson 1967).

A. Bivariate Niche Dimensions

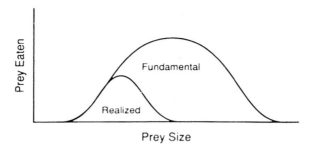

B. Multivariate Niche Dimensions

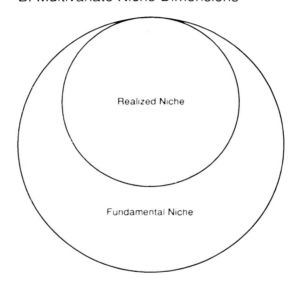

Figure 12.1 The concepts of the fundamental and realized niches (Hutchinson 1958) displayed as a single-niche axis (A) and in multiple-niche dimensions (B). Note that the realized niche in (A) may be limited by several factors such as predator size distribution, size distribution of available prey, and available prey species.

The theory of limiting similarity undergirds the theory of island biogeography. It is also known as the species packing hypothesis: as the number of species in an assemblage increases, greater partitioning of resources occurs, resulting in niche compression. At some level resources cannot be partitioned further because there is a limit to the degree of niche similarity among species.

The theory of island biogeography applies not only to actual islands, but to isolated patches of habitats as well. Freshwater habitats are typically islands of water surrounded by a sea of land. The number of species supported by an "island" is a function of accessibility to colonists, as well as patch size and habitat diversity. These two factors affect immigration and extinction rates, thus limiting the number of species that an island can support. One prediction from this is that if repeated introductions are made into an aquatic system, extinction rates will be high and the fish community will become unstable and difficult to manage (Magnuson 1976).

The logical extension of these two theories is that a species-rich fauna will be more resistant to invasion than one that is species poor (Diamond and Case 1986), but this generality is far from universal. For instance, some depauperate native fish communities in undisturbed California streams have repeatedly resisted invasions by introduced species (Baltz and Moyle 1993), whereas the extraordinarily rich fish fauna of Lake Victoria, Africa, has been devastated by a single introduction (see 12.3.5). The reasons that the two theories have limited reliability is that they presume steady state conditions and assume that competition among species is the most important factor structuring communities. These theories also do not take into account that moderately disturbed environments often have the highest diversity of species because random environmental fluctuations prevent the most efficient competitors from becoming dominant. Thus, mixed assemblages of native and introduced species are more likely to coexist where there are intermediate levels of disturbance. Highly disturbed environments are most likely to contain only a few tolerant species, typically human symbionts like common carp, goldfish, fathead minnow, and mosquitofish.

12.3.3 Competition

Competition occurs when two organisms require the same resource that is in limited supply. Competition is frequently cited as a major reason why introduced fishes replace native fishes, but most of the evidence is anecdotal or inferential and does not demonstrate conclusively that there is some limiting resource (Fausch 1988; Ross 1991). The most dramatic examples occur among territorial salmonids, where aggressive behavior leads to the dominance of one species over another (interference competition). For instance, Fausch and White (1986) found that introduced adult brown trout displaced native adult brook trout from the best habitats, making the brook trout more vulnerable to fishing and other forms of predation.

Exploitation competition has not been as well documented as interference competition because it is not conspicuous. However, it may be extremely common. It occurs when one species uses resources more quickly and more efficiently than the other. Thus, the redside shiner largely replaced juvenile rainbow trout in the littoral zone of a Canadian lake because they more efficiently exploited invertebrates associated with beds of aquatic macrophytes (Johannes and Larkin 1961). Declines in kokanee in lakes of western North America have been caused by the near elimination of large zooplankton species by introduced opossum shrimp (Lasenby et al. 1986), whereas declines and extinctions of whitefish in the Great Lakes have been at least partially caused by the removal of large zooplankton species by introduced alewife (Crowder and Binkowski 1983). Less characteristically, introduced creek chub forced brook trout in Quebec Lakes to switch from feeding on benthos to feeding on zooplankton, apparently causing reduced growth rates (Mangan and Fitzgerald 1984).

12.3.4 Parasite–Host Interactions

Introduced species can be sources of introduced diseases that severely deplete native populations. The transfer of these organisms worldwide during the 20th

Century has been without precedent (Ganzhorn et al. 1992). The European crayfish *Asticus* spp. was virtually eliminated from northern Europe by a disease brought in with signal crayfish *Pascifasatcus leniusculus* from North America. Ironically, the demand this created for imported crayfish in Scandinavian countries caused the development of fisheries for crayfish in California. In the former Soviet Union, attempts to introduce a new species of sturgeon into the Caspian Sea failed, but did succeed in introducing a sturgeon parasite that devastated the populations of native sturgeons.

A more subtle effect of parasites is their "use" by an immune host as a weapon against a competitor by introducing an energy drain and a source of additional mortality. Coexistence between native and introduced species may depend on relative degrees of immunity from reciprocal parasites. Thus, whitefish are restricted to benthic prey in the presence of cisco, and they are further disadvantaged because of their increased susceptibility to parasites hosted by benthic invertebrates (Holmes 1979). One of the causes of the decline of the woundfin, an endangered species in Utah, is heavy infestations by an Asiatic tapeworm. This tapeworm accompanied the introduced red shiner, which seems to be replacing the woundfin in part because it is more resistant to the tapeworm. The red shiner, in turn, picked up the tapeworm in its own native range from introduced grass carp (Deacon 1988).

12.3.5 Predator–Prey Interactions

Predation is a powerful evolutionary force and there are many studies that demonstrate that top carnivores can determine not only the kinds and numbers of potential prey species but also the kinds and numbers of species at lower trophic levels. However, the most dramatic effects are often on their prey species, and the introduction of a predator into a system containing prey species not evolved to counter its particular style of predation can lead to dramatic changes in the numbers and diversity of the prey assemblage (Zaret and Paine 1973; Li et al. 1987; Arthington 1991; Holcik 1991). There are many examples of how introduced predators have altered biotic communities of inland waters of North America, but in this section we will focus on the introduction of Nile perch into Lake Victoria in East Africa. We do this because it has been the most devastating introduction in modern times and because the introduction was based in part on the advice of western fisheries biologists.

The absence of large piscivorous fishes from most African rift lakes is at least partially responsible for their incredibly high species richness, mostly in small cichlids. These cichlids, with more than 300 species in Lake Victoria alone, are one of the world's marvels of natural history and evolution. In the 1950s, the introduction of the large, predatory Nile perch into Lake Victoria was proposed because (from a western perspective) the abundant, but small, cichlids did not provide an adequate fishery for the native peoples; they were eaten locally but were not considered suitable for export. Gee (1965) considered the introduction desirable because it would "control" the haplochromine cichlids and convert them to more useful, larger fish. While the debate on whether or not to introduce the Nile perch was occurring, it mysteriously appeared in the lake, justifying further introductions. Until about 1975, the Nile perch populations remained small. Suddenly, the populations exploded and in the process extirpated most of the cichlids and 40 species of noncichlids (Hughes 1986). Today, three species dominate the fish fauna: Nile perch, an introduced cichlid, and a native zooplank-tivore. The prey of the Nile perch has shifted from small cichlids to shrimp and its own young (Hughes 1986).

Initially, Nile perch did not find favor with the local peoples because they were too large to preserve by traditional methods of sun drying and difficult to catch by traditional methods. However, people soon adjusted their gear to catch the Nile perch and learned to process them for oil, which they used for cooking the fish. The fishery has greatly expanded and some of the Nile perch are now exported. However, one consequence of the new processing methods is an increased demand for charcoal, which is needed to rend the fish oil. This is causing local woodlands to be cut, with unknown effects on the terrestrial biota. Because of the apparent success of the fishery, however, Nile perch may be planted in other rift lakes with similar devastating effects on local ecosystems.

The saga of the Nile perch is similar to what happened in the Laurentian Great Lakes following the invasion of the sea lamprey: drastic declines of fishes not adapted to lamprey predation. It is likely that the large native fishes would have disappeared from the Great Lakes altogether if a massive program of lamprey control had not been initiated and sustained at considerable cost. Predators need not be large in size, like Nile perch or sea lamprey, to have a major impact; predation on eggs and larvae of native fishes by introduced zooplanktivores can also cause major population declines (Crowder 1980; Kohler and Ney 1980).

12.3.6 Complex and Indirect Effects

More often than not, the effects of introductions are more complicated than previously described. The effects may be noticeable only as a gradual, indirect, restructuring of the recipient biotic community (Ross 1991). Two interrelated types of effects are most common: habitat modification and cascading trophic interactions.

The success of the common carp is at least partly the result of its ability to modify the shallow-water habitats it favors. It roots up the bottom and aquatic macrophytes, making ponds and shallow lakes more turbid. This decreases the abundance of visual feeding predators and competitors. Similarly, declines of sport fisheries have been associated with the invasion of the rusty crayfish *Orconectes rusticus* in lakes, where it eliminates aquatic plant beds used for cover by juvenile fishes and as a source of invertebrates for larger fishes (Lodge et al. 1985).

Some introductions are made because the species can alter habitats. The best example of this is the grass carp, which can eliminate large beds of troublesome aquatic macrophytes. A study by Rowe (1984) on a New Zealand lake revealed some of the possible indirect effects of macrophyte removal. After the introduction of grass carp, the following sequence of events occurred: (1) removal of macrophytes; (2) increase in phytoplankton production; (3) increase in zooplank-ton production causing an increase in growth rates of rainbow trout; (4) increase in predation on trout by cormorants because of lack of cover for the fish; and (5) shifts in feeding habits, relative density, and growth rates of other resident fishes.

When a top predator is introduced into a lake and severely reduces the populations of dominant planktivorous fishes, the effect cascades down the food

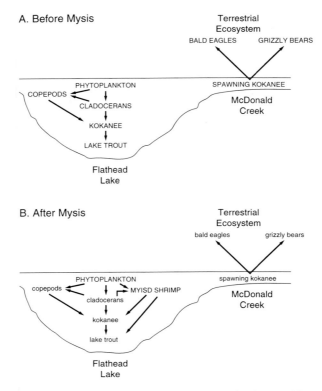

Figure 12.2 Effects of mysid introductions to the Flathead Lake trophic network. Arrows denote the direction of prey consumption. The lowercase lettering in (B) denotes the diminution of affected populations. (Modified from Spencer et al. 1991.)

chain, resulting in alterations of zooplankton and phytoplankton abundances and species. This in turn affects the abundances of other fishes, including benthic species. Such effects have been called cascading trophic interactions (Carpenter et al. 1985). Thus, Scavia et al. (1986) suggest that the introduction of predators, coupled with some climatic changes, have drastically altered the trophic structure of Lake Michigan. The likelihood that an introduced species would produce such cascading effects seems to be related to ecosystem productivity (Li and Moyle 1981). All other things being equal, more eutrophic environments can support more species and be less susceptible to disruption by introductions. For example, large cladocerans (zooplankton) and introduced opossum shrimp coexist in mesotrophic lakes but not in oligotrophic lakes (Lasenby et al. 1986).

Occasionally, the effects of an introduction can cascade into terrestrial systems. For example, introduction of opossum shrimp into Flathead Lake, Montana, ultimately caused increases in mortality of bald eagles *Haliaeetus leucocephalus* (Spencer et al. 1991). The shrimp competed with kokanee for zooplankton resulting in fewer salmon carcasses available for eagles and grizzly bears *Ursus arctos* to scavenge upon in spawning streams (Figure 12.2). Eagles then shifted their foraging efforts to scavenging road-killed animals, with unfortunate consequences to both motorists and birds (J. A. Stanford, University of Montana, personal communication).

12.3.7 Hybridization

Introduced fishes commonly hybridize with closely related native species, usually to the detriment of the native species. For example, this has resulted in elimination of cutthroat trout by rainbow trout in much of the Great Basin. In recent years, an even bigger problem than hybridization between species has been hybridization between genetically distinct stocks of the same species. For example, indiscriminate introductions of Florida strain largemouth bass and northern strain largemouth bass may result in a loss of fitness of both strains in their native ranges (Philipp 1991). The same concern is shared by salmonid biologists on the Pacific Coast. Various genetic stocks may be highly adapted to local conditions so the interbreeding with other strains, especially hatchery strays, may reduce their ability to respond to fluctuations in environmental conditions (Utter 1981; Meffe 1992).

12.3.8 Evolutionary Overview

Ultimately, the degree to which alien organisms affect indigenous communities is dependent upon the evolutionary pedigree of the donor community in relation to that of the recipient community. The evolutionary trajectory of each community will determine the degree of preadaptation of the alien organism to its new surroundings and that of the recipient community to the newcomer. For instance, adaptability to hydrologic conditions can affect interactions among native and alien fishes and determine whether or not alien species will become established in a new environment. Coexistence of the Sonoran topminnow with introduced mosquitofish in the desert streams of Arizona is predicated upon periodic flash floods (Meffe 1984). The topminnow has evolved behaviors to cope with sudden increases in discharge, but the alien mosquitofish has not. In the absence of periodic flash flooding, predation by mosquitofish would extirpate the Sonoran topminnow within a few years. Similarly, the fish community of a small mountain stream seems to shift between two equilibria, one dominated by native fishes and one dominated by introduced brown trout, with the dominant species determined by the recent hydrological history of the stream (Strange et al. 1992). It would be mistaken to assume that interbasin transfers are safer introductions than exotic species from another country. Brown and Moyle (1991) suggested that the interbasin transfer of squawfish may result in the elimination of at least one native species in the Eel River of California. In short, evolutionary mismatches cause most of the disastrous losses of biodiversity cited throughout this chapter. Such disasters can be avoided in the future only if introductions are evaluated in both an ecological and evolutionary context.

12.4 ALTERNATIVES TO INTRODUCTIONS

To paraphrase Aldo Leopold (1938), introductions can serve as the perfect alibi for postponing the practice of fisheries management. Although most fishes are introduced with the best of intentions and may live up to the manager's expectations at the local level, at least for short periods of time, unexpected negative effects of introductions often outweigh the positive effects. This phenomenon was called the "Frankenstein effect" by Moyle et al. (1986) after the central figure of Mary Shelley's 1818 novel. Dr. Frankenstein thought he was creating an improved version of man, but instead created a monster. There are many examples of the Frankenstein effect in North America, including many not mentioned previously: common carp have reduced the ability of waterfowl refuges to support ducks by eliminating aquatic macrophytes the ducks feed upon; introduced catfishes and cyprinids prey on the eggs of endangered fishes in the Colorado River, negating most efforts to restore the natives; introduced trout eliminate frogs from mountain regions.

Because the side effects of an introduction are difficult to predict, it is prudent to first consider alternative management strategies when an introduction is suggested as a solution to a management problem. Such strategies include (1) use of native fishes, (2) better water management, (3) habitat protection, and (4) use of sterile fishes.

12.4.1 Use of Native Fishes

Often a fish species is introduced simply because a fisheries manager is familiar with it. Native fishes may be underused because of ignorance. Often native species are superior to introduced species for aquatic pest control and for forage species because they are better adapted to local conditions. Ahmed et al. (1988) provided a system for rating the suitability of native fishes for aquatic pest control that could be modified for other purposes. Intensive investigation of native species should be conducted to see if they fit the management need before any introduction is made. For example, cisco and lake trout form a coadapted predator–prey complex in many of the Canadian boreal lakes. Therefore, cisco introductions to lakes with lake trout have long-lasting positive benefits, enhancing the production of lake trout (Matuszek et al. 1990).

For improving fisheries, there are many situations where increased fishing opportunities can be provided simply by educating the public about the value of underused species such as large cyprinids and suckers. Often these fishes already support small fisheries by people who know how to catch and prepare them. Expanding such fisheries can be accomplished by educating anglers on angling techniques and new ways of preparing the fish to eat. For example, suckers are highly acceptable to North American palates if smoked or pickled.

12.4.2 Better Water Management

Increasingly, North American waters are dammed, diverted, polluted, or otherwise made less suitable for fishes, especially high-value species like trout and salmon. Fisheries managers are given responsibility for managing the fish populations in such altered waters and are often blamed if the fishing is poor. Introductions of tolerant species or of hatchery-reared fish are then made as a way of providing better fishing because reversing environmental degradation is so difficult. Yet introductions into such altered waters are often undesirable because, while they may temporarily provide better fishing, they may also prevent the public from realizing the seriousness of the water quality and water management problems. For example, the unique fishes of the Colorado River survived the environmental changes of numerous impoundments, but introduction of predatory fishes may have guaranteed their extinction or at least made them forever dependent on artificial propagation (Minckley 1991). If possible, attention should be focussed on the environmental causes of lost fisheries and the blame for poor fishing should be shifted onto the shoulders of water managers and polluters. They in turn should have the responsibility for providing the conditions necessary to support resident sport fishes. It is likely that such improvements would benefit not only fisheries, but also entire communities of aquatic organisms of which the "important" species are part.

A good example of the need for better water management is found in most reservoirs, which are often built using fisheries values as a partial justification. Reservoir water levels and outflows are rarely managed specifically for fish, even for short periods of time. For example, stabilizing water levels during the spawning period of centrarchids can greatly enhance their populations; releases from reservoirs to enhance streamflows can improve survival of migrating salmonids.

12.4.3 Habitat Protection

It is axiomatic that native fishes and most high-value sport fishes thrive in undisturbed habitats or in habitats that most closely resemble those in which they evolved. Habitat protection and, if necessary, restoration should therefore be cornerstones of fisheries management. These measures are often slow to produce results and are difficult to accomplish; however, they can result in more predictable and more manageable fisheries resources than are likely to occur through repeated introductions of new species.

Habitat is the template for community structure because control by geomorphic, climatic, energetic, and hydrologic factors set boundary conditions for community interactions (Steedman and Regier 1987). As such, its role should be carefully examined before introductions are made. The near extinction of lake whitefish after the introduction of walleye in Canadian lakes occurred because the role of habitat was not well understood. Mistakes such as this can be avoided in the future because the faunal composition can now be predicted from area, mean depth, water transparency, and nutrient availability; certain lakes are best suited for northern pike, others for lake trout (Marshall and Ryan 1987). The impact of rainbow smelt introductions was less severe in Lake Champlain than in other Laurentian Great Lakes because the lake's morphometry allowed for greater thermal, spatial, and temporal segregation of smelt from other fishes (Colby et al. 1987). A major reason that degraded habitat is more subject to invasions by exotics is that it is often less complex and diverse, and opportunities for ecological segregation are reduced (Evans et al. 1987).

12.4.4 Use of Sterile Fishes

There are many situations, mainly in highly disturbed or artificial habitats, where special management problems (mosquitos, weeds, heavy fishing pressure) make the introduction of fish the best management strategy. If at all possible, the fish planted in such situations should not be able to escape and populate habitats where they are not wanted. The best way to ensure this is to plant fish that cannot reproduce. The high demand for grass carp for weed control, coupled with the widespread fear of the ecological damage they can cause, has lead to the development of sterile, triploid stocks for planting in open waters. Another method that is being explored for grass carp, tilapia, and other fishes is to produce fish of only one sex (Shelton 1986). The biggest problem with planting fish incapable of reproduction is assuring that all fish are in fact sterile or of the same sex. This becomes increasingly difficult as sterile fish are mass-produced in hatcheries.

12.5   EVALUATING POTENTIAL INTRODUCTIONS

Despite the many problems created by introduced species, further introductions seem inevitable. For example, the newsletter of the Introduced Fish Section of the American Fisheries Society exists in part to inform its readers of new and proposed introductions in North America. Given the recent advances in ecological theory and our increased knowledge of the effects of introductions, no introduction today should be made without a detailed evaluation of its potential impact. Lasenby et al. (1986) pointed out that many of the problems created by opossum shrimp introductions would have been avoided if evaluation procedures such as those suggested by Li and Moyle (1981) had been followed. Kohler and Stanley (1984) provided a detailed protocol, including a review and decision model, for evaluating potential introductions.

The following guidelines need to be considered when an introduction is proposed.

1. No introductions should be made into the few aquatic systems left that show little evidence of human disturbance. Oligotrophic, nutrient-poor, or open marine systems should, in general, be considered to be poor sites for introductions.

2. Introductions should be considered mainly for systems that have been so altered by human activity that original fish communities have been disrupted or eliminated.

3. Introductions should be considered mainly for bodies of water that are sufficiently isolated so that uncontrolled spread of the introduced species is unlikely. Because most problem waters are not isolated, the best alternative is to evaluate potential effects of the introduction on all connected waters, no matter how distant. Nearby unconnected waters also have to be evaluated, as anglers are fond of moving fish around.

4. Any system being considered for an introduction should have its biota thoroughly inventoried and a list of species developed that might be sensitive to the introduction. Special consideration should be given to rare species or species ecologically most similar to the species proposed for introduction.

5. From the inventory, species should be categorized according to functional groups by habitat and trophic position. Food webs should be constructed using whatever information is available, and the potential effect of the introduction on trophic structure evaluated. This would essentially provide

an overview of the possible interactions among native and introduced species.

6. If major gaps of understanding emerge from the above exercise, then further research should be conducted on the system. Particularly recommended are experiments with the proposed introduction in isolated ponds or laboratory systems.

7. The potential of a proposed introduction for bringing new parasites or diseases into a recipient system should be thoroughly investigated. Ideally, a proposed species should be raised under quarantine conditions for several generations before being introduced.

8. The life history characteristics of the potential introduction should be thoroughly understood before the introduction is made. The ideal introduced species will be fairly specialized and have coevolved with many members of the assemblage to which it is being added. The broader the diet of the species, the more likely it will be to create unexpected problems. It should have low vagility in case it escapes from the original site of introduction so that its spread will be easier to control.

9. Each proposed introduction should be evaluated by an independent review panel of scientists familiar with ecological principles and aquatic systems. It is important not to be too hasty with an introduction, as most are irreversible.

12.6 CONCLUSIONS

The management of natural resources ultimately is based on value systems. Political, social, economic, and aesthetic values are powerful engines that drive management policies. Human values have permitted and even encouraged introductions as a management tool. This is likely to be the case for some time, but the indiscriminant methods of the past are no longer acceptable. The American Fisheries Society has adopted a position statement (Kohler and Courtenay 1986) that advises caution and restraint with respect to introduced species.

Ecological theory is well enough developed to say that introduced species will alter the communities into which they are introduced but not well enough developed to predict precisely what the changes will be in most cases. Therefore, extreme caution should be exercised in every proposed introduction to prevent irreversible damage to natural systems. The Frankenstein effect should always be kept in mind.

Fisheries managers also need to be cautious because values are changing, especially in the face of expanding human populations and declining natural habitats. Few fisheries managers in the 1950s or 1960s had any inkling of the present public concern for saving endangered species, for preserving natural diversity, for maintaining water quality, for protecting wild lakes and streams, or even for being able to angle for native fishes. Accordingly, we need to keep as many options open as possible for future fisheries managers; poorly considered introductions significantly reduce these options.

12.7 REFERENCES

Ahmed, S. S., A. L. Linden, and J. J. Cech, Jr. 1988. A rating system and annotated bibliography for the selection of appropriate indigenous fish species for mosquito and weed control. Bulletin of the Society of Vector Ecologists 13:1–59.

Arthington, A. H. 1991. Ecological and genetic impacts of introduced and translocated freshwater fishes in Australia. Canadian Journal of Fisheries and Aquatic Sciences 48(Supplement 1):33–43.

Arthington, A. H., and D. S. Mitchell. 1986. Aquatic invading species. Pages 34–53 in R. H. Groves and J. J. Burdon, editors. Ecology of biological invasions, an Australian perspective. Australian Academy of Science, Canberra.

Balon, E. K. 1974. Domestication of the carp, *Cyprinus carpio* L. Royal Ontario Museum, Miscellaneous Publications, Toronto.

Baltz, D. M. 1991. Introduced fishes in marine systems and seas. Biological Conservation 56:151–178.

Baltz, D. M., and P. B. Moyle. 1993. Invasion resistance to introduced species by a native assemblage of California stream fishes. Ecological Applications 3:246–255.

Brown, L. R., and P. B. Moyle. 1991. Changes in habitat and microhabitat partitioning with an assemblage of stream fishes in response to predation by Sacramento squawfish (*Ptychocheilus grandis*). Canadian Journal of Fisheries and Aquatic Sciences 48:849–856.

Bryan, J. E., and P. A. Larkin. 1972. Food specialization by individual trout. Journal of the Fisheries Research Board of Canada 29:1615–1624.

Carlton, J. T. 1985. Transoceanic and interoceanic dispersal of coastal marine organisms: the biology of ballast water. Oceanography and Marine Biology an Annual Review 23:313–371.

Carpenter, S. R., J. F. Kitchell, and J. R. Hodgson. 1985. Cascading trophic interactions and lake ecosystem productivity. BioScience 35:635–639.

Colby, P. J., P. A. Ryan, D. H. Schupp, and S. L. Searns. 1987. Interactions in north-temperate lake fish communities. Canadian Journal of Fisheries and Aquatic Sciences 44 (Supplement 2):104–128.

Courtenay, W. R., Jr., and C. C. Kohler. 1986. Review of exotic fishes in North American fisheries management. Pages 401–413 in Stroud (1986).

Courtenay, W. R., Jr., and C. R. Robins. 1989. Fish introductions: good management, mismanagement, or no management? Reviews in Aquatic Sciences 1:159–172.

Crosby, A. W. 1986. Ecological imperialism: the biological expansion of Europe, 900–1900. Cambridge University Press, Cambridge, UK.

Crowder, L. B. 1980. Alewife, rainbow smelt, and native fishes in Lake Michigan: competition or predation? Environmental Biology of Fishes 5:225–233.

Crowder, L. B., and F. P. Binkowski. 1983. Foraging behaviors and the interaction of alewife, *Alosa pseudoharengus,* and bloater, *Coregonus hoyi.* Environmental Biology of Fishes 8:105–113.

Deacon, J. E. 1988. The endangered woundfin and water management in the Virgin River, Utah, Arizona, Nevada. Fisheries (Bethesda) 13(1:)18–29.

Diamond, J., and T. Case. 1986. Overview: introductions, extinctions, exterminations, and invasions. Pages 65–79 in J. Diamond and T. Case, editors. Community ecology. Harper and Row, New York.

Eck, G. W., and E. H. Brown, Jr. 1985. Lake Michigan's capacity to support lake trout (*Salvelinus namaycush*) and other salmonines: an estimate based on the status of prey populations in the 1970's. Canadian Journal of Fisheries and Aquatic Sciences 42:449–454.

Evans, D. O., B. A. Henderson, N. J. Bax, T. R. Marshall, R. T. Oglesby, and W. J. Christie. 1987. Concepts and methods of community ecology applied to freshwater fisheries management. Canadian Journal of Fisheries and Aquatic Sciences 44 (Supplement 2):448–470.

Fausch, K. D. 1988. Tests of competition between native and introduced salmonids in streams; what have we learned? Canadian Journal of Fisheries and Aquatic Sciences 45:2238–2246.

Fausch, K. D., and R. J. White. 1986. Competition among juveniles of coho salmon, brook trout, and brown trout in a laboratory stream, and implications for Great Lakes tributaries. Transactions of the American Fisheries Society 115:363–381.

Ganzhorn, J., J. S. Rohovech, and J. L. Fryer. 1992. Dissemination of microbial pathogens through introductions and transfers of finfish. Pages 175–192 in A. Rosenfield and R. Mann, editors. Dispersal of living organisms into aquatic ecosystems. University of Maryland, Maryland Sea Grant Program, College Park.

Gee, J. M. 1965. The spread of Nile perch (Lates niloticus) in East Africa with comparative biological notes. Journal of Applied Ecology 2:407–408.

Giles, R. H. 1978. Wildlife management. Freeman, San Francisco.

Herbold, B., and P. B. Moyle. 1986. Introduced species and vacant niches. American Naturalist 128:751–760.

Holcik, J. 1991. Fish introductions in Europe with particular reference to its central and eastern part. Canadian Journal of Fisheries and Aquatic Sciences 48(Supplement 1):13–23.

Holmes, J. C. 1979. Parasite populations and host community structure. Pages 27–46 in B. B. Nikol, editor. Host-parasite interfaces. Academic Press, New York.

Hughes, N. F. 1986. Changes in the feeding biology of the Nile perch Lates nilotica (L.) (Pisces: Centropomidae) in Lake Victoria, East Africa since its introduction in 1960 and its impact on the native fish community of the Nyanza Gulf. Journal of Fish Biology 29:541–548.

Hutchinson, G. E. 1958. Concluding remarks. Cold Spring Harbor Symposia on Quantitative Biology 22:415–427.

Johannes, R. E., and P. A. Larkin. 1961. Competition for food between redside shiners (Richardsonius balteatus) and rainbow trout (Salmo gairdneri) in two British Columbia lakes. Journal of the Fisheries Research Board of Canada 18:203–220.

Karasev, G. L. 1974. Reconstruction of the fish fauna in transBaikalia. Journal of Ichthyology 14:164–186.

Kohler, C. C., and W. R. Courtenay, Jr. 1986. American Fisheries Society position on introduction of aquatic species. Fisheries (Bethesda) 11(2):39–42.

Kohler, C. C., and J. J. Ney. 1980. Piscivory in a land-locked alewife (Alosa pseudoharengus) population. Canadian Journal of Fisheries and Aquatic Sciences 37:1314–1317.

Kohler, C. C., and J. G. Stanley. 1984. A suggested protocol for evaluating proposed exotic fish introductions in the United States. Pages 387–406 in W. R. Courtenay, Jr., and J. R. Stauffer, editors. Distribution, biology and management of exotic fishes. Johns Hopkins University Press, Baltimore, Maryland.

Kornberg, H., and M. H. Williamson, editors. 1986. Quantitative aspects of the ecology of biological invasions. Philosophical Transactions of the Royal Society of London B 314:501–742.

Lasenby, D. C., T. G. Northcote, and M. Furst. 1986. Theory, practice and effects of Mysis relicta introductions to North American and Scandinavian lakes. Canadian Journal of Fisheries and Aquatic Sciences 43:1277–1284.

Leopold, A. S. 1938. Chuckaremia. Outdoor America 3:3.

Li, H. W., and P. B. Moyle. 1981. Ecological analysis of species introductions in aquatic systems. Transactions of the American Fisheries Society 110:772–782.

Li, H. W., P. B. Moyle, and R. W. Garrett. 1976. Effect of the introduction of the Mississippi silverside (Menidia audens) on the growth of black crappie (Pomoxis nigromaculatus) and white crappie (Pomoxis annularis) in Clear Lake, California. Transactions of the American Fisheries Society 105:404–408.

Li, H. W., C. B. Schreck, C. E. Bond, and E. Rexstad. 1987. Factors influencing changes in fish assemblages of Pacific Northwest streams. Pages 193–202 in W. J. Matthews and D. C. Heins, editors. Community and evolutionary ecology of North American stream fishes. University of Oklahoma Press, Norman.

Lodge, D. M., A. L. Beckel, and J. J. Magnuson. 1985. Lake bottom tyrant. Natural History 94(8):32–37.

MacArthur, R. H., and R. Levins. 1967. The limiting similarity, convergence and divergence of coexisting species. American Naturalist 101:377–385.

MacArthur, R. H., and E. O. Wilson. 1967. The theory of island biogeography. Princeton University Press, Princeton, New Jersey.

Magnuson, J. J. 1976. Managing with exotics—a game of chance. Transactions of the American Fisheries Society 105:1–10.

Mangan, P., and G. J. Fitzgerald. 1984. Mechanisms responsible for the niche shift of brook char, *Salvelinus fontinalis* Mitchill, when living sympatrically with creek chub, *Semotilus atromaculatus* Mitchill. Canadian Journal of Zoology 62:1543–1555.

Marshall, T. R., and P. A. Ryan. 1987. Abundance patterns and community attributes of fishes relative to environmental gradients. Canadian Journal of Fisheries and Aquatic Sciences 44 (Supplement 2):198–215.

Matuszek, J. E., B. J. Shuter, and J. M. Casselman. 1990. Changes in lake trout growth and abundance after introduction of cisco into Lake Opeongo, Ontario. Transactions of the American Fisheries Society 119:718–729.

Meffe, G. K. 1984. Effects of abiotic disturbance on coexistence of predator and prey fish species. Ecology 65:1525–1534.

Meffe, G. K. 1992. Techno-arrogance and halfway technologies: salmon hatcheries on the Pacific coast of North America. Conservation Biology 6:350–354.

Meng, L., and J. J. Orsi. 1991. Selective predation by larval striped bass on native and introduced copepods. Transactions of the American Fisheries Society 120:187–192.

Minckley, W. L. 1991. Native fishes of the Grand Canyon region: an obituary? Pages 124–177 *in* Colorado River ecology and dam management. National Academy Press, Washington, D.C.

Mooney, H. A., and J. A. Drake, editors. 1986. Ecology of biological invasions of North America and Hawaii. Springer-Verlag, New York.

Moyle, P. B. 1986. Fish introductions into North America: patterns and ecological impact. Pages 27–43 *in* H. A. Mooney and J. A. Drake, editors. Ecology of biological invasions of North America and Hawaii. Springer-Verlag, New York.

Moyle, P. B., H. W. Li, and B. A. Barton. 1986. The Frankenstein effect: impact of introduced fishes on native fishes in North America. Pages 415–426 *in* Stroud (1986).

Nesler, T. P., and E. P. Bergersen, editors. 1991. Mysids in fisheries: hard lessons from headlong introductions. American Fisheries Society Symposium 9.

Nichols, F. H., J. K. Thompson, and L. Schemel. 1990. Remarkable invasion of San Francisco Bay (California, U.S.A.) by the Asian clam *Potamocorbula amurensis*. II. Displacement of a former community. Marine Ecology Progress Series 66:95–108.

Noble, R. L. 1981. Management of forage fishes in impoundments of the southern United States. Transactions of the American Fisheries Society 110:738–750.

Nordeng, H. 1983. Solution to the "char problem" based on arctic char (*Salvelinus alpinus*) in Norway. Canadian Journal of Fisheries and Aquatic Sciences 40:1372–1387.

Ogutu-Ohwayo, R., and R. E. Hecky. 1991. Fish introductions in Africa and some of their implications. Canadian Journal of Fisheries and Aquatic Sciences 48(Supplement 1):8–12.

Philipp, D. P. 1991. Genetic implications of introducing Florida largemouth bass, *Micropterus salmoides floridanus*. Canadian Journal of Fisheries and Aquatic Sciences 48(Supplement 1):58–65.

Roberts, L. 1990. Zebra mussel invasion threatens U.S. waters. Science (Washington, D.C.) 249:1370–1372.

Ross, S. T. 1991. Mechanisms structuring stream fish assemblages: are there lessons from introduced species? Environmental Biology of Fishes 30:359–368.

Rowe, D. K. 1984. Some effects of eutrophication and the removal of aquatic plants by grass carp (*Ctenopharyngodon idella*) on rainbow trout (*Salmo gairdneri*) in Lake Parkinson, New Zealand. New Zealand Journal of Marine and Freshwater Research 18:115–137.

Scavia, D., G. L. Fahnensteil, M. S. Evans, D. J. Jude, and J. T. Lehman. 1986. Influence of salmonine predation and weather on long-term water quality in Lake Michigan. Canadian Journal of Fisheries and Aquatic Sciences 43:435–443.

Scavia, D., G. A. Lang, and J. F. Kitchill. 1988. Dynamics of Lake Michigan plankton: a model evaluating nutrient loading competition and predation. Canadian Journal of Fisheries and Aquatic Sciences 45:165–177.

Shapiro, J., and D. I. Wright. 1984. Lake restoration by biomanipulation: Round Lake, Minnesota, the first two years. Freshwater Biology 14:371–383.

Shelton, W. L. 1986. Reproductive control of exotic fishes—a primary requisite for utilization in management. Pages 427–434 in. Stroud (1986).

Spencer, C. N., B. R. McClelland, and J. A. Stanford. 1991. Shrimp stocking, salmon collapse, and eagle displacement. BioScience 41:14–21.

Steedman, R. J., and H. A. Regier. 1987. Ecosystem science for the Great Lakes: perspectives on degradative and rehabilitative transformations. Canadian Journal of Fisheries and Aquatic Sciences 44 (Supplement 2):95–103.

Strange, E. M., P. B. Moyle, and T. C. Foin. 1992. Interactions between stochastic and deterministic processes in stream fish community assembly. Environmental Biology of Fishes 36:1–15.

Stroud, R. H., editor. 1986. Fish culture in fisheries management. American Fisheries Society, Fish Culture Section and Fisheries Management Section, Bethesda, Maryland.

Tabor, R. A., R. S. Shively, and T. P. Poe. In press. Predation on juvenile salmonids by smallmouth bass and northern squawfish in the Columbia River near Richland, Washington. North American Journal of Fisheries Management.

Tomcko, C. M., R. A. Stein, and R. F. Carline. 1984. Predation by tiger muskellunge on bluegill: effects of predator experience, vegetation, and prey density. Transactions of the American Fisheries Society 113:588–594.

Utter, F. M. 1981. Biological criteria for the definition of species and distinct intraspecific populations of anadromous salmonids under the United States Endangered Species Act of 1973. Canadian Journal of Fisheries and Aquatic Sciences 38:1625–1635.

Werner, E. E. 1986. Species interactions in freshwater fish communities. Pages 344–358 in J. Diamond and T. J. Case, editors. Community ecology. Harper and Row, New York.

Wydowski, R. S., and D. H. Bennett. 1981. Forage species in lakes and reservoirs of the western United States. Transactions of the American Fisheries Society 110:764–771.

Zaret, T. M., and R. T. Paine. 1973. Species introduction in a tropical lake. Science (Washington, D.C.) 182:449–455.

Chapter 13

Stocking for Sport Fisheries Enhancement

ROY C. HEIDINGER

13.1 HISTORY

Fish are currently being raised for stocking in sport fisheries by federal and state agencies and by private fish culturists. In order to understand the history of fish stocking in the United States, one must become familiar with the major developments in fisheries management.

At the time the United States was settled, fish were abundant, but as population centers developed, certain stocks of fish were reduced by overharvest and environmental degradation. Spawning runs of some fish stocks fluctuated widely from year to year. Because the commercial harvest of many of these stocks occurred during the spawning run, processing and shipping facilities were either underused, or they could not handle the number of fish caught. Such events eventually led to public concern for the resource.

By 1871, 10 U.S. coastal states had established fisheries commissions (McHugh 1970). Trout hatcheries were started in the mid-1800s, and there was a general belief that most of the problems associated with low catch rates were due to a failure in reproduction that could be corrected by stocking. Dr. Theodatus Garlick, who is considered the father of fish culture in the United States, published his book on artificial propagation in 1857 (Thompson 1970). At that time fisheries management was dominated by fish culture. In fact, what we now know as the American Fisheries Society was founded in 1870 as the American Fish Culturist's Association (Clepper 1970). The name was changed to the American Fisheries Society in 1884. In 1871, Congress authorized the formation of the U.S. Fish Commission, and Spencer Fullerton Baird was named the first commissioner. A milestone was reached in 1872 when Congress appropriated $15,000 for the production and introduction of American shad, salmon, and other valuable species (Clepper 1970). This was the beginning of federal support for fisheries management. In the meantime, Seth Green had opened a private trout hatchery in 1864, and by 1870 there were approximately 200 private individuals practicing fish culture (Thompson 1970).

During the 1930s and 1940s some researchers began to question the merits of wholesale stocking. Drs. Swingle and Smith at Auburn University determined that the basic management problem in small ponds was not too few fish, but rather too many small- and intermediate-size fish (Swingle 1970). The failure of many of the early stocking programs led to a period from the late 1940s to the 1960s when there was a reduction in the number of hatcheries and a corresponding reduction in stocking, especially for species in habitats where they were not self-sustaining.

The Federal Aid in Fish Restoration Act, commonly called the Dingell–Johnson Act, passed by Congress in 1950, significantly increased the number of dollars that could be used by states for fisheries management (see Chapter 4). The Wallop–Breaux Aquatic Resources Trust Act of 1984 increased the funds from approximately $35 million to more than $200 million in fiscal year 1991.

In the 1970s, the U.S. Congress reexamined Fish and Wildlife Service (FWS) responsibilities for fish stocking. Concern was expressed about the role of the FWS in providing fish for stocking private farm ponds. In the late 1970s FWS terminated this practice except in eight southeastern states (Chandler 1985). During the early 1980s the FWS identified 11 species of fish or groups of fish as having national concern, so with a new emphasis on only nationally significant species, FWS closed 31 of their 104 hatcheries in fiscal year 1983 (Chandler 1985).

Similarly, from the 1970s to the present there has been a general reevaluation of state stocking programs. According to Keith (1986), fish stocking is a major form of fisheries management in some states, but plays only a minor role in others. In most cases agencies' philosophies have developed over a long period of time through routine monitoring and informal observation; political expediency has also played a major role. More recently, emphasis has been placed on deliberate, biologically justifiable stockings designed to optimize recreational fishing in many states (Smith and Reeves 1986).

13.2 CURRENT PHILOSOPHIES

Even with the same information available to them, biologists do not always agree on the best use of a resource. For example, many biologists would argue that it was a mistake to stock salmon in the Great Lakes. At the extremes there are two broad approaches to fisheries management in North America: that of the purist and that of the pragmatist. The purist's philosophy places considerable emphasis on trying to maintain a pristine state, especially when dealing with natural bodies of water such as trout streams and natural lakes. Thus, their goal is to maintain or return the fish community to the way it was before it was disturbed by human intervention. The pragmatist's philosophy takes the waters as they are—some of them humanly created and all of them humanly altered in some way—and seeks to establish a fish community that is suitable for the existing environment and for the intended purpose. In many cases, this is done by stocking with either predator or prey species. The views of most biologists fall somewhere between these two philosophies.

One should remember that not every body of water lends itself to the same management protocol and, therefore, not every body of water should be managed the same way. Also, because the expectations of users differ, some bodies of water should be managed for different user groups. The idea of something for everyone makes sense from both a biological and public relations point of view.

13.3 GOALS, OBJECTIVES, AND CRITERIA

Noble (1986) discussed the relationship among goals, objectives, and criteria as these terms relate to stocking as a tool in fisheries management. Goals of stocking

Table 13.1 Some of the more commonly stocked warmwater species. (After Smith and Reeves 1986.)

Taxa	Number of states	Number stocked/hectare	
		Fry	Fingerlings
Largemouth bass	41	124–247	12–494
Smallmouth bass[a]	22		
Bluegill	29	618–3,706	variable
Redear sunfish[b]	14		
Crappie	11		
Striped bass	19	2–124	2–124
Hybrid striped bass	26	up to 247	12–49
Channel catfish	35		25–1236
Flathead catfish	8		
Blue catfish	10		12–124
Grass carp	8		2–37
Forage species[c]	8		

[a]Number stocked usually dependent upon hatchery allocation.
[b]15–30% of all sunfish stocked in the reporting states.
[c]Fathead minnow, threadfin shad, gizzard shad, golden shiner, blueback herring, creek chubsucker, and spottail shiner.

programs have tended to be amorphous, so it is difficult to determine if they have been attained. The goal of stocking to enhance a fishery is difficult to quantify. Objectives, in contrast to goals, should be expressed in explicit terms, with achievement being in distinct, measurable form. For example, a quantifiable objective would be to maintain a catch rate of 1 fish/h or an electrofishing rate of 100 largemouth bass/h in a given body of water. Implementation of a stocking program implies a benefit, but this benefit may accrue at the population, community, fishery, agency, or society level. It is not necessary that all levels will benefit (Noble 1986).

Potter and Barton (1986) reviewed a number of agencies' stocking goals and criteria for coldwater fisheries. Explicit, quantifiable objectives tended not to be included; instead, more general goals were used, such as to maintain quality fisheries, protect quality and diversity, maintain a healthy aquatic environment, meet the demand for angling, and provide a diversity of fish species.

13.4 CURRENT STOCKINGS IN NORTH AMERICA

The extent of fish stocking and the kinds of fish being stocked, especially by the U.S. states and federal government, have been reviewed. Stocking programs reviewed were subdivided into the following groups: (1) warmwater species (Smith and Reeves 1986); (2) coolwater species (Conover 1986); (3) coldwater species (Wydoski 1986); (4) anadromous species (Moring 1986); and (5) marine fisheries (Richards and Edwards 1986).

Smith and Reeves (1986) surveyed all 50 U.S. states; 43 states reported they used warmwater species in their stocking program (Table 13.1). In a similar survey, of the 47 states and 9 maritime provinces that responded, 48 reported stocking coolwater species (Conover 1986).

In a 1984 survey of 50 states and 10 provinces, Conover (1986) found coolwater taxa that were stocked included walleye, sauger, yellow perch, muskellunge, tiger muskellunge (northern pike × muskellunge) and saugeye (sauger × walleye).

Table 13.2 Life stages of four taxa of coolwater fish stocked in the United States and Canada in 1983–1984. (After Conover 1986.)

Taxa	Fry (%)	Fingerlings (%)
Walleye	98	2
Northern pike	93	7
Muskellunge	87	13
Tiger muskellunge	7	93

Approximately 1,157 million coolwater fish were stocked in the 1983–1984 season. Walleye accounted for 95% of these fish, 98% of which were stocked as fry; 4% were northern pike. The remaining taxa contributed only 1% of the total stockings. The majority of walleye, northern pike, and muskellunge are stocked as fry, while most tiger muskellunge are stocked as fingerlings (Table 13.2). The reason that tiger muskellunge are stocked as fingerlings instead of as fry reflects the stocking preference of fisheries biologists in the few states (e.g, Michigan, New York, and Pennsylvania) that stock the majority of the hybrids, rather than a biological difference between the two parental species and the hybrid. Also, because the hybrid accepts a prepared diet more readily than the muskellunge, it is more practical to rear the tiger muskellunge to the fingerling stage.

Of the coldwater species, approximately 256 million trout and salmon were stocked by federal and state hatcheries in 1983 (Wydoski 1986). Of these fish, 51% were stocked as fingerlings, 30% as catchables, and 19% as fry (Table 13.3).

Large numbers of Pacific salmon continue to be stocked on the West Coast, and the anadromous Atlantic salmon and striped bass are stocked on the East Coast (Table 13.4). Several thousand adult American shad have been stocked into rivers in the New England states (Moring 1986).

Interest in marine fish stocking is increasing in the United States and throughout

Table 13.3 Inland coldwater fish, by life stage, produced by federal and state hatcheries in 1983. (After Wydoski 1986 based on data from U.S. Fish and Wildlife Service 1984.)[a]

Species	Fry	Fingerlings	Catchables	Total
Salmon				
Atlantic	144.8	7,125.4	232.6	7,502.8
Coho		4,889.3	33.0	4,922.3
Chinook	491.1	13,269.3		13,760.4
Kokanee	25,342.7	3,417.0	56.7	28,816.4
Trout				
Brook	2,817.7	8,087.5	6,102.2	17,007.4
Brown	3,382.5	12,830.0	8,894.2	25,116.7
Cutthroat	5,599.8	8,948.0	7,868.6	22,411.4
Golden	15.0	53.0		68.0
Lake	290.7	10,444.7	159.2	10,894.6
Rainbow	9,668.2	58,123.9	5,484.6	122,638.7
Steelhead	2.3	2,553.4	5.0	2,560.7
Splake		290.8	139.0	429.8
Grayling	214.5	32.5		247.0
Others	32.0	3.1	7.2	42.3
Total	47,996.3	130,067.9	78,344.3	256,408.5

[a]In thousands of fish.

Table 13.4 Selected anadromous species stocking (after Moring 1986).

Species	State, federal, or provincial stocking agency	Number (million)	Year
Atlantic salmon	Fish and Wildlife Service	1.7	1984
	Maritime provinces	1.1	1982
Striped bass	11 coastal states	3.8	1983
Pacific salmon	Washington	328.0[a]	1983
	Oregon	54.0[a]	1983
	British Columbia	449.0	1982
	Alaska	47.0[b]	

[a]Various agencies involved.
[b]Alaska's salmonid stocking goal.

the world (Richards and Edwards 1986). Fifty-six million red drum were stocked into bays along the coast of Texas from 1975 to 1982; 15% were stocked as eggs, 80% as fry, and 5% as fingerlings. The state of Florida is stocking snook. As of 1978 Japan had stocked over 9 million marine fish of 12 different species (Richards and Edwards 1986).

Fish are being stocked by private fish culturists in at least 46 of the 50 states (Davis 1986). In over one-half of the states that Davis surveyed, the respondents replied that there was no record of the extent of the private fish culture industry in their states. However, based on the positive responses, private production of channel catfish, rainbow trout, largemouth bass, and sunfish for stocking exceeds 660 million fish annually. Smaller numbers of many other species are also raised.

13.5 STOCKING PROGRAMS

There is a lack of uniform stocking terminology with management implications (Laarman 1978; Radonski and Martin 1986; Smith and Reeves 1986). For the purposes of this chapter, stocking programs are divided into two categories: introduction and enhancement; enhancement stocking is subdivided into maintenance and supplemental.

Introductory stocking of species into waters where they do not exist has been practiced for many years in North America. The classic combination of largemouth bass, bluegill, and channel catfish for stocking new or renovated farm ponds is just one example that is perceived as successful by the angling public (Chapter 20). Most of the freshwater sport fishes presently found in California were introduced. The Pacific coast striped bass populations have resulted from stocking programs. Nonnative species have been introduced into many bodies of water (Chapter 12). In newly constructed lakes and reservoirs and in renovated bodies of water, these introductions are usually successful in terms of relatively high survival of the stocked fish, because of a high density of vulnerable food organisms and low density of predators. Introduced fishes become self-sustaining only if all limiting environmental factors fall within the species' ranges of tolerance. For example, it is necessary that suitable habitat, food supply, and water temperatures exist for spawning. Even if habitat and environmental

Figure 13.1 Fingerling and adult fish are transported in specially designed insulated tanks supplied with oxygen.

conditions are favorable and spawning takes place, predators can reduce or prevent recruitment.

When recruitment is limited, it may be necessary to augment the population through enhancement stocking (Figure 13.1). If the stocking is done to enhance a weak year-class, this is called supplemental stocking. Maintenance stocking is used if recruitment fails completely, which in many cases is anticipated (e.g., when striped bass are stocked into a reservoir without a major headwater stream or when certain hybrids or triploid fish are stocked). Laarman (1978) evaluated walleye stocking in 125 bodies of water over a 100-year period. Approximately 48% of the introductory stockings and 32% of the maintenance stocking were successful, but only 5% of the supplemental stocking succeeded.

Keith (1986) reviewed introduction and enhancement stocking in reservoirs. In stocking coldwater reservoirs, some effort was made to exclude nonnative species, while warmwater reservoirs were more intensively stocked with non-native species. He concluded:

> Impacts on prey populations have varied from unnoticeable to severe depletion, and effects on native predator fishes have been judged largely insignificant. However, displacement of ecologically similar species, predation on desirable native species and production of intergrade populations have been documented.

The enhancement strategy of stocking sub-harvestable-size fish is called put grow and take. Under certain conditions, harvestable-size fish are stocked that are not expected to reproduce or even grow significantly before they are caught. These put-and-take fisheries are characteristically used where fishing pressure is extremely high, such as urban areas, or in bodies of water that will support fish for only a limited time during the year. For example, trout are stocked during the fall and spring in temporary streams or in lakes where they die because of low oxygen and high temperatures in the summer. Some stockings, such as rainbow trout in the White River below Bull Shoals Dam in Arkansas, initially created a put-grow-and-take fishery, but as the fishing pressure increased, it became a put-and-take fishery. Even though rainbow trout grow 2.5 cm/month in the White River, most are removed by fishing within a few weeks after stocking. In general, states can use Wallop–Breaux monies for put-grow-and-take stocking, but not for put-and-take. On average, fish have to be 110% of stocked size when harvested for states to use federal aid to support the program.

13.6 STOCKING TECHNIQUES

Why stock? What species? What size? What number? When and where? What quality? These questions should be asked before a stocking program is implemented. It may not be possible to answer all of the questions, but they should be asked. In addition, the biologist should always keep in mind what is biologically feasible and desirable, versus what is economically feasible and desirable.

13.6.1 Why Stock?

Normally, fish are stocked to fulfill perceived needs. These needs range from a carefully conceived plan to fill a niche with a predator or prey species, to frivolous stocking of fish because hatcheries happen to have a surplus on hand. Quite frequently, political or social issues override the biological ones. Ideally, when considering why to stock, the fisheries manager states objectives and implements evaluation procedures. Some of the most common reasons for stocking are (1) to introduce a new sport or commercial species, (2) to introduce a new forage species, (3) to establish a biological control, (4) to start or maintain a put-grow-and-take fishery, (5) to supplement a year-class, (6) to satisfy public and political pressure to stock, (7) to mitigate an existing situation, (8) to start or maintain a put-and-take fishery, (9) to reestablish a species that has been lost from an area (some endangered species plans call for stocking), (10) to establish a trophy fishery, (11) to increase angler satisfaction, and (12) to redistribute fishing pressure.

A major concern at this stage of program development is the possible negative effects of the stocking (see Chapter 12). Unfortunately, if negative long-term community changes are not considered, it usually costs much more to critically evaluate the results of a stocking program than to stock the fish in the first place. Even though this is a fact, it is not a compelling argument for no evaluation.

13.6.2 What Species?

When considering an endangered species or supplementing a year-class the biologist has little choice of species to stock. Mitigation stocking is also usually targeted to a specific species. Choices may or may not be available when an agency attempts to satisfy public pressure for stocking. Often such pressure comes from organizations that are interested in a species. Normally, choices are available when the goal is to fill a niche by introducing new species that will be self-sustaining, or to start put-grow-and-take and put-and-take fisheries. In many areas of North America there is a tremendous latent demand for fishing, which can be released by starting a put-and-take or put-grow-and-take fishery. However, once the fishery has become established, public pressure often demands continuation of the stocking program. This can develop into a public relations problem if the hatchery system cannot meet the demand for fish.

Another difficult situation can arise if a body of water is being used for collecting brood fish or eggs. For example, when a lake is being used as a source of walleye brood fish that are captured in gill nets, and a dense population of hybrid striped bass is established, the hybrids may become a tremendous nuisance in the nets when the walleye are being collected. A public relations problem can also develop when fish populations, such as walleye or striped bass, are established in lakes and rivers that have a commercial fishery based on entanglement gear, such as gill nets or trammel nets. When anglers observe sport species trapped in entanglement gear, they often demand closure of the commercial fishery.

It is beyond the scope of this chapter to list all of the species that are being stocked in North America and all the biological characteristics that should be considered when choosing a fish for stocking. Table 13.5 is presented to give the reader information on the range of biological characteristics that influence the choice of a fish to stock.

13.6.3   What Size?

In general, there is a positive correlation between the size of fish stocked and survival; however, it costs more to rear fingerlings than it does to rear fry, and normally a hatchery system has more fry than fingerlings available for stocking. Thus, the decrease in vulnerability and increase in relative survival of larger fish must be weighed against the higher cost of producing larger fish. Once the relative cost of production and relative survival of the different sizes of fish have been determined for various environmental conditions, the optimal stocking size can be calculated. Unfortunately, little information has been published on this subject. One problem has been marking fry and fingerlings without inducing size-selective mortality so that relative survival rates could be calculated. With the advent of better marking techniques, such as coded wire tags, chemical marks, and genetic markers, studies can now be conducted to determine the optimum size of fish to stock. An alternative to stocking different sizes of fish the same year is to stock different sizes of fish in alternate years and then evaluate year-class strength. The problem with this approach is that many factors other than size of fish stocked can contribute to year-class strength, and the effect of size cannot be determined.

In general, when fry are stocked into a new or renovated body of water that is free of predator fish, their chance of survival is much higher than when they are stocked into established fish populations. However, in such situations an abundant food supply for the fry must be present. Even if the small fish do not die of starvation, a low food supply reduces their growth rate and increases their period of vulnerability.

Eggs are stocked only on a limited basis. Red drum eggs have been stocked into bays along the coast of Texas and threadfin shad have been successfully stocked using eggs (Maxwell and Easbach 1971). The success of red drum stockings has not been evaluated. Some agencies are recommending that sportsmen's groups "plant" trout eggs. Harshbarger and Porter (1982) found that direct intergravel plants produced twice the hatch and 3.5 times more swim-up fry than eggs planted in Whitlock–Vibert boxes. Two of five plantings of muskellunge eggs in Wisconsin

Table 13.5 Selected list of taxa that are stocked with some of the pertinent biological characteristics to consider when choosing a fish for stocking.

Taxa	Characteristics
Largemouth bass	Sport fish; Florida subspecies cannot survive in coldwater as well as northern subspecies
Smallmouth bass	Grows well on insects and crayfish as forage; grows well at warm temperatures, but does not recruit in southern states in ponds with largemouth bass and sunfish present
Bluegill	Becomes stunted in small ponds and can limit bass recruitment; readily accepts prepared diets
Redear sunfish	Harder to catch than bluegill; capable of eating molluscs; does not readily accept a prepared diet
Green sunfish	Very vulnerable to largemouth bass predation
Black crappie	Easier to handle and transport than white crappie; tends to predominate over white crappie in northern and southern portions of United States; does not readily accept a prepared diet
White crappie	Tends to overpopulate in small ponds and lakes or does not recruit; tends to dominate over black crappie in turbid water; does not readily accept a prepared diet
Hybrid *Lepomis*	Grows faster than parentals; certain F_1s are predominately males; F_1s tend to be fertile
Channel catfish	Requires cavity in which to spawn; very vulnerable to largemouth bass below 15–20 cm; may not recruit in small ponds; will readily accept prepared diets
Black and yellow bullhead	Used in urban fisheries; tends to reproduce at small size (15-cm); heavy populations capable of keeping a pond muddy in areas of colloidal clay
Striped bass	Pelagic sport fish capable of eating large shad; floating eggs require large headwater stream to recruit; some populations are maintained by stocking fry
Hybrid striped × white bass	Cross using female striped bass grows larger than reciprocal; easier to train to take prepared diet than parentals
Gizzard shad	Very fecund forage species; not desired in small pond or lake when managing for sunfish; spawns at 2 years of age
Threadfin shad	Very fecund forage species; young-of-the-year spawn; winter-kills at temperatures below 10°C
Common carp	Commercial species; capable of eating infauna; fecund; long lived; wide temperature tolerance
Grass carp	Used as biological control of vegetation; stocked at 12 to 74/hectare; triploids are available; not approved in all states; commonly reaches 14 kg; very vulnerable to largemouth bass predation below 20 cm
Fathead minnow	Forage fish; so vulnerable that it tends to be eliminated by largemouth bass
Golden shiner	Has been stocked in small lakes and ponds as forage for largemouth bass; tends to be more successful in northern part of largemouth bass range; in Midwest may overpopulate and limit bass recruitment
Inland silverside	Forage fish; winter-kills, but can tolerate colder temperatures than threadfin shad; young-of-the-year reproduce in Midwest

Table 13.5 Continued.

Taxa	Characteristics
Walleye	Sport fish; some populations can be maintained by fry stocking; others require fingerling stocking
Saugeye	Currently being investigated, but there is some indication it may produce a fishery where walleye stocking has failed
Muskellunge	Trophy sport fish; fry are very vulnerable to fish predation
Tiger muskellunge	Accepted as trophy sport fish by most muskie anglers; easier to raise to advanced fingerling stage on prepared diet than parentals
Coho salmon	Requires approximately 18 months in hatchery to reach a size for imprinting; tends to school more densely than chinook; 6–12 months in Great Lakes fishery; many unit stocks on West Coast
Chinook salmon	Requires 6 months in hatchery to reach size for imprinting; 24–26 months in Great Lakes fishery; many unit stocks on West Coast
Rainbow trout	Tolerates low oxygen better than most trout; very vulnerable to bass and striped bass predators; used frequently in two-story fish populations; highly vulnerable to sportfishing; can grow to a harvestable size on zooplankton
Brown trout	Harder to catch and longer life than rainbow or brook trout; many strains available; provides a trophy component, more tolerant of warm water
Brook trout	May out-compete other trout in cold headwater areas; very vulnerable to fishing; more acceptable to anglers at a small size
Kokanee	Small freshwater form of sockeye salmon that feeds on zooplankton during its entire life; very susceptible to competition
Atlantic salmon	Unlike Pacific salmon, all do not die after spawning; high sport fish quality; requires a forage fish to obtain large size

resulted in measurable year-classes. Lake trout eggs have been hatched (73%) between layers of artificial turf suspended over a reef in Lake Superior (Swanson 1982).

One of the best known put-grow-and-take fisheries is the salmon fishery that has been made possible by smolt stocking in the Great Lakes (Chapter 22). Salmon that are stocked as smolts imprint and return to the streams where they were released. In put-grow-and-take fisheries in smaller inland waters, there are only a few species that are routinely stocked as larvae with any degree of success. These include the striped bass, hybrid striped bass, walleye, and saugeye. The larvae of all of these species are pelagic, and their survival may be due to reduced predation in the open area of a lake as opposed to the vegetated littoral area. Even so, in many lakes larval stocking of walleye and striped bass has not been successful, whereas the stocking of fingerlings has developed and maintained a sport fishery.

Channel catfish are frequently stocked into farm ponds and small lakes that contain largemouth bass (see Chapter 20). In general, there is a positive correlation between size of channel catfish stocked and relative survival (Krummrich and Heidinger 1973).

Two-story rainbow trout populations in reservoirs may have to be maintained

Figure 13.2 Results of a put-and-take rainbow trout fishery in the White River, Arkansas.

by stocking large, less vulnerable trout. Often 40–50% of the stocked fish are creeled when 20- to 23-cm trout are stocked, and no fishery may develop if 5- to 10-cm fish are stocked. By its nature a put-and-take fishery requires what is usually called a catchable-size fish for stocking. The operational definition of catchable is a fish large enough to be acceptable to the angling public.

Put-and-take trout fisheries are often based on fish larger than 178 mm in total length (Figure 13.2). In 1983, states stocked 54,000 km of streams and 550,000 hectares of lakes with more than 53 million catchable-size trout that cost almost $37 million to produce. Rainbow trout was the most frequently stocked species (78%), followed by brook trout (11%), brown trout (10%), and other species (1%). Some fresh- and warmwater urban put-and-take fisheries are based on 0.2-kg bullheads while others use 1.1-kg common carp and 0.5-kg channel catfish. The fisheries manager always has to guard against superimposing his or her personal recreational bias upon the angling public.

13.6.4 What Number?

Obviously, the number of fish stocked should be based on biologically sound principles and the objectives associated with a specific management plan. In reality, few hatchery systems produce enough fish of the needed size to meet the demand. The management decision then becomes: Should these fish be distrib-

Table 13.6 Stocking densities of selected taxa in two urban put-and-take fisheries.

Taxa	Number/hectare	Reference
Carp, bullhead, and channel catfish	373[a]	Haas 1984
Bullhead	415–1,532	Lange 1984

[a]Stocked 10 times during the year at this density.

uted at low density into all the waters for which they have been requested, or should only a relatively few of the bodies of water be stocked at the desired density? Many state agencies address this problem by prioritizing the bodies of water. The priority order in many states—based on ownership—is: state, public, federal, and private bodies of water. State biologists then proceed down the priority list, stocking fish until the supply has been exhausted. A priority list can lead to stocking when conditions are less than optimal, such as during periods of low zooplankton or ichthyoplankton abundance.

When maintenance stocking is involved and the brood fish are collected from the wild, the bodies of water from which the brood fish are being taken often are given first priority. Normally, brood fish of only a relatively small number of species are kept in hatcheries. Typically, bluegill, redear sunfish, channel catfish, crappie, and trout are kept in the hatchery; largemouth bass and smallmouth bass may be; and the rest, such as salmon, walleye, striped bass, shad, northern pike, and muskellunge, are usually collected on their spawning run. Some recommended stocking densities for warmwater and coolwater fish are listed in Tables 13.1 and 13.6–13.9. Many states stock more fish per hectare in a small body of water than in a large body of water. This sliding scale reduces the absolute number of fish required. The tremendous range of stocking densities for various species reflects the biology of the species in terms of normal standing stock, life stage at which fish are stocked, the management program being followed, and the productivity of the water. Obviously, more fry than fingerlings are stocked per hectare, which reflects the relatively low survival of the fry versus fingerlings. This general pattern of decreasing mortality when large fish are released has been demonstrated among stream trout populations (Hume and Parkinson 1988).

Some species such as the muskellunge are at a high trophic level. Their standing stock may only be one legal size fish for every 0.4 to 4.0 hectares (Brege 1986). Fish that produce a trophy fishery are usually stocked at a relatively low density. Is it possible to stock too many fish? Hume and Parkinson (1987) have shown that mortality increases when steelhead are stocked at high densities in British Columbia streams. One concern in this type of trout stocking is that the fry do not move extensively from the stocking site. This fish behavior can lead to overstocking when the number to be stocked has been calculated for the total area of the stream or river. The only solution to this problem is to disperse the fish at the time of stocking. Unfortunately, it is not always economically or physically feasible to do so. Currently, a major concern of biologists working on the Great Lakes is that the forage base may be depleted if additional salmon are stocked annually.

The biology of the species is not the only criterion that influences the number of fish to be stocked. The actions of the fishing public can be as important. A phenomenon known as "truck following" is often associated with put-and-take trout stocking and put-and-take urban fisheries with catfish, common carp, and

Table 13.7 Part of Embody's revised planting table for trout streams. (After Everhart et al. 1975.)

Stream width meter	Number of 7.6-cm fingerlings/kilometer[a]								
	Pool grade A			Pool grade B			Pool grade C		
	Food grade			Food grade			Food grade		
	1	2	3	1	2	3	1	2	3
0.3	89	72	56	73	56	29	56	39	22
1.5	447	364	280	364	280	196	280	196	112
3.0	895	727	559	727	559	391	559	391	224
4.6	1,342	1,091	839	1,091	839	587	839	587	336
6.1[b]	1,611	1,309	1,087	1,309	1,007	705	1,007	705	403

[a]The number of other size trout to stock can be calculated by multiplying the number of 7.6-cm fish indicated in the table by 12, 1, 0.75, 0.6, and 0.3 for 2.5-, 7.5-, 10.1-, 15.2-, and 25.4-cm fish, respectively.

[b]For streams over 6.1 m in width, the formula is $X = 1.642\ WN_1 + 8N_1$, where N = number of 7.6-cm fingerlings for 0.3-m wide stream, W = average width of stream in meters, and X = number to be stocked/km.

bullheads. Anglers go to extremes to learn when and where fish will be released, and large numbers of them deplete the vulnerable fish almost as soon as they are stocked. This situation is aggravated when large numbers of fish are infrequently stocked. Frequent stockings of small numbers of fish reduce truck following and tend to level out the catch per unit effort.

One of the earliest systems to estimate the number of trout that should be stocked in streams was proposed by Embody in 1927 (Everhart et al. 1975). In Embody's system, the number of trout to be stocked depends upon stream size, suitability of the stream to support trout, food supply, and size of trout (Table 13.7). Embody's tables have been modified by various managers to reflect their specific goals and environmental conditions (Smith and Moyle 1944; Cooper 1948).

A number of states have written detailed trout stocking procedures for the management of stream and lake fisheries. Pennsylvania's plan is an example; it is detailed and includes goals and many testable objectives. In addition, it integrates the available resources with management techniques and user pressure to protect and maintain the resource (Tables 13.8 and 13.9).

13.6.5 When and Where?

Trout, especially catchables, must be stocked frequently to avoid truck following, to equalize fishing effort, and to maintain an acceptable harvest rate. In the southern states that have two-story fish populations in reservoirs with trout as the coldwater species, the trout are not usually stocked in the summer to avoid the possibility of temperature shock. Attempts have been made to stock trout in the summer through a tube that extends into the metalimnion. One reason for summer stocking is to give a hatchery system more flexibility. In a steelhead stream, Hume and Parkinson (1988) confirmed a pattern of decreasing mortality with later release dates.

One of the most common stocking combinations for small lakes and ponds includes largemouth bass, bluegill, redear sunfish, and channel catfish (see Chapter 20). In many states the bluegill, redear sunfish, and channel catfish are

Table 13.8 Partial summary of trout stream stocking policy for Pennsylvania. (From Pennsylvania Fish Commission 1986.)

Wild trout biomass (kg/hectare)	Recreational[a] use potential	Mean width (m)	Human population (persons/km²)	Stocking intensity (trout/hectare/year)		
				Preseason	Inseason	Total
High yield strategy						
<20	High	4–20	NA[b]	80	30–30–30	170
Optimum yield strategy						
Brook trout 20–30	High	4–20	NA	50	30–30	110
Brook trout plus brown trout 20–40						
20–40	Good	4–20	≥125	70	50–40	160
20–40	Good	4–20	40–125	40	40–40	120
20–40	Good	4–20	<40	40	40	80
20–40	High	<4	NA	30	30	60
Low yield strategy						
NA	Good to low	<4	NA	30	0	30
Rivers strategy						
NA	High	>30	≥125			75
NA	High	>30	40–125			65
NA	High	>30	<40			60
NA	Good	>30	≥125			45
NA	Good	>30	40–125			35
NA	Good	>30	<40			30
NA	Low	>30	NA			20

[a]Based on percentage of streams in private or public ownership, proximity to road, and available parking.
[b]NA = Not applicable.

stocked in the fall, and the largemouth bass are stocked the following spring. To a great extent, this facilitates efficient use of hatchery space and does not necessarily reflect biological restraints. Dillard (1971) concluded that in Missouri there was little difference in the results of stocking farm ponds with largemouth bass in the summer and bluegill and channel catfish in the fall, compared to stocking all three species in the fall.

Most warm- and coolwater species spawn in the spring; therefore, the fry must be stocked in the spring. Ideally, fry should be stocked when their natural zooplankton food is abundant. The density of zooplankton can change drastically within 2 weeks. Realistically, because the fry cannot be held for more than a few days without food, they are stocked when they are available.

The difference between water temperatures may be troublesome when the hatchery and the body of water to be stocked are geographically separated. When fry are moved north from a southern hatchery the water temperature at the stocking site is often below optimum. Similarly, when they are moved south, water temperature at the stocking site may be too warm for the fry to survive. Water temperature can be an extremely acute problem when power plant cooling

Table 13.9 Partial summary of trout lake stocking policy for Pennsylvania (Pennsylvania Fish Commission 1986).

Human population density rank[a]	Stocking rate Trout/hectare/year	
	Optimum	Alternate[b]
Lake ≤ 8 hectares		
1	250	120
2	240	110
3	220	100
4	210	80
Lake >8 ≤ 20 hectares		
1	200	70
2	190	60
3	170	50
4	160	40
Lake >20 ≤ 40 hectares		
1	120	60
2	110	50
3	100	40
4	80	30
Lake >40 ≤ 81 hectares		
1	80	40
2	70	30
3	60	20
4	50	10
Lake >81 hectares		
1	16	
2	12	
3	8	
4	4	

[a]Ranges from urban (1) to rural (4).
[b]Reduced stocking rates when access or some social, biological, or chemical factor limits suitability for intensive management.

lakes are stocked. In some cases hatching temperatures can be adjusted to produce fry for stocking at more optimum conditions.

It would seem that the hatchery manager would have much more timing flexibility when stocking fingerlings, and this is true if the fingerlings are being fed a prepared diet. But most sport species, except channel catfish, some sunfish, and salmonids, do not accept prepared diets without an intensive training period. Ideally, walleye, striped bass, and hybrid striped bass fingerlings should be stocked into a strong ichthyoplankton pulse. In most cases, these fish tend to be raised in fertilized rearing ponds. At stocking densities of 125,000 fry/hectare and 50% survival, walleye deplete the zooplankton by the time the fish reach 5 cm in total length (Nickum 1986), and if the fish are not harvested within a few days, cannibalism will rapidly reduce the density. Thus, the fish culturist who is saddled with a numbers quota will want to move the fish out of the hatchery as soon as the zooplankton community is depleted to avoid extensive cannibalism.

The optimum location for stocking and the best time of day or night to stock various species have not been thoroughly investigated. We know zooplankton density is not uniform throughout a reservoir, and that they are patchy within any area of the reservoir. Also higher densities of ichthyoplankton and young-of-the-year fish tend to occur in the back end of large coves. Since the larvae of walleye, striped bass, and hybrid striped bass inhabit open water, it may be beneficial to stock them away from shore instead of at convenient boat docks. Alternatively, fingerling northern pike, largemouth bass, and muskellunge are probably better adapted for shoreline stocking than pelagic stocking.

In Nebraska, Miller (1971) stocked 5- to 8-cm northern pike at a rate of 62 to 155/km into small streams that emptied into larger bodies of water. Of 15 such stockings, five were judged successful, five were partially successful, and five were considered failures.

As indicated earlier, some salmon and trout fry and fingerlings do not disperse well in streams if spot stocked. Evidently this is not the case for all salmonid fingerlings. Marshall and Menzel (1984) found no difference in survival or growth of spring spot-stocked versus scatter-stocked brown trout fingerlings (9 cm) in six northeast Iowa streams (put-grow-and-take fisheries). Summer survival ranged from 3 to 22% and overwinter survival ranged from 2 to 14%. Similar overwinter survival (7%) of brown trout has been reported in a Wisconsin stream by Brynildson et al. (1966) and in a New York stream (6%) by Schuck (1943).

In some cases fry or fingerlings are not stocked directly into the body of water where the fishery occurs. For example, in the north central states northern pike have been reared in managed marshes (Gregory et al. 1984). Production is quite variable, but 1,000 to 3,000 fingerlings (75 to 150 mm total length) per hectare is normal. Both fry and adult brood fish have been stocked into the marshes. Royer (1971) found that adult brood fish stockings were much more successful than fry stocking. However, fry stockings have been successful in other marshes (Fago 1977). A variation of the managed spawning marsh is the use of winterkill lakes to rear fish for stocking. In 1967–1968, Minnesota reared about 500,000 northern pike weighing 136,000 kg in shallow winterkill lakes by stocking adult fish in the spring and harvesting in the late fall or winter (Johnson and Moyle 1969). Walleye can also be raised in winterkill lakes.

Some states construct nursery ponds adjacent to large reservoirs in order to produce fingerlings of various species for stocking into the reservoir. Arkansas built a number of these nursery ponds during the 1960s (Keith 1970). Production of fish in such nursery areas can be quite high (Table 13.10).

Certain species are being raised to larger sizes in cages or pens suspended in the body of water in which they are to be released. Channel catfish have been raised to larger sizes in cages in order to reduce their vulnerability to bass predation when they are stocked (Collins 1971). Alternate crops of catfish (blue and channel) and rainbow trout have been raised to large size (0.2 to 0.4 kg/fish) in net-pens. Arkansas's Bull Shoals net-pen facility produced 44% (by number) of all catchable catfish and 15% of all trout stocked in the state in 1986 (Oliver and Rider 1986). Trout stocking density in the cages ranges from 106 to 176 fish/m^3 and

Table 13.10 Fish produced in four Arkansas nursery ponds (Keith 1970).

Year	Species	Number/hectare	Length (mm)
	Norfork (3.2 hectare)[a]		
1964	Walleye	12,500	63
1965	Walleye	23,400	63
1966	Channel catfish	12,500	228
1967	Northern pike	1,200	253
	Striped bass	1,600	114
1968	Crop failure		
1969	Muskellunge	400	228
	Bull Shoals (4.0 hectare)		
1966	Walleye	30,000	63
1967	Walleye	75,000	63
	Channel catfish	4,500	177
	Largemouth bass	1,200	88
1968	Walleye	2,200	177
1969	Walleye	25,000	101
	Beaver (10.9 hectare)		
1968	Walleye	45,900	63
1969	Northern pike	22,900	114
	Greeson (8.1 hectare)		
1968	Channel catfish	2,800	b
	Striped bass	2,100	126
1969	Walleye	37,000	63

[a]Size of nursery ponds in parentheses.
[b]Mean weight 0.23 kg.

catfish were usually stocked at approximately 140 fish/m^3. One prerequisite of this technique is that the cultured species will accept a prepared diet.

13.6.6 What Quality?

In addition to factors discussed above, other parameters contribute to the success of a stocking program including genetic characteristics, parasite infestation, physiological integrity, and handling history (Murphy and Kelso 1986).

Kutkuhn (1981) defined the term stock as a randomly mating group of individuals having temporal, spatial, or behavioral integrity. This is somewhat similar to the ecological definition of a population. Accordingly, the stock, and not the species or subspecies, should be considered the operational unit of interest. In actuality this concept may be quite difficult to apply.

The existence of genetic stocks of salmon and trout has been recognized for many years. In their review, Kincaid and Berry (1986) found 143 genetic stocks of rainbow trout, 43 of brook trout, 42 of brown trout, 39 of cutthroat trout, and 26 of lake trout. These include both natural and domestic stocks. Different breeding stocks of salmon spawn at various times in different reaches of the streams. These strains can have considerable differences in biological parameters, such as growth rate, fecundity, disease resistance, stress resistances, age to maturity, time of spawning, and location of spawning. For an example of characteristics of specific rainbow trout strains see Kincaid and Berry (1986). Initially, biologists attempted to identify these strains using classical meristic techniques, but achieved only limited success. The more recent gel electrophoresis technique is more powerful. To the applied fisheries manager and culturist, differences in allozyme patterns become significant when they are correlated with some behavioral or physiological parameter. For example, the southern subspecies of the largemouth bass is not as tolerant of cold water as the northern subspecies of largemouth bass. Undoubtedly, there are differences even among populations of a subspecies, but these differences may or may not have biological, social, economic, or political significance.

The degree of domestication (number of generations in a hatchery) of a species can change the characteristics of a species. For example, in 1959 Boles and Borgeson (1966) stocked equal numbers of brown trout from the Convict Lake strain (wild), the Mount Whitney strain (several generations in a hatchery) and the Massachusetts strain (many generations in hatchery). After 2 years, the return to creel of the Massachusetts, Mount Whitney, and Convict Lake strains were 63, 37, and 25%, respectively. Other studies have shown that hatchery-reared salmonids, including brook, brown, and rainbow trouts, may be more vulnerable to angling than are wild fish (Boles 1960; Hunt 1979). Other differences have also been noted between wild and "domestic" strains of salmonids. Symons (1969) determined that wild strains of Atlantic salmon smolts dispersed more widely than domestic strains, and Flick and Webster (1976) found that hatchery-reared wild brook trout lived longer than domestic strains of brook trout when released into natural ponds. Reisenbichler and McIntyre (1977) found that a hatchery strain of steelhead had a higher growth rate than wild steelhead in hatchery ponds but not in streams; however, the wild trout × hatchery trout hybrid had the highest growth rate in streams.

Stocks can be identified with morphological characters, DNA–DNA hybridization, cytogenetic analysis, electrophoretic characteristics of specific protein molecules, immunological analyses of specific protein molecules, mitochondrial DNA analysis, and nuclear DNA analysis (Philipp et al. 1986). It is possible that additional, even more sensitive, techniques will be available in the future. Even though the genetic stock concept is important, care must be taken that it is not overemphasized. With sufficient analytical sensitivity every population could probably be shown to be distinctive. In fact, every individual is distinctive. From a practical point of view, in terms of relocating genetic stocks through stocking, the unit that makes up a given stock must have measurable, meaningful differences in such things as growth rate, reproductive rate, tolerance of environmental conditions, and behavior. The argument over what is a meaningful difference will continue for a long time, since in many cases the point is philosophical, rather than biological.

Goede (1986) discussed the potential effects of stocking fish infected with parasites. He pointed out that stocking diseased or carrier fish can have serious consequences, including imminent mortality or reduced performance and increased sensitivity to stressors. An equally serious problem is the possible establishment of a reservoir of infection. Most fisheries agencies have attempted to deal with whether or not to stock infected fish through a set of internally developed policy guidelines. For example, salmon infected with hematopoietic necrosis virus or *Myxosoma cerebralis,* which induces whirling disease, are not usually stocked, especially in waters containing salmonid populations with no known history of these diseases. On the other hand, fish that have been infected with bacteria, such as *Aeromonas,* or with protozoans, such as *Ichthyopthirius multifilia,* are frequently stocked. The latter category of parasites is found in almost all bodies of water. In fact, *Aeromonas* is a nonobligatory fish parasite.

The physiological integrity of hatchery-reared fish can be compromised during the rearing process. For example, it is possible to produce large numbers of fish that do not have inflated swim bladders. These fish do not survive when released into the wild. This is known to occur in a number of both saltwater and freshwater species (e.g., snook, striped bass, and walleye). Under certain conditions, cultured fish have been shown to have overinflated swim bladders.

A phenomenon even less understood than noninflated swim bladders is whether or not piscivorous sport fish that have been raised on a prepared diet have a survival rate as high after they are released as the same taxa reared on natural forage. Biologists have mixed opinions on this subject. If fish do not survive as well on a prepared diet, the problem may relate to the quality of the diet or to a change in the behavior of the fish.

Even if all of the previously discussed criteria, such as species, genetic stock, size, health, inflated swim bladder, etc., have been met, fish must be handled correctly during harvest and hauling to the stocking site or high mortality will result. Some of the fish may be dead when they are stocked (initial mortality) or they may die within several days or weeks after stocking (delayed mortality). Many management failures have probably resulted from fish that were predestined to die from handling, hauling, and stocking stress. Often the people who stock fish are not trained biologists. If the fish are moved too fast from one temperature to another, thermal shock could occur. For information on this subject see Piper et al. (1982).

13.7 EVALUATION OF STOCKINGS

In a review of the economic benefits and costs associated with stocking fish, Weithman (1986) indicated that factors usually considered to determine success or failure of stocking programs include (1) survival, (2) reproduction, (3) angler acceptance, (4) days and quality of fishing provided, (5) percentage of stocked fish that are harvested, and (6) economic benefits to anglers and state and local governments. In actuality, public perception plays a major part in determining the success of a program.

Post-stocking fish surveys are usually conducted to evaluate survival and reproduction. Days of fishing provided and percentage of stocked fish harvested are obtained from creel surveys. The fish community is sampled to document the presence of stocked fish, obtain fish for age and growth analyses, and determine population structure.

Economic evaluation of the success of a stocking is usually based on benefit:cost ratio. Monetary cost of stocking is theoretically easy to calculate, but biological cost is often hard to identify and more difficult to quantify. Likewise,

Table 13.11 Selected examples of four types of economic cost analysis. (See Weithman 1986.)

Species	State	Cost/ angler-day ($)	Cost/ fish creeled ($)	Benefit:cost Dollar value of fishing day	Benefit:cost Yearly value to anglers	Reference
Rainbow trout	Virginia	0.69				Applegate (1963)
Rainbow trout	New Mexico	0.54				Ferkovich (1969)
Channel catfish	Missouri	0.24				Haas (1984)
Channel catfish[a]	Virginia		18.00			Bryson et al. (1975)
Walleye	Wisconsin		1.46			Hauber (1983)
Walleye	Michigan		4.00			Laarman (1981)
Largemouth bass, bluegill, and channel catfish	Missouri			90:1		Novinger (1977)
Coho and Chinook salmon	Oregon			8:1		Brown and Hussen (1976)
Channel catfish	Kansas			39:1		Stevens (1982)
Salmon and Trout	Great Lakes				18:1	Talhelm and Ellefson (1973)
Muskellunge	Missouri (Pomme de Terre Lake)				9:1	Belusz and Witter (1986)

[a] Average length at stocking.

the biological, social, and economic benefits of stocking may be difficult to quantify. Various methods are used to calculate economic cost and benefits. These include (1) agency cost/angler-day, (2) agency cost to provide fish in the creel, (3) benefit:cost ratio expressed as agency cost for stocking versus days of fishing multiplied by the dollar value of a day's fishing, and (4) benefit:cost ratio expressed as agency cost versus benefit to anglers, and state and local economies (Weithman 1986). Examples are compared in Table 13.11.

It is questionable whether benefit:cost ratios have included all of the costs. Attempts have been made to compare the cost of fish produced by government agencies to the cost of those produced by the private sector. A comparative cost analysis showed that fingerling rainbow trout, largemouth bass, and bluegill were probably more expensive to buy from federal hatcheries than from private hatcheries. Salmon were not available from the private sector and channel catfish were probably cheaper to buy from private producers (USFWS 1982). Catchable rainbow trout were cheaper to buy from commercial producers in some regions of the United States, but more expensive in others.

The daily cost of fishing, if it is not known for a specific fishery, is often based on national or state averages. Some economists feel that the true value of a fishing day is how much a person is willing to spend to go fishing; others argue that the dollars spent on recreational fishing are "displaced." This latter philosophy concludes that if the money were not spent on fishing, it would be spent for other forms of recreation and, therefore, it would really not be lost to the total economy. An objective evaluation of stocking is not always easy. It can be measured in terms of survival rate, year-class strength, and contribution to the harvest.

13.8 THE FUTURE

Radonski and Martin (1986) considered the history of stocking and concluded that:

> The arc of the fish culture pendulum has come full swing: from early consideration as a universal fisheries management panacea through a transitional period of questioning and disrepute, to final recognition as an indispensable tool when appropriately integrated with other equally essential fisheries management protocol.

Due to the continued increase in numbers of anglers without a concurrent increase in lake construction, stocking will continue to be used as a tool in sport fisheries management. In particular, it will be used more frequently in large rivers and coastal areas. Such examples as transplanting the snail darter (Hickman 1981) and the progress made in recovery of native fishes in the southwestern United States (Rinne et al. 1986) will lead to stocking as an acceptable tool in the management of both endangered species and nonsport fishes.

Through no fault of their own, managers of hatchery systems often become locked into production based on numbers, which may have little to do with the success of stocking programs. An alternative to the "head count" is needed. Quotas (Hatchery Benefit Units) can be set in more biologically significant terms once the relative survival and cost of different size individuals of various species is known, and a weighting system is set up among species (Heidinger et al. 1987).

The economics of rearing various taxa to larger sizes will change as more biological information is accumulated and as more fish are reared for stocking by private fish culturists. The considerable resistance by state and federal agencies to buying fish from private fish culturists for stocking will eventually be reduced due to legal and political pressure and the maturing of the private sector.

No matter how it is done, the actual cost of stocking makes it an expensive management tool. In some cases, part of the cost is recovered by requiring a special license or stamp to fish for the stocked fish. This is most frequently done for salmon and trout put-and-take and put-grow-and-take fisheries. The requirement for a special fishing license or stamp will likely be extended to other sport fisheries.

One of the most important problems associated with stocking programs in artificial lakes and reservoirs is the uncontrolled loss of stocked species over the spillway. We need to start thinking of total mortality as equaling fishing mortality plus natural mortality and spillway emigration.

Fish genetics, in terms of identifying stocks, hybridization, sterilization, and strain improvement, will increasingly become more important. With the advent of gene transfer technology, there is tremendous potential for engineering fish communities well suited for perturbed or artificial bodies of water. In the future, genetic stocks will be preserved in gene banks. The philosophical difference between those who advocate no stocking and those who advocate complete "laissez faire" will initially widen, but as more information is obtained, they will be reconciled to a much greater degree than they are today.

13.9 REFERENCES

Applegate, R. L. 1963. Evaluation of trout stocking practices. Virginia Commission of Game and Inland Fisheries, Federal Aid in Fish Restoration, Project F-13-R-1, Completion Report, Richmond, Virginia.

Belusz, L. C., and D. J. Witter. 1986. Why are they here and how much do they spend? A survey of muskellunge angler characteristics, expenditures, and benefits. Pages 39–45 in G. E. Hall and M. J. Van Den Avyle, editors. Reservoir fisheries management: strategies for the 80's. American Fisheries Society, Southern Division, Reservoir Committee, Bethesda, Maryland.

Boles, H. D. 1960. Experimental stocking of brown trout in a California Lake. Proceedings of the Annual Conference Western Association of State Game and Fish Commissioners 40:334–349.

Boles, H. D., and D. P. Borgeson. 1966. Experimental brown trout management of Lower Sardine Lake. California Fish and Game 52:166–172.

Brege, D. A. 1986. A comparison of muskellunge and hybrid muskellunge in a Southern Wisconsin Lake. American Fisheries Society Special Publication 15:203–207.

Brown, W. G., and A. Hussen. 1976. A production economic analysis of the Little White Salmon and Willard National Fish Hatcheries. Oregon State University, Agricultural Experiment Station, Special Report 428, Corvallis.

Brynildson, O. M., P. E. Degurse, and J. W. Mason. 1966. Survival, growth, and yield of stocked domesticated brown and rainbow trout fingerlings in Black Earth Creek. Wisconsin Conservation Department Research Report (Fish) 18, Madison.

Bryson, W. T., R. T. Lackey, J. Cairns, Jr., and K. L. Dickson. 1975. Restocking after fish kills as a fisheries management strategy. Transactions of the American Fisheries Society 104:256–263.

Chandler, W. J. 1985. Inland fisheries management. Pages 93–129 in R. L. Di Silvestro, editor. Audubon Wildlife Report 1985.

Clepper, H. 1970. A century of fish conservation. American Forests 76(11) 16–19, 54–56.

Collins, R. A. 1971. Cage culture of catfish in reservoirs and lakes. Proceedings of the Annual Conference Southeast Association of Fish and Wildlife Agencies 24(1970): 489–496.

Conover, M. C. 1986. Stocking cool-water species to meet management needs. Pages 31–39 in Stroud (1986).

Cooper, G. P. 1948. Fish stocking policies in Michigan. Transactions of the North American Wildlife Conference 13:187–198.

Davis, J. T. 1986. Role of private industry in increasing production capabilities. Pages 243–248 in Stroud (1986).

Dillard, J. 1971. Evaluation of two stocking methods for Missouri farm ponds. Pages 203–204 in R. J. Muncy and R. V. Bulkley, editors. Proceedings of the North Central Warmwater Fish Culture-Management Workshop. Iowa State University, Ames.

Embody, G. C. 1927. An outline of stream study and the development of a stocking policy. Cornell University Aquaculture Laboratory, Ithaca, New York.

Everhart, W. H., A. W. Eipper, and W. D. Youngs. 1975. Principles of fishery science. Cornell University Press, Ithaca, New York.

Fago, D. M. 1977. Northern pike production in managed spawning and rearing marshes. Wisconsin Department of Natural Resources Technical Bulletin 96.

Ferkovich, P. 1969. Cost:harvest ratios from southeastern area and northeastern area trout waters. New Mexico Department of Game and Fish, Federal Aid in Fish Restoration, Project F-22-R-10, Completion Report, Santa Fe.

Flick, W. A., and D. A. Webster. 1976. Production of wild, domestic and interstrain hybrids of brook trout (*Salvelinus fontinalis*) in natural ponds. Journal of the Fisheries Research Board of Canada 33:1525–1539.

Goede, R. W. 1986. Management considerations in stocking of diseased or carrier fish. Pages 349–355 in Stroud (1986).

Gregory, R. W., A. A. Elser, and T. Lenhart. 1984. Utilization of surface coal mine waste water for construction of a northern pike spawning/rearing marsh. U.S. Fish and Wildlife Service FWS/OBS-84103.

Haas, M. A. 1984. The Missouri urban fishing program. Pages 275–259 *in* L. J. Allen, editor. Urban fishing symposium, proceedings. American Fisheries Society, Fisheries Management Section and Fisheries Administrators Section, Bethesda, Maryland.

Harshbarger, T. J., and P. E. Porter. 1982. Embryo survival and fry emergence from two methods of planting brown trout eggs. North American Journal of Fisheries Management 2:84–89.

Hauber, A. B. 1983. Two methods for evaluating fingerling walleye stocking success and natural year class densities in Severn Island Lake, Wisconsin, 1977–1981. North American Journal of Fisheries Management 3:152–155.

Heidinger, R. C., J. H. Waddell, and B. L. Tetzlaff. 1987. Relative survival of walleye fry versus fingerlings in two Illinois reservoirs. Proceedings of the Annual Conference Southeastern Association of Fish and Wildlife Agencies 39(1985):306–311.

Hickman, G. D. 1981. Is the snail darter transplant a success? Pages 338–344 *in* L. A. Krumholz, editor. The warmwater streams symposium: a national symposium on fisheries aspects of warmwater streams. American Fisheries Society, Southern Division, Bethesda, Maryland.

Hume, J. M. B., and E. A. Parkinson. 1987. Effect of stocking density on the survival, growth and dispersal of steelhead trout fry (*Salmo gairdneri*). Canadian Journal of Fisheries and Aquatic Sciences 44:271–281.

Hume, J. M. B., and E. A. Parkinson. 1988. Effects of size at and time of release on the survival and growth of steelhead fry stocked in streams. North American Journal of Fisheries Management 8:50–57.

Hunt, R. L. 1979. Exploitation, growth, and survival of three strains of domestic brook trout. Wisconsin Department of Natural Resources, Report 99, Madison.

Johnson, F. H., and J. B. Moyle. 1969. Management of a large shallow winter-kill lake in Minnesota for the production of pike (*Esox lucius*). Transactions of the American Fisheries Society 98:691–697.

Keith, W. 1970. Preliminary results in the use of a nursery pond as a tool in fishery management. Proceedings of the Annual Conference Southeastern Association of Game and Fish Commissioners 23(1969):501–511.

Keith, W. E. 1986. A review of introduction and maintenance stocking in reservoir fisheries management. Pages 144–155 *in* G. E. Hall and M. J. Van Den Avyle, editors. Reservoir fisheries management: strategies for the 80's. American Fisheries Society, Southern Division, Reservoir Committee, Bethesda, Maryland.

Kincaid, H. L., and C. R. Berry, Jr. 1986. Trout broodstocks in management of national fisheries. Pages 211–222 *in* Stroud (1986).

Krummrich, J. T., and R. C. Heidinger. 1973. Vulnerability of channel catfish to large-mouth bass predation. Progressive Fish-Culturist 35:173–175.

Kutkuhn, J. H. 1981. Stock definition as a necessary basis for cooperative management of Great Lakes fish resources. Canadian Journal of Fisheries and Aquatic Sciences 38:1476–1478.

Laarman, P. W. 1978. Case histories of stocking walleyes in inland lakes, impoundments, and the Great Lakes—100 years with walleyes. Pages 254–260 *in* R. L. Kendall, editor. Selected coolwater fishes of North America. American Fisheries Society Special Publication 11.

Laarman, P. W. 1981. Vital statistics of a Michigan fish population with special emphasis on the effectiveness of stocking 15-cm walleye fingerlings. North American Journal of Fisheries Management 1:177–185.

Lange, R. E. 1984. Fishing in the big apple: a demonstration program for New York City. Pages 263–274 *in* L. J. Allen, ed. Urban fishing symposium proceedings. American Fisheries Society, Fisheries Management Section and Fisheries Administrators Section, Bethesda, Maryland.

Marshall, S. A., and B. A. Menzel. 1984. Recovery, growth and habitual utilization of spring-stocked fingerling brown trout (*Salmo trutta*) in six northeast Iowa streams. Proceedings of the Annual Conference Western Association of Fish and Wildlife Agencies 64:411–421.

Maxwell, R., and A. R. Easbach. 1971. Eggs of the threadfin shad successfully transplanted and hatched after spawning on excelsior mats. Progressive Fish-Culturist 33:140.

McHugh, J. L. 1970. Trends in fishery research. Pages 25–56 *in* N. G. Benson, editor. A century of fisheries in North America. American Fisheries Society Special Publication 7.

Miller, E. 1971. Some success and failures using very small streams as northern pike stocking areas. Pages 84–88 *in* R. J. Muncy and R. V. Bulkley, editors. Proceedings of the North Central Warmwater Fish Culture-Management Workshop. Iowa State University, Ames.

Moring, J. R. 1986. Stocking anadromous species to restore or enhance fisheries. Pages 59–74 *in* Stroud (1986).

Murphy, B. R., and W. E. Kelso. 1986. Strategies for evaluating fresh-water stocking programs: past practices and future needs. Pages 306–313 *in* Stroud (1986).

Nickum, J. G. 1986. Walleye. Pages 115–126 *in* R. R. Stickney, editor. Culture of nonsalmonid freshwater fishes. CRC Press, Boca Raton, Florida.

Noble, R. L. 1986. Stocking criteria and goals for restoration and enhancement of warmwater and coolwater fisheries. Pages 139–159 *in* Stroud (1986).

Novinger, G. D. 1977. An assessment of Missouri's pond program. Missouri Department of Conservation, Federal Aid in Fish Restoration, Project F-1-R-25, Study I-19, Job 1, Completion Report, Columbia.

Oliver, M. L., and L. L. Rider. 1986. Net-pen aquaculture in Bull Shoals Reservoir. Pages 287–300 *in* Stroud (1986).

Pennsylvania Fish Commission. 1986. Management of trout fisheries in Pennsylvania waters. Pennsylvania Fish Commission, Division of Fisheries, Harrisburg.

Philipp, D. P., J. B. Koppelman, and J. L. Van Orman. 1986. Techniques for identification and conservation of fish stocks. Pages 323–338 *in* Stroud (1986).

Piper, R. G., I. B. McElwain, L. D. Orme, J. P. McCraren, L. G. Fowler, and J. R. Leonard. 1982. Fish hatchery management. U.S. Fish and Wildlife Service, Washington, D.C.

Potter, B. A., and B. A. Barton. 1986. Stocking goals and criteria for restoration and enhancement of coldwater fisheries. Pages 147–159 *in* Stroud (1986).

Radonski, G. C., and R. G. Martin. 1986. Fish culture is a tool, not a panacea. Pages 7–13 *in* Stroud (1986).

Reisenbichler, R. R., and J. D. McIntyre. 1977. Genetic differences in growth and survival of juvenile hatchery and wild steelhead trout, *Salmo gairdneri*. Journal of the Fisheries Research Board of Canada 34:123–128.

Richards, W. J., and R. D. Edwards. 1986. Stocking to restore or enhance marine fisheries. Pages 75–80 *in* Stroud (1986).

Rinne, J. N., J. E. Johnson, B. L. Jensen, A. W. Ruger, and R. Sorenson. 1986. The role of hatcheries in the management and recovery of threatened and endangered fishes. Pages 271–285 *in* Stroud (1986).

Royer, L. M. 1971. Comparative production of pike fingerlings from adult spawners and from fry planted in a controlled spawning marsh. Progressive Fish-Culturist 33:153–155.

Schuck, H. A. 1943. Survival, population density, growth, and movement of wild brown trout in Crystal Creek. Transactions of the American Fisheries Society 73:209–230.

Smith, B. W., and W. C. Reeves. 1986. Stocking warmwater species to restore or enhance fisheries. Pages 17–29 *in* Stroud (1986).

Smith, L. L., Jr., and J. B. Moyle. 1944. A biological survey and fishery management plan for the streams of the Lake Superior north shore watershed. Minnesota Department of Conservation Technical Bulletin 1.

Stevens, V. 1982. Channel catfish fishery maintenance for public fishing waters. Kansas

Fish and Game Commission, Federal Aid in Fish Restoration, Project F-25-D, Completion Report, Pratt.

Stroud, R. H., editor. 1986. Fish culture in fisheries management. American Fisheries Society, Fish Culture Section and Fisheries Administrators Section, Bethesda, Maryland.

Swanson, B. L. 1982. Artificial turf as a substrate for incubating lake trout eggs on reefs in Lake Superior. Progressive Fish-Culturist 44:109–111.

Swingle, H. S. 1970. History of warmwater pond culture in the United States. Pages 95–106 in N. G. Benson, editor. A century of fisheries in North America. American Fisheries Society, Special Publication 7.

Symons, P. F. K. 1969. Greater dispersal of wild compared with hatchery-reared juvenile Atlantic salmon released in streams. Journal of the Fisheries Research Board of Canada 26:1867–1876.

Talhelm, D. R., and P. V. Ellefson. 1973. Michigan's Great Lakes trout and salmon fishery (1969–1972). Bureau of Sport Fisheries and Wildlife, and the National Marine Fisheries Service, Fisheries Management Report 5, Great Lakes Fish Resource Development Study AFSC-8, Ann Arbor, Michigan.

Thompson, P. E. 1970. The first fifty years—the exciting ones. Pages 1–11 in N. G. Benson, editor. A century of fisheries in North America. American Fisheries Society Special Publication 7.

USFWS (U.S. Fish and Wildlife Service). 1982. Comparative costs of alternative sources of fish for federal management needs. USFWS, Division of Hatcheries and Fishery Resource Management, Washington, D.C.

USFWS (U.S. Fish and Wildlife Service). 1984. Propagation and distribution of fishes from National Fish Hatcheries for fiscal year 1983. USFWS, Fish Distribution Report 18, Washington, D.C.

Weithman, A. S. 1986. Economic benefits and costs associated with stocking fish. Pages 357–264 in Stroud (1986).

Wydoski, R. D. 1986. Informational needs to improve stocking as a coldwater fisheries management tool. Pages 41–57 in Stroud (1986).

Chapter 14

Management of Undesirable Fish Species

ROBERT W. WILEY AND RICHARD S. WYDOSKI

14.1 HISTORY

During the second half of the 19th Century many species of fish were transplanted throughout the United States to mitigate for losses of fish stocks associated with industrial growth. However, the success of such introductions (i.e., good growth and survival) required the control of other fish that were predators or competitors, as well as providing adequate sources of food. Occasionally, introduced predator or forage fish were more deleterious than beneficial (see Chapter 12). Although unsuccessful, the earliest attempted control of fish in the United States was the application of copper sulfate in a Vermont lake (Titcomb 1914).

By the 1930s, fisheries management techniques included various strategies for the manipulation (i.e., control) of fish populations to improve sport fishing. Netting, seining, trapping, and chemical control described by Hubbs and Eschmeyer (1938) are still in use today. Since the 1930s, more emphasis on the control of undesirable fishes has appeared in the literature. Clearly, as the demand for desirable sport or commercial fishes increased, the perceived need to control undesirable fishes has also increased.

14.2 WHAT IS AN UNDESIRABLE FISH SPECIES?

To develop an understanding of current strategies for the management of undesirable fishes, a questionnaire was mailed to all state conservation agencies and Canadian provinces in 1988. The agencies that responded (90%) provided a broad range of perceptions of undesirable fish species. Their responses can be concisely summarized by stating that an undesirable species of fish is virtually any species that does not meet human needs.

There is no specific rating system to aid in determining whether or not a fish species is undesirable; professional judgement must be used to make the final decision. When a fish species has been determined to be a pest, a control program should be designed based on an understanding of the biology and habitat of the species, a consideration of all effective methods of control, and an understanding of the level of control that is needed and possible (Binning et al. 1985).

Undesirable fishes are not considered beneficial to humans—analogous to

---

Box 14.1   Criteria to Determine if Fish are Undesirable

1. The species does not contribute to the sport fishery (may not be available or acceptable to anglers or commercial fishers) or forage base.
2. The species inhibits development or maintenance of desirable fish through predation or direct competition with sport or commercial fishes.
3. The species is detrimental to the biological balance of the aquatic system (e.g., large gizzard shad sometimes constitute most of the biomass in reservoirs).
4. The species may serve as a potential reservoir for pathogenic organisms in a hatchery water supply.
5. The species may interfere with other wildlife management practices (e.g., common carp cause turbidity inhibiting the growth of aquatic plants that are used as food by waterfowl).

---

weeds and gardeners (Box 14.1). Early fisheries administrators in the United States (e.g., Spencer F. Baird) were instrumental in the introduction of common carp into North America because it was considered to be a good food fish. Within 20 years (1877–1896) after its introduction, the common carp was disdained because it failed to provide sport for anglers and was not used for food (Cumming 1975). In addition to control of nongame species, stunted stocks of game fish are sometimes controlled because they are not of a size that is acceptable to anglers.

14.2.1   Families of Fish Targeted for Control

Control programs have targeted virtually all species of freshwater fishes (Lennon et al. 1970; Dunst et al. 1974). Most (54%) of the current control efforts are directed at three families of fishes: herrings (primarily gizzard shad), minnows (almost exclusively common carp), and sunfish (Table 14.1). An additional 22% of the control efforts involved suckers (11%) and bullhead catfishes (11%). Although the need was recognized as early as 1900, conservation agencies did not routinely implement control measures until after 1930 (Figure 14.1).

The management of undesirable fish species in North America appears to be evolving from control of a single species toward an approach that considers relationships with companion species of fish. Modified stocking practices, changes in regulations, and development of strains of stocked fish are being used as strategies to manage undesirable species.

14.3   FACILITATING MANAGEMENT OF UNDESIRABLE FISH STOCKS

Three methods are used to control undesirable stocks of fish: chemical, biological, and mechanical. Chemical and biological means are favored over mechanical controls for most families of fishes (Table 14.1). Chemical control of fish stocks is summarized in Johnson and Finley (1980) and Schnick et al. (1986). Biological control of fish stocks appears to be increasing, perhaps as a result of

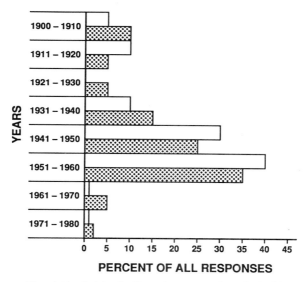

Figure 14.1 Recognition (stippled bar) of need to control undesirable species of fish and implementation (clear bar) of control programs by conservation agencies. (Data for this and subsequent figures provided by state and provincial agencies in response to a 1988 questionnaire.)

increased consideration of fish communities. North American fisheries managers reported that mechanical control methods are not often used because they are labor intensive and generally ineffective (Dunst et al. 1974). Criteria used in selecting a method to control problem fishes include cost, size of water body, water temperature and quality, target species, public opinion, ownership of water, environmental concerns, and location of the water body (Figure 14.2).

All control programs should involve an awareness of the entire fish community because targeting a single species without regard to its relationship to others in the

Table 14.1 Families of North American freshwater fishes that have been targeted for control. Values given for the three control methods are percentages.

Family	Number of states and provinces	Control method		
		Biological	Chemical	Mechanical
Amiidae	1			100
Castostomidae	11	28	41	31
Centrarchidae	19	36	38	26
Characidae	1	50		50
Cichlidae	1	25	50	25
Clupeidae	13	47	32	21
Cyprinidae	22	28	50	22
Escocidae	1		67	33
Ictaluridae	11	21	48	31
Lepisostiidae	1		33	67
Percidae	9	39	26	35
Petromizontidae	2		60	40
Salmonidae	7	21	32	47
Scianidae	1	33	33	33

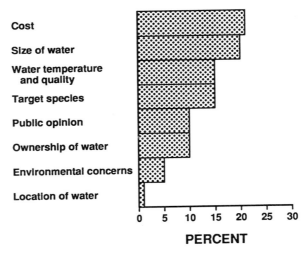

Cost
Size of water
Water temperature
and quality
Target species
Public opinion
Ownership of water
Environmental concerns
Location of water

0   5   10   15   20   25   30

PERCENT

Figure 14.2   Criteria used to determine method employed to control undesirable fishes.

population can result in additional problems. These may include overpopulation and eventual stunting of predator fish or virtual annihilation of forage fish by effective predators.

14.4   CHEMICAL METHODS FOR MANAGING UNDESIRABLE FISH SPECIES

Chemical control is the most popular method to control undesirable fishes because of the ease of application, the short time required to achieve lasting results, and lower cost when compared with other control measures. Generally, a complete removal of all fish is the goal of a treatment project, although partial treatments can also be effective. Partial treatments with chemicals include (1) treatment of spawning sites of fish, (2) treatment of particular sections of lakes, or (3) "thinning" of stunted stocks of game fish (Lennon et al. 1970; Bradbury 1986). Partial kills are usually followed by introductions of predatory game fishes to feed on forage species or to control less desirable fishes. Restocking following any chemical treatment should consider available food sources. For example, stocked fish may perform poorly if zooplankton abundance is low. Partial treatments have met with varying success depending on size of the water body treated, water quality, water temperature, and target species (Bradbury 1986).

14.4.1   Properties of an Ideal Fish Toxicant

Chemical fish toxicants must have properties that are consistent with the needs of fisheries managers and that meet requirements of government regulations. The toxicant selected should be specific to the species of fish targeted, easy and safe to apply, degrade to harmless constituents in a limited time without the aid of a detoxicant, be harmless to nontarget organisms (plant and animal), be effective over a broad range of water quality conditions, and be registered for use in the aquatic environment (Lennon et al. 1970). No currently registered fish toxicant

---

Box 14.2   Items for Consideration in Planning Chemical Treatment

1. Determine the need for chemical treatment to restore the sport fishery based on pretreatment surveys of the fish population.
2. Obtain and evaluate complete water quality and fishery statistics.
3. Determine the volume (lake or pond) or length and volume (stream) of water to be treated.
4. Determine the amount of toxicant required to obtain desired treatment (amounts of toxicant may be decreased if lake levels can be lowered or the flow of regulated streams reduced).
5. Determine if the chemical must be detoxified (some break down to nontoxic components quickly due to water temperature, sunlight, etc.); accurately determine the amount of material required to detoxify the specific concentration of the toxicant.
6. Provide an opportunity for the public to become informed and allow them an opportunity to comment on the treatment.
7. Ensure that the treatment will not contaminate potential sources of drinking water.
8. Evaluate the potential adverse impacts on environmentally sensitive species (including threatened and endangered species).
9. Develop a detailed operational plan that completely describes all aspects of the project.
10. Carefully consider fish species to be used in restocking waters to ascertain that suitable environmental conditions are present.

---

meets all of these criteria. Therefore, fisheries managers must carefully evaluate the benefits of using toxicants with the potential adverse environmental effects.

14.4.2   Planning a Chemical Treatment Project

Environmental assessments provide a mechanism to plan a project and to select the best alternative that will accomplish the goals and objectives (Box 14.2). In the United States, environmental assessments are required by the National Environmental Policy Act (42 U.S.C. Sections 4321–4361; see Chapter 4). Environmental assessments must include a description of the proposed treatment, why the treatment is proposed, a description of the environment, environmental impacts of the proposed treatment, mitigating measures to offset adverse impacts of the proposed treatment, discussion of unavoidable adverse impacts, discussion of irreversible and irretrievable commitments of resources, documentation of public and agency interest, and alternatives to accomplish the proposed work.

Environmental assessments are reviewed by the Fish and Wildlife Service, other federal and state agencies, and the public. If no adverse comments are received or if benefits clearly exceed adverse impacts, the treatment can proceed. (If there are significant adverse environmental impacts or potential public health hazards, an environmental impact statement must be prepared.) Pretreatment studies may be necessary to obtain information needed to prepare an environmental assessment. A comparison of the alternative actions and the potential

---

Box 14.3   Guidelines Suggested for Handling Toxicants in Fisheries Management

1. Always read and follow directions on the container label (information on the label includes product name, type of chemical, formulation, statement of ingredients, net contents, name and address of manufacturer, Environmental Protection Agency (EPA) registration number, number identifying where the product was produced, reference to toxicity to humans, cautionary statements, whether product is for general or restricted use, directions for use, statement on restricted use, statement on misuse, instructions for storage and disposal, and limitations or restrictions on use).
2. Store chemical toxicants only in the original labeled containers.
3. Avoid smoking in areas where toxicants are handled or used.
4. Avoid inhaling fumes or dusts from toxicants and wear protective clothing if so instructed.
5. Avoid toxicant spills; if skin contact occurs wash immediately.
6. Dispose of empty toxicant containers following EPA guidelines.
7. Always wash thoroughly, including clothing, following use of toxicants.
8. Prevent access to any toxicants by children, pets, or irresponsible adults.

---

impacts are vital to the success of the project and such planning should be axiomatic in any chemical control program.

14.4.3   Safe Use of Chemical Toxicants in Fisheries Management

Safety precautions must always be carefully followed in using chemicals (Sowards 1961; Binning et al. 1985). Following proper protocol and using common sense when planning and executing chemical control projects will ensure the safety of workers, the public, and the environment (Box 14.3).

14.4.4   Chemicals Registered for Use as Piscicides

Four toxicants are registered for use as piscicides in the United States (Schnick et al. 1986); rotenone, antimycin, TFM, and Bayluscide. Fish killed using these chemicals should not be used as food. The Fisheries Act of Canada does not permit the application of any deleterious substance (such as toxicants) in fish-bearing waters. The Pesticide Chemicals Act of Canada allows use of a registered powdered form of rotenone, but only in small landlocked lakes. Accordingly, the use of toxicants has been very limited in Canada since 1980. In the United States, regulations governing piscicide use are administered by the federal government (Federal Insecticide, Fungicide, and Rodenticide Act of 1972, as amended, 7 U.S.C. Sections 136–136y) and by the respective states. Most states require that pesticides (including piscicides) be used by certified applicators. Use of chemicals to control fish are sometimes referenced in state conservation codes. Information on use of piscicides in Mexico is very limited.

14.4.5 Properties of Registered Fish Toxicants

Two of the toxicants (rotenone and antimycin) are registered for general use and are used on a nationwide basis, and two (Bayluscide and TFM) are registered as lampricides with primary use in northeastern and central North America.

14.4.5.1 Rotenone

Rotenone is the most commonly used fish toxicant and was first used in North America in 1934 (Lopinot 1975). This chemical has been found to be safe when applied by certified applicators following label instructions (Sousa et al. 1987). In the United States efforts are underway to obtain approval from the Environmental Protection Agency and the Food and Drug Administration for human consumption of fish killed by rotenone. It is presently registered for nonfood use as a general fish toxicant. The chemical affects the oxygen transfer systems in fish and results in physiological suffocation. Rotenone is available from several suppliers in powder or liquid form. Powder is less expensive than the liquid, and when mixed into a slurry before application, is very effective. Powdered derris root can be mixed with sand (1 part to 3 parts), a small amount of gelatin, and water to form a paste for use in heavily vegetated waters or where waters are deep.

Rotenone is commonly dispersed using pump sprayers, mixed with water in the propwash of boats, pumped into deep waters in lakes, and applied from aircraft. Constant flow drip stations are often used in treating streams. Biologists should always consult the container label when calculating the amount of toxicant to be used. Rotenone is applied at various rates—depending on water chemistry—that may affect the success of fish removal. Davies and Shelton (1983) provide information on the use of rotenone in lakes and streams including calculations of amount of toxicant to use, equipment needed for a project, species sensitivity, use of a detoxifier, and how to carry out a treatment project.

Rotenone is environmentally nonpersistent. The emulsified form of the chemical causes avoidance reaction in fish, is relatively nontoxic in birds and mammals, and does not kill fertilized fish eggs, but may be absorbed by bottom sediments and aquatic plants. Toxicity is affected by water temperature, light, dissolved oxygen, turbidity, and alkalinity. Some species of fish that can tolerate low oxygen levels (such as common carp and bullhead catfish) are relatively resistant to rotenone.

Rotenone should be applied at water temperatures of 20°C and above to obtain optimum fish kills and to facilitate detoxification (Davies and Shelton 1983). Rotenone kills fish quickly and breaks down rapidly in warm water. Biologists in Wisconsin (Roth and Hacker 1988) reported using rotenone applied shortly before freeze-up (water temperature 4.4°C or less) to obtain complete kills of common carp and bullhead catfish. At low water temperatures (especially if ice cover forms shortly after application), rotenone can remain toxic for periods up to 3 months.

Rotenone will detoxify naturally within 2 days to 2 weeks in late summer or early fall. Detoxification rates are accelerated by warm water temperatures, high alkalinity, and in clear waters with high light penetration. Detoxification is inhibited by turbidity and deep water because of decreased light penetration. Where chemically induced detoxification is necessary, such as near potable water supplies or to protect downstream fishes, potassium permanganate is usually

Figure 14.3 A potassium permanganate drip station used to detoxify rotenone in streams at the downstream end of reaches treated to control undesirable fish species.

added in an amount equal to the rotenone used plus the permanganate demand of the water (Figure 14.3; Davies and Shelton 1983).

14.4.5.2 Antimycin

Antimycin (Fintrol is the registered name) is an antibiotic and is the only other compound registered as a general fish toxicant. It is less commonly used than rotenone because of limited availability. Fish absorb the chemical through the gills and are killed by interruption of respiration within the body cells. Antimycin is available only in a liquid form from a single supplier (Aquabiotics Corporation, Bainbridge Island, Washington).

Antimycin is an effective fish toxicant in soft-water streams, lakes, and ponds and will kill fertilized fish eggs. The first reported use of antimycin was in 1963 (Lopinot 1975). Antimycin does not elicit an avoidance response in fish and it is highly toxic to some rotenone-resistant fish. Because antimycin is available only in a liquid formulation, it is not as effective in deep lakes as the sand-based compound that was formerly available. Some biologists apply the liquid to sand to treat deep lakes and ponds.

14.4.5.3   Lampricides

Two lampricides, TFM (3-triflouromethyl–4-nitrophenol) and Bayluscide (Bayer 73) are currently registered in the United States. Both toxicants are used only in control of the sea lamprey, largely by the Great Lakes Fishery Commission (see Chapter 22).

TFM. This selective toxicant, which was first tested as a lampricide in Michigan in 1958 (Cumming 1975), is applied as a liquid or soluble bar formulation to control larval lampreys (ammocetes) in spawning and nursery areas. The mode of action is not well understood but TFM appears to cause general collapse of the circulatory system. The toxicant TFM is restricted for nonfood use as a lampricide and can be used only by applicators certified by the Fish and Wildlife Service (Schnick et al. 1986). The effectiveness of TFM is depressed at water temperatures near freezing and at pH values above 8.0. The chemical is environmentally nonpersistent, does not affect birds, mammals, or aquatic plants (though photosynthesis may be reduced), has a varied affect on invertebrates depending upon treatment location and species, and may depress associated fish populations.

Bayluscide (Bayer 73). Bayluscide is a granular or powdered toxicant that was developed for the control of mollusks (snails) in tropical areas and first used as a lampricide in 1963 (Cumming 1975). The mode of action of Bayluscide is thought to be similar to that of TFM, but the exact mode is unknown (National Research Council of Canada 1985). Bayluscide is registered as a survey tool for sea lamprey populations and the synergistic combination of Bayluscide–TFM (not more than 2% TFM by weight) is registered as a lampricide. A formulation applied to sand has been found effective for control of larval lamprey, especially at the mouths of rivers entering lakes.

Although Bayluscide is used as a lampricide, it is highly toxic to associated fishes and must be used with care. The compound is environmentally nonpersistent, is pH-sensitive, moderately toxic to mammals, and very toxic to mollusks and aquatic annelids.

14.4.6   Evaluation of Chemical Control of Undesirable Fishes

Cost–benefit information on the chemical control of fishes is not readily available. Numerous states report annually spending large sums to control undesirable fish stocks but few have summarized actual cost–benefit data. The cost per hectare to treat with rotenone ranged from $2.00 to $240.00 and averaged $40.00 in the late 1980s. Costs varied in relation to volume of water treated. Rotenone is such an important fish toxicant than the expenditure of nearly $3 million and investment of 6 years of testing were justified to reregister the toxicant for fishery use (Sousa et al. 1987).

Information on costs associated with other registered fish toxicants is not adequate to permit comparisons. Much time and money have been invested in the control of sea lamprey in the Great Lakes. The increase in the number of lake trout is, at least partly, attributable to chemical control of the sea lamprey. It

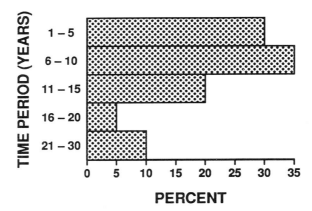

Figure 14.4 Length of time that chemical treatment has been found to be effective. Effective time depends on the physical and chemical factors of the treated waters as well as the target species.

should be obvious that programs not considered to be beneficial would be discontinued. Cost alone cannot serve as the measure of the value of a treatment program. Common sense must play a large role.

Trimberger (1975) reported that increases in the number of game fish available to anglers and increases in fishing pressure followed most chemical reclamation projects in Michigan. Fishing pressure on formerly marginal trout waters after chemical treatment has equalled the pressure on the best trout waters. Trimberger also reported that chemical treatments of warmwater lakes consistently resulted in more angler days per acre than were attributable to other management strategies. Similar statements have been made by fisheries managers throughout North America.

Periods of effectiveness range from less than 1 to 30 years (Figure 14.4). When benefits are substantial, treatment programs can be scheduled at regular intervals. Annual treatments of particular waters are usually considered too costly for benefits realized.

14.5 BIOLOGICAL METHODS FOR MANAGING UNDESIRABLE FISH SPECIES

Biological control methods can be grouped into three categories: (1) grazing and predation by protozoa, zooplankton, fish, birds, insects, snails, crayfish, turtles, and mammals; (2) use of pathogens (viruses, bacteria, and fungi); and (3) biomanipulation, in which interrelationships among plants, animals, and their environment are adjusted to achieve the desired control and ecological balance (Schuytema 1977). Biomanipulation is the most promising biological control technique for managing fish populations in that it may minimize competition and establish balanced predator–prey populations (Box 14.4). The use of pathogens for fish control is risky; therefore, pathogens are not generally used in aquatic environments.

Mechanical and chemical control of nuisance aquatic plants is generally temporary and expensive. Biological control of such vegetation by using grazers

---

Box 14.4 Important Factors that Must be Considered in Choosing a Suitable Predator or Prey Species

1. Numbers of undesirable or unexploited fishes.
2. Size of the prey fish species.
3. Potential for use of forage organisms by sport fish predators.
4. Size of the water body.
5. Species and strains of predators available.
6. Size of predators at stocking and estimated survival based on size.
7. Stocking rates as well as timing and frequency of stocking.
8. Habitat suitability for the predator.
9. The desirability of the predator as a sport species.
10. Angler harvest of stocked game fish.
11. Control of the predator in the stocked water.
12. Native fish species (especially threatened and endangered fishes) that may be impacted by the predator (e.g., hybridization, competition and predation).
13. Potential for transmission of disease or parasites.
14. Fisheries management plan for the basin where introduction is to take place.
15. Public preference.

---

such as sterile triploid grass carp has been very effective. However, many biologists have opposed the use of grass carp because of the perceived potential for adverse environmental impacts (see Chapter 12).

14.5.1 Predatory Fish

Biologists control fast-growing populations of forage fish or stunted sport fish by stocking sport-fish predators. Various predator–prey combinations are used in farm ponds to manipulate the carrying capacity for desired fishes. Variations in predator and prey growth by latitude requires different approaches to pond fisheries management as suggested for largemouth bass–bluegill (see Chapter 20).

In special situations, stocking of predators may improve sport fishing in waters that contain overabundant and stunted fish populations. To be effective as a management tool, predator stocking should (1) be cost-effective with respect to culture cost, stocking rates, and survival; (2) result in survival to sufficient numbers so anglers have a reasonable chance to catch them; (3) grow large enough to be of interest to anglers; and (4) produce consistent fisheries so biologists can reasonably predict the outcome of such stockings.

Almost all states and Canadian provinces use predators for biological control of abundant forage fish. Some of the encountered problems include: (1) the predator is not stocked at adequate density, (2) the predator reduces the desirable forage fish resulting in reduced growth of the predator, (3) the predator becomes overpopulated and stunted, (4) the predator causes declines of other sport species, (5) the predator species is so exploited as a sport fish that it never becomes abundant enough or large enough to control the targeted forage fish, and (6) the predator is ineffective as a control agent.

Table 14.2 Fishes managed using biological control. Values given for each rank are percentages.

Species	Rank[a]		
	1	2	3
Gizzard shad	39		4
Yellow perch	18	3	4
Sunfish	9	36	29
Common carp	6	18	17
Other cyprinids	9	11	4
Suckers	6	11	8
Bullheads		11	
Crappie		3	4
White perch	3	7	4
Threadfin shad	3		4
Alewife	3		4
Brook trout			13
Tilapia	3		4

[a]Ranked in order of importance by state and provincial agencies in response to a 1988 questionnaire.

Quantitative results of predator stocking as a biological control are scarce (Hall 1985). Most predator–prey interaction studies in large reservoirs have been confined to single species or a few species. The complex interactions of total fish communities has not been adequately studied.

Typically, aquatic organisms are introduced to solve local or regional problems. However, recent studies demonstrate that introduced fish can eliminate native species, reduce survival and growth rates of established species, or change the structure of the fish community (see Chapter 12). Furthermore, introductions of nonnative species may impact endangered species (see Chapter 15).

Gizzard shad, yellow perch, sunfish, common carp, and other minnows, are the primary fishes being controlled by biological methods (Table 14.2). The primary species that are used as predators to control undesirable species or forage species include largemouth bass, walleye, northern pike, striped bass, white bass × striped bass hybrids, and muskellunge (Table 14.3).

Table 14.3 Species that are used as predators in biological control programs. Values given for each rank are percentages.

Species	Rank[a]		
	1	2	3
Largemouth bass	26	27	27
Walleye	23	7	25
Northern pike	13	7	4
Striped bass	10	13	7
Striped bass hybrids	10	20	13
Muskellunge	6	7	17
Trout	6	7	7
Salmon	3		4
Channel catfish	3	7	
Flathead catfish		6	
Smallmouth bass		6	
Peacock bass			4

[a]Ranked in order of importance by state and provincial agencies in response to a 1988 questionnaire.

14.5.2 Controlling Reproduction

Other biological controls include stocking hybrids that have little or no potential to reproduce or altering the sex of fish by chemicals and heat to produce fish that will never develop sexually or that will produce monosex individuals. Natural reproduction of introduced fish species is sometimes considered undesirable. In such cases, controlling reproduction by genetic manipulation is sometimes warranted. In farm ponds, hybrid sunfish (limited reproductive capability) can be established by two methods: (1) stocking F_1 hybrid sunfish directly into the ponds, or (2) stocking parental fish of the correct sex to produce the desired hybrids (Kurzawski and Heidinger 1982). In such situations, largemouth bass can control any F_2 hybrid sunfish since recruitment of F_2 hybrids are minimal (Brunson and Robinette 1986).

14.5.3 Forage Fish

Prey introductions are sometimes made in efforts to increase the growth of stunted predators that are too small to be acceptable to anglers. However, problems have also been caused by well-intentioned decisions to improve the forage base for game fish. A classical example is the competitive interaction that resulted from the introduction of redside shiners as possible forage for rainbow trout in British Columbia lakes (Johannes and Larkin 1961). The introduction failed because the behavior of the shiners resulted in ecological separation from trout large enough to be effective predators while the shiners competed directly for food with small trout.

Fish that are considered undesirable in some locations may serve as acceptable forage in others. For example, Tui chubs provide the primary forage for a rainbow trout subspecies in Eagle Lake, California (Burns 1966). In Pyramid Lake, Nevada, Lahontan cutthroat trout prey upon Tui chub and the chub is not a problem there. However, in other waters of the intermountain and western states, minnows such as the Tui chub and Utah chub compete with stocked fingerling rainbow trout. Consequently, they are controlled by stocking strains of trout that are piscivorous, stocking warm or coolwater predators, or through chemical rehabilitation.

Some prey species are used more by predator species than others. For example, rainbow smelt are considered to be better forage organisms for landlocked Atlantic salmon in New England than alewife because of being more spatially available. In some cases, the introduction of a prey species to serve as forage for one game fish may result in competition with another. For example, cisco stocked as prey for lake trout in Lake Opeongo, Ontario, competed with young small-mouth bass (Emery 1975). Therefore, the biology of the species to be introduced should be carefully reviewed. An ideal forage organism should be prolific, stable in abundance, trophically efficient, vulnerable to predators, nonemigrating, and innocuous to other species (Ney 1981).

14.5.4 Underexploited Fishes

Some conservation agencies have initiated programs to encourage the use of less desirable fishes for food and sport. The common carp is one of the most important problem fish species in North America because it is very prolific and sought by few anglers. Miller (1972) offered suggestions on how to fish for

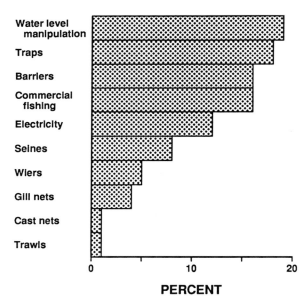

Figure 14.5 Mechanical methods used to manage undesirable fish species.

common carp, when to fish for them, baits that are effective, and tips on dressing and cooking them. The American Fisheries Society published a book to stimulate greater interest in common carp (Cooper 1987). Similar programs may be developed for other underexploited species. For example, the Utah Division of Wildlife Resources published brochures on fishing techniques, fishing locations, and cooking mountain whitefish and bullhead catfish. Angling for mountain whitefish provides stream fishing opportunities in winter when the trout fishery is closed. In addition, many conservation agencies adopt liberal bag limits and expand seasons to encourage increased harvest of underexploited species.

14.6 MECHANICAL METHODS FOR MANAGING UNDESIRABLE FISH SPECIES

The most frequently used (81%) mechanical control methods include water level manipulation, traps, barriers, commercial fishing, and electricity (Figure 14.5). Other methods include using seines, weirs, gill nets, cast nets, and trawls. Mechanical control methods are employed on a variety of species in North America (Table 14.4).

Water level manipulations in reservoirs can be achieved by drawdowns, by increasing water levels, or a combination of the two (see Chapter 10). Planned drawdowns can be used to control nuisance aquatic vegetation, to increase predator utilization of forage fishes resulting in faster growth of the predators, to release nutrients from bottom sediments, and to stimulate increased reproduction and survival of young fish following reflooding. Significant increases in water levels timed to occur just before, during, or for a short time after the spawning season for spring spawners can be used to provide more littoral habitat and

Table 14.4 Species that are controlled by mechanical methods. Values given for each rank are percentages.

Species	Rank[a]		
	1	2	3
Common carp	23	6	17
Other cyprinids		21	9
Suckers	12	6	4
Sunfish	17	16	17
Yellow perch	9	16	4
White perch		9	4
Gizzard shad	6	13	4
Crappie	6		4
Trout	6	3	4
Tilapia	6		
Gar	3		4
Sacramento blackfish	3		
Goldfish	3		
Northern pike		3	
Sea lamprey			9
Freshwater drum			4
Largemouth bass			4

[a]Ranked in order of importance by state and provincial agencies in response to a 1988 questionnaire.

increased productivity. However, water level manipulations may conflict with other reservoir uses such as domestic or irrigation water supply, power generation, flood control, and recreation.

Water manipulation has some limitations because this method can only be used on reservoirs with water control structures. Drawdowns will reduce the invertebrate biomass in the fluctuating littoral zone and, if done in late summer or fall, may provide conditions for winterkill. The reduction in invertebrate biomass can be compensated by revegetating the fluctuating littoral zone to increase productivity during the following spring.

Game fish populations may benefit from commercial fishing of undesirable species (including competitors or predators) if the numbers are reduced significantly. For example, nearly 80 million kilograms of commercial fish (mostly freshwater drum) were removed by trap nets and trawls from Lake Winnebago, Wisconsin, between 1955 and 1966 (Priegel 1971). This removal coincided with an increase in populations of walleye, sauger, white bass, yellow perch, and black crappie (all desirable species). Commercial fishing is an acceptable way of removing less desirable fish species at no cost to the managing agency; however, most inland commercial fisheries do not remove enough of the undesirable species to benefit game fisheries (Grinstead 1975). Intensive harvest of preferred commercial species usually results in reduction of average size and catch rate that ultimately makes fishing unprofitable. Other undesirable species may remain unharvested because of low market value. Energy transfer (i.e., productivity) is complex in fish communities and some biologists believe that suppression of game fish by other less desirable species may be greatly exaggerated (Marrin and Erman 1982).

Trapping or netting of fish is practiced to control fish in small waters (ponds or lakes) or for partial control in larger waters (Figures 14.6 and 14.7). Mechanical controls are effective only if there is a significant decrease in the target species.

Figure 14.6   Seining operation for rough fish removal from a lake.

Netting must occur regularly to be effective. Vulnerability of the target species is an important consideration in netting and trapping efforts. Common carp, buffalo fish, freshwater drum, and bullhead catfishes can be effectively harvested with nets and traps during periods of spawning aggregations. Certain species of fish (such as common carp) assemble into huge schools just before freeze-up and can be effectively removed at that time. Nevertheless, netting and trapping programs have produced varying results. Rose and Moen (1953) reported a sixfold increase in game-fish stocks when competing species were reduced to about 35% of former abundance in Lake Okoboji, Iowa. However, removal of over 27,000 longnose suckers over a 7-year period did not improve the rainbow trout fishery in Pyramid Lake, Alberta (Rawson and Elsey 1950). Such controls are temporary because fish exhibit compensatory survival, growth, and reproduction.

Mechanical devices such as fishways and screens may have potential for controlling fish. Fishways are usually used to pass desirable fish species over obstacles such as dams, falls, or rapids. By careful design of fishways, passage of undesirable species of fish can be prevented (Broach 1968). Fish screens are used

Figure 14.7   Trap net set as part of a rough fish removal program.

to keep fish out of particular reaches of streams but little is known about the fundamental design needs to make them perform efficiently. The purposes for which screens are constructed are important in determining where they should be used, what types should be used, and if the costs are justified (Huber 1974). Designs for screens, bypasses, and fishways must take into account the amount of water diverted, availability of water to attract fish, the swimming ability of the fish, the behavior and sizes of the fish, and the quantity and size of debris that may reach the screen.

Electricity and sound are methods sometimes considered in efforts to control or guide fish. Albertson et al. (1965) and McLain (1956) reported limitations in controlling sea lampreys with electricity. A voltage gradient sufficient to control small fish would narcotize, injure, or even kill larger fish in the electrical field. Water resistance, content of dissolved substances, and water temperature also influence the effectiveness of electricity. Portable and boat electroshockers can be used to remove fish from confined areas but do not work well in open, large, or deep waters. Electrical fields have been used to repel fish such as gizzard shad and alewives from water intakes of hydroelectric plants. Although the use of sound has not been thoroughly studied, some preliminary investigations indicate that sound may not be useful in guiding or repelling fish.

It may be helpful to use combinations of control methods to capture or eliminate target species under different conditions when they are most vulnerable to particular methods. For example, copper sulfate has been used to increase fish movement that resulted in larger fyke-net catches (Brown 1964). Electrofishing equipment can be used in streams to force fish into gill nets, trammel nets, or trap nets.

Successful use of nets and traps has been limited usually because an inadequate portion of the offending population is removed. The sizeable escapement from nets in large lakes and reservoirs requires continuous fishing. Some nets (e.g., fyke nets) are effective only in shallow waters since fish often swim over the leads in deeper water. Seines and trawls may have some promise for control in special situations but those gears require a relatively smooth bottom, free of snags (a condition rare in most waters), to be fished properly. In small farm ponds with smooth bottoms, seines can be used effectively. Wing nets and trap nets are effective for crappies, bullheads, sunfish, and gizzard shad in shallow waters. Passive gear such as gill nets and trammel nets are effective on various species of fish when the fish are active. Weirs can be effective in streams when fish are migrating, but require cleaning of waterborne debris from the mesh.

The main drawback in using nets is that they require a great deal of human effort, and are rather expensive equipment that must be repaired or replaced fairly often. Nets may also cause some mortality of desirable game fish. Commercial fishers use nets designed to catch certain sizes of fish, but are seldom designed for control of a given species. Usually catches will be largest at the beginning of the netting operation. As the fish population declines, the catch per unit effort will drop making netting inefficient.

14.7 CONCLUSION

If a fish species is considered undesirable, a variety of management methods (e.g., chemical, biological, or mechanical) should be considered to achieve carefully planned and predetermined objectives. The goals and objectives that are determined for control of undesirable species must be realistic in terms of success that may be achieved and costs related to benefits. Complete control of undesirable fish is seldom possible and partial controls must remove enough of the undesirable fish population so that competition or predation on desired fish are reduced. Control programs should be carefully evaluated to determine if reasonable predictability of the outcome of the control method is possible.

Fisheries managers must be flexible to increase or decrease management efforts depending on the response of the target fish population. Control, using the most effective method(s), should be applied at a time and location when the pest species is most vulnerable. At the same time, control should be implemented when there is the least risk of adverse impacts (Berryman 1972). Often the concept of integrated management (using a combination of methods) may be necessary to achieve selective manipulation of fish populations.

14.8 REFERENCES

Albertson, L. M., B. R. Smith, and H. H. Moore. 1965. Experimental control of sea lampreys with electricity on the south shore of Lake Superior, 1953–60. Great Lakes Fishery Commission Technical Report 10.

Berryman, J. H. 1972. The principles of predator control. Journal of Wildlife Management 36:395–400.

Binning, L., and six coauthors. 1985. Pest management principles for the commercial applicator, aquatic pest control. University of Wisconsin, Cooperative Extension Service, Madison.

Bradbury, A. 1986. Rotenone and trout stocking: a literature review with special reference to Washington Department of Game's lake rehabilitation program. Washington Department of Game, Fisheries Management Report 86-2, Olympia.

Broach, D. 1968. A small-capacity spillway modified to prevent re-entry of undesirable fishes. Progressive Fish-Culturist 30:38.

Brown, E. H., Jr. 1964. Fish activation with copper sulfate in relation to fyke-netting and angling. Ohio Department of Natural Resources, Division of Wildlife, Publication W-71, Columbus.

Brunson, M. W., and H. R. Robinette. 1986. Evaluation of male bluegill × female green sunfish hybrids for stocking Mississippi farm ponds. North American Journal of Fisheries Management 64:156–167.

Burns, J. W. 1966. Rough fish management. Pages 492–498 in A. Calhoun, editor. Inland fishery management. California Department of Fish and Game, Sacramento.

Cooper, E. L., editor. 1987. Carp in North America. American Fisheries Society, Bethesda, Maryland.

Cumming, K. B. 1975. History of fish toxicants in the United States. Pages 5–21 in Eschmeyer (1975).

Davies, W. D., and W. Shelton. 1983. Sampling with toxicants. Pages 119–213 in L. A. Nielsen and D. L. Johnson, editors. Fisheries techniques. American Fisheries Society, Bethesda, Maryland.

Dunst, R. C., and nine coauthors. 1974. Survey of lake rehabilitation techniques and experiences. Wisconsin Department of Natural Resources Technical Bulletin 75.

Emery, A. R. 1975. Stunted bass: a result of competing cisco and limited crayfish stocks. Pages 154–164 in H. Clepper, editor. Black bass biology and management. Sport Fishing Institute, Washington, D.C.

Eschmeyer, P. H., editor. 1975. Rehabilitation of fish populations with toxicants: a symposium. American Fisheries Society, North Central Division, Special Publication 4, Bethesda, Maryland.

Grinstead, B. G. 1975. Response of bass to removal of competing species by commercial fishing. Pages 475–479 in H. Clepper, editor. Black bass biology and management. Sport Fishing Institute, Washington, D.C.

Hall, G. E. 1985. Reservoir fishery research needs and priorities. Fisheries (Bethesda) 10(2):3–5.

Hubbs, C. L., and R. W. Eschmeyer. 1938. The improvement of lakes for fishing. University of Michigan, Institute for Fishery Research, Bulletin 2, Ann Arbor.

Huber, E. E. 1974. Fish protection at intake structures and dams: guidance, screens, and collection devices: a selected bibliography with abstracts. Oak Ridge National Laboratory, ORNL-EIS-74-67, Oak Ridge, Tennessee.

Johannes, R. E., and P. A. Larkin. 1961. Competition for food between redside shiners (Richardsonius balteatus) and rainbow trout (Salmo gairdneri) in two British Columbia lakes. Journal of the Fisheries Research Board of Canada 18:203–221.

Johnson, W. W., and M. T. Finley. 1980. Handbook of acute toxicity of chemicals to fish and invertebrates. U.S. Fish and Wildlife Service Resource Publication 137.

Kurzawski, K. F., and R. C. Heidinger. 1982. The cyclic stocking of parentals in a farm pond to produce a population of male bluegill × female green sunfish F_1 hybrids and male redear sunfish × female green sunfish F_1 hybrids. North American Journal of Fisheries Management 2:188–192.

Lennon, R. E., J. B. Hunn, R. A. Schnick, and R. M. Burress. 1970. Reclamation of ponds, lakes, and streams with fish toxicants: a review. FAO (Food and Agriculture Organization of the United Nations) Fisheries Technical Paper 100.

Lopinot, A. C. 1975. Summary of the use of toxicants to rehabilitate fish populations in the midwest. Pages 1–4 in Eschmeyer (1975).

Marrin, D. L., and D. C. Erman. 1982. Evidence against competition between trout and nongame fishes in Stampede Reservoir, California. North American Journal of Fisheries Management 2:262–269.

McLain, A. L. 1956. The control of the upstream movement of fish with pulsated fish current. Transactions of the American Fisheries Society 86:269–284.

Miller, G. 1972. Time out for carp. Nebraska Game and Parks Commission, Lincoln.

National Research Council of Canada. 1985. TFM and Bayer 73: lampricides in the aquatic environment. Environmental Secretariat, Publication NRCC 22433, Ottawa.

Ney, J. J. 1981. Evolution of forage-fish management in lakes and streams. Transactions of the American Fisheries Society 110:725–728.

Priegel, G. R. 1971. Evaluation of intensive freshwater drum removal in Lake Winnebago, Wisconsin, 1955–1966. Wisconsin Department of Natural Resources Technical Bulletin 47.

Rawson, D. S., and C. A. Elsey. 1950. Reduction in the longnose sucker population of Pyramid Lake, Alberta, in an attempt to improve angling. Transactions of the American Fisheries Society 78:13–31.

Rose, E. T., and T. Moen. 1953. The increase in game-fish populations in East Okoboji Lake, Iowa, following intensive removal of rough fish. Transactions of the American Fisheries Society 82:104–114.

Roth, J., and V. Hacker. 1988. Fall use of rotenone at low concentrations to eradicate fish populations. Wisconsin Department of Natural Resources, Research and Management Findings 9, Madison.

Schnick, R. A., F. P. Meyer, and D. L. Gray. 1986. A guide to approved chemicals in fish production and fishery resource management. University of Arkansas, Cooperative Extension Service Bulletin MP-241, Little Rock.

Schuytema, G. S. 1977. Biological control of aquatic nuisances—a review. U.S. Environmental Protection Agency, Corvallis Environmental Research Laboratory, Corvallis, Oregon.

Sousa, R. J., F. P. Meyer, and R. A. Schnick. 1987. Re-registration of rotenone: a state/federal cooperative effort. Fisheries (Bethesda) 12(4):9–13.

Sowards, C. L. 1961. Safety as related to the use of chemicals and electricity in fishery management. U.S. Fish and Wildlife Service, Washington, D.C.

Titcomb, J. W. 1914. The use of copper sulfate for the destruction of obnoxious fishes in ponds and lakes. Transactions of the American Fisheries Society 44:20–26.

Trimberger, E. J. 1975. Evaluation of angler-use benefits from chemical reclamation of lakes and streams in Michigan. Pages 60–65 in Eschmeyer (1975).

Chapter 15

Endangered Species Management

CLARENCE A. CARLSON AND ROBERT T. MUTH

15.1 INTRODUCTION

Several nontechnical accounts of declining fish faunas and fisheries have been published (e.g., Brown 1982; Mowat 1984; Warner 1984; Ashworth 1986). Recurrent themes run through these stories; species of value to humankind were overexploited, "worthless" species were destroyed for the presumed good of other commercially valuable species, and various human activities (e.g., damming rivers, polluting water and air, and introducing nonnative species) led to depletion of other species. Many more tales of fish or fisheries depletions that are likely to be read by nonbiologists could be mentioned. Accurate reports like these are important; their examples of biotic declines and extirpations conform to a general trend that is not limited to North American fishes, and they help to increase public awareness of serious environmental problems. The ultimate decline, that which culminates in extinction, may be occurring worldwide at an unprecedented rate and in many forms of life as a result of human actions.

15.1.1 Accelerated Extinction

Extinction usually occurs when populations are unable to persist as their environments change. Extinction may be local, when a given population fails, or global, when an entire species is eliminated. Extinction is a normal process, and global extinction has been the fate of most species produced during the Earth's history (Ehrlich and Ehrlich 1981). Estimates of the rate of species loss have been made in spite of difficulties inherent in such estimation and lack of data on biotic diversity. Myers (1979a, 1979b) estimated that at least 90% of all species that ever existed have become extinct, mostly as a result of natural processes. By 1600, humans became capable of driving animals to extinction. The human-induced extinction rate increased to about one species per year and held relatively constant from 1600 to 1900. It increased again to perhaps one species per day by 1979 as humans exploited ecologically diverse moist tropical forests. Myers (1985) expected humans to accelerate the extinction rate to one species per hour by the late 1980s and to dozens per hour by the end of the 1990s. He compared these rates to the maximum rate of extinction of the dinosaurs (one species every 10,000 years) to emphasize his point that there has never been such a period of massive and compressed extinction as that which we now face. Predictions of the percentage of today's species that will be lost by the end of this century range from 15 to 50 (Shen 1987).

Many see a need for human intervention to slow the accelerated rate of extinction, but attempts to reduce the extinction rate are unlikely to succeed unless causes of extinction are understood. Humphrey (1985) listed three biological circumstances that lead to extinction: natural rarity, lack of habitat, and inbreeding depression (loss of genetic diversity). Persistent rarity often arises from a position high on a grazing food chain or from large body size. An organism that has evolved to maximize its fitness in a particular limited habitat is a candidate for extinction if that habitat is threatened. Liability to extinction is decreased if a species occurs in a large number of semiautonomous populations over a wide geographical range. Without sufficient numbers of individuals, genetic diversity in biological populations may be decreased to harmful levels. The concept of minimum viable population—a level of abundance above which persistence, without loss of fitness, for a given length of time is assured—has recently come into vogue (Soule 1987).

Environmental change resulting in extinction may be physical or biological in nature. Habitat disruption, a type of physical change, is the foremost cause of plant and animal extinctions, and it is usually difficult or impossible to rectify. Wolf (1987) stressed the significant effects of fragmenting once-continuous habitat into island-like refuges; animals typically die out in a manner referred to as faunal collapse. Habitat destruction in tropical forests, where up to half the species on earth reside (Myers 1986), is widely recognized as the greatest threat to wild plants and animals. Extinction from biological causes may come about from improved abilities of predators (including people) or competitors, which evolve over time, or invasion by nonnative species. Humans have contributed to extinction through commercial hunting, predator and pest control, and collection of organisms for medical research, zoos, and sources of houseplants. Impacts of alien organisms introduced by humankind may combine with effects of human exploitation to devastate native species. Ehrlich and Ehrlich (1981) provided excellent reviews of how species endangerment by humans occurs directly (e.g., by overexploitation and predator control) or indirectly (e.g., in their words, through paving over, plowing under, spewing, cutting down, transporting, and recreation).

15.1.2 Reasons to Minimize Extinction

Ehrlich and Ehrlich (1981) grouped reasons to care about extinction into (1) direct economic benefits to humans; (2) indirect (ecological) benefits to humans; and (3) compassion, aesthetics, fascination, and ethics. Direct benefits include allowing us to meet future agricultural, pharmaceutical, and industrial needs. Potentially useful species may become extinct before humans discover and exploit them as sources of food, fiber, shelter, recreation, medicines, lubricating oils and waxes, energy, biological pest controls, or research subjects.

The ecological benefits of maintaining diverse natural systems are less obvious, more difficult to convey to nonbiologists, and sometimes the subject of controversy among ecologists. Myers (1979a) emphasized the role of biological diversity in contributing to stability of ecosystems. Ehrenfeld (1976) described the diversity–stability concept as the most vexing and embarrassing example of conservationists being provoked into exaggerating and distorting alleged values of "nonresources" (e.g., species, biotic communities, and ecosystems). He explained the issue as a case of inverted cause and effect; the most diverse communities simply have occupied the most stable environments for the longest periods of time. Stability within a community depends on species diversity (richness and composition) and trophic complexity (number and strength of species interactions) of the community, the type of perturbation, and the species being perturbed (Kikkawa 1986). Goodman (1975) concluded that, even if the diversity–stability hypothesis were completely false, its basis of natural community preservation is correct, and the conservation principles it promotes are sound.

Ehrlich and Ehrlich (1981) summarized ecosystem services, or ways in which ecosystems support human life. They stressed that maintaining genetic diversity is important to the continued functioning of ecosystems; only genetic variability allows evolution in response to environmental change. The world's gene pool (i.e., its species diversity) should be considered one of earth's most important and irreplaceable resources. Continuation of ecosystem services depends on the capacity of ecosystems to evolve, their resilience, and human willingness to lessen the assault on them. If the assault continues unabated, accelerated extinction may be counterbalanced by increased speciation, and a disproportionate number of opportunistic, r-selected species (which we often consider to be pests) may occur in the biota our descendants inherit from us (Myers 1979a).

Others have argued for preservation of biotic diversity on moral grounds. Ehrenfeld (1976) counseled against assigning economic values to nonresources. He coined the term "Noah principle" for the notion that communities and species should be conserved simply because they exist and have existed for a long time. Callicott (1986) reviewed arguments that humans have a moral obligation not to extirpate species, i.e., species have intrinsic value and therefore, a right to exist. He stated that the ethical argument for preservation of species may be most compelling but has not been well articulated.

15.1.3 Slowing the Extinction Rate

As the major causal agents of the current extinction spasm, we humans have a responsibility to reduce the rate of extinction. Fostering and applying a conservation or evolutionary ethic may be a good starting point, but practical means should also be used. Before they can be, the fundamental political, economic, and psychological resources of humankind must be mobilized (Ehrlich and Ehrlich 1981). The best way to safeguard living diversity—through reserves and parks—has been hampered by lack of knowledge on which to base decisions about location and size of reserves. However, the extinction rate may be slowed by such recent developments as work of the Nature Conservancy, which tries to preserve diversity by preserving selected land areas; UNESCO's system of biosphere reserves; and emergence of a Society for Conservation Biology (Ehrlich 1987). Habitat in damaged ecosystems can sometimes be rehabilitated to enhance chances for survival of endangered or threatened species. However, restoration should not be considered a substitute for efforts to preserve natural ecosystems.

Zoos, aquariums, and botanical gardens are useful for preserving organisms that cannot survive in the wild. The major disadvantages of depending on zoos as reservoirs of organic diversity are their limited capacity and their need to display as many animals as possible to attract visitors and fulfill their educational objectives. The few species that are chosen to be bred in captivity may not be subjected to the same selection pressures that affect natural populations. Possible reduction of genetic diversity leads to questions about the value of such innovations as gene banks, frozen sperm banks, seed banks, and embryo banks.

Protection of biotic diversity is obviously dependent on future research and political action. National and international conservation programs must share knowledge about biological diversity, extinction rates, and location and size of reserves and use this information as a basis for legislation. Extinction rates also can be slowed by treaties and laws that protect rare organisms in nature.

15.2 ENDANGERED SPECIES LAWS AND TREATIES

15.2.1 The U.S. Endangered Species Act

As the environmental movement of the 1960s progressed, the American public became increasingly concerned about accelerated extinction. The U.S. Congress responded by passing the Endangered Species Preservation Act of 1966 and the Endangered Species Conservation Act of 1969. These laws did not provide comprehensive authority for protecting endangered wildlife; in particular, they failed to prohibit taking of endangered species or require federal agencies to comply with their intent.

The 1966 and 1969 Acts were replaced by the more comprehensive and prohibitive Endangered Species Act (ESA) of 1973 (16 U.S.C. Section 1531 et seq.), an attempt to slow the rate of extinction by singling out animals and plants thought to be near extinction and giving them special protection. Its stated purposes were to provide (1) conservation of the ecosystems upon which endangered and threatened species depend, and (2) a program for the conservation of such species. The Act declared that it was the policy of Congress that all federal departments and agencies seek to conserve endangered and threatened species and use their authorities in furtherance of the Act (see Box 15.1).

The 1978 amendments weakened the ESA by making it more difficult to list species as endangered. Before a species could be listed, it was necessary to describe economic impacts and boundaries of habitat and to hold public hearings within a 2-year time period. Many conservationists also considered creation of the Endangered Species Committee an invitation to agencies to avoid seeking acceptable alternatives to their actions that would be less likely to jeopardize endangered species. However, the 1978 amendments may fairly be said to have made the Act more flexible and the consultation and public participation processes stronger. The Endangered Species Act Amendments of 1982 streamlined listing and exemption processes, and the review board was replaced by the Secretary of the Interior. More active management of listed species was facilitated by provision for experimental populations, especially by introduction to apparently suitable or formerly occupied habitats. The 1988 amendments were the result of a 4-year struggle for reauthorization of the Act by the scientific and environmental communities.

Because the ESA will continue to be amended, interested individuals need means of keeping abreast of changes in the Act. The United States Code, last revised in 1983, contains general and permanent laws of the United States as amended, including the ESA as amended in 1982. A minor problem is presented by differences between Section numbers in the Code and the original Act; numbers in Box 15.1 will resolve this problem. Original Section numbers are used in Fish and Wildlife Service (FWS) regulations and guidelines and in common parlance. Changes in the ESA and in lists and regulations are reported in the *Federal Register* and the *Endangered Species Technical Bulletin*. The *Environmental Reporter* of the Bureau of National Affairs is another source of up-to-date information on the ESA.

15.2.2 Implementing the ESA

Recovery plans are the primary tools for restoring listed species to self-sustaining components of their ecosystems. In recovery planning, priority is given to species most likely to benefit from such plans; priorities are based on the degree of threat facing a species, its taxonomic uniqueness and recovery potential, and its likelihood to encounter conflicts with development. Plans are intended to guide various conservation programs of federal, state, and local agencies and other organizations. Regional directors of the Fish and Wildlife Service are responsible for preparing plans for species in their region, but they may assign the preparation to FWS personnel, a volunteer recovery team, a state or federal agency, a conservation organization, or knowledgeable individuals. Approved plans are reviewed periodically and revised as new information is compiled or the status of listed species changes. One of the most important things to remember about recovery plans is that they are only guides to conservation of listed species; they do not mandate action, and no one is legally obligated to carry out the conservation measures they contain.

Kohm (1991) presented a good review of what has been learned since the ESA was passed and discussed how future efforts should be directed. Ehrlich and Ehrlich (1981) described the ESA, as originally written, as a powerful weapon on the environment's behalf and considered the modified version (prior to 1982) a potentially strong weapon in defense of species and environmental integrity. Enforcement of the ESA has prevented the extinction of some species and slowed the downward trends of others. Implementation of the ESA has, however, been widely criticized (Yaffee 1982). Rates and apportionments of species listings have been condemned by environmentalists and industry alike. During the first decade under the Act, the Office of Endangered Species favored listing of vertebrates over insects, mollusks, and plants. By 1984, nearly 60% of listed species were vertebrates, but plants and invertebrates outnumbered vertebrates nine to one on the list of organisms being considered for listing. By 1990, 565 U.S. species were listed as endangered or threatened; 39% were plants, 13% were invertebrates, and 14% were fishes (USFWS 1990). By January 1990, 256 plans for 307 U.S. endangered and threatened species were completed and approved (USFWS 1990), yet few of these species are making progress toward recovery. Conservation measures spelled out in recovery plans are not mandated, and citizen action, in the form of a lawsuit, is often required to accomplish recovery of a species. In that sense, the ESA affords far less protection than most people assume. Nongovern-

Box 15.1   The U.S. Endangered Species Act of 1973 and its Amendments

I. Major provisions of the Endangered Species Act (ESA) are summarized here by Section. Numbers in parentheses correspond to Sections of Title 16, Chapter 35 of the 1982 edition of the United States Code, which incorporated ESA amendments through 1982.

Section 2 (§1531) covered findings, purposes, and policy of Congress in passing the Act.

Section 3 (§1532) defined endangered species (one in danger of extinction throughout all or a significant part of its range) and threatened species (likely to become endangered in the foreseeable future in all or a significant portion of its range). The Directors of the Fish and Wildlife Service (FWS) and the National Marine Fisheries Service (for marine species) are responsible for determining which species are threatened or endangered.

Section 4 (§1533) required the Secretaries of the Interior and Commerce to establish lists of threatened or endangered species (and/or subspecies and/or isolated populations) and publish them in the *Federal Register*. The Secretaries were also required to develop and implement recovery plans for the conservation and survival of listed species.

Section 5 (§1534) gave the FWS authority to acquire lands for endangered species. Such lands become part of the National Wildlife Refuge System.

Section 6 (§1535) was designed to overcome potential problems between states and the federal government in implementing the Act by providing federal matching funds through cooperative agreements for states that enact laws for protection of endangered species and cooperate in research programs that contribute to better management of such species.

Section 7 (§1536) required all federal agencies to ensure that actions they authorized, funded, or carried out did not jeopardize the continued existence of threatened or endangered species or result in destruction or adverse modification of their critical habitats. Federal agencies were required to consult with the FWS on actions likely to jeopardize listed species or their critical habitats.

Section 8 (§1537) authorized funding to help other countries develop and manage programs for conservation of threatened or endangered species and provided for funding of CITES.

Section 9 (§1538) prohibited certain actions (including taking or importation into or exportation from the United States) involving listed species. Taking was defined as to harass, harm, pursue, hunt, shoot, wound, kill, trap, capture, or collect.

Section 10 (§1539) allowed for exceptions to prohibited actions under special circumstances (e.g., biologists may obtain collection permits for scientific purposes or to enhance propagation or survival of listed species).

Section 11 (§1540) provided for penalties for violating the Act and opportunity for anyone to sue (in federal courts) anyone else who violates the Act.

Box 15.1 Continued.

II. Amendments of ESA have resulted from required and regular reauthorization by Congress.

1978 Amendments (P.L. 95-632) improved consultation stipulated in Section 7 and provided a review process to resolve conflicts which persist after consultation. Applications for exemption from requirements of the Act may be approved by a review board and an endangered species committee. Decisions of the committee are subject to judicial review. Species and critical habitat were redefined, and listing of species and critical habitat were, to the maximum extent possible, required to be simultaneous.

1979 Amendments (P.L. 96-159) were of a relatively minor nature.

1982 Amendments (P.L. 97-304) included changes in Section 4 to ensure that listing decisions are based solely on biological criteria and to authorize listing without the requirement for concurrent critical habitat designation in some cases. Section 7 was changed to allow early initiation of the consultation process and extension of the normal consultation period. Section 10 was also modified to allow release of experimental populations outside the range of a listed species if this would further the conservation of the species.

1988 Amendments (P.L. 100-478) increased funding for ESA-related activities of the Departments of the Interior, Commerce, and Agriculture. They also required the Secretary of the Interior to report to Congress every 2 years on progress to develop and carry out recovery plans and on the status of species for which recovery plans have been developed. The FWS must also more closely monitor the status of listing candidates. A Cooperative Endangered Species Conservation Fund was established to facilitate cooperative state programs, protection of endangered and threatened plants was strengthened, and fines for willful violation of the Act were more than doubled.

III. For further information, consult Shea (1977), Cadieux (1981), Ehrlich and Ehrlich (1981), Campbell (1982), USFWS (1982), Yaffee (1982), Ono et al. (1983), SCEPW (1983), USDI (1984), Rinne et al. (1986), and Kohm (1991).

mental environmental organizations with interests in conserving endangered or threatened species can and do play important roles in enforcing the ESA.

15.2.3 State Endangered Species Programs

States began enacting laws and enforcing protection of endangered species and natural habitats in the late 1960s. Some state laws protect only species on federal lists, and other states prepare their own lists. California was first to develop a state list of endangered and threatened species. Many states have potentially excellent programs to aid listed species, but their success has been linked to funding levels. The Endangered Species Act encouraged state participation in conservation of endangered species through cooperative agreements which allow federal matching funds for state projects involving listed species. This funding, however, has been minuscule compared to other federal aid for wildlife. In many states, endangered species programs are linked with nongame wildlife programs. State fish and game agencies have become the front-line protectors of federally listed and other rare species in their purviews.

15.2.4 Canadian and Mexican Endangered Species Programs

Canada has no legislative equivalent of the ESA; it has a national committee that evaluates and assigns status to species at risk. The Committee on the Status of Endangered Wildlife in Canada (COSEWIC), established in 1977, includes representatives of federal, provincial, and territorial governments and national conservation organizations such as the Canadian Nature Federation, Canadian Wildlife Federation, and World Wildlife Fund Canada (Cook and Muir 1984). Scientific subcommittees, corresponding to major taxonomic categories, arrange for and approve status reports on species that are candidates for a national list of species regarded as extinct, extirpated, endangered, threatened, rare, or not in any category (Cook and Muir 1984). Approved status reports are copied for limited distribution, and printed summary sheets are widely distributed to the public. Committee decisions (lists) have been reported in the *Canadian Field-Naturalist*. Animal species assigned threatened or endangered status are published in the *Canada Gazette* to provide official notice of such action. The committee has no legislative or management role but has close connections to the Convention on International Trade in Endangered Species of Wild Fauna and Flora (CITES).

Mexico is the only North American country that has not developed an official list of organisms in danger of extinction (Contreras Balderas 1987). Development of a Mexican list has been prevented by lack of biological information and specific legislation. Contreras Balderas (1991) described legislation to protect the environment and ecological equilibrium which indirectly affords protection to freshwater habitats and fishes.

15.2.5 International Treaties

Laws and treaties that restrict shipments of wildlife between countries are not a recent phenomenon. The Lacey Act of 1900 was the first U.S. law applied to control of wildlife imports, and the 1940 Convention on Nature Protection and Wildlife Preservation in the Western Hemisphere was the first international treaty that protected all animals and plants. As international trade in wildlife and wildlife products increased in the 1960s and early 1970s, broader and more effective controls were needed (Ehrlich and Ehrlich 1981). In 1975 CITES was implemented to regulate international trade in endangered, rare, and protected species among signatory nations. Appendix I of CITES lists species or other taxa that are in danger of extinction. Trade in these species is restricted: an export permit from the country of origin and an import permit from the destination country must be approved by management and scientific authorities in both countries. One hundred and seven nations were members of CITES in 1990.

Trade in endangered wildlife has flourished in spite of CITES. Countries vary significantly in their ability to enforce the convention, and only signatories are bound by it. However, the United States forbids imports of endangered species without regard to their country of origin.

15.3 ENDANGERED FISHES OF THE UNITED STATES AND CANADA

Serious threats to many unique, native fishes in North America have been recognized since the early 1960s. Miller (1972) made the first attempt to list native threatened fishes of the United States. Over 300 kinds of fishes were listed to enhance chances for their survival through protective legislation and stronger concern for natural resources.

The American Fisheries Society Endangered Species Committee (Deacon et al. 1979) prepared a list of endangered, threatened, and special-concern fishes of North America from available literature in 1976 and had it reviewed by many biologists. Endangered and threatened were defined as in the ESA, and the special-concern category was developed for species that could become threatened or endangered by relatively minor disturbances to their habitat or for which additional information was required to determine their status. Deacon et al. (1979) based their list on biological considerations throughout the range of each taxon without regard for jurisdiction or politics; it contained 251 taxa appropriate to one of their three categories, regardless of whether they appeared on any official list. Threats and percentages of the listed taxa affected by them were habitat modification (98%), other natural or human-induced factors (37%), restricted range (16%), overexploitation (3%), and disease (2%). The authors considered this to be evidence that well over 90% of the endangered and threatened fishes of North America could be restored by nationwide programs of habitat restoration and protection.

Compilers of data on endangered and threatened species generally do not attempt to chronicle fish extinctions, probably because it is difficult to demonstrate that a fish species is extinct. Miller et al. (1989) reported documented extinction of 3 genera, 27 species, and 13 subspecies of fishes from North America during the past 100 years. Physical habitat alteration and detrimental effects of introduced species were the most commonly cited factors contributing to these extinctions. Ono et al. (1983) presented a list of North American fishes that in their view were endangered, threatened, or extinct. Seventy-eight percent of the endangered and threatened fishes on their list inhabited only the United States, 10% were from Mexico, and 1% were from Canada. Almost 80% of the extant U.S. fishes on their list were confined to the Southwest and Southeast. Over 60% of the species they listed inhabit streams, and 90% of those live in warmwater streams. North American fishes then made up about 70% of the known endangered and threatened fishes of the world. The work of Ono et al. (1983) was unique in presenting accounts of the circumstances attending the plight of listed species grouped by geographic regions in North America.

McAllister et al. (1985) discussed 37 Canadian fishes classed provisionally or by COSEWIC as rare, endangered, extirpated in Canada, or extinct. Campbell (1988) discussed status reports in preparation and approved by the COSEWIC Fish and Marine Mammals Subcommittee; by 1987, 38 fish species had been classified, and many more status reports were under review.

Johnson (1987), on behalf of the American Fisheries Society Endangered Species Committee, assembled lists of native U.S. and Canadian fishes receiving legal protection or of special concern due to low numbers, limited distribution, or recent declines. The protected fishes included threatened and endangered species on FWS lists and those proposed for listing. Species of special concern are not protected by law but were considered either more secure than protected species or too poorly studied to determine if they need full protection. Johnson listed 517 taxa, including described and undescribed species and subspecies, that had legal protection and 709 of special concern in the United States and Canada. By 1986, over 90% of the fishes listed by Deacon et al. (1979) as in need of assistance had at least special-concern status, and 62% were fully protected by state or federal law. Only four states and six provinces had no protected fishes or fishes of special concern. A sobering statistic was that 56% of the described fish species of the United States and Canada were receiving some degree of protection and 32% were fully protected by one or more governments under endangered species laws in all or a portion of their ranges.

The most up-to-date and authoritative information on endangered, threatened, and special concern fishes of North America is that of Williams et al. (1989). The authors, members of the American Fisheries Society Endangered Species Committee, updated the list compiled by Deacon et al. (1979) by adding 139 new taxa and removing 26. Williams et al. (1989) listed 364 fishes in North America that deserved endangered (103), threatened (114), or special-concern status. Twenty-two of the listed fishes were endemic to Canada, 254 to the United States, and 123 to Mexico. Fifty-six of the new taxa on the list were Mexican fishes added because of new information on their status and the serious degradation of aquatic habitats in that country. Fishes of the southwestern and southeastern United States showed most adverse impacts of habitat loss and introductions of exotic species. No fish was removed from the 1979 list because of successful recovery efforts. Although recovery plans have been approved for 47 of the 82 species of U.S. fishes listed as threatened or endangered by FWS (1990), the American Fisheries Society lists far more fishes than does the FWS. Williams et al. (1989) stated that recovery efforts had been locally effective for some species but that North America's fish fauna has generally deteriorated. Future management for conservation of entire ecosystems and establishment of long-term monitoring programs for fish populations and aquatic habitats were encouraged.

15.4 CASE HISTORIES

15.4.1 Causes of Fish Endangerment

Ono et al. (1983) presented many case histories on rare and endangered fishes. North American fishes have been forced to the brink of extinction by several human activities. Our case histories and those presented by Ono et al. illustrate that habitat alteration, introduction of nonnative species, and overexploitation have had the greatest impacts on native fishes. Lake sturgeon, shortnose sturgeon, ciscoes of the Great Lakes, Atlantic (Acadian) whitefish, and the cui-ui have declined significantly because of their desirability as food fish. Introduction of nonnative fishes has significantly impacted western trouts and fishes of such diverse habitats as North American deserts, the Colorado River system, Pacific coast streams, and Texas springs. However, there is no doubt that human capacity to change habitats has been the most significant threat to native North American fishes. In their case histories, Ono et al. (1983) reviewed situations in

Figure 15.1   Snail darter. (Photograph courtesy of R. Behnke.)

which fish habitat was modified by pollution of surface or groundwaters, water diversion, impoundment, stream channelization, dredging, mining, overgrazing, timber cutting, and recreation. Emphasis on western fishes in the following case histories reflects our interests and the geographical distribution of endangered fishes in North America. Minckley and Deacon (1991) provided excellent information on the precarious status of native western fishes.

15.4.2   The Snail Darter

No discourse on endangered fishes would be complete without considering the celebrated case of the snail darter, an early test of the ESA which resulted in significant changes in the Act in 1978. The snail darter (Figure 15.1) was discovered in 1973 in a part of the lower Little Tennessee River known as Coytee Springs (Ono et al. 1983). Shortly thereafter, the ESA was signed into law by President Nixon. Convinced that the snail darter was new to science and very rare, Dr. David Etnier, one of the discoverers, submitted a status report on the fish to the FWS. The report suggested that the 7- to 8-cm-long fish, which lives in fast waters and feeds on snails in winter and caddisfly and black fly larvae in summer, was endangered, and its existence was jeopardized by a Tennessee Valley Authority (TVA) dam (Ono et al. 1983). The snail darter may have lived throughout much of the Little Tennessee before the advent of TVA, but dams had apparently restricted its habitat to the last 24 km of the river before it joined the Tennessee River. By the time the snail darter was described, construction of TVA's controversial Tellico Project, a multipurpose water resource development project, had been approved by Congress (in 1966), begun (in 1967), delayed by a court injunction (for lack of an environmental impact statement), and resumed (Ono et al. 1983). Controversy stemmed from TVA's assertion that the benefits of the project would outweigh any disadvantages. Environmentalists countered that the dam would flood productive farm land, destroy an excellent trout stream, and inundate a sacred place of the Cherokee Indians. The snail darter's habitat, above the site of the proposed Tellico Dam, was certain to be destroyed by closure of the dam.

The Fish and Wildlife Service was petitioned to list the snail darter as endangered in January 1975, a proposal to do so was published in the *Federal Register* in June, and the fish was listed in October. That same year, TVA began transplanting snail darters to the Hiwassee River, another tributary of the Tennessee River, without informing state authorities or the FWS (Ono et al. 1983). Clearly a federal agency in violation of the ESA, TVA rushed to complete construction of the dam. In 1976, the Environmental Defense Fund, whose help had been enlisted by dam opponents, sued TVA for violating the ESA, and the FWS declared the lower portion of the Little Tennessee River critical habitat for the snail darter. The District Court in which trial was held found that the snail darter would be eradicated by the dam but denied a request for a permanent injunction on completion of construction. In early 1977, a Circuit Court of Appeals reversed the District Court's decision and terminated all work on the dam that threatened the snail darter or its critical habitat. News media began to criticize the decision to stop a multimillion-dollar project because of a small, inconsequential fish; the story was often presented as a case of environmentalism carried to extremes (Ehrlich and Ehrlich 1981).

After apparent defeat, TVA asked the FWS to delist the snail darter and requested permission to move all snail darters from the Little Tennessee River; both requests were refused. The Tennessee Valley Authority also appealed the Appeals Court's decision to the Supreme Court, which in June 1978 upheld the injunction against the dam but ". . . virtually invited Congress to amend the law to allow exceptions" (Ehrlich and Ehrlich 1981). Congress did just that, passing the 1978 amendments to the ESA's Section 7 which allowed an Endangered Species Committee to resolve conflicts between endangered species and projects that threaten them. The committee established to resolve the snail darter-Tellico Dam conflict considered evidence of the dam's benefits, alternatives to completing the dam, values that would be lost if it were finished, and new economic analyses which showed that electricity would be produced at a deficit (Ehrlich and Ehrlich 1981). It voted unanimously in early 1979 not to exempt the Tellico Project because it was ill-conceived and uneconomical (Yaffee 1982; Ono et al. 1983). Later that year, the committee's decision was reversed and the conflict finally ended when an amendment exempting the Tellico Project from provisions of all federal laws was attached to the water projects appropriation bill in the House of Representatives and passed without debate or opposition. The Senate tried without success to kill the amendment, and President Carter cited political problems and difficulties in vetoing a multipurpose bill; he signed the bill (with regret, he said) in September 1979 (Ehrlich and Ehrlich 1981; Yaffee 1982). The Tellico Dam was completed and closed early in 1980. The snail darter has recently been said to be doing well in the Hiwassee River (Hickman 1981) and several other places where new populations had been found (Ono et al. 1983). It was downlisted from endangered to threatened in 1983.

The snail darter case has been called a classic boondoggle and a prime example of how pork-barrel politics threaten the environment and efforts to preserve species. However, Yaffee (1982) cited the case as an example of how the judiciary, Congress, the media, and administrative agencies interact to determine how the ESA, a statute intended to be prohibitive, is implemented.

15.4.3 Death Valley Fishes

Another case of vanishing fishes which involved intervention by the Supreme Court is that of the desert fishes of Death Valley. Scientists disagree on historic drainage patterns in the Death Valley region of southwestern Nevada and southeastern California, but they agree that much of the area was covered by large lakes interconnected by rivers during the Pleistocene epoch (Soltz and Naiman 1978). Today, Death Valley is one of the hottest and driest places in the United States, and its waters have largely shrunk to a few small spring pools and intermittent streams. The Owens and Mohave rivers receive runoff from within the basin, but much of the basin's water enters through two groundwater systems, the Pahute Mesa and Ash Meadows groundwater basins. Only a few native fishes existed in Death Valley at the turn of the century. Among them were killifishes (several species of *Cyprinodon* and two of *Empetrichthys*), minnows (speckled dace and tui chub), and the Owens sucker (Pister 1974, 1981). Many lived in very restricted habitats subject to destruction by a single catastrophic event (Ono et al. 1983).

Human activities have posed potential and realized catastrophes in Death Valley. Early in the 1900s, construction of the Los Angeles Aqueduct greatly modified habitats in the Owens River drainage. New dams reduced flooding that had created killifish habitat, marshes were drained, and game fishes (e.g., brown trout and largemouth bass) and mosquitofish were introduced. The game fishes used small native fishes as food, and the mosquitofish competed with some of them. Use of groundwater for crop irrigation disrupted fish habitat and lowered water tables on which fish depended in the Amargosa Valley and Ash Meadows. Introduction of nonnative fishes to waters of Ash Meadows also presented problems for its native fishes. These threats were particularly devastating for killifishes. The Devils Hole pupfish, which exists only in a small spring pool called Devil's Hole in Ash Meadows, was particularly affected by groundwater withdrawals (Figure 15.2). Lowering water levels gradually reduced photosynthesis in its limited habitat and exposed a shelf on which it depended for spawning substrate and its algal food supply.

To forestall threats to Death Valley fishes, wildlife agency personnel met in 1969 to develop an integrated plan for their preservation. The Desert Fishes Council, which continues to be active today, was established at a 1970 meeting at the Death Valley National Monument headquarters. Consisting largely of state, federal, and academic biologists, it has assumed a strong role in coordinating preservation efforts and assisting government agencies with preservation programs.

Various strategies have been employed to preserve Death Valley fishes. Several fishes were transferred to natural or artificial refugia. The Devils Hole pupfish, for example, was first transferred to five refugia in California and Nevada and to an artificial refugium built by the Bureau of Reclamation below Hoover Dam. The latter transfer resulted in phenotypic changes in the fish (Williams et al. 1988). Later, a fiberglass shelf was suspended below the surface of its pool and artificially lighted to enhance algal production. The fish was also reared in aquaria,

Figure 15.2 Biologists touring the Devil's Hole pool, home of the Devils Hole pupfish. (Photograph courtesy of J. Hawkins.)

but none of these efforts met with success. Since it was apparent that only maintenance of groundwater levels would protect the Devils Hole pupfish and other Ash Meadows fishes, legal action to prohibit removal of groundwater in the area was initiated. In July 1972, before the ESA was passed, the people of the United States, through the Department of Justice, sued the land developer and the state of Nevada as codefendants on a point of water law (Pister 1979). A District Court issued a temporary injunction to prohibit pumping in April 1973. The injunction was made permanent by a unanimous ruling of the Supreme Court in June 1976. Since land ownership and land-use control best assure habitat integrity, attempts were then made to buy Ash Meadows. After the Fish and Wildlife Service (supported by the Nevada Department of Wildlife) tried and failed, the land was bought by a real estate developer, who started a housing development. In 1984, Ash Meadows National Wildlife Refuge was established (Deacon and Williams 1991). Fish Slough (Inyo and Mono counties, California),

Figure 15.3 Razorback sucker. (Photograph courtesy of E. Wick.)

the primary refugium of the endangered Owens pupfish and Owens tui chub, was saved from development by a 1982 Act of Congress which allowed a land exchange between the Bureau of Land Management and a developer (Pister 1985).

What remains of the native fishes of Death Valley? Of the endemic Owens River fishes, the Owens pupfish was nearly wiped out, speckled dace populations were greatly diminished, and tui chub hybridized with bait minnows (Pister 1974). Only the Owens sucker was largely unaffected. In Ash Meadows, interactions with introduced fishes led to extinction of the Ash Meadows killifish, Raycraft Ranch killifish, and Pahrump Ranch killifish. The Pahrump killifish survives perilously close to extinction in artificially maintained refugia, where it is threatened by bullfrogs and introduced fishes. The Ash Meadows speckled dace and Ash Meadows Amargosa pupfish were seriously threatened by groundwater pumping (Ono et al. 1983). Of the native fishes of Death Valley, Mohave tui chub, Owens tui chub, Ash Meadows speckled dace, Devils Hole pupfish, Owens pupfish, Ash Meadows Amargosa pupfish, and Pahrump killifish are endangered. Recent habitat preservation measures lead one to limited optimism about their future, but most native fishes of Death Valley remain endangered in every sense of the word. Ono et al. (1983) noted that 70% of the fishes on federal threatened and endangered lists were from desert environments.

15.4.4 Colorado River Basin Fishes

The indigenous fish fauna of the Colorado River basin includes some of the most unique freshwater fishes in North America. Because the basin has been isolated for millions of years, its native fishes are morphologically different and evolutionarily and geographically distant from their nearest relatives. In terms of number of species or subspecies, cyprinids, catostomids, and cyprinodontids head the list of native fishes. Among the better-publicized natives are the Colorado squawfish, a large predatory minnow, and humpback chub and razorback sucker (Figure 15.3), both distinguished by a prominent dorsal hump at the nape. In all, about 54 fishes (presently recognized species and subspecies) are native to the basin, and most (83% of the total number) are also endemic. Stanford and Ward (1986b) and Carlson and Muth (1989) summarized information on within-basin distributions of native fishes.

Because of severe abundance and range reductions, most native fishes of the Colorado River are currently in jeopardy. Two are extinct (Pahranagat spinedace and Las Vegas dace), 16 are listed as federally endangered, 6 are listed as federally threatened, and 13 have either been proposed for federal listing or are under review for federal listing. In all, 35 native fishes (65% of the total number) receive legal protection under the ESA; except for one, the Gila topminnow, all are endemic to the system. Several of the remaining native fishes are variously protected by one or more basin states. Of the federally protected fishes, 29 occur only in the lower Colorado River basin (Arizona, California, Nevada, and New Mexico), 2 occur only in the upper Colorado River basin (Colorado, Utah, and Wyoming), and 4 occurred in both the upper and lower basins. The two extinct fishes occurred in the lower basin. Researchers have attributed the decline of native fishes to (1) modification and loss of habitat and (2) introduction of nonnative species (Stanford and Ward 1986b). Typically, these two factors occur together and have synergistic adverse effects.

Closure of Hoover Dam in 1935 marked the end of the free-flowing Colorado River. Since then, the basin has become one of the most altered and controlled river systems in the United States. Dam construction and regulation have drastically altered physical and biological features of the river system (Stanford and Ward 1986a), and, as a result, native fishes inhabiting the larger river channels have been severely impacted. Biologists are just beginning to recognize the detrimental effects of the spread of nonnative riparian vegetation on stream widths and, therefore, fish habitat in the basin.

Since the late 1800s, approximately 67 fish species have been successfully introduced into the basin, raising the total number of species to over 100 (Carlson and Muth 1989). Most of these introductions resulted from efforts to establish sport fisheries in and downstream of newly created reservoirs. Several authors have noted adverse impacts of introduced nonnative species on native fishes of the basin. Invoking the island biogeography theory, Molles (1980) proposed that, because the Colorado River basin is an insular system, the native fish fauna is especially vulnerable to invasion by nonnative fish species. Using local extirpations of native fishes in the lower basin as examples, he suggested that successful invasions can cause removal of native fishes through competitive replacement or predation.

Of the basin's 16 federally endangered fishes, the bonytail, Pahranagat roundtail chub, Virgin River roundtail chub, woundfin, Colorado squawfish, razorback sucker, desert pupfish, and Gila topminnow are currently being maintained at Dexter National Fish Hatchery, New Mexico (Johnson and Jensen 1991). Since 1974, the Dexter hatchery has served as a refuge and propagation facility for several rare and imperiled native fishes of the American Southwest. Personnel from the Dexter facility have assisted in development of similar programs elsewhere in North America and in other countries.

Despite the severely perturbed state of the basin, threats of additional water development, and in some cases years of research and volumes of data, efforts to recover the federally protected fishes by direct management of wild populations or their habitats have been limited. In the lower basin, where river habitat degradation and fish extirpation has been especially severe, recovery work has emphasized acquisition of habitats and brood stocks, propagation, and reintroduction. Hatchery brood stocks of the threatened Apache trout and endangered Gila trout now exist, fish are being reintroduced, and several new stream populations have been established. Since 1981, young razorback sucker have been reintroduced annually in the Gila, Salt, and Verde river drainages, Arizona. Reintroduction of Colorado squawfish in the Salt and Verde river drainages began in 1985, and desert pupfish and Gila topminnow have been stocked at several reintroduction sites in Arizona. Success of these reintroductions has yet to be fully evaluated, but the programs are expected to continue and expand to other lower-basin locations.

To date, most work on native fishes in the upper basin has been oriented toward research and population monitoring, in part because river habitat degradation is less severe, and wild populations of federally protected and rare fishes are still present. Plans for reintroducing upper-basin endangered fishes incorporate and depend upon understanding population genetics and on habitat enhancement or restoration.

A program has been developed to recover, delist, and manage humpback chub, bonytail, Colorado squawfish, and razorback sucker in the upper basin while allowing for additional water development (Wydoski and Hamill 1991). Elements of the recovery program are (1) provision of instream flow—determine, procure, and maintain instream flows required to protect and recover the species; (2) habitat development and maintenance—increase the amount of favorable habitat for reproduction, production, and overwintering; (3) artificial propagation and stocking of rare and endangered fish species; (4) nonnative species and sportfishing management—regulate stocking of nonnative fishes, selectively remove nonnative species, regulate sportfishing in areas where rare and endangered fishes occur, and educate the fishing public on rare and endangered fishes; and (5) research, monitoring, and data management. The time frame for this $60 million recovery program is 15 years, and an Upper Colorado River Recovery Implementation Committee oversees recovery efforts.

Even with operating recovery strategies in place, the future of the Colorado River's native fishes appears bleak. Carlson and Muth (1989) concluded that conflicts between development and natural ecosystems in the basin will continue and probably worsen as demand for water increases. Stanford and Ward (1986a) stated that the river's future depends on whether (1) there will be enough water to maintain desirable ecosystem values and (2) native and nonnative fishes can coexist. They concluded that the endangered fishes are incompatible with stream regulation and nonnative species and that future water shortages will preclude allocations for them and other ecological concerns.

15.4.5 Western Trouts in the Genus *Oncorhynchus*

Since the turn of the century, most native western trouts have experienced declining populations, and some species have been pushed to near extinction (Ono et al. 1983). Behnke (1992) classified native western trouts into four species and 22 subspecies. Five are federally listed as threatened (Little Kern golden trout, Apache trout, Lahontan cutthroat trout, Paiute cutthroat trout, and greenback cutthroat trout), and one is federally listed as endangered (Gila trout). Other native western trouts are variously protected by several western states and often appear on unofficial lists of threatened and endangered fishes. Some of these, such as Colorado River cutthroat trout and Bonneville cutthroat trout, are under review for federal listing.

Several factors have contributed to the decline of native western trouts. One major threat has been extensive introduction to areas outside their native ranges (Behnke and Zarn 1976); this has led to hybridization and resulted in loss of genetically pure populations, increased competition for food and space, and increased predation on young trout. Allendorf and Leary (1988) considered introgression, dilution of native alleles, the most important effect of introductions of nonnative western trout. Widespread introgression can lead to replacing the diversity of taxa produced by evolution with a single taxon. Such homogenization could result in loss of locally adapted populations and, less importantly, outbreeding depression (hybrids with reduced fitness). Another factor blamed for the decline of native western trouts is habitat destruction caused by water diversion and removal, channelization, water pollution, and overgrazing of rangeland and damage to streambanks by livestock.

Originally, the Apache trout and greenback, Lahontan, and Paiute cutthroat trouts were federally listed as endangered, but they were reclassified as threatened to facilitate recovery and management efforts (Behnke and Zarn 1976). Federal recovery plans have been approved for several native western trouts, and management programs have been successful in rehabilitating certain populations. Status and management of Lahontan and greenback cutthroat trouts were considered in Gresswell (1988); in general, elimination of nonnative trout, reintroduction of genetically pure native stocks, habitat protection and improvement, and restricted harvest have been used to enhance populations of these fishes. These actions conform to the steps in a native trout management program outlined by Behnke and Zarn (1976): (1) collection of specimens from suspected genetically pure populations, (2) identification of pure populations by taxonomic study of collected specimens, (3) protection and possible improvement of existing habitat, (4) introduction of native trout into isolated waters devoid of nonnative trouts, and (5) establishment of special-regulation fisheries where applicable.

Native western trouts are some of the most aesthetically pleasing fishes in North America and have high sport value. Behnke and Zarn (1976) stated that low-cost quality fisheries can be based on native trouts if they are properly managed and publicized. Therefore, odds favor native western trouts if any of North America's threatened or endangered fishes have a chance of being recovered and delisted. Public awareness and support are needed if native western trouts are to be saved.

15.5 MANAGEMENT FOR ENDANGERED FISHES

Fisheries biologists can use many approaches to alleviate threats to rare native fishes. In general, they should strive to maintain populations of threatened or endangered fishes by recognizing their status and the importance of preserving biological diversity, contributing to knowledge about them, applying management strategies (and provisions of the ESA), or instructing others on the merits of such actions.

When conservation of fishes is possible, Moyle and Cech (1988) recommended the following steps: (1) inventory fishes; (2) monitor habitats, communities, and species; (3) conduct research on best management of resource and nonresource fishes; (4) plan regional management following natural, rather than political, boundaries; and (5) manage fish communities and habitats as units. We see a special need for biological surveys to determine the status of rare fishes and gain legal protection for those near extinction, particularly in Mexico and, to a lesser extent, Canada. Williams et al. (1989) stressed the need for long-term monitoring programs to provide baseline data on the status of rare fishes and allow future assessment of both fish populations and their habitats. Managers of rare fishes must proceed cautiously, however, because the field is relatively new and much remains to be learned (Meffe 1986). We must, for example, avoid over-monitoring populations of threatened or endangered fishes; it is possible to study some species to extinction. Because of political and biological urgency regarding recovery, aspects of basic biology (e.g., genetics, reproduction, growth, age structure, sex ratio) have often been neglected in research on rare fishes. Many suggestions of Mlot (1989) and Soule and Kohm (1989) regarding research directions for conservation biology should be applied to rare fishes; emphasis on ecosystem fragmentation, biology of small systems, reproduction, reintroduction, and effects of stress and disease is justified. Population dynamics and habitat needs, with particular emphasis on early life stages, must be understood for effective management. Single-species approaches to recovery efforts must and have begun to give way to considering endangered fishes as components of complex ecosystems. Biologists must also recognize their obligations as informed citizens to see that laws that protect species or their habitats are passed, modified as necessary, and enforced.

15.5.1 Managing Fish and Habitat

Specific practices that can be applied to management of endangered fishes include (1) maintaining and enhancing historic populations; (2) protecting, expanding, or restoring habitat; (3) moving specimens to refuges; (4) rearing and stocking in new or formerly occupied areas; (5) minimizing introductions and undesirable effects of introduced nonnative organisms; and (6) controlling exploitation. First, factors leading to species rarity must be identified and, if possible, controlled.

We have listed maintaining and enhancing historic native populations first as a reminder of their importance. This could involve some or all of the other management practices considered in this section. One approach to avoiding extinction, however, might be to simply leave some populations and their habitats alone. Saving entire species from extinction is of paramount importance.

Habitat should be a primary consideration in recovery of rare fishes. Land acquisition as accomplished in the Death Valley drainage is the most effective means of protecting habitat for endangered species. Other approaches involve participation in pollution-control efforts, land-use planning, and legal actions to prevent habitat destruction. Definition and development of optimal habitats is far more effective in enhancing endangered fish populations than imposition of restrictive regulations (Deacon et al. 1979). Johnson and Rinne (1982) noted that protection and enhancement of habitats is the almost universal goal of fish recovery plans, but recovery teams have generally left implementation to the discretion of land managers. They suggested that existing habitats be protected through increased federal management, prioritizing protected habitats, and use of ecosystem-based recovery teams. To make protection more efficient, only optimal

Figure 15.4 Low log dams in streams are often used to enhance pool habitat for western trouts. (Photograph courtesy of K. Fausch.)

habitats, rather than all areas occupied by a species, need to be identified and designated for preservation. Recovery planning on an ecosystem basis has proven more effective than the conventional species-by-species approach.

Restoration of damaged or depleted habitat is also possible. Earlier, we discussed attempts to restore the habitat of the Devils Hole pupfish by shelf construction. Re-creating backwaters, wetlands, embayments, and other habitat types has been suggested to aid recovery of the protected native fishes of the Upper Colorado River system. Use of fishways and other remediation measures has also been suggested to preserve those species. Rinne (1982) described effects of log habitat-improvement structures on Gila trout growth and dispersal; similar techniques have been applied to streams containing other rare western trouts (Figure 15.4). Restoration of aquatic habitat might also include removal of dams, returning channelized streams to their original courses, silt removal, and restoration of riparian vegetation.

Our case histories included examples of transferring threatened fishes to refugia. One of the earliest transfers was that of the Owens pupfish. Thought to be extinct when it was described in 1948, it was later rediscovered in Fish Slough in Mono County, California (Miller and Pister 1971). To protect the remnants of the species from introduced predatory fishes and competition from mosquitofish, the Owens Valley Native Fish Sanctuary was built by the California Department of Fish and Game in 1969. Owens River fishes were moved to the sanctuary in 1970. Two other refugia were later constructed to receive the Owens pupfish and other native Owens Valley fishes, and all have contributed to their continued existence. Fish in such refugia generally are not representative of endemic native species (Ono et al. 1983). New selection pressures are expected to adapt them to their refugia and alter their gene pools. Turner (1984), however, found no apparent loss

Figure 15.5 Ponds at Horsethief Canyon State Wildlife Area, Colorado, are used as refugia for endangered fishes of the upper Colorado River basin. (Photograph courtesy of D. Gates.)

of average genic heterozygosity in refugium populations of desert pupfish; he considered use of refugia to be a valid practice for endangered fish management and conservation (Figure 15.5). Similarly, Ammerman and Morizot (1989) found that captive Colorado squawfish did not differ significantly from wild populations in allele frequencies at loci that affect overall individual fitness. Nonetheless, there is general agreement that fishes should be transplanted to refugia only when their extinction is imminent and they should be reintroduced to native habitats as soon as possible.

Allendorf and Leary (1988) stated that the primary goal of any conservation program must be to ensure maintenance of existing genetic variation. Loss of genetic variation is generally expected to increase the probability of extinction by adversely affecting such survival-related characteristics as growth, fertility, longevity, normal development, disease resistance, and ability to adapt to changing environmental conditions. Long-term conservation of a large range of species is possible only in natural communities, and we may have a responsibility to keep evolutionary options open. To conserve our genetic estate (biological heritage), Frankel (1974) urged that geneticists play a role in planning ecological reserves for conservation of such communities and that the genetic principles of conservation be explored and clarified. Smith and Chesser (1981) emphasized the importance of genetic variability in maintaining stable, complex ecosystems and species. Since management practices may reduce that variability below normal levels, they called for conservation efforts to maintain existing levels of genetic variability in natural fish populations. Meffe (1986) reviewed the nature of potential genetic problems in management of endangered fishes and suggested means to minimize genetic deterioration of endangered stocks. Management of endangered fishes, or any endangered species, should first avoid extinction and then maintain the ability of fishes to adapt to changing environments and their capacity for continued speciation. Maintaining genetic variability and evolutionary flexibility in small populations depends on maximizing within-population and among-population variance. To accomplish this for endangered fishes, Meffe (1986) recommended the best approximation to the following set of management guidelines: (1) genetic monitoring should be done to determine how variation is distributed within a species and how to preserve it, (2) the largest feasible genetically effective population size should be maintained in captive breeding programs, (3) if a large population cannot be maintained, inbreeding may be avoided by selective mating, (4) stocks should be kept in hatchery environments for as short a time as possible, and (5) separate stocks of isolated populations should be maintained to preserve interpopulation variance.

Meffe (1987) argued that fisheries biologists must become more aware of conservation genetics and incorporate it into fish conservation programs if they expect to make a contribution to the preservation of biotic and genetic diversity. He also urged that a conservative approach to conservation genetics be followed. Conservation of fish gene pools is limited by lack of genetic information on fishes, facilities for storing fish genomes, experience in breeding and rearing many species, understanding of the effect of genetic heterozygosity on individual fitness, and information on how fish life histories can affect the genetic structure of populations. Soule (1985) noted that the new field of conservation biology, which is devoted to preserving biological diversity and addresses biology of species, communities, and ecosystems perturbed by human activities, is often a crisis discipline. In such disciplines, decisions must often be made in haste and without benefit of knowing all the facts. Because the science of conservation genetics is also new, and there is often uncertainty about necessary and sufficient conservation methodology, retention of maximum global genetic diversity for future research and management represents the conservative approach (Meffe 1987).

Although fish seem to have been somewhat ignored by conservation biologists (Allendorf 1988), a few recent studies emphasize the importance of genetic analyses in their conservation. Meffe and Vrijenhoek (1988) developed two models of isolation and gene flow to address how the population genetic structure of desert fishes relates to programs for their recovery and conservation. The authors urged use of three types of experimental research on population genetics and fitness to allow prediction of survival and long-term success of remnant populations of endangered and threatened fishes. Echelle et al. (1989) observed that most between-sample gene diversity in their work on Pecos gambusia was due to differences among populations in the four primary areas where the fish occurred. Maintaining the species in all four areas will ensure maximum allelic diversity and also protect genetic diversity in other members of the communities of which Pecos gambusia are a part. The review by Simons et al. (1989) of plans to downlist the Gila topminnow also demonstrated the importance of maintaining diverse relict populations.

A different approach is needed in conservation programs when most alleles are distributed throughout the range of a species and little genetic divergence exists between regions within its general range. Allendorf and Leary (1988) found that conservation of some subspecies of cutthroat trout requires maintenance of many populations, but other subspecies can be conserved by protecting only a few populations. Their data suggested that some subspecies of cutthroat trout have long evolved separately from others and that taxonomic revision of the species is needed to (1) reflect these distinct evolutionary histories, and (2) allow development of good conservation plans for the diverse groups now considered a single species.

Rearing and reintroducing endangered fishes have generally met with limited success. Ono et al. (1983) discussed several failed attempts to propagate rare fishes. But, for some species such as the endangered cui-ui of Pyramid Lake, Nevada, artificial propagation is the key to survival. State and federal agencies at first seemed reluctant to embrace reintroduction because of its potential to limit other water uses. Nonetheless, reintroductions of rare fishes began in the southwestern United States in the early 1980s. Elimination of the requirement that critical habitat be designated when species were listed and provision of experimental populations by the 1982 ESA amendments initially stimulated increased interest in reintroduction. However, regulations finalized in 1984 were so complicated that many reintroduction plans were disrupted again (Rinne et al. 1986). Williams et al. (1988) examined recovery plans for 39 endangered and threatened U.S. fishes and found that 82% of them called for reintroductions to establish new populations, begin artificial propagation, or set up educational exhibits.

Many fishes have been introduced to North America, primarily to provide sport and food (Courtenay and Kohler 1986), and others have been moved within North American countries to areas outside their native ranges. Effects of such introductions may include habitat alteration, introduction of parasites or diseases, reduction in growth or survival (or even elimination) of native fishes, and changes in community structure. Adverse effects of introduced exotics have contributed to 66% of fish extinctions in North America in the past century (Miller et al. 1989). When an introduction seems the only solution to a problem, guidelines to facilitate sound decision making should be followed (see Chapter 12).

Artificial barriers have been used to separate endemic rare fishes from potential predators, competitors, or sources of hybridization (Ono et al. 1983). Threatened fishes may be the objective of properly managed sport fisheries. Kucera et al. (1985) and Gresswell (1988) described introductions of threatened greenback and Lahontan cutthroat trouts to establish sport fisheries. Sport fisheries should not be developed at the expense of natural populations or habitats. Although exploitation of endangered fishes is controlled as mandated by the ESA, regulations may be necessary to restrict their incidental catch or limit the extent to which they are taken by mistake. Such regulations are often considered a form of people management.

15.5.2 Managing People

Economic, political, and social constraints often have strong bearing on decisions related to managing endangered fishes. The snail darter controversy and other case histories illustrate this point. Conflicts between endangered fishes and construction projects or other human activities with significant economic, political, and social support are certain to continue. Simply gaining funding for research on and management of endangered fishes requires consideration of these factors. Fish and game agencies continue to struggle to support nongame programs with funding gained primarily from license fees. However, the people have mandated management and preservation of all fish and wildlife (Pister 1976). The rapidly increasing demand for fish- and wildlife-oriented recreation will not be met unless agency perspectives continue to move away from emphasis on game species and improved fishing. Fisheries management programs need to recognize the inherent values of native fishes, whether they are harvestable or not. Programs will evolve toward that recognition only if practicing fisheries professionals are willing to broaden their perspectives and fisheries students are exposed to new ideas. Moyle et al. (1986) urged fisheries managers to study recent developments in ecology and support ecologically orientated research. Meffe (1987) stated that education of the public and professionals in values of biotic diversity and conservation of biological resources would help to launch educational programs in fish conservation genetics. The same can be said regarding educational programs in management of endangered fishes in general. O'Connell (1992) stressed the need for scientists to increase their involvement in legislating sound science and develop a bit of political sophistication. Biologists should collaborate more with social scientists, be prepared to present social justification of their work to the public, and quibble less with one another. Public disagreements among scientists and overemphasis of what is not yet known have led society to believe that science is uncertain about some matters that are widely accepted or quite predictable. Johnson and Rinne (1982) stressed the need for cooperation among biologists and management agencies if recovery actions are to succeed, and Carlson and Muth (1989) have argued that rancor needs to be replaced by cooperation and communication in research on and management of Colorado River fishes. All must recognize, however, that the conflict inherent in scientific pursuit of truth cannot be avoided.

15.6 REFERENCES

Allendorf, F. W. 1988. Conservation biology of fishes. Conservation Biology 2:145–148.
Allendorf, F. W., and R. F. Leary. 1988. Conservation and distribution of genetic variation in a polytypic species, the cutthroat trout. Conservation Biology 2:170–184.
Ammerman, L. K., and D. C. Morizot. 1989. Biochemical genetics of endangered Colorado squawfish populations. Transactions of the American Fisheries Society 118:435–440.
Ashworth, W. 1986. The late, Great Lakes. Knopf, New York.
Behnke, R. J. 1992. Native trout of western North America. American Fisheries Society Monograph 6.
Behnke, R. J., and M. Zarn. 1976. Biology and management of threatened and endangered western trouts. U.S. Forest Service General Technical Report RM-28.
Brown, B. 1982. Mountain in the clouds: a search for the wild salmon. Simon and Schuster, New York.
Cadieux, C. L. 1981. These are the endangered. Stone Wall Press, Washington, D.C.
Callicott, J. B. 1986. On the intrinsic value of nonhuman species. Pages 138–172 *in* B. G. Norton, editor. The preservation of species: the value of biological diversity. Princeton University Press, Princeton, New Jersey.
Campbell, F. T. 1982. The Endangered Species Act: facing extinction? Environment 24:6–13, 39–42.
Campbell, R. R. 1988. Rare and endangered fish and marine mammals of Canada: COSEWIC Fish and Marine Mammal Subcommittee status reports IV. Canadian Field-Naturalist 102:81–86.
Carlson, C. A., and R. T. Muth. 1989. The Colorado River: lifeline of the American

Southwest. Canadian Special Publication of Fisheries and Aquatic Sciences 106:220–239.

Contreras Balderas, S. 1987. Threatened and endangered fishes of Mexico. Proceedings of the Desert Fishes Council 16–18:58–65.

Contreras Balderas, S. 1991. Conservation of Mexican freshwater fishes: some protected sites and species, and recent federal legislation. Pages 191–197 in Minckley and Deacon (1991).

Cook, F. R., and D. Muir. 1984. The Committee on the Status of Endangered Wildlife in Canada (COSEWIC): history and progress. Canadian Field-Naturalist 98:63–70.

Courtenay, W. R., and C. C. Kohler. 1986. Exotic fishes in North American fisheries management. Pages 401–413 in Stroud (1986).

Deacon, J. E., G. Kobetich, J. D. Williams, and S. Contreras. 1979. Fishes of North America endangered, threatened, or of special concern: 1979. Fisheries (Bethesda) 4(2):29–44.

Deacon, J. E., and C. D. Williams. 1991. Ash Meadows and the legacy of the Devils Hole pupfish. Pages 69–87 in Minckley and Deacon (1991).

Echelle, A. F., A. A. Echelle, and D. R. Edds. 1989. Conservation genetics of a spring-dwelling desert fish, the Pecos gambusia (Gambusia nobilis, Poeciliidae). Conservation Biology 3:159–169.

Ehrenfeld, D. W. 1976. The conservation of non-resources. American Scientist 64:648–656.

Ehrlich, P. R. 1987. Population biology, conservation biology, and the future of humanity. BioScience 37:757–763.

Ehrlich, P., and A. Ehrlich. 1981. Extinction: the causes and consequences of the disappearance of species. Random House, New York.

Frankel, O. H. 1974. Genetic conservation: our evolutionary responsibility. Genetics 78:53–65.

Goodman, D. 1975. The theory of diversity–stability relationships in ecology. Quarterly Review of Biology 50:237–266.

Gresswell, R. E., editor. 1988. Status and management of interior stocks of cutthroat trout. American Fisheries Society Symposium 4.

Hickman, G. D. 1981. Is the snail darter transplant a success? Pages 338–344 in L. A. Krumholz, editor. The warmwater streams symposium: a national symposium on fisheries aspects of warmwater streams. American Fisheries Society, Southern Division, Bethesda, Maryland.

Humphrey, S. R. 1985. How species became vulnerable to extinction and how we can meet the crises. Pages 9–29 in R. J. Hoage, editor. Animal extinctions: what everyone should know. Smithsonian Institution Press, Washington, D.C.

Johnson, J. E. 1987. Protected fishes of the United States and Canada. American Fisheries Society, Bethesda, Maryland.

Johnson, J. E., and B. L. Jensen. 1991. Hatcheries for endangered freshwater fishes. Pages 199–217 in Minckley and Deacon (1991).

Johnson, J. E., and J. N. Rinne. 1982. The Endangered Species Act and Southwest fishes. Fisheries (Bethesda) 7(4):2–8.

Kikkawa, J. 1986. Complexity, diversity and stability. Pages 41–62 in J. Kikkawa and D. J. Anderson, editors. Community ecology: patterns and process. Blackwell Scientific Publications, Melbourne, Australia.

Kohm, K. A., editor. 1991. Balancing on the brink of extinction—the Endangered Species Act and lessons for the future. Island Press, Washington, D.C.

Kucera, P. A., D. L. Koch, and G. F. Marco. 1985. Introductions of Lahontan cutthroat trout into Omak Lake, Washington. North American Journal of Fisheries Management 5:296–301.

McAllister, D. E., B. J. Parker, and P. M. McKee. 1985. Rare, endangered and extinct fishes in Canada. National Museums of Canada, National Museum of Natural Sciences Syllogeus 54, Ottawa.

Meffe, G. K. 1986. Conservation genetics and the management of endangered fishes. Fisheries (Bethesda) 11(1):14–23.

Meffe, G. K. 1987. Conserving fish genomes: philosophies and practices. Environmental Biology of Fishes 18:3–9.

Meffe, G. K., and R. C. Vrijenhoek. 1988. Conservation genetics in the management of desert fishes. Conservation Biology 2:157–169.

Miller, R. R. 1972. Threatened freshwater fishes of the United States. Transactions of the American Fisheries Society 101:239–252.

Miller, R. R., and E. P. Pister. 1971. Management of the Owens pupfish, *Cyprinodon radiosus*, in Mono County, California. Transactions of the American Fisheries Society 100:502–509.

Miller, R. R., J. D. Williams, and J. E. Williams. 1989. Extinctions of North American fishes during the past century. Fisheries (Bethesda) 14(6):22–38.

Minckley, W. L., and J. E. Deacon, editors. 1991. Battle against extinction—native fish management in the American West. University of Arizona Press, Tucson.

Mlot, C. 1989. The science of saving endangered species. BioScience 39:68–70.

Molles, M. 1980. The impacts of habitat alterations and introduced species on the native fishes of the upper Colorado River basin. Pages 163–181 *in* W. O. Spofford, A. L. Parker, and A. V. Kneese, editors. Energy development in the Southwest: problems of water, fish and wildlife in the upper Colorado River basin, volume 2. Resources for the Future, Washington, D. C.

Mowat, F. 1984. Sea of slaughter. Atlantic Monthly Press, Boston.

Moyle, P. B., and J. J. Cech. 1988. Fishes: an introduction to ichthyology, 2nd edition. Prentice-Hall, Englewood Cliffs, New Jersey.

Moyle, P. B., H. W. Li, and B. A. Barton. 1986. The Frankenstein effect: impact of introduced fishes on native fishes in North America. Pages 415–426 *in* Stroud (1986).

Myers, N. 1979a. The sinking ark: a new look at the problem of disappearing species. Pergamon Press, New York.

Myers, N. 1979b. Conserving our global stock. Environment (Washington, D.C.) 21(9): 25–33.

Myers, N. 1985. A look at the present extinction spasm and what it means for the future evolution of species. Pages 47–57 *in* R. J. Hoage, editor. Animal extinctions: what everyone should know. Smithsonian Institution Press, Washington, D.C.

Myers, N. 1986. Tropical deforestation and mega-extinction spasm. Pages 394–409 *in* M. E. Soule, editor. Conservation biology: the science of scarcity and diversity. Sinauer, Sunderland, Massachusetts.

O'Connell, M. 1992. Response to: "Six biological reasons why the Endangered Species Act doesn't work and what to do about it." Conservation Biology 6:140–143.

Ono, R. D., J. D. Williams, and A. Wagner. 1983. Vanishing fishes of North America. Stone Wall Press, Washington, D.C.

Pister, E. P. 1974. Desert fishes and their habitats. Transactions of the American Fisheries Society 103:531–540.

Pister, E. P. 1976. A rationale for the management of nongame fish and wildlife. Fisheries (Bethesda) 1(1):11–14.

Pister, E. P. 1979. Endangered species: costs and benefits. Environmental Ethics 1:341–352.

Pister, E. P. 1981. The conservation of desert fishes. Pages 411–445 *in* R. J. Naiman and D. L. Soltz, editors. Fishes in North American deserts. Wiley, New York.

Pister, E. P. 1985. Desert pupfishes: reflections on reality, desirability, and conscience. Environmental Biology of Fishes 12:3–11.

Rinne, J. N. 1982. Movement, home range, and growth of a rare southwestern trout in improved and unimproved habitats. North American Journal of Fisheries Management 2:150–157.

Rinne, J. N., J. E. Johnson, B. L. Jensen, A. W. Ruger, and R. Sorenson. 1986. The role of hatcheries in the management and recovery of threatened and endangered fishes. Pages 271–285 *in* Stroud (1986).

SCEPW (Senate Committee on Environment and Public Works). 1983. The Endangered Species Act as amended by Public Law 97-304 (The Endangered Species Act Amendments of 1982). S. Prt. 98-9, U.S. Government Printing Office, Washington, D.C.

Shea, K. 1977. The Endangered Species Act. Environment 19:6–15.

Shen, S. 1987. Biological diversity and public policy. BioScience 37:709–712.

Simons, L. H., D. A. Hendrickson, and D. Papoulias. 1989. Recovery of the Gila topminnow: a success story? Conservation Biology 3:11–15.

Smith, M. H., and R. K. Chesser. 1981. Rationale for conserving genetic variation of fish gene pools. Ecological Bulletins (Stockholm) 34:13–20.

Soltz, D. L., and R. J. Naiman. 1978. The natural history of native fishes in the Death Valley System. Natural History Museum of Los Angeles County Science Series 30.

Soule, M. E. 1985. What is conservation biology? BioScience 35:727–734.

Soule, M. E., editor. 1987. Viable populations for conservation. Cambridge University Press, New York.

Soule, M. E., and K. A. Kohm, editors. 1989. Research priorities for conservation biology. Island Press, Washington, D.C.

Stanford, J. A., and J. V. Ward. 1986a. The Colorado River system. Pages 353–374 *in* B. R. Davies and K. F. Walker, editors. The ecology of river systems. Dr. W. Junk, Dordrecht, The Netherlands.

Stanford, J. A., and J. V. Ward. 1986b. Fishes of the Colorado system. Pages 385–402 *in* B. R. Davies and K. F. Walker, editors. The ecology of river systems. Dr. W. Junk, Dordrecht, The Netherlands.

Stroud, R. H., editor. 1986. Fish culture in fisheries management. American Fisheries Society, Fish Culture Section and Fisheries Management Section, Bethesda, Maryland.

Turner, B. J. 1984. Evolutionary genetics of artificial refugium populations of an endangered species, the desert pupfish. Copeia 1984:364–369.

USDI (U.S. Department of the Interior). 1984. Endangered and threatened wildlife and plants: experimental populations. Federal Register 49:33885–33894.

USFWS (U.S. Fish and Wildlife Service). 1982. President signs amendments to Endangered Species Act. Endangered Species Technical Bulletin 7(11):1, 7–8.

USFWS (U.S. Fish and Wildlife Service). 1990. Box score—listings and recovery plans. Endangered Species Technical Bulletin 15(1):12.

Warner, W. W. 1984. Distant waters: the fate of the North Atlantic fisherman. Penguin Books, New York.

Williams, J. E., D. W. Sada, and C. D. Williams. 1988. American Fisheries Society guidelines for introductions of threatened and endangered fishes. Fisheries (Bethesda) 13(5):5–11.

Williams, J. E., and seven coauthors. 1989. Fishes of North America endangered, threatened, or of special concern: 1989. Fisheries (Bethesda) 14(6):2–20.

Wolf, E. C. 1987. On the brink of extinction: conserving the diversity of life. Worldwatch Paper 78.

Wydoski, R. S., and J. Hamill. 1991. Evolution of a cooperative recovery program for endangered fishes in the upper Colorado River basin. Pages 123–139 *in* Minckley and Deacon (1991).

Yaffee, S. L. 1982. Prohibitive policy: implementing the federal Endangered Species Act. MIT Press, Cambridge, Massachusetts.

Chapter 16

Managing Fisheries with Regulations

RICHARD L. NOBLE AND T. WAYNE JONES

16.1 INTRODUCTION

Successful fisheries management must incorporate the use of appropriate and effective regulations along with other management tools to meet defined management objectives. Although we generally think of regulations as being imposed upon the angler, they are broadly applicable to anyone who might influence fisheries, whether directly or indirectly.

Today's managers must be able to identify fisheries that will respond to either more restrictive or more liberal regulations. In doing so, they must consider ecological data, angler opinions, enforcement ability, judicial systems, and economic impacts.

Regulations are tools to be used in conjunction with other management practices such as population and habitat manipulation. Regulations may seem to be an appealing management tool because, in comparison with population or habitat manipulation, they are relatively inexpensive to implement. Nevertheless, unlike the other management approaches, publicity and enforcement can become major costs.

16.1.1 Responsibilities for Regulating Fisheries

Responsibility for regulating fisheries in public waters rests primarily with state or provincial fisheries agencies. However, other agencies may have authority over habitat protection or alterations that are critical to fish communities. Nevertheless, a major responsibility of fisheries agencies is the review of proposed regulations and applications for permits by potential resource users other than anglers. Depending on locale, fisheries on private property—particularly those in small impoundments—may be exempt from agency regulations. If so, owners and managers are free to establish their own regulations to achieve specific objectives.

The role of the federal government is typically limited to managing endangered species, anadromous fishes, federal waters, and international fisheries. Much of the management of the Great Lakes is undertaken through an international commission which coordinates regulations among the bordering states and provinces (see Chapter 22).

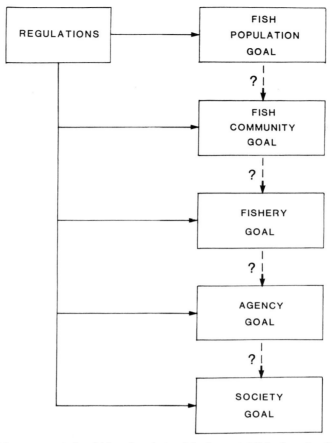

Figure 16.1 Management should be aimed at achieving established goals. (Modified from Noble 1986.)

16.2 ATTAINING OBJECTIVES THROUGH REGULATIONS

Generally, the purpose for regulations is to protect or enhance a fishery for the benefit of the user. Consequently, regulations may be implemented for biological, ecological, sociological, and economic reasons. Sound management should be based upon a strong scientific base, and regulations may be used to obtain information that strengthens that base. In any case, regulations should be enacted to meet specific measurable objectives and established management goals (Figure 16.1; see Chapter 2).

16.2.1 Biological and Ecological Objectives

Regulations have been used to protect fish populations from overexploitation and to distribute the catch among anglers. As fishing pressure increased, stocks normally declined and fishing success decreased. Therefore, regulations became increasingly stringent on the assumption that reduced harvest would prevent overfishing. Concern that fishing was removing too many adult fish, thereby reducing the spawning population to inadequate levels, has frequently been the justification for protection of spawners, particularly during the spawning period when they may be more vulnerable to capture.

As the dynamics of fish populations and communities have become better understood, regulations have been accepted as a means of enhancing, as well as protecting, stocks. Fishing can be regulated to adjust size composition of stocks so more fish are in the desirable size range. Our understanding of predator–prey systems has improved to the point that we can use regulations to manipulate predator size and abundance, thereby having a top-down or trophic-cascade effect (Carpenter et al. 1985) on the prey populations, and sometimes upon populations of undesirable species.

Productivity of fisheries depends on quality of habitat. Although most regulations directed at the user have little effect upon the habitat, a series of regulations, based primarily upon fishery requirements, have been promulgated for protection of aquatic habitat. These regulations primarily affect agricultural, industrial, and municipal users of surface waters and watersheds. Habitat protection or enhancement may be the single most important regulatory process with which the professional fisheries manager will deal.

16.2.2 Sociological Objectives

Sociological objectives are primarily aimed at providing the resource user—typically the angler—with a valid expectation for good fishing success. In most inland waters, the fishery resource is preferentially reserved for recreational fishing; consequently sociological objectives tend to be qualitative. Experiences are measured by satisfaction rather than catch rates or value of catch; therefore, regulations may be aimed at achieving optimal sustainable yield of inland recreational fisheries by incorporating biological, sociological, and economic considerations (Anderson 1975). There is generally an excess of demands upon our resource and a variety of types of anglers to satisfy. Regulations can be used to divide the resource among users; without such regulations a few anglers harvest most of the allowable catch. Regulations may also be used to provide options for anglers according to their desired experiences. Such regulations reduce conflicts between user groups, while simultaneously protecting or enhancing the aesthetic qualities of the fishing experience. Finally, it may be necessary to protect the anglers from themselves. In the fervor of the quest for fishing success, anglers may continue their pursuit even though the circumstances may be precarious.

The professional biologist has traditionally been trained to deal with population dynamics and community ecology, and is readily prepared to address ecological objectives. Today's fisheries manager must also develop an appreciation for sociological objectives.

16.2.3 Informational Objectives

A rather ubiquitous regulation is that most anglers must have a license to fish, at least in public waters. People involved with the sale and transport of fish are also typically licensed. Although sale of such licenses has economic justification, the angler statistics generated by such sales are equally important. In many states, recreational fisheries management programs are supported primarily with federal aid funds. Statistics are used to calculate federal aid allotments, which are allocated partially on the basis of number of licensed anglers. Lists of licensed anglers can define human populations to be surveyed for fishing effort and harvest. In some closely monitored or experimental fisheries, a special regulation may require that each angler report all fish caught or harvested, as well as size information on each fish, to replace costly creel census and fish population sampling.

The exceptions to licensing create severe limitations on the validity of statistics generated. Exemptions for youth, senior citizens, Native American anglers, and local residents differ, making statistical comparisons among jurisdictions and over time difficult. Nevertheless, statistics on licensed anglers can provide important trends which can be useful in evaluating changes such as effects of license fee increases.

16.3   TYPES OF REGULATIONS

Many types of regulations exist, most of which are unfamiliar to biologists entering the profession. Discussions usually emphasize regulations that affect harvest or gear type. Ability to use regulations effectively depends upon an understanding of fish population dynamics and how the populations can be expected to respond under the existing habitat conditions.

16.3.1   Licenses and Permits

Licensing varies with agency but generally includes basic fishing rights for a period of time. The sale of annual licenses provides revenue for the agency and a data base for angler surveys. Additional privilege licenses or stamps may be issued for certain bodies of water or species. The sale of special licenses pays for specific programs which are too expensive for agencies to provide from the sale of basic licenses. In most instances the cost of special licenses is considered a user fee, and proceeds are applied to the specific resource whose use generates the fees. Some special fees are short term, such as a one-time surcharge to pay for hatchery construction or to acquire angler easements.

Most agencies issue short-term licenses to both residents and nonresidents. These provide angling opportunities for users who only wish to try their skill occasionally. Short-term licenses add revenue to the program and provide information to agencies about visitation by nonresidents. In some cases, daily permits are available. Daily stocking rates for intensive put-and-take fisheries, such as Virginia's Big Tumbling Creek trout program, may be determined on the basis of daily permits sold.

Fisheries biologists are subject to regulations that govern their work. Scientific collecting permits are usually required of teachers and researchers who use special equipment to collect fishes or who collect fishes protected by law. Detailed reporting is required so that numbers of each species that are removed for scientific purposes are documented.

16.3.2   Size Limits

Historically, minimum size limits were implemented to prevent overharvest and depletion of fish stocks, and were frequently set to protect juvenile fish until maturity. Today, a variety of size limits, including both minimum and maximum size limits, as well as protected size ranges are employed to maintain favorable fish populations, community structure, and quality of fishing.

Minimum size limits, which prohibit harvest of fish below some specified length, are generally imposed to lower both angling and total mortality in highly vulnerable populations and to reduce exploitation of fish before they reach sexual maturity. For example, Hunt (1970) stated that a minimum size limit for brook trout, if wisely applied, is the best single regulation for preventing excessive harvest. Clark et al. (1981) demonstrated that as size limits increased, the following general relationships occurred for "quality" trout fisheries: (1) harvest of trout in terms of numbers and weight decreased, (2) catch-and-release of trout in numbers and weight increased, (3) total catch in numbers and total yield in weight increased, and (4) numbers of large trout harvested increased. In contrast, Austen and Orth (1988) found that a 305-mm minimum size limit on smallmouth bass in Virginia provided no benefits, apparently due to high exploitation of new recruits, slow growth, and illegal harvest of sub-legal size fish. Minimum size limits have been effective in many fisheries where harvest is high or recruitment is low. However, in some heavily fished, productive waters with minimum size limits, the size composition may shift to a modal size just below the minimum limit, a situation termed stockpiling (e.g., Johnson and Anderson 1974). Intra-specific competition then can cause stunting and reduced recruitment to legal size. In waters where harvest has little impact on total mortality, minimum size limits may be of little value.

Maximum size limits are rare, but in situations with relatively few sexually mature adults or where large numbers of smaller fish exist and the manager intends to increase growth rates, maximum size limits can be beneficial. The Yellowstone National Park's cutthroat trout and Idaho's Snake River sturgeon are two examples in which maximum size limits have been imposed. Care must be taken to clarify implementation of maximum size limits in areas where minimum size limits have been promoted previously. A proposal for a maximum size limit may be misunderstood by constituents, law enforcement officials, and the judicial system. Consider the problem of an agency which for years has promoted minimum size limits for a species and then imposes a maximum size limit to protect a dwindling stock of spawning fish. While this change may have biological validity, judicial systems may be reluctant to convict violators due to a lack of understanding.

Slot limits prohibit harvest from a designated intermediate size range (Anderson 1980). Harvest below the slot may take advantage of a surplus of recruits, and allow additional energy to be channeled into mid-size fish, which then survive and grow out of the protected range and into the harvestable-size range. In addition to the increased harvest of large fish, the fishery provides harvest of some relatively young fish and substantial catch-and-release of protected fish. For slot limits to be successful, harvest of the surplus of young must be high enough to sustain the objective structure of the population (Figure 16.2). The amount of harvest needed depends upon the body of water, species vulnerability to angling, longevity, and rates of reproduction and recruitment. Satisfactory growth rates must be sustained so that fish will not accumulate in the slot range.

Minimum size limits may be combined with slot limits, and may be associated with limits on the numbers of individuals allowed to be harvested. For example, North Carolina established a 178-mm minimum size limit for salmonids on a

Black Bass Slot Limit

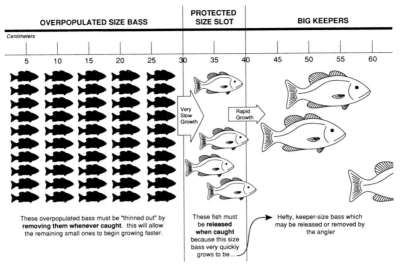

Figure 16.2 Managers should attempt to educate anglers about slot limits, which are often difficult to understand.

stream and imposed a protected slot of 254 to 406 mm, allowing only one fish of the four-fish creel (catch) to be harvested above the slot. Such complexity, although designed to achieve a specific objective, makes compliance difficult. The managers must be certain that there is backing from constituents and law enforcement officials. The general fishing public is often unwilling to remove small fish and is inclined not to return larger, slot-size fish. A regulation of this nature has a reduced chance of accomplishing its objectives without an effective education program and angler compliance.

16.3.3 Creel Limits

Creel limits are usually defined as the allowable fish, either in number or weight, that can be harvested during a particular time period. Generally, this is in the form of number of fish per day. Creel limits may be established by individual species or for species in aggregate such as salmonids, black bass, crappies, and sunfish. Creel limits are instituted so that harvest may be more equitably divided; they also have sociological value, since they give the angler a specific target and the satisfaction of catching a limit. Although creel limits sometimes may be justified as a means of preventing overexploitation, they are generally ineffective in reducing exploitation since few anglers actually harvest their limit.

Creel limits are readily understood by constituents, law enforcement officers, and judicial systems. Daily creel limits have a high degree of voluntary compliance and are easily enforced. Agencies that have personnel and monies available to monitor creel limits on specific bodies of water have a distinct advantage in managing fish populations.

Possession limits, usually one or two times the daily limit, are established at the request of law enforcement officials to aid enforcement of daily creel limits, e.g., when anglers fish all night. Creel limits usually require fish to be kept in a manner so number, species, and length (when applicable) can be determined.

Seasonal or annual creel limits may be developed to limit total harvest by individual anglers, but such regulations are ineffective unless checking stations are available to record catches. Due to lack of enforcement, voluntary compliance is usually nonexistent and such regulations only serve as additional red tape.

Catch-and-release regulations have gained popularity in recent years (Barnhart and Roelofs 1977). The principal objective is to reduce fishing mortality, thereby maintaining high catch rates and increasing the catch of larger fish. Achievement of this objective is dependent upon minimization of hooking mortality, so some gear restrictions may be incorporated. Catch-and-release regulations are also sociological in nature since anglers must fish only for fun. Catch-and-release regulations may be enacted on waters where pesticide accumulations make the fish inedible. Rather than completely closing such waters to fishing, agencies post health advisories and keep the water open to catch-and-release fishing, thereby providing angling opportunities on what otherwise might be a lost resource.

The ideal catch regulation is one that controls total harvest by fish size group, based upon knowledge of the population dynamics of the target species. Such a regulation is termed a quota, which is typically applied on an annual basis. After the established annual quota has been harvested, the fishery for that species is closed to further harvest. Unfortunately most agencies do not possess the necessary personnel to collect up-to-date harvest information from which to determine when a quota is reached. It is also difficult to inform the public of closure. However, on private waters with controlled access, an annual quota is readily applicable and can be effective in preventing overfishing. For vulnerable species, an annual quota can be filled quickly. For example, Ming (1974) reported that annual quotas of largemouth bass in a 9.5-acre public fishing lake were taken in an average of 4.2 days.

16.3.4 Seasons

Although many underexploited fisheries are managed with continuous open seasons, many agencies close seasons during peak spawning activities to protect spawning fish. Closure during this period can have the additional effect of conserving the limited stock for later harvest by a larger clientele. If timing of spawning runs is predictable, it may be possible to open the season after the spawning is well underway. This timing allows sufficient reproduction to sustain the population, while still giving anglers access to a readily available resource. Such an approach has been used successfully for rainbow trout runs from the New York finger lakes, where closure through March protects the early spawners. Their reproduction is adequate to produce enough fingerlings for the limited nursery habitat.

Seasons are sometimes established for the safety of users. In the high latitudes, fishing seasons are frequently closed on lakes and reservoirs during springtime ice breakup. This regulation helps keep people off the ice during this dangerous period. These closures may be species-specific or pertain only to certain areas.

Diel seasons, which comprise closed hours, may also have some justification on the basis of protecting the resource or the user. More likely, however, the justification is enforceability. If fishing is closed during certain hours, such as at night, any activity during that period can more easily draw the attention of enforcement officers.

Although the purpose is not generally recognized, season closures for fisheries are sometimes established in order to divert enforcement manpower to monitor traditionally high periods of hunting activity. On the surface this may appear unreasonable; but if adequate protection cannot be guaranteed for the fishery resource, closure may be the only logical solution to the problem.

16.3.5 Closed Areas

Areas are sometimes closed to fishing, generally to protect the user. Fishing is often prohibited within a certain distance below dams to protect anglers from extreme turbulence and sudden releases of water. In some instances, waters have been closed to fishing to reduce human disturbance in areas where threatened or endangered plants and animals exist. Other areas are closed when contaminant accumulations make the fish inedible. As previously discussed, a prohibition on harvest may make more sense than complete closure for the latter situation.

To a limited extent, areas may be closed to protect or enhance the resource (Hill and Shell 1975). During periods of fish concentration on spawning or nursery grounds, fish may be very vulnerable to harvest. Restriction of anglers from those areas would prove to be beneficial. Management of put-and-take fisheries may include closed areas where stocking is done. Exclusion of anglers from the immediate area of stocking prevents harvest of the concentrated fish and provides opportunity for a steady dispersal into adjacent waters. Pennsylvania has used this method in some of their trout waters. This approach compromises maximization of catch, but extends benefits to a greater number of users.

The seasonal closure of areas to protect certain species brings up the question of whether anglers who fish for other species should be denied access. If enforcement is adequate, a regulation prohibiting harvest of species of concern would appear to be more viable than closure of the area to all anglers.

Closure of areas may be necessary as fish populations are initially established or are reestablished following reclamation. Closure will allow a fish community to develop and to be subsequently managed effectively. Fishing during the initial establishment and growth period, when fish are particularly vulnerable to angling, could upset the community balance and preclude sustained management success.

When managing waters that are open to both recreational and commercial fishing, segregation of the two types of users can be attained by restricting one or both to certain areas. Such zoning can be an effective approach to minimizing conflicts between users, although segregation itself is unlikely to have any direct effects on the resource.

16.3.6 Gear Restrictions

Restrictions on gear are imposed to promote a diversity of fishing experiences and inefficiency of harvest. Gear restrictions, unless used to decrease mortality, should only be considered to meet social, as opposed to biological, objectives in sport fishing. If proper size limits, creel limits, and seasons are established, most restrictions on conventional recreational fishing gears are not usually necessary for biological purposes.

Although gear restrictions are used to decrease user conflicts and promote a

Figure 16.3 Designated areas for specialized fishing can add to the quality of the fishing experience.

diversity of angling opportunities, this approach is not without liability. Implementation of new gear restrictions usually creates a great furor. On the surface it would appear that gear types that result in increased mortality or overexploitation should be regulated. Consider, however, the individual who has been using these gear types for a number of years and suddenly discovers that his favorite fishing spot is off limits unless he changes gear. Anticipated public resistance does not mean that managers should not consider gear restrictions. A manager has the responsibility to meet social, as well as biological objectives. The important point is to realize the difference between the two.

Fly-fishing is a form of angling that has gained widespread popularity in recent years. Because of some special requirements of fly-fishing, especially adequate spatial segregation from other anglers, some areas are designated for fly-fishing only (Figure 16.3). This type of regulation must be considered social in nature, and provides no biological gain (Latta 1973). Public demand may be sufficient to warrant gear restriction; however, such a regulation should be weighed against diversification of fishing to determine whether angling opportunity would increase or decrease.

Gear restrictions are also applicable when a large number of fish are to be released after capture, such as when slot limits, minimum size limits, or catch-and-release regulations apply. Single, barbless hooks are frequently required on the premise that minimal physical damage is done to the fish, thereby increasing their chances of survival to be caught again. Multiple points on hooks (e.g., treble hooks) may also be prohibited in fisheries where high catch rates of protected fish are anticipated. The value of such regulation should not be overestimated. Studies have shown that hooking mortality is minimal unless fish are deeply hooked.

Number of hooks per line and number of lines tended by each angler are commonly limited. Likewise, regulations may require continuous tending of lines. These regulations reduce losses of hooked fish, and are consistent with an ethic of not wasting or maiming our fishery resources.

Restriction on types of bait can also be considered a form of gear regulation. Bait restrictions are implemented to reduce hooking mortality, to decrease angling efficiency, and to prevent contamination of the habitat by unwanted species. A widespread bait restriction to reduce vulnerability is prohibition of the use of salmon eggs. During spawning runs, this bait is so effective for salmonids that its use is considered to give anglers an unfair advantage. Use of exotic and otherwise undesirable fish is prohibited to prevent their spread into new areas. Despite the effectiveness of species such as common carp, goldfish, blue tilapia, and sea lamprey as live bait, their use can only contribute to further expansion of the ranges of these potentially deleterious fishes. For different reasons, use of game fish or other recreational species as bait is normally prohibited. Although it is conceivable that collection of sport fishes for bait could have detrimental effects on stocks, their prohibition as bait makes enforcement of creel and size limits much more explicit.

Inland fisheries are primarily reserved for recreational rather than commercial use. When commercial fishing is allowed, regulations carefully limit the gears and harvest (Fritz and Wight 1986). If nets are used, mesh sizes are regulated to minimize catches of sport fishes. Agencies sometimes contract with commercial fishers to harvest underexploited species. These limited-entry fisheries are becoming increasingly popular, and some agencies have gone exclusively to this program for commercial fishing on inland waters. Although studies have shown that the incidental catch of sport fishes is negligible, public reaction is typically negative if game fish are caught by commercial fishers.

16.3.7 Reporting

Total catch and effort data are valuable to fisheries managers and can be provided by the angler. Nevertheless, regulations that mandate anglers to keep catch and effort records are tenuous at best. No matter what the objective, the effectiveness of such a regulation depends upon voluntary compliance unless the agency has the personnel and funds for enforcement. As an example, all Atlantic salmon anglers in the Canadian provinces are required to report catch and effort information. In order to obtain complete information, the Canadian authorities send follow-up letters to those anglers who did not report their catch as required. Even if total response is obtained, the accuracy of the information relies ultimately upon the integrity of the angler.

Record-keeping regulations appear to have their greatest effectiveness for waters with a single point of access. At that location, the value of total catch records can be publicized to encourage accurate reporting. Reporting forms should be readily available and a secure, well-marked depository for reports should be established. Single access points that have limited access hours also can be staffed, either by a creel clerk or, as is frequently the case with single access sites, by a concessionaire.

In some fisheries, record-keeping in the form of diaries kept by selected anglers may provide valuable information on annual trends in size composition and catch rates. Such records have the benefit of measuring changes among a fixed subset of anglers who use similar fishing techniques each year. Integrity of these anglers can be anticipated to be such that more reliable data are acquired than can be expected from the general public. Individual record-keeping has been used frequently by private pond owners and fishing clubs and can be a cost-effective method of assessment. It is important to communicate regularly with the record keepers to show them the results of their efforts and to reinforce the value of their data.

Throughout most of the United States, procedures for voluntary registration of state record fish have been established. Additional citation programs for trophy fish are also common (Quinn 1987). Regulations governing the procedures to follow when submitting an application for a state record fish or a citation fish are usually simple. Most agencies require identification of the fish by a professional and verification of weight on certified scales. This type of regulation is a service to the angler; standardization is designed to control cheating. In addition, the program may provide valuable information to the fisheries agency.

16.3.8 Regulation of Commercial Ventures

Since inland fisheries are primarily recreational, commercial fisheries are highly regulated, if not totally prohibited, by fisheries agencies. Gear restrictions are set to minimize catch of sport fishes; size limits, seasons, and designated areas are also likely to be stringently regulated. A regulation that prevents sale of sport fish species, whether caught by recreational or commercial fishing, is frequently implemented to keep recreational fishing from becoming commercial, and to limit commercial fishing to nonsport fishes.

Federal treaty rights for Native Americans frequently exempt tribal fisheries from state agency regulations. Commercial harvest from fisheries which are otherwise designated by agencies as recreational has been a source of conflict between recreational anglers and Native Americans. This is a good illustration of the conflict that can arise when sociological, economic, and ecological objectives are incompatible.

Because of the economic values of fish beyond those associated with commercial fishing, commercialization of fish production and recreational fishing has occurred. To avoid conflicts with conventional fisheries and to facilitate better enforcement, agencies may develop specific regulations on commercial activities such as fishing tournaments and aquaculture.

One form of commercialization of sport fishing has resulted in what can be described as tournament mania. Organizations have developed around the concept of competitive fishing and conduct a regular series of tournaments with high stakes, which have ranged from large cash prizes to expensive fishing equipment and accessories. Private businesses and civic organizations have used fishing tournaments as promotional events. The debate over the effects of fishing tournaments on fish communities is ongoing. Some agencies have enacted regulations that require tournaments to be permitted. However, most states do not regulate tournaments in any fashion (Duttweiler 1985). Many well-organized tournament sponsors develop and impose their own regulations governing handling of fish and release of fish after check-in. The intent of such regulations is good; however, only the highly organized tournaments are involved. Tournaments between local clubs and among groups of anglers typically have few self-imposed restrictions. Because these tournaments are so numerous, spontaneously organized, and rarely publicized, their regulation by agencies is impractical. Until a feasible method is found to regulate this type of tournament, the effectiveness of tournament regulations is in doubt. However, without regulation, this commercial use of public fisheries resources will continue to be a source of conflict with the sport-fishing public.

Most agencies require commercial production and distribution operations to be licensed. This provides information regarding the number of hatcheries and the species raised, and allows monitoring to prevent diseases and undesirable species from entering the area. Generally, regulations prohibit the sale, transport, and release of any undesirable species, nonindigenous strains, or any fish not certified to be free of pathogens. Further regulations govern quantity and quality of water intake and release by production facilities. As demand for stocking private waters for recreation has grown, private production and distribution of sport fishes has increased. Licensing of commercial operations which deal with sport fishes minimizes enforcement problems related to discrepancies in possession and size limits otherwise applicable to anglers. Some complications have arisen as a result of bans on the sale of sport fishes, intended primarily to protect wild stocks. When aquaculture enterprises have gotten into production and sale of sport fish species, either for stocking private waters or as foodfish, special provisions have become necessary.

16.3.9 Regulation Evaluation

Effects of regulation changes should always be evaluated. The extent of the evaluation may vary from a complex research study to simply polling law enforcement officers or constituents on the effectiveness of the regulation. The results of evaluations, whether positive or negative, should be made available to the public. The manager should have valid data for periods prior to and after implementation to evaluate the effectiveness of the change. Since results may not be immediate, post-implementation data should be collected over an ample time period for effects to be detected.

Regulations should always be evaluated relative to the specific biological, sociological, informational, or economic objectives which were established. Evaluation methodologies should be developed and incorporated into the initial action plan when the regulatory action is designed. Too often an agency's decision to evaluate is an afterthought; consequently there is likely to be little comparative data collected prior to the new regulation.

16.4 REGULATIONS FOR SPECIFIC FISHERIES

As levels of governmental agencies became established and acquired regulatory authority, the number and complexity of regulations increased. With this came a shift from specific regulations to more general regulations enacted to cover entire systems. This generic approach had two advantages: it was easily administered and easily understood by constituents. However, failure to tailor regulations to specific situations and objectives often led to resources being managed for mediocrity. In contrast, there were resources which, when placed under more restrictive regulations, responded positively. Conversely, there were bodies of water where generalized regulations were too restrictive and negative results were obtained. Consider if a statewide minimum size limit was imposed on salmonids in infertile waters where density was high and growth rates were slow. Such a regulation would essentially prevent harvest.

The Norris Reservoir study (Eschmeyer 1945) provided the impetus toward more liberal regulations. Agencies began removing closed seasons on reservoirs and dropping restrictions on other types of water (Fox 1975; Redmond 1986). This trend continued into the early 1960s, when acquisition of data on fish populations and harvest substantially improved due to widespread use of assessment techniques such as electrofishing, tagging, and creel census. Subsequent experimentation with slot limits, higher minimum size limits, and more restrictive creel limits led to the imposition of water-specific and species-specific regulations.

16.4.1 Coldwater Fisheries

Regulations are particularly applicable to oligotrophic, cold waters, where exploitation may exceed annual production. Consequently, we see the most widespread use of them in such habitats. Due to low fertility and cold temperatures, growth rates are slow, thereby extending the time to maturation and recruitment to the fishery. Furthermore, low productivity and standing crops of fish make their populations subject to impacts of fishing. Therefore, regulations are frequently promulgated to protect coldwater fisheries from overexploitation. In contrast, many coldwater streams are in remote mountainous areas which are not readily accessible to anglers. Recognizing this, some agencies have liberalized regulations for such areas.

Management of coldwater streams which are marginal for trout or are subject to excessive fishing pressure frequently includes put-and-take or put-grow-and-take stocking, which may require management through regulations. In these cases of more intensive management, regulations are usually set with the objectives of maximizing catch and dividing the resource among users.

It is also in cold waters that much emphasis has been placed upon diversification of fishing experiences. Therefore, regulations are frequently established for sociological needs. Catch-and-release and fly-fishing-only are commonly used regulations in these waters.

Coldwater habitats are so sensitive to disturbance that emphasis is placed on strict regulations governing habitat degradation and modification. In lakes, discharge regulations are imposed to minimize eutrophication, which could easily upset the fragile food chains. In streams, regulations emphasize minimal instream disturbances of flow and substrate, as well as limited riparian disturbances by forestry practices, road building, and livestock grazing.

16.4.2 Coolwater and Warmwater Streams and Rivers

With some exceptions, coolwater and warmwater streams are seldom over-exploited as recreational fishing resources due to limitations in access. Consequently, regulations to protect or divide the resources are unnecessary. These waters are often opened to commercial fishing, even for commercial harvest of sport fishes, and regulations are apt to be imposed to minimize conflicts.

One exception to liberalized regulations is the smallmouth bass fishery in streams. Much like trout, smallmouth bass have fairly stringent habitat requirements. As a result, a variety of size and catch limits are imposed for this species. Because of slow growth rates in some areas, regulations have often been counterproductive, except for providing increased numbers of strikes due to release rather than harvest of fish.

Another exception to liberal regulations in stream fisheries applies to anadromous stocks such as striped bass. Until recently, liberalized regulations applied to anadromous fishes in inland waters; the spawning runs allowed an opportunity for inland exploitation of a perceived inexhaustible, ocean-produced resource. However, most fisheries have declined due to environmental degradation of water quality, spawning grounds, and migration routes, and overharvest of stocks while at sea. Stringent harvest regulations are now in place for most anadromous fisheries; these apply to both recreational and commercial users, although division of the limited resource continues to be a major controversy. Included in this controversy are the rights of Native American anglers to exploit a fishery under treaty rights which are not subject to limitation by state and provincial agencies.

Decisions on habitat regulations commonly confront the stream fisheries manager. Instream industrial activities such as gravel dredging or diversion of irrigation waters are subject to regulation on the basis of fisheries and water quality. With the widespread construction of dams on rivers and streams, instream flow requirements have increasingly become recognized as a basis for regulation of consumptive users of lotic water resources. Although agencies other than the state or provincial fisheries management agency will likely be responsible for promulgation of habitat regulations, the biologist must be prepared to contribute objectively and quantitatively to development of such regulations on the basis of impacts upon fisheries.

16.4.3 Coolwater and Warmwater Lakes

Natural lakes with moderate productivity, which characterizes most coolwater and warmwater lakes, have highly evolved and complex communities. Consequently, the principal approach to management of these fisheries has been through regulations. A common objective is to prevent overharvest of game fishes, particularly top carnivores. However, as eutrophication and overharvest have altered community dynamics, wide annual fluctuations in harvest have become more common. These fluctuations are usually associated with variations in recruitment due to variations in year-class strength. Effective use of regulations under such conditions is a challenge to the fisheries manager; although restrictive regulations may be desirable to rebuild depleted stocks, flexibility to change regulations is needed to allow harvest of surpluses when they occur.

Although most management and regulatory attention in lake fisheries has been directed at game fishes, effort is also directed at pan fish such as yellow perch, crappies and sunfish, which frequently serve as prey for game fishes. Until recently, liberal harvest regulations were applied to these fishes. As fishing pressure has increased and data on exploitation rates have been obtained, the need for more restrictive harvest regulations have been recognized.

Because there may be more than one piscivore of interest to sport fisheries, it may be desirable to direct management at one species to optimize its fishery. In such cases, stringent regulations are likely to be imposed on the primary fishery while liberal regulations govern the secondary fishery. An example is New York's liberalized regulations on walleye in Chautauqua Lake, where primary interest is in a trophy muskellunge fishery (Mooradian et al. 1986).

16.4.4 Reservoirs

Reservoir management has been characterized by liberal harvest regulations (Redmond 1986). In contrast to natural lakes, reservoirs have poorly evolved predator–prey systems and frequently have large populations of potential prey such as gizzard shad (Noble 1986). In addition, some reservoirs are subject to extreme habitat variations such as water level fluctuations, which limit the effectiveness or applicability of regulations as a management tool. As fishing effort has increased, exploitation rates of some reservoir sport fishes have been excessive, especially during the early post-impoundment years and on smaller reservoirs. Recent years have brought more restrictive harvest regulations with the objective of reducing fishing mortality and improving size structure. In particular, size and catch limits have been implemented in reservoirs with limited recruitment in attempts to improve size structure, thereby providing higher quality fish to the creel. Missouri has implemented such regulations on crappie to accomplish this objective.

Because of their size and dendritic shape, which results from impounding stream systems, reservoirs may offer an opportunity for achieving varied fisheries objectives through spatial within-reservoir zoning. Closed areas have been shown to serve as nurseries which can supply catchable-size fish to open areas, while preventing lakewide overfishing. Likewise, it should be possible to apply different regulations to different areas, thereby creating differing size structures and varied fishing experiences. Occasionally lakes have been zoned to achieve specific management objectives. An example is Missouri's Table Rock Lake, where one arm of the reservoir is managed with a more restrictive regulation for crappie. Such an approach is applicable only to relatively sedentary species.

16.4.5 Private Ponds

Privately owned ponds are conducive to management with harvest regulations. Fish populations of small impoundments are particularly susceptible to harvest (e.g., Funk 1974). However, small size and control over users allows implementation of a variety and complexity of regulations. These can be closely integrated with other forms of management to meet well-defined objectives. Although enforcement through conventional agency personnel is limited, if at all applicable, the limited fishing constituency of any pond can be readily reached with educational programs and materials. This control ensures a high degree of compliance to self-imposed regulations (Ashley and Buff 1988). If regulations are developed with the objective of adjusting size structure of fish populations through fishing, adequate fishing pressure must be exerted. In large ponds with limited access, it may be necessary to promote a directed fishery to attain adequate fishing pressure. Likewise, total control of access and compliance with special regulations is necessary. A few uninvited guests or renegade members of a fishing club can quickly offset the accomplishments achieved through the dedicated efforts of the remaining anglers.

Increased demands for urban fishing opportunities have been addressed through a variety of programs (Allen 1984). Among these is intensive put-and-take fishing in small, closely controlled ponds. Although there is no need for regulations to protect the stocks under these highly artificial conditions, special regulations can serve important sociological purposes. Limits on gear can improve human safety and minimize injury of unharvested fish. Regulations may include sociological zoning, providing independent fishing opportunities for youth, senior citizens, handicapped individuals, and the general public. Such highly structured regulations are unlikely to be possible in any other setting.

16.5 REGULATORY PROCESS

The regulatory process varies with each agency, yet follows a generalized procedure. In response to an identified objective, a proposal is formulated. A subsequent period of review by the biological staff will include evaluation of effects of the regulation on the fish stock and the fishery, and feasibility of implementation and enforcement (Figure 16.4). This may involve a formal presentation or review in written format. The next step varies by agency but some period of public comment, either before or following evaluation by a board of

Figure 16.4 Effective enforcement is critical to the success of fishing regulations. (Photograph courtesy of North Carolina Wildlife Resources Commission.)

Figure 16.5 Public hearings offer the opportunity to learn angler reactions to proposed regulation changes and simultaneously to inform them of the anticipated benefits. (Photograph courtesy of North Carolina Wildlife Resources Commission.)

commissioners or directors, is usually established. In some agencies ratification by the legislature is necessary depending upon the nature of the regulation.

An important factor ignored by many biologists is the public comment period (Figure 16.5). Regulations, especially new or radically different ones, often meet with opposition regardless of their merit. The manager must be prepared to sell an idea to peers, the agency governing board, and constituents. In many instances, this requires a carefully planned educational process in which the manager must play the key role. The manager must anticipate questions and concerns that will be expressed by the public and others so that sound logical answers can be readily provided. Depending upon the awareness level of the audience, a number of different versions of the same presentation may be required. Passage of regulations in many instances depends on the "selling" done by the biologist regardless of the validity or merit of the regulation (see Chapter 3).

The responsibility of today's professional does not end with passage of the regulation. Only by working with the public for acceptance and compliance with the regulation can the objective be met. Professionals cannot rely solely on law enforcement officers to ensure compliance with the enacted regulation. Acceptance and compliance evolve after passage, so the manager may need to continue to work for years toward that goal.

This brings us to one of the most complicated aspects of regulations. Unpopular regulations have been proposed and enacted many times. In many instances, these were necessary to protect the resource; however, cases exist where the regulations were of dubious merit. The following scenario will serve as an example.

In a statewide angler opinion survey a large percentage of largemouth bass anglers indicated they would prefer to catch larger bass, even if it meant a decrease in creel limits and an increase in size limits. You have a reservoir which meets all the biological requirements and you propose a regulation to increase the minimum size limit and reduce the creel limit. This proposal meets with opposition from the anglers who regularly fish the reservoir and are satisfied with the *status quo*. Should that regulation be enacted? The professional will face issues like this constantly. The answer is found only by evaluating each case on its own merits and with the knowledge that mistakes can and will be made.

16.6   FUTURE ROLE OF REGULATIONS

There can be little doubt that the role of regulations in fisheries management will increase in the future. As reservoir construction diminishes, we can expect intensified fishing effort on the fixed inland aquatic resources. In addition, the efficiency of angling will continue to increase as more sophisticated methods are employed. Large, powerful boats now propel many anglers to the most remote coves of our biggest lakes and reservoirs. Fish-finders not only chart the bottom topography, but also the distribution of fish. Physical conditions such as temperature, oxygen, pH, and color are being monitored by some anglers to increase their probability of success. Without appropriate regulations, overharvest will occur on most of our major fisheries. Likewise, without appropriate regulations, the resource would be allocated inadvertently to the best-equipped individuals rather than divided among the various users.

Pressure on habitat will also continue to increase. The public sector demands water for many uses, not just recreation. Although legislation protects our water from pollution and our waterways from alteration more than ever before, water withdrawals will continue to become demanding influences on our fisheries. The fisheries manager must be able to provide objective inputs to decision making on the basis of impacts on fisheries. In particular, effects of water flow and water level fluctuations will need to be considered. With better knowledge and some imagination, regulations may be developed which allow net benefits to fisheries from industrial and municipal use of water, rather than primarily negative impacts.

To meet the increased demand for fishing, both in quantity, quality, and diversity, it will be necessary to more closely tailor regulations to specific situations. That will mean abandonment of many regional regulations in favor of resource-specific regulations which are based upon such factors as fish population dynamics, characteristics of the habitat, proximity to population centers, other fishing alternatives in the area, angler preferences, and enforcement limitations (e.g., Brousseau and Armstrong 1987). The benefit of this approach will be establishment of more precise management objectives and higher chances of attaining those objectives through integration of all available management tools. There is general resistance to site-specific regulations from the public and those responsible for enforcement, who understandably prefer the simplicity of standardized regulations. Nevertheless, the major obstacle is likely to be personnel to adequately assess fisheries to determine management needs. Site-specific regulations already are common in highly populated states with popular trout fisheries. With the recent recognition that overexploitation of warmwater fisheries can commonly occur, and the increasing interest in providing greater diversity and better quality of fishing, greater emphasis on site-specific regulations is justified. However, in those areas where regional standard regulations have been employed, public education will be an integral part of the development of site-specific regulations.

Regardless of whether new regulations are being proposed or old ones defended, the fisheries manager will be more accountable. Justification for regulations will be measured increasingly by their anticipated effectiveness in the achievement of defined objectives in management plans, where regulations are incorporated with other management approaches. Although biologists must start from the standpoint of what is good for the resource, they must be cognizant of the fact that fisheries are often a part of a multiple-use resource. Therefore, public decisions are made according to the political process. Consequently, the manager will need to be able to justify and evaluate sociologically as well as biologically based objectives. Furthermore, attention will need to be given to public education, thereby developing a better appreciation of objectives which will, in turn, raise the level of accountability.

The process by which regulatory change is developed is through proper assessment of need and opportunity by the fisheries manager. The process by which implementation is effective is through understanding and compliance by the user and the ability of the law enforcement officers to enforce the change.

16.7 REFERENCES

Allen, L. J., editor. 1984. Urban fishing symposium, proceedings. American Fisheries Society, Fisheries Management Section and Fisheries Administrators Section, Bethesda, Maryland.

Anderson, R. O. 1975. Optimum sustainable yield in inland recreational fisheries management. Pages 29–38 in P. M. Roedel, editor. Optimum sustainable yield as a concept in fisheries management. American Fisheries Society Special Publication 9.

Anderson, R. O. 1980. The role of length limits in ecological management. Pages 41–45 in S. Gloss and B. Shupp, editors. Practical fisheries management: more with less in the 1980's. American Fisheries Society, New York Chapter, Ithaca, New York.

Ashley, K. W., and B. Buff. 1988. Pond owner utilization of management recommendations made by state fishery personnel. Fisheries (Bethesda) 13(6):12–14.

Austen, D. J., and D. J. Orth. 1988. Evaluation of a 305-mm minimum length limit for smallmouth bass in the New River, Virginia and West Virginia. North American Journal of Fisheries Management 8:231–239.

Barnhart, R. A., and T. D. Roelofs, editors. 1977. A national symposium on catch-and-release fishing. Humboldt State University, Arcata, California.

Brousseau, C. S., and E. R. Armstrong. 1987. The role of size limits in walleye management. Fisheries (Bethesda) 12(1):2–5.

Carpenter, S. R., J. F. Kitchell, and J. R. Hodgson. 1985. Cascading trophic interactions and lake productivity. BioScience 35:634–639.

Clark, R. D., Jr., G. R. Alexander, and H. Gowing. 1981. A history and evaluation of regulations for brook trout and brown trout in Michigan streams. North American Journal of Fisheries Management 1:1–14.

Duttweiler, M. W. 1985. Status of competitive fishing in the United States: trends and state fisheries policies. Fisheries (Bethesda) 10(5):5–7.

Eschmeyer, R. W. 1945. The Norris Lake fishing experiment. Tennessee Department of Conservation, Division of Fisheries, Nashville, Tennessee.

Fox, A. C. 1975. Effects of traditional harvest regulations in bass populations and fishing. Pages 392–398 in H. Clepper, editor. Black bass biology and management. Sport Fishing Institute, Washington, D.C.

Fritz, A. W., and H. L. Wight. 1986. Commercial fishing as a reservoir management tool. Pages 196–202 *in* Hall and Van Den Avyle (1986).

Funk, J. L., editor. 1974. Symposium on overharvest and management of largemouth bass in small impoundments. American Fisheries Society, North Central Division, Special Publication 3, Bethesda, Maryland.

Hall, G. E., and M. J. Van Den Avyle, editors. 1986. Reservoir fisheries management: strategies for the 80's. American Fisheries Society, Southern Division, Reservoir Committee, Bethesda, Maryland.

Hill, T. K., and E. W. Shell. 1975. Some effects of a sanctuary on an exploited fish population. Transactions of the American Fisheries Society 104:441–445.

Hunt, R. L. 1970. A compendium of research on angling regulations for brook trout conducted at Lawrence Creek, Wisconsin. Wisconsin Department of Natural Resources, Research Report 54, Madison.

Johnson, D. L., and R. O. Anderson. 1974. Evaluation of a 12-inch length limit on largemouth bass in Philips Lake, 1966–1973. Pages 106–113 *in* J. L. Funk, editor. Symposium on overharvest and management of largemouth bass in small impoundments. American Fisheries Society, North Central Division, Special Publication 3, Bethesda, Maryland.

Latta, W. C. 1973. The effects of a flies-only fishing regulation upon trout in the Pigeon River, Otsego County, Michigan. Michigan Department of Natural Resources, Research Report 1807, Ann Arbor.

Ming, A. 1974. Regulation of largemouth bass harvest with a quota. Pages 39–53 *in* J. L. Funk, editor. Symposium on overharvest and management of largemouth bass in small impoundments. American Fisheries Society, North Central Division, Special Publication 3, Bethesda, Maryland.

Mooradian, S. R., J. L. Forney, and M. D. Staggs. 1986. Response of muskellunge to establishment of walleye in Chautauqua Lake, New York. Pages 168–175 *in* G. E. Hall, editor. Managing muskies, a treatise on the biology and propagation of muskellunge in North America. American Fisheries Society Special Publication 15.

Noble, R. L. 1986. Predator–prey interactions in reservoir communities. Pages 137–143 *in* Hall and Van Den Avyle (1986).

Quinn, S. P. 1987. The status and usefulness of angler recognition programs in the United States. Fisheries (Bethesda) 12(2):10–16.

Redmond, L. C. 1986. The history and development of warmwater fish harvest regulations. Pages 186–195 *in* Hall and Van Den Avyle (1986).

COMMON MANAGEMENT
PRACTICES

Chapter 17

Coldwater Streams

J. S. GRIFFITH

17.1 CHARACTERISTICS OF COLDWATER STREAMS

Although coldwater streams come in a variety of sizes and shapes and hold numerous fish species, for the purposes of this chapter they will be defined as streams where game fish populations are predominantly salmonids. These are streams that maintain salmonid populations by sustaining spawning and rearing, rather than just serving as migratory pathways or receptacles for hatchery-produced fish.

Fluvial salmonids in North America may be divided into three categories:

1. Anadromous species that move out of streams and into the sea (chum salmon and pink salmon) or into lakes (sockeye salmon) immediately after emergence (see Chapter 23).
2. Anadromous species that spend 1 to 4 years in streams before moving to the sea. These include chinook salmon, coho salmon, and Atlantic salmon, as well as anadromous races of rainbow trout (steelhead), cutthroat trout, brown trout, brook trout, and Dolly Varden.
3. Species that spend their entire lives in streams and large river systems, such as some populations of bull trout and resident races of rainbow trout, cutthroat trout, brown trout, and brook trout.

All of these fishes, except brown trout, are native to North America. In addition to those mentioned above, other species, such as Arctic char and Arctic grayling, may be locally important. The mountain whitefish is abundant in western rivers, but differs from other salmonids in life history characteristics and is not considered in this chapter.

Coldwater fish communities in streams are the product of a highly dynamic environment. Beginning at the headwaters and moving downstream, the physical variables within most stream systems provide a continuous gradient of conditions such as width, depth, velocity, and temperature. The river continuum concept (Vannote et al. 1980) has been proposed as a framework for describing the structure and function of communities along a river system and their change in response to fluvial geomorphic processes (see Chapter 19).

Salmonids are present at predictable positions along a stream continuum. Their environmental requirements vary considerably among species and life stages, but a defined range of tolerable physical, chemical, and biological characteristics exists. Water temperature may be the primary factor dictating the lower limit of salmonid distribution in terms of elevation and latitude. Although some populations—like those in the Firehole River in Yellowstone National Park—may spend periods of time at 26°C, summer stream temperatures for most coldwater fishes do not exceed 22°C. The scope for activity (the difference between active metabolism and standard metabolism that indicates the amount of oxygen physiologically available for activity) is greatest for wild rainbow trout at 20°C and for hatchery cutthroat trout at 15°C. Growth for most salmonids rapidly declines above 20°C. Dissolved oxygen concentration, typically near saturation in unpolluted coldwater streams, should remain at a minimum of 8 mg/L for rearing and 10 mg/L for egg and larval development.

Flow in coldwater streams may range from a few liters per second to hundreds of cubic meters per second. Although salmonids may be found in streams that range from first-order to sixth-order or larger, salmonid abundance and diversity in the Salmon River, Idaho, was greatest in fourth- and fifth-order streams (Platts 1979). A similar relationship holds for stream gradient. Despite the fact that some salmonid populations may exist in streams with gradients exceeding 20%, a range of 0.5 to 6.0% is typical.

As stream size increases, the influence of riparian vegetation declines. Shading of the stream lessens, and the relative contribution of allochthanous input in the form of litterfall (and terrestrial insects) is reduced. Water temperatures usually increase as stream size increases, providing more thermal units for production. The maximum diel temperature range increases from a few degrees in second-order headwaters to a maximum of about 10°C in about fifth-order streams, and then the range decreases with further increases in stream order.

With these concepts in mind regarding physical and biological changes along the length of a stream, it is important to consider some basic differences between one stream system and another. Fluvial environments vary greatly in their basic nature, and the contrast, for example, between a glacial stream and one that is spring fed is striking. Conditions at high latitudes are marked both by extreme photoperiods and temperatures, and as such these streams represent special cases.

One key characteristic of a coldwater stream is its flow pattern (Poff and Ward 1989). An annual hydrograph (Figure 17.1) depicts average daily flows as a time series and provides information regarding the extent and timing of minimum, maximum, and average stream flows. Three hydrographs show a range of typical patterns. For a high-elevation, unregulated stream, as much as 75% of the precipitation in the drainage basin may come from snow. Its hydrograph has a peak flow in spring that may be from ten to several hundred times that of the average summer low flow. A spring-fed stream may show a much flatter hydrograph. Coastal systems, such as those on the west side of the Cascade Mountains in Oregon and Washington, may be characterized by a hydrograph that reflects 250 cm of rainfall, most of which falls between October and March. Another type of hydrograph may be typical of a regulated stream below an impoundment, where flows are drastically modified by reservoir storage.

The geology of a drainage basin determines water chemistry parameters such as dissolved solids and conductivity. In a similar manner, the geomorphic characteristics of the watershed dictate critical features of a stream, such as its ability to transport sediment. In a process such as the transport of sediment, there are eight major interacting variables: channel width, depth, slope, roughness of bank and bed, discharge, form, sediment concentration, and sediment size. A change in any

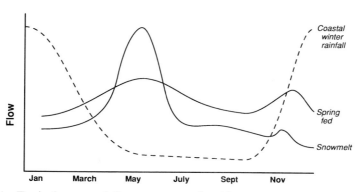

Figure 17.1 Typical seasonal flow patterns of coldwater streams. The coastal winter rainfall hydrograph represents the Alsea River, Oregon; the spring-fed hydrograph, Silver Creek, Idaho; and the snowmelt, Falls River, Idaho, where it exits Yellowstone National Park.

of these factors may bring a response from the others as the system moves toward a new equilibrium. For example, channel substrate systematically changes with gradient (Figure 17.2).

The transport and deposition of fine sediment have particularly important consequences for fluvial salmonids. Excessive turbidity may impair the health of fish, and levels of sediment that heavily embed the substrate reduce the survival of embryos and of juvenile salmonids that overwinter in the substrate.

Interactions with other fish species have shaped the niches of coldwater salmonids. Biologic diversity of native species in coldwater streams ranges from simple systems with limited diversity (such as in some western streams, where only six to eight species of *Cottus, Catostomus,* and *Rhinichthys* are found with one or two salmonids) to more elaborate biologic communities typical of streams in eastern North America (with perhaps 20 species of cyprinids, cottids, catostomids, and darters interacting with salmonids).

17.2 DEMANDS FOR COLDWATER STREAM FISHERIES

Coldwater streams are valuable resources that are in demand by a number of current and potential users, some of whose interests conflict. Streams are highly

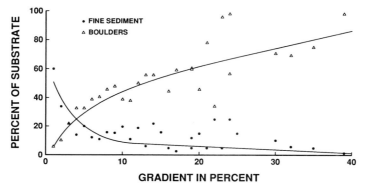

Figure 17.2 Change in stream channel substrate with gradient in tributaries of the Salmon River, Idaho. (Data from Platts 1979.)

valued by anglers, and are preferred over lakes by the majority. In Idaho, for example, a recent survey indicated that twice as many trout anglers preferred stream fishing over lake and reservoir fishing, even though streams total only about 20% of the state's surface water.

The economic value of stream fisheries is significant. The net economic value (over and above actual expenditures) of a day of stream fishing in Montana in 1985 was $102. Angling effort is predicted to reach an average of 300 h/hectare for public lakes, streams, and reservoirs in the contiguous 48 states by the year 2000 (McFadden 1969). If so, each hectare of coldwater habitat (a 400-m length of a 25-m-wide stream) will provide $7,500 of net economic value per year. Some popular trout fisheries now provide more than 2,500 h of angling per hectare each year, and effort in some has doubled in the past decade. Many land-use practices directly compete with salmonid fishes and anglers for coldwater stream habitat. Examples are described in Chapters 8 and 9.

17.3 LIFE HISTORY AND BEHAVIORAL PATTERNS OF SALMONIDS

17.3.1 Territoriality

As juvenile salmonids absorb their yolk sacs and emerge from the gravel in which they developed, three changes in behavior occur. They become positively phototactic, they start to nip at and chase neighboring fish, and they begin to take particles of food from the drift. Defending a territory of adequate size increases the probability that sufficient food (aquatic insects and other macroinvertebrates) will be carried past them by the current. Agonistic (threat and defense) interactions between fish become more subtle with age as nipping and chasing are replaced with ritualized threat postures and changes in body shading that serve the same purposes. These behavioral signals appear to be effectively communicated among species of fluvial salmonids. The ability to establish and hold a territory is largely dependent on body size, with a fish that is even slightly larger than another being dominant in most encounters.

In a conceptual sense, individual fish remain at a point in the territory referred to as the focal point where they may spend the majority of their time during daylight hours. The focal point may be near the center of a teardrop-shaped territory, and the fish darts out from the focal point to feed or to threaten intruders. The distance moved to feed is usually equivalent to about twice its body length. Because the amount of drift that passes a given point in the water column is approximately linear with water velocity (Figure 17.3), the fish scans a high-velocity area for drift. However, because the energy it expends increases exponentially as its swimming speed increases, the fish selects a focal point with reduced water velocity, often 10–20 cm/s, to maximize net energy gain (Fausch 1984). Territory size increases with fish size. A trout 10 cm long may defend a 0.1 m^2 area and a 30-cm-long fish may defend a territory of 1.0 m^2.

In reality, there are numerous variations on the basic theme described above. Juvenile coho salmon, for example, may be territorial in riffles, but are not territorial in pools although aggression may be displayed toward other pool inhabitants (Puckett and Dill 1985). Other individuals, called floaters, never defend an area and exist in spaces between the territories of other fish. Territorial

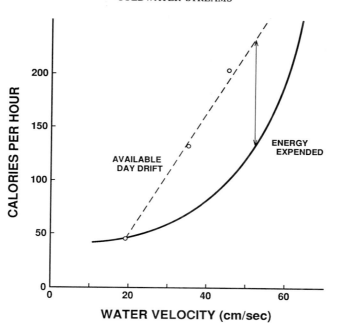

Figure 17.3 Energy expended by a 100-g trout at 15°C by swimming, and energy available from drifting invertebrates, at a range of water velocities. If the fish did not move several body lengths from its focal point to feed, net energy gain would be maximized at a velocity of 52 cm/s. (Data from Feldmeth and Jenkins 1973.)

fish have a net energy intake advantage because of reduced costs in searching for and pursuing food and reduced agonistic activity costs (Puckett and Dill 1985). In a study of wild adult brown trout, Bachman (1984) developed a size-dependent linear dominance hierarchy of individuals with overlapping home ranges. No fish had exclusive use of any home range and no clearly defined territories were observed. The brown trout spent 86% of daylight hours in a sit-and-wait state, searching the water column for drifting food.

Practical implications of these patterns of salmonid behavior are that only a portion of a stream may actually be used by a given life stage of a particular species, and that an ontogenetic shift occurs as the habitat used changes with fish growth. Age-0 rainbow trout, for example, use a combination of habitat characteristics such as water depth and velocity, substrate, and cover that define their microhabitat. Sets of probability-of-use curves (e.g., Bovee 1978) developed for many fluvial salmonids illustrate differences in habitat used by species and life stages (Figure 17.4).

17.3.2 Population Dynamics

Fish production is defined as the total elaboration of fish tissue during any time interval, including what is formed by individuals that do not survive to the end of the interval. The components of production are growth and survival.

Growth of salmonids in streams is typically less than the maximum possible because of periods of suboptimal temperature, limited food and space, and other

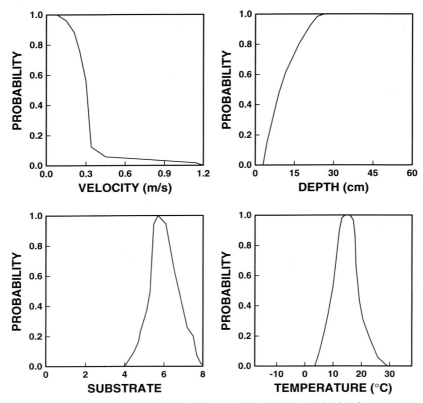

Figure 17.4 Probability-of-use curves for adult brook trout. Typical substrate categories are 2 = mud, 4 = sand, 6 = cobble, and 8 = bedrock. (From Bovee 1978.)

factors. Growth is most rapid during the first summer of life, when fish may reach sizes of 100–150 mm (Figure 17.5). In subsequent years, growth rate declines but may continue at 20–40 mm per year during the last few years of life. When carrying out investigations of population dynamics, care must be taken in assigning ages to salmonids. Recent studies using otoliths and daily growth rings have shown that age determination based on traditional scale reading may have underestimated true age. Furthermore, in severe environments, salmonids that grow little during their first year may not develop first-year annuli. Biologists conducting age and growth studies should be aware of these sources of bias.

Natural mortality of stream-dwelling salmonids varies considerably with age. Mortality during the first few months of life may exceed 90%, although part of this loss may actually reflect emigration. Factors such as intraspecific competition and starvation that cause mortality of age-0 fish are density dependent. Others, such as spring floods and high summer temperatures, operate regardless of fish density. Predation may be substantial, but is difficult to accurately assess. An extensive study on the Au Sable River in Michigan could not identify the causes of mortality during the first year of life for brook trout and brown trout (Alexander 1979). Predation on yearling and older trout was significant by great blue herons *Ardea herodias,* common mergansers *Mergus merganser,* otters *Lutra canadensis,* mink

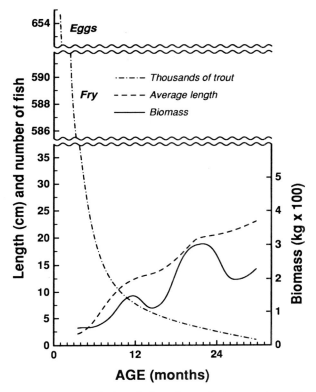

Figure 17.5 Summary of some vital statistics of a typical year-class of brook trout for the first 31 months of life in 3.5 hectares of spring-fed Lawrence Creek, Wisconsin. (From McFadden 1961.)

Mustela vison, and brown trout (Table 17.1). These predators killed 79% and 45% of the age-1 brook trout and brown trout present, respectively, and 25% and 44% of the older brook trout and brown trout, respectively. In a Vancouver Island stream in British Columbia, common merganser broods consumed sufficient wild coho salmon fry during summer months to account for 24–65% of the potential smolt production (Wood 1987).

Overwinter mortality of salmonids in streams appears to approximate 50% of the fish present in the fall, although data often include spawning-related mortality and sometimes mortality from spring floods. Death may result from starvation, predation, or physical damage. Because of the decrease in metabolic activity at cold temperatures, starvation of wild salmonids is unlikely. Healthy trout survive several months of fasting in cold water. Predation in winter is generally regarded as minimal, although it has not been carefully evaluated.

Physical injury appears to be the major cause of winter mortality in stream reaches where snow bridging does not occur. During nights when clear skies maximize radiant heat loss from streambed and water, and when air temperatures are low enough to supercool the stream, crystals of ice, called frazil ice, form in the water column. The frazil ice drifts downstream and attaches to the first object it contacts, where it may accumulate as a mat of anchor ice. Anchor ice tends to

Table 17.1 Percentage of total mortality of trout due to predation in the Au Sable River, Michigan (data from Alexander 1979). Mortality estimates were made from the fall of one year to the fall of the next for the age-groups indicated.

Predator	Brown trout		Brook trout	
	Age 0–1	Age 1–8	Age 0–1	Age 1–4
Common merganser	10.7	12.8	3.6	10.5
Great blue heron	6.8	14.1	7.7	3.3
Belted kingfisher	3.2	0	4.0	0.1
Brown trout	16.1	0.2	58.0	5.1
Mink	7.7	11.4	5.0	4.8
Otter	0.2	5.4	0.6	1.2
Anglers	6.8	46.0	6.6	43.7
Unknown	48.5	10.1	14.5	31.3

build up where velocity is greatest, flow is most turbulent, and where there is no surface ice. As the mats build, dams form and the stream becomes elevated and is forced out of its bed. Fish may lose their orientation and swim aimlessly into the impounded areas. During the day, heat from the sun melts the ice dams and the stream quickly returns to its former bed, often stranding fish and other aquatic organisms in the overflow areas. This sequence may be repeated frequently throughout the winter.

Salmonids display one or more of a range of responses to the onset of winter, depending upon fish species and life stage, minimum temperatures experienced, and habitat quality. The two basic options are to move elsewhere, or to remain in the same areas used in summer. Fish that move may enter thermal refuges such as groundwater seeps or move to off-channel habitat such as ponds and side channels. Movements generally begin when water temperatures drop below 7–10°C. In many Idaho streams, Chapman and Bjornn (1969) suggested that 50% or more of the juvenile chinook salmon and steelhead that left the stream did so in fall and winter months. Adults of resident species characteristically move to the deepest pools available, leaving small tributaries to enter main river areas.

Once in the area to be used in winter, juvenile salmonids typically either aggregate in the water column in deep, slow water beneath cover or conceal themselves in the substrate. Movement into or beneath cobble, small boulders, or large woody debris to depths of 30 cm has been described by Chapman and Bjornn (1969) for juvenile chinook salmon and steelhead and more recently by others for most other fluvial salmonids. These fish enter a semi-torpid state and expend a minimum of energy. Where cobble was not available in heavily sedimented portions of an Idaho stream, juvenile chinook salmon overwintered in low velocity areas among dense growths of sedges and grasses which draped over the bank (Hillman et al. 1987). A recent observation that has significant implications for winter ecology is that fish which are in the substrate during the day may emerge at night. In small Washington streams, Campbell and Neuner (1985) found that juvenile and adult resident rainbow trout were concealed during the day in winter, but these fish moved inshore into shallow water at night.

Overwintering within the substrate appears to be adaptive for high-elevation or more northerly conditions where winters are characteristically severe (Cunjak and Power 1986). Here streams freeze over early and remain frozen throughout the winter, and fish survival might be maximized by remaining in the substrate without expending energy for swimming. At lower latitudes or at lower elevations, streams display alternating freezing and thawing, variable discharge, and anchor ice formation. Under such conditions, an active existence might be most adaptive for survival.

After reaching sexual maturity, natural mortality may reduce fluvial populations of trout by 50% or more annually, with males typically showing higher mortality than females. Therefore, odds are against extended longevity, especially if the population is heavily fished. "Old" fish might be 4–5 years old, depending on the species and the environment. For brook trout in Lawrence Creek, Wisconsin, some year-classes never had members surviving past age 3 (McFadden 1961).

17.3.3 Spawning

The attainment of sexual maturity appears to be more related to fish size than to age. In some populations of brook trout, males become sexually mature at the end of their first growth season, when less than 100 mm in length, and females mature as yearlings. Other species or other populations of brook trout may not mature until age 4 or older. Trout may reproduce over several years or they may reproduce only once before death. Egg production varies with size of female, and ranges from a few hundred eggs to several thousand. Spawning occurs as water temperature approaches 10°C, either in the fall for species such as brown trout and brook trout or in spring for rainbow trout and cutthroat trout. Spawning habitat is predictable in terms of water velocity, water depth, and substrate size. Eggs develop in the gravel until they accumulate approximately 300 degree-days over 0°C for resident trout species. The eggs of fall-spawning species remain in the gravel during the winter when they are prone to damage from intergravel ice formation under severe conditions, but fall-spawned fry emerge from the gravel before spring-spawned fry.

17.4 DETERMINANTS OF COLDWATER STREAM FISH POPULATIONS

17.4.1 Species Composition

The present distribution of salmonid species in North America bears little resemblance to original ranges. Efforts began over 100 years ago to introduce virtually every species into every drainage across the continent. Now few coldwater stream systems exist that do not hold brook trout, brown trout, or rainbow trout. The ranges of coho salmon, chinook salmon, and steelhead have been increased dramatically in the past decade by introductions into the Great Lakes. Only Atlantic salmon, cutthroat trout, bull trout, and Dolly Varden have not been spread widely out of their original ranges. Cutthroat trout have been transplanted around the world, but nearly all of these efforts have been unsuccessful.

Within the continuum of physical variables in a coldwater stream, it is possible that salmonid species might occupy separate stream portions that do not overlap. Some segregation has been documented, such as in the Salmon River drainage in Idaho where first-order streams were used only by bull trout, and fifth-order streams were not used by brook trout and bull trout (Platts 1979; Figure 17.6).

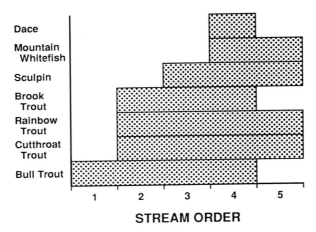

Figure 17.6   Relationship between stream order and fish species composition for tributaries of the Salmon River, Idaho. (Data from Platts 1979.)

Rainbow trout were the only salmonids found in stream gradients over 16%. Bull trout were most abundant in gradients of between 6% and 10%, cutthroat trout between 8% and 14%, and brook trout between 2% and 5%. Stream gradient has been shown to have a strong negative influence on brook trout abundance in Wyoming streams (Chisholm and Hubert 1986); however, stream-dwelling salmonids are generalists, both in terms of habitat use and diet. Their coexistence in multi-species situations depends upon their ability to specialize when influenced by interspecific interactions such as predation, hybridization, and competition.

Competition occurs when a number of individuals (of the same or different species) use common resources which are in short supply or, if the resources are not in short supply, the individuals harm one another in the process. Competition among fluvial salmonids usually translates into attempts by individuals of the same or different species to secure adequate space and, therefore, food and cover. Continued interactions between two species should either produce a niche shift in one or both species, the extinction of one species, or fluctuating coexistence as the environment alternately favors one species over the other. Coexistence of sympatric species may result from interspecific differences in habitat use which reflect species differences in agonistic behavior, innate habitat preference, timing of emergence, morphology, or a combination of these factors. However, there is evidence that the presence of introduced salmonids has detrimentally impacted native species. Examples include the decline of inland subspecies of cutthroat trout following range expansion of brook trout (Griffith 1988), decline of brook trout in the southeastern portion of its range following invasion by rainbow trout (Kelly et al. 1980), and replacement of brook trout by brown trout in Minnesota (Waters 1983).

17.4.2   Typical Density, Standing Stock, and Production

Annual production of salmonids, either in mixed- or single-species situations in coldwater streams, usually falls within the range of 15 to 50 kg/hectare for soft-water streams or for those at higher latitudes, and around 100 kg/hectare for

Table 17.2 Frequency of annual production values for resident salmonids in streams in North America, Europe, and New Zealand. (Modified from Waters 1977.)

Species	Production, kg/hectare				
	<50	51–100	101–150	151–200	>200
Brook trout	3	3	2	1	3
Brown trout	5	4	6	0	2
Rainbow trout	4	0	2	0	0
Multi-species	0	2	2	2	1

more productive streams (Waters 1977; Chapman 1978; Table 17.2). Often more than half of the production is contributed by the youngest age-group. The turnover ratio, or ratio of production to mean standing stock over a time interval, is commonly 1.0 to 1.5 for fluvial salmonids (Chapman 1978), but ranges from about 0.5 to 1.0 for small, high-elevation Colorado streams (Scarnecchia and Bergersen 1987).

Midsummer estimates of salmonid biomass and density were summarized by Platts and McHenry (1988) for 313 interior streams in the western United States with pristine or lightly altered habitat. Biomass exhibited great variability, ranging from 0 to 819 kg/hectare with an average of 54 kg/hectare. Salmonid density was less variable, ranging from 0 to 420 fish/100 m^2 with an average of 25 fish/100 m^2 for all sites. Brown trout density and biomass were significantly greater than those of other species, and multi-species communities had densities and biomasses that were not significantly different from those of streams occupied by a single species.

Densities of juvenile salmon and steelhead in fully-seeded stream-rearing habitat may exceed 700 fish/100 m^2 for brief periods. By late summer, however, densities of greater than 50 to 370 fish/100 m^2 are typical for age-0 Atlantic salmon, and such a range is also realistic for Pacific salmon and steelhead. Average annual smolt production for Atlantic salmon in North American streams may reach 6 to 10 fish/100 m^2 (Bley 1987).

17.4.3 Predictors of Abundance

A number of models have been developed to predict the biomass of resident trout in streams based on measurements of habitat. The habitat quality index (HQI) model II (Binns and Eiserman 1979) predicted trout biomass in Wyoming streams using nine attributes: late-summer streamflow, annual streamflow variation, maximum summer stream temperature, nitrate nitrogen, cover, eroding streambanks, submerged aquatic vegetation, water velocity, and stream width. The model explained 96% of the variation in trout biomass for the 36 streams from which it was developed, and 87% for 16 Wyoming streams in a follow-up study by Conder and Annear (1987).

Attempts to apply the HQI to populations of salmonids in streams outside of Wyoming have generally not been successful, indicating that trout populations in different areas respond to different sets of factors. Correlative models based on the HQI have been assembled for areas ranging from Colorado to Ontario (Table 17.3). Most recently, geomorphic variables such as basin relief and drainage density have been shown to correlate with trout biomass in Wyoming streams (Lanka et al. 1987). This linkage may enable the use of simple measures of

Table 17.3 Models other than the HQI that predict trout biomass in streams based on measurement of habitat. See Fausch et al. (1988) for a thorough review.

Study	Location	Trout species	Biomass range, kg/hectare	Significant variables	r^2
Wesche et al. (1987)	Southeast Wyoming	Brown	2–211	overhead bank cover	0.31
Scarnecchia and Bergersen (1987)	Northern Colorado	Brook, cutthroat, rainbow, brown	39–282	width and depth, width, alkalinity	0.82
Bowlby and Roff (1986)	Southern Ontario	Brook, brown, rainbow	0.5–150	suspended microcommunity, biomass, mean maximum temperature, benthic biomass, pool area, piscivore presence	0.62

drainage basin geomorphology, perhaps in combination with stream habitat variables, to predict potential habitat quality for salmonids.

17.4.4 Limiting Factors

An examination of the stability of fluvial salmonid populations over time provides insight into their limiting factors. Populations in spring-fed streams, where flows are relatively uniform throughout the year and floods and droughts are uncommon, are most stable. An example is Lawrence Creek, Wisconsin, where brook trout populations in September varied from 55 to 111 kg/hectare over a 5-year period in the mid-1950s (McFadden 1961) and annual production varied only 20% over 11 consecutive years (Hunt 1974).

At the opposite extreme are Great Basin (Nevada) and Rocky Mountain streams for the period 1975–1985, when both near-record low flows and high flows, and the most severe winter on record, occurred (Platts and Nelson, 1988). Biomass of bull trout in the South Fork Salmon River, Idaho, fluctuated fourfold (range 11–39 kg/hectare) in those 11 years, and brook trout in three small Idaho streams also fluctuated fourfold in an 8-year period. In the Great Basin, biomass of cutthroat, brown, and rainbow trout also showed a three- or fourfold fluctuation (cutthroat trout range 33–98 kg/hectare, brown trout 20–80 kg/hectare, rainbow trout 4–14 kg/hectare) in the same time period. Trout number fluctuated more than did biomass, often eightfold. Platts and Nelson suggested that for these and many other western streams, abiotic factors may be the causal factors limiting salmonid abundance.

Most salmonid streams are intermediate between the two extremes described above. Abiotic events such as floods and severe winters may commonly occur, but not frequently enough to affect every generation of trout. When they do occur, populations would be depressed by the climatic factors so that resources such as food would be abundant and competition between individuals minimal. However, between these periods, "ecological crunches" as described by Wiens and Rotenberry (1981) would be interspersed when populations increase and resources become limited. Interspecific interaction then becomes important, and the niches of one or both species may be modified as a result. For fluvial trout, interspecific competition should be most significant in late summer in those years when population levels are at or above normal. At these times, space and food are minimal, and niche shifts have been observed. Also, when population levels are at or above normal levels, density-dependent mortality of salmonids during the first few months of life is an important population regulating process.

In a practical evaluation of limiting factor theory, Mason (1976) provided juvenile coho salmon in a small Vancouver Island stream with supplemental food in the form of marine invertebrates. As a result, the density effect on survival was eliminated and growth increased, producing a coho salmon biomass that was six- to sevenfold above natural levels at the end of the summer. However, coho salmon remain in the rearing stream through their first winter of life. By February, Mason's expanded populations had declined to densities typical of most natural coho salmon populations, and the net result was output of about the same number of smolts that the stream would have produced without summer feeding. To accomplish a successful "lifting of the lid," Mason would have had to increase appropriate winter habitat as well as provide summer food.

17.5 OPTIONS FOR COLDWATER STREAM MANAGEMENT

17.5.1 History

A highlight movie reel depicting fisheries management of coldwater streams during the past 100 years would show numerous shifts in emphasis during that period. The interval from 1870 through the early 1900s was the golden era of introductions. During this time exotics such as brown trout (MacCrimmon and Marshall 1968) and common carp were widely distributed. At the same time, most salmonids native to North America were transplanted throughout the continent. Brook trout, for example, were brought into 12 central and western states before 1900, and into five other states and three provinces 30–50 years later (MacCrimmon and Campbell 1969).

In an attempt to maintain high catch rates while stream habitat was noticeably shrinking in the middle of the 20th Century, hatchery programs proliferated. Not until the early 1970s were questions being raised about impacts of stocked trout on wild populations. A gradual change in attitude, with increased emphasis on wild fish in a quality setting, began and continues to grow. Currently, emphasis is being placed upon maintaining or restoring species or subspecies of special concern, adopting special regulations, and protecting and rehabilitating instream and riparian habitat.

17.5.2 Management Goals

For anadromous salmonids rearing in freshwater, the management goal is simple: produce as many smolts as possible. The issue becomes the quality of the smolts, and in particular whether those of hatchery origin are adequate.

For resident salmonids that support sport fisheries, the traditional goal has been maximization of harvest. However, this is shifting to a more challenging goal of optimum sustainable yield where yield includes fish harvested, as well as fish caught and released as they contribute to the quality of the angling experience. It is important that managers understand what constitutes a successful angling

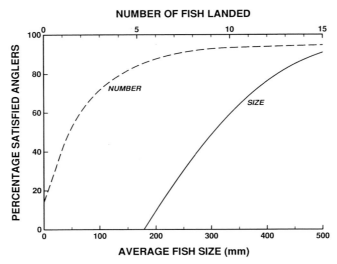

Figure 17.7 Satisfaction of anglers in Yellowstone National Park as determined by number of fish landed and average size of fish landed. (From Varley 1984.)

experience for their constituents. To establish what makes a satisfying trip for anglers throughout Yellowstone National Park, the Fish and Wildlife Service interviewed over 20,000 anglers (Varley 1984). The percentage of satisfied individuals increased, but at a decreasing rate, as the number and size of trout landed increased (Figure 17.7). About one person in five was satisfied even though no fish were landed. Satisfaction increased greatly when one fish was caught, and about 80% of the anglers were satisfied after catching three fish.

Use of a standardized length-categorization system (Gabelhouse 1984) would enable a manager to compare the length-frequency distribution of a population with the sizes of fish desired by anglers. Minimum lengths for "preferred," "memorable," and "trophy" lengths correspond to certain percentages of the world-record length for that species. For example, preferred rainbow trout would exceed 49 cm, and memorable and trophy individuals would exceed 65 and 81 cm, respectively, while trophy brook trout would exceed 59 cm.

It has been facetiously said that management of coldwater streams boils down to three basic components: protect habitat, preserve habitat, and restore habitat. Actually, there are other components (specifically, to manipulate the biotic community and to regulate the catch), nonetheless, habitat is the key. The need to maintain quality habitat for coldwater streams is equivalent to the overriding need to maintain a predator–prey balance in ponds and reservoirs. Predator–prey balance in salmonid streams is of lesser concern because invertebrate food sources are generally abundant, and their overall quantity is typically not impacted by trout to a significant extent.

The limiting factor concept may help coldwater stream managers in allocating their limited resources. It is important to distinguish between deleterious signs and problems. Deleterious signs can be seen (perhaps too easily), whereas the underlying problems may be difficult to detect. By examining the situation from a basinwide perspective and critically reviewing the requirements of each life-stage of the species concerned, it may be possible to identify the factors that most urgently require management action. For example, an evaluation of steelhead and coho salmon populations and their habitat in Fish Creek, Oregon, provided information on factors limiting their production in the basin (Everest et al. 1985). For steelhead, spawning habitat was more than adequate, but rearing habitat was inadequate. For coho salmon, smolt production was limited only by the numbers of adults returning to the basin. Following this analysis, specific management plans could be implemented to improve steelhead rearing habitat and to increase escapement of adult coho salmon.

17.6 HABITAT MANAGEMENT

When considering management actions, the general strategy is to document existing habitat conditions and then to assess habitat improvement potential (or document habitat degradation) using a model such as the HQI. Chapter 9 describes techniques for the restoration of coldwater stream habitat. The placement of instream and streambank structures, currently in vogue, needs to be done after careful consideration. More than $100 million will be spent on enhancement of salmonid habitat in the western United States in the next decade. The ability of this program to provide cost-effective results must be carefully examined; in many cases, placing structures to increase fish numbers may not be justifiable, but implementing measures to improve watershed conditions or to deal with specific habitat problems may have large pay offs. Follow-up evaluation is critical, and it is important to realize that the benefits from instream structures may not be detectable for 5 years or more.

If streams have been damaged by livestock, they are usually rehabilitated more easily and cheaply by relieving the grazing pressure (by fencing or other compatible strategies) and then allowing streambanks to restore themselves than by building artificial stream structures. Also, it is unlikely that the habitat structures would survive in the presence of heavy grazing.

Habitat management should be based on a need to provide habitat that is as complex as possible. The need for management of large woody debris in salmonid nursery streams is clearly recognized. In Oregon, House and Boehne (1986) examined stream segments in old growth coniferous forests where large woody debris was abundant, and compared them with previously logged segments that held little large debris. The presence of large woody debris had profound effects on channel morphology. Although channel gradient was the same in test and control segments, large woody debris "stairstepped" the channel, caused the formation of secondary channels, meanders, and undercut banks, and caused the tipover of trees to expose rootwads. Spawning gravel was trapped in the stairstepped sections, and pool quality and quantity were increased. Significantly greater salmonid biomass was found in the old-growth segments, and the number of coho salmon was positively correlated with the number of large woody debris intrusions into the stream channel.

In the absence of complete ecological understanding, management for diversity will usually keep the stream in sound condition for biological resources. The best situation has plunges, backwaters, large woody debris and streamside vegetation, cobble, gravel, even sand and silt, shear lines, depth, pockets, undercuts, and an optimal relationship between this diversity and year-long trophic and habitat requirements.

17.7 MANIPULATIONS OF COMMUNITY STRUCTURE

A commonly used management approach is to modify the community structure of a stream by introducing nonnative salmonids, by stocking hatchery trout, or by removing nonsalmonid competitors. Much has been written about the role of hatchery-reared fish as they have been used to either supplement wild trout populations or to provide angling where no wild trout existed. In 1983, 43 states stocked over 50 million catchable-size trout at a production cost of about $37 million (Hartzler 1988). The fraction of these fish that were stocked into streams is unknown, but at least 30 states planted "catchables" into more than 54,000 km of streams in the United States. Rainbow trout accounted for 77% of all catchable salmonids stocked, and brook and brown trout made up 11 and 10% of the total, respectively.

Recent studies, however, show detrimental effects of hatchery-reared trout on wild salmonids. Populations of 2-year-old and older wild brown trout and rainbow trout in two Montana streams declined significantly in numbers and total weight following introduction of catchable-size hatchery rainbow trout (Vincent 1987). As Vincent points out, such stocking can reduce the number of wild trout available to anglers and may cause some genetic alteration of the wild stocks. In these situations, management of self-sustaining wild trout streams would be better directed at maintaining or enhancing riparian habitat, maintaining adequate water flows, and applying appropriate catch regulations.

Experience with an extensive system of hatcheries for anadromous salmonids in the western United States and Canada has shown that hatchery production may reduce natural production through competition, predation, interbreeding, and transmission of disease. Procedures have been developed to mitigate this effect, but most are directed toward hatcheries that release fish as smolts, and involve separation of hatchery and wild fish. Recently, interest has expanded to an alternative approach called "outplanting" (releasing hatchery fish to rear or spawn in streams, usually remote from the hatchery), which may have merit in some situations (Reisenbichler and McIntyre 1986).

"Control" of potential competitors that appear to have a negative impact on trout and salmon in streams has been a focus of attention for several decades (see Chapter 14). In streams, sculpins have been thought to impact salmonid populations through predation on their eggs and fry and through competition for food in the form of benthic invertebrates. However, Moyle (1977) suggested that sculpin impacts would be expected only under conditions where salmonid populations had been badly damaged by overharvest or environmental disturbance. After trapping and removing nontrout species from a small New York stream for 13 years, Flick and Webster (1975) could detect no change in growth of brook trout. They suggested that removal of nontrout species did not seem to be a viable management technique.

In western streams, the question of competition between trout and mountain whitefish remains unanswered. Whitefish, which may outnumber trout 10 to one in some fourth- and fifth-order streams, show a partial diet overlap with trout, especially with juvenile trout.

A program to improve the quality of angling by removing predators from a portion of the Au Sable River, Michigan, from 1964 through 1966 included reducing numbers of large brown trout by 40–66% and harassing common mergansers (Shetter and Alexander 1970). No significant differences in size or angler catch of smaller brook trout and brown trout were subsequently observed, and it was concluded that predator reduction would have to be conducted at a much more intensive level to produce the desired results.

17.8 REGULATION OF THE FISHERY

The traditional objective of catch regulation in freshwater sport fisheries has been to maximize harvest by shifting as much of total mortality as possible into angling-induced mortality (see Chapter 16). Typical annual harvest for some stream fisheries may approach 50% of the biomass present. Regulations are used to protect the fish population from depletion and to maintain an equitable distribution of the resource among the fishing public. Fairly simple regulations, usually bag limits, are set to provide the greatest allowable harvest without overexploitation. Mathematical models have been developed to evaluate the response of trout populations to these and other regulations.

However, as Anderson and Nehring (1984) have pointed out, above some level of fishing pressure, the maximum-yield approach sacrifices the "quality" aspects of the fishery because larger and older fish exist in greatly reduced numbers. This threshold of fishing pressure has been surpassed on the more popular trout streams in Colorado and perhaps across the continent.

More specialized regulations include slot limits where only fish smaller than and larger than a length interval can be killed. Another approach is to restrict harvest to a few fish per day while limiting terminal gear to flies or artificial lures only. Because of the 30% mortality expected if bait-caught resident salmonids are released, the use of bait is generally not compatible with such restrictive regulations. Yet another approach is delayed harvest, where stream sections set aside for fly and lure fishing are managed as "no-kill" for part of the year when environmental conditions are ideal for trout growth and survival. When conditions deteriorate, such as during low flow and high water temperatures, some harvest is allowed.

Complete no-kill regulation (catch-and-release) has become an accepted management technique over the past decade. The strategy here is to recycle individual medium-size and large trout as many times as possible, such as in the Yellowstone River in Yellowstone National Park where cutthroat trout were each caught an average of nearly ten times in a season (Schill et al. 1986). Catch-and-release would be expected to be successful for populations having good growth potential with low natural mortality rates and substantial longevity. Although often called no-kill regulations, catch-and-release does result in 4–6% mortality from each capture for salmonids caught with artificial lures or flies (Wydoski 1980). This could cause significant cumulative mortality for populations whose members are repeatedly caught and released, especially when water temperatures exceed the preferred range for the species. For the Yellowstone Park cutthroat trout fishery,

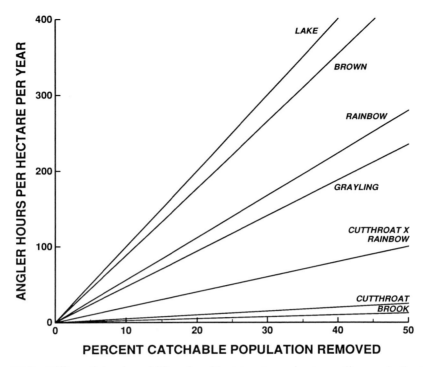

Figure 17.8 Differential vulnerability of resident trout species to angling, as indicated by the number of angling hours required to catch certain fractions of each population. (From Varley 1984.)

however, single capture mortality was estimated to be only 0.3% (Schill et al. 1986).

Where they are biologically effective in "stockpiling" larger trout, catch-and-release regulations provide opportunities for high-quality angling. For example, the catch rate of trophy-size trout in the South Platte River in Colorado was 28 times greater in the catch-and-release area than in the harvest area. Rainbow trout dominated the population in the catch-and-release area with over 500 kg/hectare present and 50% of the trout longer than 30 cm. In the general regulation area, brown trout were most abundant, biomass was about one-third of that in the catch-and-release area, and only 17% of the population exceeded 30 cm (Anderson and Nehring 1984). This example also illustrates the differential vulnerability of salmonid species to angling, a phenomenon which must be considered in any regulation of harvest.

An index of the susceptibility of salmonids to angling is the amount of angling effort that results in removal of a certain fraction of the catchable population. Varley (1984) developed a preliminary model for Yellowstone National Park trouts (Figure 17.8). Approximately half of the catchable cutthroat trout in a population would be removed by about 35 hours of fishing per hectare per year. Corresponding values for rainbow trout and brown trout were approximately 285 hours and over 400 hours, respectively.

17.9 THE FUTURE OF COLDWATER STREAMS

The future of management for coldwater streams lies with managing habitat, as discussed previously, and with managing the angler. As Larkin (1988) points out, the manager of the future may well conduct a market analysis of latent demand to help develop a management strategy. The value of this technique is shown by Sorg et al. (1985), who determined that the value of a coldwater fishing trip would increase by $8 if the number of fish caught doubled, and would increase by $13 if the size of fish caught increased by 50%.

The manager of the future will also have to deal with calls for limiting entry to the more popular publicly owned streams. Most of the current interest in limited entry comes from anglers who are concerned that the quality of their experience is reduced by high angler density. To date, there is only minimal biological evidence that catch-and-release anglers are impacting stream resources, but this may change as pressure increases.

17.10 REFERENCES

Alexander, G. R. 1979. Predators of fish in coldwater streams. Pages 153–170 *in* H. Clepper, editor. Predator–prey systems in fisheries management. Sport Fishing Institute, Washington, D.C.

Anderson, R. M., and R. B. Nehring. 1984. Effects of a catch-and-release regulation on a wild trout population in Colorado and its acceptance by anglers. North American Journal of Fisheries Management 4:257–265.

Bachman, R. A. 1984. Foraging behavior of free-ranging wild and hatchery brown trout in a stream. Transactions of the American Fisheries Society 113:1–32.

Binns, N. A., and F. M. Eiserman. 1979. Quantification of fluvial trout habitat in Wyoming. Transactions of the American Fisheries Society 108:215–228.

Bley, P. W. 1987. Age, growth, and mortality of juvenile Atlantic salmon in streams: a review. U.S. Fish and Wildlife Service Biological Report 87(4).

Bovee, K. D. 1978. Probability-of-use criteria for the family Salmonidae. U.S. Fish and Wildlife Service, Cooperative Instream Flow Service Group, Information Paper 4, Fort Collins, Colorado.

Bowlby, J. N., and J. C. Roff. 1986. Trout biomass and habitat relationships in southern Ontario streams. Transactions of the American Fisheries Society 115:503–514.

Campbell, R. F., and J. H. Neuner. 1985. Seasonal and diurnal shifts in habitat utilized by resident rainbow trout in western Washington Cascade Mountain streams. Pages 39–48 *in* F. W. Olson, R. G. White, and R. H. Hamre, editors. Symposium on small hydropower and fisheries. American Fisheries Society, Western Division and Bioengineering Section, Bethesda, Maryland.

Chapman, D. W. 1978. Production in fish populations. Pages 5–25 *in* S. D. Gerking, editor. Ecology of freshwater fish production. Wiley, New York.

Chapman, D. W., and T. C. Bjornn. 1969. Distribution of salmonids in streams, with special reference to food and feeding. Pages 153–176 *in* T. G. Northcote, editor. Salmon and trout in streams. H. R. MacMillan Lectures in Fisheries, University of British Columbia, Vancouver.

Chisholm, I. M., and W. A. Hubert. 1986. Influence of stream gradient on standing stock of brook trout in the Snowy Range, Wyoming. Northwest Science 60:137–139.

Conder, A. L., and T. C. Annear. 1987. Test of weighted usable area estimates derived from a PHABSIM model for instream flow studies on trout streams. North American Journal of Fisheries Management 7:340–350.

Cunjak, R. A., and G. Power. 1986. Winter habitat utilization by stream resident brook trout (*Salvelinus fontinalis*) and brown trout (*Salmo trutta*). Canadian Journal of Fisheries and Aquatic Sciences 43:1970–1981.

Everest, F. H., G. H. Reeves, J. R. Sedell, J. Wolfe, D. Hohler, and D. A. Heller. 1985. Abundance, behavior, and habitat utilization by coho salmon and steelhead trout in Fish Creek, Oregon, as influenced by habitat enhancement. Annual Report to Bonneville Power Administration, Project 84-11, Portland, Oregon.

Fausch, K. D. 1984. Profitable stream positions for salmonids: relating specific growth rate to net energy gain. Canadian Journal of Zoology 62:441–451.

Fausch, K. D., C. L. Hawkes, and M. G. Parsons. 1988. Models that predict standing crop of stream fish from habitat variables: 1950–85. U.S. Forest Service, Pacific Northwest Region, General Technical Report PNW-GTR-213, Portland, Oregon.

Feldmeth, C. R., and T. M. Jenkins. 1973. An estimate of energy expenditure by rainbow trout in a small mountain stream. Journal of the Fisheries Research Board of Canada 30:1755–1759.

Flick, W. A., and D. A. Webster. 1975. Movement, growth, and survival in a stream population of wild brook trout *Salvelinus fontinalis* during a period of removal of non-trout species. Journal of the Fisheries Board of Canada 32:1359–1367.

Gabelhouse, D. W. 1984. A length-categorization system to assess fish stocks. North American Journal of Fisheries Management 4:273–285.

Griffith, J. S. 1988. Review of competition between cutthroat trout and other salmonids. American Fisheries Society Symposium 4:134–140.

Hartzler, J. R. 1988. Catchable trout fisheries: the need for assessment. Fisheries (Bethesda) 13(2):2–8.

Hillman, T. W., J. S. Griffith, and W. S. Platts. 1987. The effects of sediment on summer and winter habitat selection by juvenile chinook salmon in an Idaho stream. Transactions of the American Fisheries Society 116:185–195.

House, R. A., and P. L. Boehne. 1986. Effects of instream structures on salmonid habitat and populations in Tobe Creek, Oregon. North American Journal of Fisheries Management 6:38–46.

Hunt, R. L. 1974. Annual production by brook trout in Lawrence Creek during eleven successive years. Wisconsin Department of Natural Resources, Technical Bulletin 82.

Kelly, G. A., J. S. Griffith, and R. D. Jones. 1980. Changes in distribution of trout in Great Smoky Mountains National Park, 1900–1977. U.S. Fish and Wildlife Service Technical Paper 102.

Lanka, R. P., W. A. Hubert, and T. A. Wesche. 1987. Relations of geomorphology to stream habitat and trout standing stock in small Rocky Mountain streams. Transactions of the American Fisheries Society 116:21–28.

Larkin, P. A. 1988. The future of fisheries management: managing the fisherman. Fisheries (Bethesda)13:3–9.

MacCrimmon, H. R., and J. S. Campbell. 1969. World distribution of brook trout, *Salvelinus fontinalis*. Journal of the Fisheries Research Board of Canada 26:1699–1725.

MacCrimmon, H. R., and T. L. Marshall. 1968. World distribution of brown trout, *Salmo trutta*. Journal of the Fisheries Research Board of Canada 25:2527–2548.

Mason, J. C. 1976. Response of underyearling coho salmon to supplemental feeding in a natural stream. Journal of Wildlife Management 40:775–788.

McFadden, J. T. 1961. A population study of the brook trout *Salvelinus fontinalis*. Wildlife Monographs 7.

McFadden, J. T. 1969. Trends in freshwater sport fisheries of North America. Transactions of the American Fisheries Society 98:136–150.

Moyle, P. B. 1977. In defense of sculpins. Fisheries (Bethesda) 2(1):20–23.

Platts, W. S. 1979. Relationships among stream order, fish populations, and aquatic geomorphology in an Idaho river drainage. Fisheries (Bethesda) 4(2):5–9.

Platts, W. S., and M. L. McHenry. 1988. Density and biomass of trout and char in western streams. U.S. Forest Service General Technical Report INT-241.

Platts, W. S., and R. L. Nelson. 1988. Fluctuations in trout populations and their implications for land-use evaluation. North American Journal of Fisheries Management 8:333–345.

Poff, N. L., and J. V. Ward. 1989. Implications of streamflow variability and predictability for lotic community structure: a regional analysis of streamflow patterns. Canadian Journal of Fisheries and Aquatic Sciences 46:1805–1818.

Puckett, K. J., and L. M. Dill. 1985. The energetics of feeding territoriality in juvenile coho salmon (*Oncorhynchus kisutch*). Behaviour 92:97–111.

Reisenbichler, R. R., and J. D. McIntyre. 1986. Requirements for integrating natural and artificial production of anadromous salmonids in the Pacific Northwest. Pages 365–374 *in* R. H. Stroud, editor. Fish culture in fisheries management. American Fisheries Society, Fish Culture Section and Fisheries Management Section, Bethesda, Maryland.

Scarnecchia, D. L., and E. P. Bergersen. 1987. Trout production and standing crop in Colorado's small streams, as related to environmental features. North American Journal of Fisheries Management 7:315–330.

Schill, D. J., J. S. Griffith, and R. E. Gresswell. 1986. Hooking mortality of cutthroat trout in a catch-and-release segment of the Yellowstone River, Yellowstone National Park. North American Journal of Fisheries Management 6:226–232.

Shetter, D. S., and G. R. Alexander. 1970. Results of predator reduction on brook trout and brown trout in 4.2 miles of the North Branch of the Au Sable River. Transactions of the American Fisheries Society 99:312–319.

Sorg, C., J. Loomis, D. M. Donnelly, G. Peterson, and L. J. Nelson. 1985. Net economic value of cold and warm water fishing in Idaho. U.S. Forest Service Resource Bulletin RM-11.

Vannote, R. L., G. W. Minshall, K. W. Cummins, J. R. Sedell, and C. E. Cushing. 1980. The river continuum concept. Canadian Journal of Fisheries and Aquatic Sciences 37:130–137.

Varley, J. D. 1984. The use of restrictive regulations in managing wild salmonids in Yellowstone National Park, with particular reference to cutthroat trout, *Salmo clarki*. Pages 145–156 *in* J. M. Walton and D. B. Houston, editors. Proceedings of the Olympic wild fish conference, Peninsula College, Fisheries Technology Program, Port Angeles, Washington.

Vincent, E. R. 1987. Effects of stocking catchable-size hatchery rainbow trout in two wild trout species in the Madison River and O'Dell Creek, Montana. North American Journal of Fisheries Management 7:91–105.

Waters, T. F. 1977. Secondary production in inland waters. Advances in Ecological Research 10:91–164.

Waters, T. F. 1983. Replacement of brook trout by brown trout over 15 years in a Minnesota stream: production and abundance. Transactions of the American Fisheries Society 112:137–147.

Wesche, T. A., C. M. Goertler, and W. A. Hubert. 1987. Modified habitat suitability index model for brown trout in southeastern Wyoming. North American Journal of Fisheries Management 7:232–237.

Wiens, J. A., and J. T. Rotenberry. 1981. Habitat associations and community structure of birds in shrub steppe environments. Ecological Monographs 51:21–42.

Wood, C. C. 1987. Predation of juvenile Pacific salmon by the common merganser (*Mergus merganser*) on eastern Vancouver Island. II: Predation of stream-resident juvenile salmon by merganser broods. Canadian Journal of Fisheries and Aquatic Sciences 44:950–959.

Wydoski, R. S. 1980. Relation of hooking mortality and sublethal hooking stress to quality fishery management. Pages 43–87 *in* R. A. Barnhart and T. D. Roelofs, editors. Catch-and-release fishing as a management tool. California Cooperative Fishery Research Unit, Humboldt State University, Arcata.

Chapter 18

Warmwater Streams

CHARLES F. RABENI

18.1 INTRODUCTION

Warmwater streams are those waters where temperature becomes high enough to exclude salmonids from remaining throughout the year and successfully reproducing. The designation developed because of different management needs for different species of fish (see Krumholz 1981). Warmwater streams comprise almost a half million kilometers of fishable waters. Fishing for warmwater species predominates in at least part of the streams in the vast majority of states and provinces and is the only type of stream fishing in over half of them (Funk 1970). This chapter examines features unique to warmwater streams including their considerable problems, current management efforts, and opportunities for the future.

Numerous species are sought by anglers in warmwater streams. Probably the most popular are largemouth bass, smallmouth bass, spotted bass, striped bass, muskellunge, northern pike, walleye, and channel catfish. Other regionally important fishes are rock bass, pumpkinseed, bluegill, other sunfishes, white and black crappies, various bullheads, white perch, yellow perch, and chain pickerel. Table 18.1 lists those species commonly taken from warmwater streams that are designated by states or provinces as game species.

Native species assemblages of warmwater streams are directly related to their geographic setting. Geologic and climatic events have influenced speciation and distribution and resulted in species richness of warmwater streams ranging from over 100 species in some streams of the Mississippi River drainage, to less than half a dozen species in some western and southwestern drainages. Species introductions, both intentionally and unintentionally, have greatly expanded the ranges of many bait and game fishes.

Factors such as temperature, stream gradient, bottom type, flow characteristics, dissolved oxygen, depth, cover, food abundance, and the presence of other fish have interacted to influence the distribution and abundance of each fish species. This is the reason that there are "catfish streams" in Kansas and "smallmouth bass streams" in Virginia. An idealized classification of warmwater stream types is presented in Table 18.2 to indicate the association of physical characteristics and some examples of resident sport fishes.

Although warmwater streams may differ from each other in physical appearance and resident fish faunas, there are patterns of fish abundance and distribution that most warmwater streams have in common, and which relate to longitudinal zones (Pflieger 1988). The headwaters represent the stream section with greatest

Table 18.1   Fish occurring in warmwater streams that state or provincial agencies list as game fish.

Fish species	Number of states and provinces where fish is listed as a game fish
Largemouth bass	41
Smallmouth bass	38
Walleye	36
Crappie	30
Northern pike	30
Striped bass	24
Yellow perch	24
White bass	23
Bluegill	22
Sauger	20
Muskellunge	19
Channel catfish	18
Bullhead spp.	17
Rock bass	17
Tiger muskellunge	16
Spotted bass	14
Pickerel	12
Blue catfish	12
Flathead catfish	10
Common carp	10
Black and white bass (8 variations)	10
Sucker	8
Warmouth	7
Green sunfish	7
Paddlefish	7
White perch	6
Redear sunfish	6
Pumpkinseed	5
Redbreast sunfish	3
Bowfin	3
Gar	2
Longear sunfish	2

gradient, with a succession of short pools and well-defined riffles. Substrates consist of bedrock, coarse gravel and rubble, or a considerable amount of silt and clay, but are usually of a larger mean particle size than downstream areas. During dry periods headwaters may become a series of isolated pools or be entirely dry for long stretches. Fishes inhabiting headwaters are small, and possess the ability to survive extremes of temperature, dissolved oxygen, and current velocities. Fish assemblages tend to be controlled primarily by the inherent variability of the prevailing physical factors (Matthews 1987). Headwater species are relatively tolerant of high temperatures and low dissolved oxygen and rapidly recolonize reaches that periodically become dry.

Midreach areas are more diverse in their habitat conditions. Pools are longer and deeper, riffles are short and well defined, fluctuations in flow occur regularly but drying out or elimination of flow usually is not a problem. Substrate particle size tends to be relatively large and cover in the form of undercut banks, rootwads, and downed trees is more prevalent than in headwater areas. Fish assemblages are more diverse and species richness is higher than in headwater areas. Biological factors, such as spawning-habitat requirements and trophic

Table 18.2 An idealized classification of warmwater streams based upon physical characteristics, with some representative fishes.

Physical characteristics	Type I	Type II	Type III
Predominant substrate type	Rock-cobble	Gravel–fine	Fine sand–silt
Gradient	Medium	Medium–low	Low
Mean velocities	Medium	Medium–low	Low
Channel type	Low sinuosity	Intermediate sinuosity	High sinuosity
Riffle-pool development	High	Medium	Low
Turbidity	Low	Low–medium	Medium–high
Edaphic factors	Coarse	Intermediate	Fine
Stream size	Small–large	Small–large	Medium–large
Representative sport fish	Smallmouth bass	Largemouth bass	Channel catfish
	Rock bass	Warmouth	Common carp
	Walleye	Buffalo	
	Northern pike	Bullheads	

interactions, start to become more important as regulators of fish assemblage structure.

Mainstem reaches are characterized by long pools separated by deep runs. Riffles are usually absent. Backwaters and cutoffs are more common than in upstream reaches. Substrate particles tend to be smaller and turbidity is usually highest in this section. Increasing species richness, species diversity, and body size characterize this region.

The change in species composition from headwaters to mainstem varies among streams yet is a function of species additions. Headwater species are found in other areas but their relative abundance is drastically decreased downstream. Figure 18.1 compares actual data from two stream types found in the Ozark and prairie faunal regions of Missouri, comparable to the type I and type III streams, respectively, of Table 18.2. Although absolute values are greater for the Ozark stream, the longitudinal patterns are similar. Species richness and species diversity both increase in the downstream direction.

The among-stream comparison and the within-stream continuum are intended to show that each stream and stream segment has a particular potential—for productivity, for species community structure, and for the ability to support a particular fishery. Managers of warmwater streams must strive to understand the potential to be able to formulate reasonable management objectives.

18.2 HISTORY

An historical perspective is valuable to better understand present conditions and the constraints under which a warmwater stream manager must work. Although documentation is limited, there is every indication that the fish of warmwater streams were much more abundant before there were human impacts in the watershed. Watershed land-use practices and point-source effluents have severely altered most warmwater stream systems in North America (see Chapter 8).

The history of warmwater streams in Ohio has been well recorded (Trautman 1981) and is probably typical of the magnitude of changes that have occurred to streams and their associated fish fauna throughout North America since the first

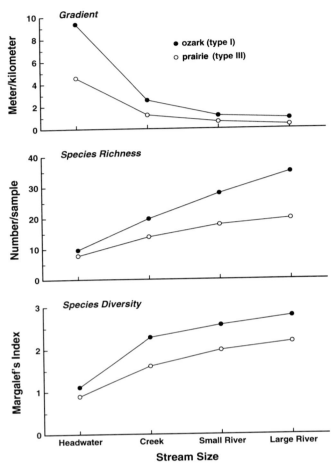

Figure 18.1 Longitudinal patterns of two warmwater stream types found in Missouri. (Data from Pflieger 1988.)

Europeans arrived (Table 18.3). Initially, around 1800, there were reports of "huge populations of the larger and better food fishes and game fishes: pikes, walleye, catfishes, suckers, drum, sturgeon, and black basses." However, within 50 years, extensive draining and filling of marshes had irrevocably altered headwater conditions so that many permanent springs and brooks became intermittent, and the shallow, ephemeral wetlands, which are so necessary for the reproduction of many species, were lost.

The hydrology of Ohio was further modified and by 1900 increasing agricultural activity and within-channel and riparian modifications created numerous clay-bottom streams, and siltation and sedimentation became the most important pollutants affecting the fish fauna. Thus, the major damage to warmwater streams probably occurred within the first 50 or so years of human habitation and the situation has been generally exacerbated since then. The most significant change, and one which is irreversible, was alteration of the water table which has dropped about 3 m in many areas. The consequences of the altered hydrologic regime to

Table 18.3 Historical factors important to warmwater streams in Ohio. (From Trautman 1981.)

Date	Estimated nonnative human population	Effects on warmwater streams
1750	0	Profusion of springs, permanent brooks, and wetlands. Banks heavily wooded, lots of trees and roots in water, gravel bottoms. Conditions conducive to production of high populations of larger and better food and game fish.
1800	45,000	Period of intensive marsh draining, stream ditching, and removal of riparian vegetation. Little or no soil erosion. Migratory fish, muskellunge, pike, lake sturgeon affected. Habitat changes favored smaller size fish.
1850	2,000,000	Intermittency increases. 500–1,000 milldams constructed—blocking fish movement and causing point-source pollution. Introduction of exotics.
1900	6,400,000	Springs rare. Drop in water table of 6 m. Siltation and sedimentation become most important pollutant. Clay bottoms common, aquatic vegetation eliminated. Increased dredging and ditching and channelization, mining wastes and industrialization.
1950	7,900,000	Increase in impounded waters. Siltation increases. Agricultural intensity increases, fence rows removed, riparian vegetation eliminated. Increase in pesticides and fertilizers.
1975	10,700,000	

the fish fauna are substantial (Figure 18.2). Historically, fish communities adapted to the regular slow springtime rises in discharge and the moderate, but adequate flows the remainder of the year. Increasing human alterations of the ecosystem have substantially reduced the water storage capacity of the earth and drastically lowered water levels. Fish species must now cope with much greater flows during limited periods of the year, and much lower flows the rest of the time.

In one major stream system of Ohio, the Maumee, 17 species of fish have completely disappeared in the last century and 26 species have become less abundant (Karr et al. 1985). Half of all species in headwater streams and 44% of species in midreach streams have disappeared. Eliminated species tended to be specialists, i.e., those requiring particularly narrow ranges of conditions for spawning or for water quality. Especially affected have been invertivores and herbivores, fish needing food types that do best on silt-free surfaces. Top carnivores, including northern pike, walleye, and smallmouth bass, have become much less abundant in the last century and less specialized species (gizzard shad, quillback, bigmouth buffalo) have increased in abundance, so that now the river system has shifted from being dominated by predatory species, to one of predominantly omnivores. This fact is important to a warmwater stream manager whose objective may be to increase the quality of fishing for top predators, while the conditions of most streams have shifted to favor the development of populations of omnivores.

Although the best documented, the changes noted for Ohio streams are not unique. In much of eastern North America stream degradation rapidly followed

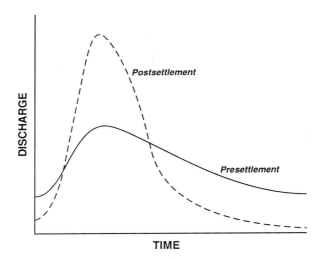

Figure 18.2   Historical alterations of stream discharge patterns.

settlement, and warmwater streams were seen as opportunities to dam for power, transport wastes, create lakes, or as obstacles to increased agricultural production. From the fisheries management perspective, sampling was difficult, free-flowing systems were considered hard to evaluate, and there was a prevailing view that warmwater streams were "self-managing" and fish were abundant for those few anglers who took advantage of the situation.

Attitudes have been gradually changing and several events since the late 1960s have encouraged a wiser use of this resource. There has been a belated realization by state and provincial agencies that fishing in warmwater streams is popular. For example, warmwater streams in Iowa account for about one quarter of all fishing trips. In Missouri, a state with extensive farm pond and reservoir resources, about 20% of all fishing occurs in warmwater streams. Oregon and Kentucky both initiated warmwater stream management programs after discovering that 25 and 35%, respectively, of all their fishing was on warmwater streams.

Federal legislation in the United States over the past few decades has significantly impacted warmwater stream management in the areas of improved water quality, increased funding for management, and increased protection for streams and their watersheds. The Federal Water Pollution Control Act Amendments of 1972 served as a major stimulus ($30 billion) for abatement of point-source pollution. There are numerous cases where, once point-source pollutants were removed, the fish rapidly returned, the stream or river was aesthetically acceptable, and anglers were once again attracted to the area. The Wallop–Breaux Aquatic Resources Trust Act of 1984, which imposed a U.S. federal excise tax on sportfishing equipment, boats, motors, trailers, and motor boat fuels, supplemented the established Dingell–Johnson Federal Aid in Sport Fish Restoration Act so that approximately $115 million are available annually for state fish and game agencies. This has provided the funding for several states to begin management programs on warmwater streams.

We now recognize the potential of a long-abused resource. There is progress in the abatement of point-source pollution, but very little has been done to address

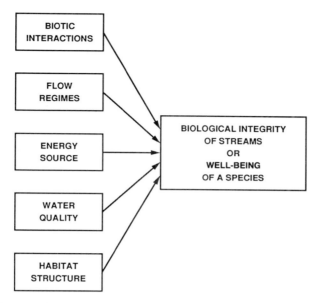

Figure 18.3 Classes of factors that influence aquatic biota and determine either the biological integrity of the stream as evidenced by the fish assemblage or the well-being of a particular species. (Adapted from Karr and Dudley 1981.)

the major problem of physical habitat degradation. An understanding of population dynamics of individual species and the interactions among species in warmwater streams lags behind comparable knowledge in other aquatic systems and limits the success of some management programs.

18.3 MANAGEMENT

18.3.1 Management Goals

The role of fisheries biologists responsible for warmwater streams is different from their coldwater counterparts in several respects: (1) warmwater streams have been more effected by human modification and often have severe habitat and water quality problems, (2) there is often a multispecies recreational fishery, and (3) basic ecological information on fish species and the associated biota is often not available.

Effective, long-term fisheries management must occur in an ecological context which takes into account complex interrelationships of the biota with its environment. A holistic view was provided by Karr and Dudley (1981), where the biological integrity of streams depends upon five primary variables: water quality, flow regime, physical habitat, energy source, and biotic interactions (Figure 18.3). Altering the processes associated with any of these variables will have a major impact on stream fish, and conversely, efforts to restore or maintain water resource quality by changing only one or a few factors will be disappointing. The responsible manager must be able to identify processes influencing each variable.

Management of warmwater stream fisheries is currently divided between efforts to manage a particular species and efforts to manage the entire fish community.

The species approach holds that sport fish, being mostly top carnivores, represent an integration of conditions in the stream environment, and maintaining their well-being will benefit the entire stream biota. The community management approach stresses that an appropriate balance of either taxonomic or ecological groups reflects the quality of the stream environment and a correct community balance will be to the best advantage of any sport fish. The two approaches are not necessarily contradictory, and the manager should view community integrity as a necessary prerequisite to species management. Managing for a variety of fish with different requirements ensures that the basic life-supporting elements within a stream are sufficient and that the results of management activities directed at any one species will have a better chance of success. Only where restoration of a fish community has been achieved, or is not necessary, should enhancement of individual species be the management focus.

Managers should always view their objective, be it the integrity of the fish community or the well-being of a particular species, in light of the controlling variables of Figure 18.3. It would be unusual, however, to mount a total effort to achieve ecological integrity. Practical considerations and economic realities require that efforts be made selectively. Physical habitat is probably the best area in which to invest, because management for physical habitat (inchannel and riparian) will pay substantial dividends in influencing water quality, flow regimes, and energy inputs.

18.3.2 Habitat Consideration

The major problem of most warmwater streams is physical habitat degradation. Some habitat problems are solvable, but in many situations the damage is permanent and restoration efforts can be only minimally successful. Streams in their unmodified condition were in a physical state of dynamic equilibrium (see Chapters 8 and 9), i.e., were in balance in terms of erosion and deposition, and as far as a fishery was concerned, steady state conditions prevailed. A fact of life as we approach the 21st Century is that with over 300 or so million people impacting our waterways, few streams will be restored to presettlement conditions. Nevertheless, there is tremendous potential for warmwater stream habitat to be manipulated to provide an increase in angling benefits.

Any habitat improvement project should focus on instream, riparian, and watershed management. While an ideal strategy would include all three areas, this is usually not practical. Warmwater stream management has been delayed and hindered by the prevailing belief among professionals that "you can't manage a stream fishery without managing the watershed." Habitat managers now know that the most economical and important stream rehabilitation or restoration takes place within and immediately adjacent to the stream channel. Efforts in such areas have the greatest potential for a successful return for effort and expense invested.

Habitat restoration or habitat manipulation must relate to the needs of the fish. Studies on midwestern streams indicate a broad use of habitat by fish with many influential variables. Stream fish efficiently partition most available space, both by species, and by size classes within species. Some species select a particular subset of habitat variables while others are forced into existing in a particular set of conditions by another member of the community, either because of competition for space or anti-predation behaviors.

Some important factors that influence how a fish community uses a warmwater stream include:

1. Horizontal heterogeneity (Gorman and Karr 1978)
2. Vertical positioning (depth) mediated by predation risk or food (Gorman 1987)
3. Structure (Angermeier and Karr 1984)
4. Stream morphology and flow regime (depth/velocity interactions) (Schlosser 1982)
5. Current velocity and cover (Felley and Felley 1987)
6. Environmental variability (Horowitz 1978; Matthews 1986)
7. Substrate type (Paine et al. 1982)
8. Predation (Bowlby and Roff 1986)
9. Food (Berkman and Rabeni 1987).

Determining preferenda for important habitat variables for multiple life stages of every species in a warmwater stream is impossible (e.g., there are over 65 species in the Current River, Missouri) and even if attainable, probably an unwise course of action. It would be much better to concentrate on a single community-level objective. Although the relations between fish communities and their physical environment are highly complex in many instances, they may often be reduced to fairly simple factors of fish types (e.g., guilds) and fish sizes and their relation to broad, similar habitat conditions (Bain and Boltz 1989). The habitat conditions most commonly cited are current velocity, depth, substrate types, and cover.

The approach recommended here is simply to increase the range of habitat variables available to fish—increase habitat diversity. If fish community members segregate and compete along continua of habitat variables, it is likely that an increase in diversity of appropriate habitat variables would offer an increase in the quality and quantity of available habitat. Nevertheless, the efficient manager must choose those important habitat elements from a wide array of choices. Given the limited resources of money and time generally available, it is suggested that most objectives can be met by concentrating on increasing diversity of depths and physical structure types. The variety of depth conditions and not just the amount of deep water should be considered. Structures useful to all sizes of fish beyond just the usual fish "cover" factors, such as substrate interstices and any other factors which increase physical heterogeneity, should also be considered. Management activities to improve instream habitat should first emphasize reestablishing some of the natural sinuosity of the stream whenever possible. Such efforts will restore the hydraulic balance, minimize downcutting and bank erosion, and in general stabilize the streambed. The next step is to maintain or reestablish two essential conditions—appropriate depths and adequate structure.

Depth is an important factor for many sport fish because it (1) serves as a cover factor, (2) serves as a refuge during low water or during winter, and (3) increases the preferred habitat for many forage fish. Greater depth can be obtained in several ways (see Chapters 8 and 9). First, if natural meandering of the channel and bank stabilization can be encouraged, increased depth will often be a natural by-product. Central Indiana streams that were flanked by woods had pools 50% deeper than pools adjacent to areas of altered land use, including residential areas, meadows, and cultivated fields. Additionally, wherever cultivated fields were separated from the stream by strips of grass or trees, the stream was considerably deeper. If instream structures are placed properly, scouring during high water will increase pool areas. The installation of gabions (rock filled baskets) or rock structures may or may not result in scouring and pool deepening, but will often act to retain water which will serve the same function.

Structure can be defined as any physical obstruction in the stream which is sufficient to provide fish with camouflage, shade, or relief from the current. Structure in warmwater streams can be achieved by many of the methods detailed in Chapter 9. In this chapter we will concentrate on the usefulness of wood as structure. The importance of woody structure (logs, branches, rootwads, snags, log jams) to salmonid stream habitat is well documented (see Chapters 9 and 17). It may be that wood is even more important in warmwater streams because there are typically fewer boulders, cobble substrate, bedrock ledges, and other objects that serve the same function.

Wood, in all its forms, provides three important functions in low-gradient, small-substrate streams: (1) it interacts with the hydraulics to increase stream channel diversity, (2) it provides fish sanctuary, and (3) it provides excellent habitat for invertebrates.

Smallmouth bass and rock bass in Ozark streams associate more closely with individual logs, log jams (consisting of trunks and branches), and rootwads than with other habitat types (Probst et al. 1984). The highest densities of muskellunge in 14 Kentucky streams were found in reaches with the greatest number of fallen trees. This factor was determined to be such an important ingredient in muskellunge management that an active program to preserve riparian vegetation is now in place (Axon and Kornman 1986).

In a medium-size stream in an agricultural area of northern Missouri, stretches without appreciable snag habitat had densities of important game fish (common carp, channel catfish, green sunfish, bluegill, smallmouth bass, and crappies), which were 25% lower than areas possessing snags (Hickman 1975). In areas without snags, there were 51% fewer catchable-size fish than in areas where snags were present.

A small stream in Illinois was longitudinally divided for a distance of 30.5 m and woody debris was removed from one side (Angermeier and Karr 1984). On the cleared side, the bottom substrate became more homogeneous and of a finer particle size. Depth on the altered channel was reduced, as was organic litter on the bottom. These changes were apparently due to the way wood altered the flow regime which had previously assisted scouring and removal of fine particles. Sixty percent of the associated fish species were more abundant on the side with woody debris, especially the larger individuals. The adaptive significance of the associations between fish and woody debris was attributed to the wood's function as camouflage rather than to increased food availability or protection from current velocity.

In streams with low gradients and fine-particle substrates, woody structure provides an important, and sometimes the only, habitat for invertebrates that are used as food by fish. The Satilla River in southeast Georgia has three major benthic habitats: shifting sand in the main channel, muddy depositional backwaters, and submerged wood (Benke et al. 1985). Woody structures (snags) make up 4% of the total habitat surface, yet support 60% of invertebrate biomass and 16%

of invertebrate production. All fish species use the snags to some extent, while four of eight major species obtain at least 60% of their prey biomass from snags.

A wood structure has many advantages. It is inexpensive, natural looking, and is important in the riparian environment.

Conditions within a stream are a direct reflection of the condition of the riparian area, and stream management requires attention to more than just the channel. The reestablishment or enhancement of riparian vegetation is a common management need in most warmwater streams, because riparian corridors have been extensively altered by channelization, increased cultivation, or livestock grazing. While few fisheries biologists dispute the value of riparian management, they do not always agree on the appropriate dimensions, both length and width, of these "buffer strips." Determining the appropriate size of a vegetated area requires taking into account: (1) objectives for the buffer strip, (2) natural situation (especially topography and soil characteristics), and (3) actual or proposed human activities.

Riparian vegetation serves a number of functions depending upon its relative position to the channel. Vegetation adjacent to and overhanging the channel contributes to bank stabilization, moderation of water temperature by shading during the summer, and an increase in allochthonous energy in the form of leaf fall and terrestrial insects. Trees that fall into the stream act as retention devices for organic matter and increase the diversity of habitat for fish and invertebrates.

Riparian vegetation which is not adjacent to or overhanging the stream channel functions to reduce the amount of sediments, pesticides, herbicides, and excessive nutrients from reaching the stream, and to increase infiltration and stabilize the water table which reduces the magnitude of hydrograph fluctuations.

The few studies that have addressed riparian zone widths on warmwater streams have concentrated on an objective of reducing sediment. Forest Service biologists have quantified soil erodability, soil drainage, and stream slope in a model which determines appropriate distances from the channel for different activities in southeastern U.S. forests. Other published guidelines or regulations are often estimates by professionals who are familiar with local conditions. For example, planning documents developed by states show recommendations for vegetation strip widths ranging from a minimum of 6 m in the lower coastal plain of Georgia to up to 152 m in some areas of Alabama.

Perhaps as important as the quantity of riparian vegetation is its quality. The ideal riparian zone would have mature stands of woods with tree spacing sufficient to allow an understory of grasses and shrubs. Trees would be close enough to the stream so that exposed roots would supply rootwad cover and allow stable undercut banks to develop. Grasses and shrubs would have deep root systems, dense well-ramified top growth, a resistance to flooding and drought, and an ability to recover growth subsequent to inundation with sediment.

The effort to establish buffer widths to protect warmwater streams emphasizes the close associations that are necessary between research (what is the truth?) and management (making decisions with the best available evidence). Clearly, more research is needed to link the well-being of aquatic biota to land-use activities so that riparian management can be conducted in an efficient, effective manner.

Increasing habitat diversity may also be achieved by eliminating factors that reduce diversity, such as siltation. In a survey of U.S. waters, excessive siltation from erosion occurred in 46% of all streams and was considered the most important factor limiting usable fish habitat (Judy et al. 1984). Similar conclusions have been reached after observing changes in fish faunas as agricultural and other human activities increased (Smith 1971; Muncy et al. 1979).

Sediments impair the biota in a number of ways. While in suspension, they limit light penetration and reduce primary production of both algae and macrophytes. Turbidity is known to affect certain fish behaviors relating to feeding, social interactions, and other life requisites. More important, when sediment comes out of suspension and settles on and within the substrate, pools fill in and the channel environment tends toward homogeneity and away from discrete riffle–pool divisions. Distinct fish assemblages are often found in riffle, run, and pool areas of a stream. As siltation increases in riffles, fish dominance changes from riffle-specific species to more run- and pool-specific species, thus decreasing overall fish diversity (Berkman and Rabeni 1987).

Siltation often degrades spawning areas and may cause behavioral changes in spawning fish, an increase in egg mortality, or a decrease in larval growth and development. The species most affected are those requiring gravel or vegetation for spawning substrate, species with complex spawning behaviors including courtship rituals, and species that construct nests from bottom materials. Sedimentation may disrupt normal trophic relations. Primary production and invertebrate secondary production can be drastically reduced by siltation. Consequently, piscivores and invertivores, which include most sport fish, are often replaced by omnivores and detritivores.

Managers must determine where sediment comes from; however, there has been no easy or even acceptable way to measure the origins of eroded materials that enter streams. Historical attempts to control sedimentation delivery to streams have emphasized the encouragement of best management practices on lands in the watershed on the assumption that reducing erosion in the watershed results in reduced sediment delivery to streams. Thus, water quality goals and resource agencies have emphasized tile outlet terraces, strip cropping, grassed waterways, and water diversion structures. Recent evidence indicates that such efforts, if directed to improve stream water quality, are not sufficient (Karr and Schlosser 1978).

It is becoming increasingly clear that a major sediment contributor in physically degraded streams is the channel itself. Several studies in Illinois (summarized by Roseboom and Russell 1985) show that about 50% of the annual sediment load of streams is from the eroding streambed and associated banks. This emphasizes, once again, that stream channel stabilization should be a management priority.

Erosion within an entire watershed was documented by examining concentrations of radioactive Cesium 137, an isotope evenly deposited onto the landscape as a consequence of the 1950s atmospheric testing of atomic weapons (Wilkin and Hebel 1982). Concentrations of the isotope in different parts of the watershed were due to erosion and redeposition. This study indicated moderate erosion from the gently sloping upland areas. However, a forested buffer between cropland and the floodplain prevented practically all erosion onto the floodplain. The amount of sediment delivery to a stream from an active floodplain could be controlled by management. Floodplains planted in row crops eroded at excessive rates and delivered all eroded soils to downstream flow. Forested floodplains, even those grazed, served as depositional areas and removed significant amounts of sediments. A floodplain in pasture or orchard had sediment movement very nearly in equilibrium where deposition equalled erosion. Thus, once streambanks are stabilized, floodplain management should be the next priority.

Channelization is the most detrimental modification to a stream ecosystem. The act of straightening a stream is, by itself, harmful to aquatic life, but it is often accompanied by clearing and snagging of instream cover and the replacement of natural riparian vegetation with a grass berm or row crops. The devastating effect of channelization on the resident fish fauna is well documented (e.g., Schneberger and Funk 1973; Stern and Stern 1980). It is not unusual for the numbers or biomass of sport fish to be reduced >90% in a channelized area. Channelization is widespread in North America but most extensive in agricultural areas. Iowa, for example, has lost over 2,500 km of streams capable of supporting permanent fish populations.

There has been little evaluation of mitigation or restoration efforts, yet there are a variety of management options available to improve fish quality in channelized stretches of streams. Minimal efforts, in some situations, can be helpful. For example, short-reach channelization (<0.5 km) in Iowa could support the same biomass of catfish as unchannelized reaches as long as brush piles and trees accumulated in the stream. This may be a special case; however, it has proved useful when the resident fish fauna is already adapted to turbid, fine-particle substrate situations.

More extensive efforts, such as installing permanent instream structures, may or may not be successful in improving a fishery. The River Styx and the Chippewa River in central Ohio have been extensively altered. Stream channels were bounded by high banks and the instream structures (mainly single and double wing deflectors) that were installed occupied a comparatively small area. The areas with structures held significantly more species and higher numbers and biomasses than did sections of streams without structures (Carline and Klosiewski 1985). Nevertheless, important game species, mainly centrarchids, did not respond as expected. While the deflectors created favorable conditions for centrarchids, the low densities and small sizes of the fish provided minimal fishing benefits. Even though some habitat conditions were improved, the erratic flows caused reproductive failures. Because of the high banked channel morphology, rising waters during rains were confined to the channel and high current velocities caused destruction of the nests and eggs of many important species. Again, the problems associated with hydraulic stability return to plague the fisheries manager.

The Olentangy River, also in Ohio, is an example of a more successful restoration (Edwards et al. 1984). It had only a small percentage of its length channelized and an unaltered section separated the mitigated area from an old channelized section. The floodplain adjacent to the channelized and unaltered sections was forested. When high flows occurred the water spread out into the floodplain and inchannel velocities were moderated. A series of artificial riffles and pools were created by constructing five equally spaced riffles, each 6.2 m wide from bank to bank, with boulders over earthen fill. Intervening pools were 250 m long. The mitigating structures served to reduce the stress on the aquatic biota that was induced by channelization. The game fish in the mitigated area responded to food availability and habitat diversity and the relative abundance and condition of sunfish, crappies, catfish, and black bass approximated that of the natural area. The diversity of macroinvertebrates and fishes in the artificial riffles and pools was intermediate when compared with the natural and channelized site. The Olentangy River project was probably more successful than the others because of hydraulic factors. Not only was adequate physical cover provided, but flow conditions were such that high flows were moderated and the structures retarded the rate of dewatering during dry periods.

The ideal fisheries management response to channelization is to restore an impacted stretch to near its original physical condition. Big Buffalo Creek, a fourth-order, chert substrate, Ozark border stream in west-central Missouri, was extensively channelized in 1948 which caused the predictable loss of stream habitat diversity and channel stability. In 1966 the stream was restored by replacing 1.3 km of channelized stream with about 2.9 km of "new" channel. High water rearranged some of the channel morphology during the next 5 years but soon stability was established and the stream came into dynamic equilibrium. Little erosion or deposition has since occurred. The standing crop of fish in the new channel increased from 0.0 kg/hectare in 1966 to 67.5 kg/hectare of which 20.6 kg were smallmouth bass in 1972. Fish populations have since remained stable. The results of this and other work demonstrate that warmwater stream habitat can be managed successfully if emphasis is placed on stabilizing hydraulic conditions that improve riffle–pool development.

The above examples illustrate that the success of a mitigation or restoration project on a channelized stream depends upon the degree of initial straightening, gradient, channel cross sectional area, riparian and floodplain land use, as well as the target species or assemblage. Species commonly occurring in moderately turbid, low-gradient streams (e.g., some ictalurids) respond better to restoration than do species requiring less turbid, high-gradient environments (e.g., centrarchids).

In summary, habitat management for stream fish should first emphasize providing a physically heterogeneous environment that is temporally stable so that the normal assemblage of species for that size stream and geographic setting can exist for an extended period. Maintaining or restoring the dynamic equilibrium, or physical stability, is the first priority. Diversity of depth and structure is important and indirectly influences other important habitat features.

18.3.3 Managing Fish and Anglers

Effective fisheries management must eventually address the triad of people, aquatic organisms, and habitat. Warmwater stream managers have had to concentrate on the fundamental problem of habitat degradation. However, there are thousands of kilometers of fishable warmwater streams where habitat restoration as a management objective has been achieved or is unnecessary. In these waters, maintenance or enhancement is the objective; the waters would benefit from manipulation of fish and fishers.

Extensive stocking of warmwater streams is not a general policy in most states because either the need has not been perceived or stocking has shown to be unnecessary. Stocking of reclaimed or restored streams where the species previously existed may have merit. Stocking native species into existing populations is usually not necessary or warranted except in special circumstances. In Missouri, smallmouth bass introductions into already established populations did not significantly alter densities or biomass. However, stocking is a promising technique in situations where inconsistent reproductive success is a problem. In

Iowa, careful examination of "walleye streams" with low production has shown a possible problem with early young-of-the-year survival, and a program of stocking fingerling walleye is underway with encouraging results. Kentucky manages several native muskellunge streams and recommends an annual supplemental stocking of 17- to 22-cm fish to maintain a more constant recruitment (Kornman 1983). Stocking is expensive and any effort should be based on biological need, be well planned, and be closely evaluated.

Special environmental situations may require an unusual approach. For example, in Kentucky acid mine drainage has devastated the local fish fauna. The introduction of redbreast sunfish into eight affected drainage basins has proven highly successful because these fish are more tolerant of acid conditions and, perhaps because of reduced competition, they have shown excellent growth and size characteristics.

In some western states, where all warmwater sport species are introduced, options for stocking are both greater and more complicated. Oregon, for example, plans to expand stocking of warmwater stream fish into waters not suited for salmonids, to introduce new warmwater species into areas where current warmwater species are not meeting management objectives, and to develop a mixed trout-warmwater species fishery. Such introductions themselves are not without problems considering that introducing nonnative sport fish into western and southwestern North American waters has been implicated as a significant factor in the reduction or extirpation of many threatened or endangered native species (Minkley and Deacon 1968).

Sportfish regulations from the states and provinces are an indication of how the management of warmwater streams lags behind coldwater stream management. While states with trout and salmon often have regulations on a stream-by-stream or at least a region-by-region basis, fish in warmwater streams are generally either ignored, or given identical treatment to populations in lentic waters.

No fisheries biologist doubts that fish population dynamics are affected by the basic differences between lentic and lotic systems, yet only four states and provinces have extensive state or provincial warmwater stream regulations. About a quarter of the states or provinces have warmwater stream regulations applying to a very restricted set of streams, usually less than three, and often associated with a tailwater. The most common regulated fishes are black bass, walleye, and the esocids.

The majority of states with warmwater stream fisheries resources have no specific regulations for their resident fish, although, as noted in Table 18.1, many species are listed as game fish, and fishing for them is encouraged. It can only be concluded that state and provincial agencies do not yet acknowledge that warmwater streams deserve or require the type of management accorded other aquatic habitats. Management, however, must be based upon a sound informational base, and appropriate management strategies will surely follow advances in our understanding of the population dynamics, environmental requirements, trophic relations, and community-level dynamics of warmwater fishes.

18.4 REFERENCES

Angermeier, P. L., and J. R. Karr. 1984. Relationships between woody debris and fish habitat in a small warmwater stream. Transactions of the American Fisheries Society 113:716–726.

Axon, J. R., and L. E. Kornman. 1986. Characteristics of native muskellunge streams in eastern Kentucky. American Fisheries Society Special Publication 15:263–272.

Bain, M. B., and J. M. Boltz. 1989. Regulated streamflow and warmwater stream fish: a general hypothesis and research agenda. U.S. Fish and Wildlife Service Biological Report 89(18).

Benke, A. L., R. L. Henry III, D. M. Gillespie, and R. J. Hunter. 1985. Importance of snag habitat for animal production in southeastern streams. Fisheries (Bethesda) 10(5):8–13.

Berkman, H. E., and C. F. Rabeni. 1987. Effect of siltation on stream fish communities. Environmental Biology of Fishes 18:285–294.

Bowlby, J. N., and J. C. Roff. 1986. Trophic structure in southern Ontario streams. Ecology 67:1670–1679.

Carline, R. F., and S. P. Klosiewski. 1985. Responses of fish populations to mitigation structures in two small channelized streams in Ohio. North American Journal of Fisheries Management 5:1–11.

Edwards, C. J., B. L. Griswold, R. A. Tubb, E. C. Weber, and L. C. Woods. 1984. Mitigating effects of artificial riffles and pools on the fauna of a channelized warmwater stream. North American Journal of Fisheries Management 4:194–203.

Felley, J. D., and S. M. Felley. 1987. Relationship between habitat selection and individuals of a species and patterns of habitat segregation among species: fishes of the Calcasieu drainage. Pages 61–68 in Matthews and Heins (1987).

Funk, J. L. 1970. Warmwater streams. American Fisheries Society Special Publication 7:141–152.

Gorman, O. T. 1987. Habitat segregation in an assemblage of minnows in an Ozark stream. Pages 33–51 in Matthews and Heins (1987).

Gorman, O. T., and J. R. Karr. 1978. Habitat structure and stream fish communities. Ecology 59:507–515.

Hickman, G. D. 1975. Value of instream cover to the fish populations of Middle Fabius River, Missouri. Missouri Department of Conservation, Aquatic Series 14, Jefferson City.

Horowitz, R. J. 1978. Temporal variability patterns and the distributional patterns of stream fishes. Ecological Monographs 48:307–332.

Judy, R. D., Jr., P. N. Seeley, T. M. Murray, S. C. Svirsky, M. R. Whitworth and L. S. Ischinger. 1984. 1982 National fisheries survey, volume 1. Technical report: initial findings. U.S. Fish and Wildlife Service FWS/OBS-84/06.

Karr, J. R., and D. R. Dudley. 1981. Ecological perspective on water quality goals. Environmental Management 5:55–68.

Karr, J. R., and I. J. Schlosser. 1978. Water resources and the land-water interface. Science (Washington, D.C.) 201:229–234.

Karr, J. R., L. A. Toth, and D. R. Dudley. 1985. Fish communities of midwestern rivers: a history of degradation. BioScience 35(2):90–95.

Kornman, L. E. 1983. Muskellunge stream investigation at Kinnicanick and Tygarts Creeks. Kentucky Department of Fish and Wildlife Research Bulletin 68, Frankfort.

Krumholz, L. A., editor. 1981. The warmwater streams symposium: a national symposium on fisheries aspects of warmwater streams. American Fisheries Society, Southern Division, Bethesda, Maryland.

Matthews, W. J. 1986. Fish faunal structure in an Ozark stream: stability, persistence and a catastrophic flood. Copeia 1986:388–397.

Matthews, W. J. 1987. Physiochemical tolerance and selectivity of stream fishes as related to their geographic ranges and local distribution. Pages in Matthews and Heins (1987).

Matthews, W. J., and D. C. Heins. 1987. Community and evolutionary ecology of North American stream fishes. University of Oklahoma Press, Norman.

Minkley, W. L., and J. E. Deacon. 1968. Southwestern fishes and the enigma of "endangered species." Science (Washington, D.C.) 159:1424–1432.

Muncy, R. J., G. J. Atchinson, R. V. Bulkley, B. W. Menzel, L. G. Perry, and R. C. Summerfelt. 1979. Effects of suspended solids and sediment on reproduction and early life of warmwater fishes: a review. U.S. Environmental Protection Agency, Environmental Research Laboratory, EPA-600/3-79-042, Corvallis, Oregon.

Paine, M. D., J. J. Dodson, and G. Power. 1982. Habitat and food resource partitioning among four species of darters (Percidae: *Etheostoma*) in a southern Ontario stream. Canadian Journal of Zoology 60:1635–1641.

Pflieger, W. L. 1988. Aquatic community classification system for Missouri. Missouri Department of Conservation, Columbia.

Probst, W. E., C. F. Rabeni, W. G. Covington, and R. E. Marteney. 1984. Resource use by stream-dwelling rock bass and smallmouth bass. Transactions of the American Fisheries Society 113:283–294.

Roseboom, D., and K. Russell. 1985. Riparian vegetation reduces stream bank and row crop flood damages. Proceedings of the North American Riparian Conference, U.S. Forest Service, Rocky Mountain Forest and Range Experiment Station, Fort Collins, Colorado.

Schlosser, I. J. 1982. Fish community structure and function along two habitat gradients in a headwater stream. Ecological Monographs 52:395–414.

Schneberger, E., and J. L. Funk. 1973. Stream channelization: a symposium. American Fisheries Society, North Central Division, Special Publication 2, Bethesda, Maryland.

Smith, P. W. 1971. Illinois streams: a classification based on their fishes and an analysis of factors responsible for disappearance of native species. Illinois Natural History Survey Biological Notes 76.

Stern, D. H., and M. S. Stern. 1980. Effects of bank stabilization on the physical and chemical characteristics of streams and small rivers: an annotated bibliography. U.S. Fish and Wildlife Service FWS/OBS-80/12.

Trautman, M. B. 1981. The fishes of Ohio, 2nd edition. Ohio State University Press, Columbus.

Wilkin, D. C., and S. J. Hebel. 1982. Erosion, redeposition and delivery of sediment to midwestern streams. Water Resources Research 18:1278–1282.

Chapter 19

Large Rivers

ROBERT J. SHEEHAN AND JERRY L.
RASMUSSEN

19.1 INTRODUCTION

Most large rivers in North America have been modified to accommodate navigation, flood control, hydroelectric power generation, and other human uses. Large Arctic rivers, such as the Mackenzie-Peace and Churchill, still exhibit long reaches where constructed barriers do not limit fish movement, but these are exceptions. Dams have transformed long river reaches into impoundments, and many of these, such as the six large Missouri River mainstem reservoirs, exhibit few riverine characteristics. Many river reaches can no longer be classified as riverine based on structure or function.

It is difficult to define a large river. Characteristics used to establish the relative size of rivers, such as length, width, average annual discharge at mouth, and size of drainage basin, are highly correlated for pristine rivers, but anomalies appear among rivers modified by man. One of the world's largest rivers based on length and drainage area, the Colorado River, has been subjected to flow diversions and reduced to a mere trickle in its lower reaches and it is no longer permanently connected to the sea (Table 19.1). Rivers can be classified into three groups based on stream order: small or headwater (orders 1–3), medium (orders 4–6), and large (orders >6). This chapter focuses on the last category and introduces the reader to the ecology and fisheries resources of large rivers, contemporary issues in river management, and potential management strategies.

19.2 DIVERSE, DYNAMIC SYSTEMS

Large rivers are diverse systems featuring a variety of lentic and lotic habitats and biotic community assemblages. Natural rivers are not fixed in space because hydraulic processes continuously create new habitats and eliminate old ones. Rivers are dynamic systems (Figure 19.1) which can be highly interactive with other aquatic systems, such as tributaries, as well as riparian and floodplain habitats (horizontal interactions). Important interactions occur between river waters, riverbeds, and groundwaters (vertical interactions), as well as between upstream and downstream river reaches, mainstem reservoirs, estuaries, and marine systems (longitudinal interactions). Many human activities interfere with one or more of these interactive pathways. The interactive nature of large rivers suggests a holistic approach to understanding and managing resources.

Table 19.1 Drainage, length, and discharge for the largest North American rivers (Czaya 1981). The Amazon River is provided for comparison (world rank order in parentheses).

River	Country at mouth	Drainage area (1,000 km)	Length (km)	Average annual discharge at mouth (m³/sec)
Amazon	Brazil	7,180 (1)	6,516 (1)	180,000 (1)
Mississippi-Missouri	United States	3,221 (3)	6,019 (3)	17,545 (8)
Mackenzie-Peace	Canada	1,805 (12)	4,250 (13)	7,500 (19)
Nelson	Canada	1,072 (17)	2,575 (31)	2,300
St. Lawrence	Canada	1,030 (19)	3,100 (20)	10,400 (16)
Yukon	United States	885 (23)	3,185 (18)	7,000 (20)
Columbia	United States	669 (31)	2,250 (37)	6,650 (21)
Colorado	Mexico	629 (33)	3,200 (17)	168
Rio Grande (Rio Bravo del Norte)	Mexico and United States	570 (35)	2,870 (25)	82
Fraser	Canada	225 (49)	1,360	3,750 (29)

19.3 ECOLOGY

The river continuum concept was developed as a model describing natural stream ecosystems (Vannote et al. 1980), including large rivers (Sedell and Richey 1989). It provides a conceptual framework from which to approach the study of large river ecosystems. This concept proposes that ecosystem structure and function varies in a predictable manner from headwaters to estuary. It is based on the idea that natural watersheds are in a state of dynamic equilibrium, and that stream channel characteristics (width, depth, velocity, sediment load) tend toward an equilibrium determined by climate, geology, and watershed. Downstream ecosystem processes are linked to upstream processes, creating resource gradients and biotic communities that are predictably organized along these resource gradients.

According to the river continuum concept, large rivers depend more on fine particulate organic matter for energy delivered from upstream reaches and from periodic inundations of the floodplain, and less on primary production, due to their greater depths and higher turbidity (Sedell and Richey 1989). Thus, large rivers are predominately heterotrophic in nature, and energy is captured in the ecosystem by collector macroinvertebrates, such as black flies, caddisflies, and filter-feeding mollusks. Consequently, fish production is largely rooted in energy flows of fine particulate organic matter by collector macroinvertebrates.

Vertical, horizontal, longitudinal, and spatio-temporal interactions are disrupted by human modification of rivers, thereby disturbing and shifting the dynamic equilibria in and between the river and its watershed (Figure 19.1). The establishment of a mainstem reservoir provides a good illustration of how dams disrupt longitudinal interactions, affecting energy flow pathways. For example, 30,000 tonnes of planktonic crustacea are swept into the Missouri River via

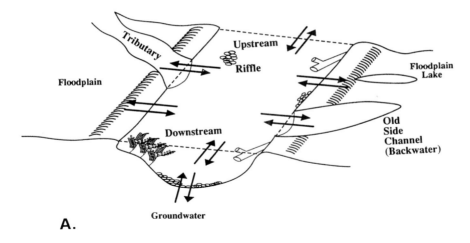

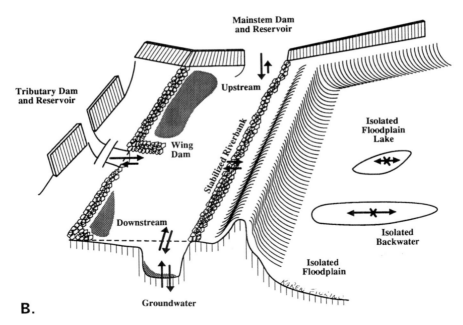

Figure 19.1 Horizontal (horizontal arrows), vertical (vertical arrows), and longitudinal (oblique arrows) physical, chemical, and biological interactions within the river complex and its drainage basin for (A) pristine (Ward and Stanford 1989) and (B) modified large rivers. Smaller arrows indicate diminished interactions in modified rivers. Channel stabilization also fixes the river in its bed, decreasing spatio-temporal interactions. Cross-hatched areas indicate deposition.

releases from one of its mainstem reservoirs (Ward and Stanford 1989). This can lead to increased fish production, but also to shifts in species composition and relative abundance. The serial discontinuity concept (Ward and Stanford 1989) has been proposed for impounded rivers in an attempt to account for the interruptions in longitudinal interactions and gradients imposed by dams. However, dams create gradients within pools; lotic conditions below dams grade into lentic conditions above the next dam. Thus, dams convert downstream reaches to conditions found in more upstream reaches (increased velocities, narrower channels, and streambed degradation). It may be appropriate to consider dams as "retrogressive discontinuities" in rivers.

19.4 CLASSIFICATION AND DESCRIPTION OF LARGE-RIVER HABITATS

Habitat use and requirements are important concerns because human activities tend to diminish habitat diversity in rivers. Factors to be considered in describing fish habitats are velocity, depth, substrate (composition, particle size, stability), structure (woody debris or snags, rock outcroppings), bank coefficient (the ratio of riverbank length to river length), presence of other biotic communities (macrophytes, mussels), salinity, and dissolved oxygen. Habitat use differs by life history, during a 24-h period, with the seasons, and with changes in discharge, making descriptions of fish habitat requirements difficult.

River habitats can be divided into either channel or extra-channel habitats. Channel habitats include riffles, runs, and pools. Sandbars and islands are important features in channel habitats because they improve habitat diversity for depth, substratum, and velocity, thereby promoting more diverse biotic communities. Islands also increase bank coefficient, which is strongly correlated with fish production.

Longitudinal gradients in physical, chemical, and biotic factors vary predictably in pristine rivers; however, dams disrupt these gradients. Physical characteristics at the upper ends of the reservoirs are often very similar to preimpoundment conditions, with prominent islands and side channels. Rivers become more lentic toward the lower ends of reservoirs; side channels, backwaters, and islands are submerged immediately upstream from the dam.

Major habitat types (Figure 19.2) in rivers modified for navigation have been described for impounded portions of the Mississippi River. This river has been highly modified to accommodate navigation via a series of 28 locks and dams and other navigation aids. A similar classification scheme for river habitats could be applied to other large river systems.

19.5 MULTIPLE USE

Large-river managers deal with multiple-use conflicts because fisheries resources have been considered secondary to other uses. We know relatively little about large-river fisheries in North America because anthropogenic alterations in hydrogeomorphology and ecology have come at a faster rate than scientific knowledge. Large Canadian rivers, though relatively free of human modifications, have not received a great deal of attention from fisheries researchers. Large rivers typically form political boundaries, obscuring responsibility for their management and protection. Effective exploitation of the resource often requires management of tributaries and the land masses they drain. No amount of stream habitat management will compensate for pollution from the watershed.

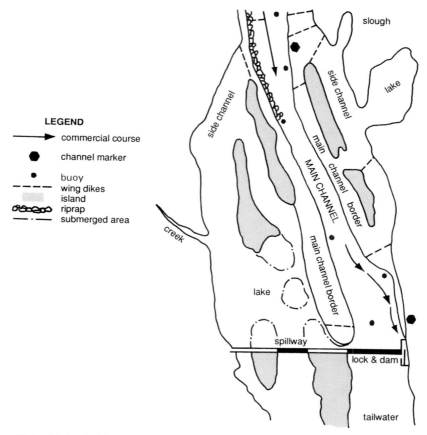

Figure 19.2 Major habitat types in the upper Mississippi River (Rasmussen 1979).

There have been attempts to return modified rivers to more natural conditions; the Kissimmee River in Florida is one example. However, from a pragmatic standpoint, impoundment and many other modifications for navigation and flood control can be considered as *fait accompli*. It is unlikely that there will be any large-scale dismantling of the inland waterway system. Therefore, biologists should consider using innovative approaches to managing large rivers under existing conditions. Management efforts should be directed at maintaining a diverse ecosystem and optimizing exploitation of renewable resources, such as sport and commercial fishes. Some management goals may necessitate artificial manipulations, such as environmental engineering of habitats or manipulations of species composition. These measures are appropriate because most large rivers no longer behave entirely as natural systems. Emphasis on conservation, rather than preservation, thus appears to be the more realistic management approach in rivers modified by humans. However, protection of unique natural habitats (such as the wild and scenic reach of the Missouri River in Montana) and threatened and endangered species, must be considered viable management practices for these resources.

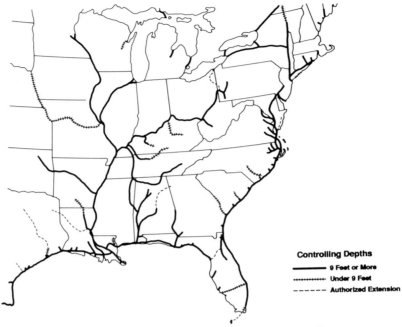

Figure 19.3 Rivers modified for navigation in the eastern United States.

19.5.1 Impoundment

All large rivers of the United States have been impounded in at least some reaches. The agency charged with maintenance of the U.S. inland navigation system is the Army Corps of Engineers. Waterways can be roughly divided into four interconnecting systems (Figure 19.3): (1) Great Lakes and connecting rivers; (2) Atlantic Seaboard; (3) Gulf States; and (4) at the center, the Mississippi River and its tributaries. The Mississippi River system comprises more than one third of the 40,000 km of navigable inland waters (Nielsen et al. 1986).

Establishment of the Mississippi River navigation system illustrates the historical use pattern in large rivers. The Ohio River became a traffic route during the earliest days of settlement in the Mississippi River valley. The first steamboat appeared in 1811. Rapidly swelling river traffic provided impetus for the Rivers and Harbors Act of 1824 which authorized snag removal from the Ohio River between the city of Pittsburgh and its confluence with the Mississippi River, marking the beginning of sanctioned habitat modifications.

Shoals and riffles of the natural river impeded boats except during times of high waters, so river training structures were designed to provide deeper navigation channels in the Mississippi River system. A series of wing dams were constructed by 1828 on the Kanawha River, an Ohio River tributary, to establish a 1-m deep navigation channel. This was followed by the completion of a system of 10 low-lift locks and dams in 1898, creating a 2-m deep channel for 142 km upstream from its confluence with the Ohio River. The lock system was replaced by four high-lift structures between 1931 and 1937, establishing a 2.75-m deep navigable depth. Between 1837 and 1866, 47 back-channel dikes and 111 training dikes were

Figure 19.4 Navigation lock (foreground) and dam. Dams such as this one are operated to maintain navigable depths in upstream pools and for flood control; locks permit vessel passage. Upstream fish movements are restricted to "locking through." (Photograph by S. Scherck.)

constructed on the Ohio River. This was followed by five low-lift locks and dams between 1875 and 1900 which provided a 2-m deep navigation channel. The completion of 51 low-lift locks and dams from 1900 to 1930 increased the Ohio River navigation channel to 2.75 m deep. The dams permitted fish and vessel passage during periods of high water, but segmented the river into a series of stair-stepped pools. Modernization of the Ohio system since 1930 has involved replacing the structures with 13 high-lift locks and dams. These high-lift dams elevated navigation pools as much as 10.5 m above downstream channels, restricting upstream fish passage.

Modification of the Mississippi River followed later, but paralleled the Ohio River. A 2.75-m deep channel was established with the construction of 28 high-lift locks and dams beginning about 1930 (Figure 19.4). Lock-and-dam systems convert rivers from a free-flowing condition to a series of impoundments. Impoundments widen some reaches and greatly increase habitat complexity due to inundation of floodplain habitat. The immediate effect of impoundment on some fishes, such as centrarchids, is beneficial by providing more habitat.

Channel degradation (deepening) occurs downstream from dams because water velocities can be greater due to upstream impoundment and constriction of the channel. Streambed degradation can occur for hundreds of kilometers downstream of dams, as on the Missouri River. Deepening of the channel causes adjacent extra-channel habitats (side channels and backwaters) to become dewatered.

Impoundment attenuates sediment deposition by reducing hydraulic gradient and flow velocities. Consequently, the boon to river fish populations from inundation of floodplain habitats can be relatively short-lived. Most backwaters of the Mississippi River will be lost to sedimentation in the next 50–100 years, a very short time on the geological time scale.

Roughly one-third of the Missouri River has been transformed to reservoirs and one-third has been channelized or stabilized. Impoundments and river-training structures have reduced most of the Missouri River to a single channel. Islands, side channels, backwaters, and sloughs have been eliminated. Hesse et al. (1989) estimate that 41,000 hectares of Missouri River habitats have been lost.

The Illinois River—near its confluence with the Mississippi River—consisted of a main channel and a network of side channels, backwaters, wetlands, and sloughs immediately after impoundment, but river fish no longer have access to much of that habitat due to sedimentation and levee construction. Soon all that will be left in many reaches will be the main channel. Together with the effects of pollution, aquatic habitat loss has led to a decline in the Illinois River commercial harvest from 10% of the entire U.S. freshwater catch in 1908 to virtually nothing in the 1980s (Karr et al. 1985).

Ecological succession is a natural process that eventually leads to the conversion of extra-channel aquatic habitats to dry land. However, river training and bank stabilization have attenuated the process. Also, agricultural practices and construction in watersheds, along with dredge-disposal programs, have accelerated aggradation in large rivers. Unfortunately, large rivers have been purposefully fixed in their beds, and natural fluvial processes that create new aquatic habitats have been all but eliminated.

The immediate effects of navigation dams and their associated floodplain levees on fishes can be highly variable, but some generalizations are possible: (1) habitats upstream of dams become favorable to lentic species, (2) riffles become inundated so fishes dependent on this habitat are impacted or eliminated, (3) species dependent on seasonal flooding of terrestrial habitats for reproduction are selected against, (4) the abundance of fish increases above dams because surface area is increased, and (5) navigation dams can block longitudinal fish migrations except for limited "locking through." Many of the positive effects are short-lived and fish species diversity and abundance can be expected to eventually decline below preimpoundment levels.

19.5.2 Navigation Aids and Flood Control Systems

The lower Mississippi River has been modified for navigation by the construction of deflection dikes to accommodate navigation during low-flows. Wing dikes constructed perpendicular to the river's flow divert water into the navigation channel (Figure 19.5) and deepen it through scour. Shorelines opposite wing dams are often armored to prevent erosion. Closing dikes are also used to divert flows from side channels to the navigation channel.

Most rivers modified for navigation are extensively leveed as a flood-control measure (Welcomme 1979). Levees block horizontal interactions, such as transfers of organic matter from the floodplain to the river. For example, levees have reduced the floodplain by more than 90% in the lower Mississippi River. Levees can limit reproduction of fish that depend on inundated floodplain habitats for spawning. They also can cut off many of the river's natural meanders and isolate floodplain lakes that provide spawning and nursery areas.

Dikes and channel stabilization structures protect the streambank from erosion, maintain channel alignment, and fix the river in its bed. The lower Mississippi River has been channelized and stabilized by more than 1,300 km of revetments.

Figure 19.5 L-head wing dam. Wing dams divert water to the channel to facilitate navigation. Wing dams increase channel border habitat and create slackwater habitat on their downstream sides but often diminish flows through side channels. Most wing dams jut out perpendicularly from the bank in a straight line, creating a scour hole at their tips. The scour hole is lost, but stratification becomes more stable on the downstream side with the L-head configuration. (Photograph by S. Scherck.)

The effect has been to halt the creation of new river habitat and to promote the loss of off-channel habitat through reductions in flow and accompanying sediment deposition.

19.5.3 Navigation Traffic

Virtually every biological component of the river ecosystem is affected by barge traffic, yet it is held by some that navigation-use-related impacts are imperceptible (Wright 1982). Towboats (which actually do more pushing than pulling) and barges (Figure 19.6) displace water during passage. One effect of boat-induced wave action and turbulence on lotic ecosystems is to increase suspended solids. A study of the Illinois River during and after a 2-month period when boat traffic was greatly reduced indicated that the return of boat traffic increased suspended solids by 30 to 40%. The suspension of sediments during vessel passage can impact fish. Alabaster and Lloyd (1980) concluded that any increase in suspended solids above a low level may cause a decline in the value and status of a freshwater fishery and that the risk of damage increases with concentration.

Most of the sediment suspended by boat passage originates in the watershed, but some may be generated by the erosional forces exerted by boats. Tows produce three successive hydraulic effects during passage: (1) a slight rise in water as the bow passes, (2) a dewatering of the shoreline and small tributaries, and (3) the return of the water in a series of waves as the stern of the towboat passes. The vertical drop in water level can be substantial (on the order of 0.5 m), and if shore or extra-channel areas have a shallow slope, a considerable portion of aquatic habitat can be dewatered. Nearby backwaters can also be dewatered if the channel is small. Dewatering of shoreline and off-channel habitats due to boat passage can kill larval fish. The proportion of damaged fish eggs has been shown

Figure 19.6   Tow boat pushing a barge string. The specific impacts of navigation on river biota remain controversial. (Photograph by S. Scherck.)

to increase subsequent to barge passage in the upper Mississippi River. Wave heights created by both recreational and commercial boats are often sufficient to cause erosion. Barges traveling upstream in the river can double or triple water velocities, while downstream barges can reverse the river's flow.

19.5.4   Electric Power Industry

The electric power industry also affects fish populations and river ecology. Use by this industry can be divided into two broad categories, direct use of a river's energy for generating hydroelectric power, and withdrawal of river water for the dissipation of waste heat from fossil fuel and nuclear power plants. Water is either returned to the river at a higher temperature or passed to cooling towers and lost as steam in the latter case.

Environmental concerns over power plants focused on effects of heated effluents in the 1960s and 1970s. These are less important in large rivers compared to other surface waters because of the large volume of water available for heat dissipation; however, power plant shutdowns can lead to thermal shock of fishes wintering in heated effluents. Increases in river temperatures can also lead to supersaturation of dissolved gases and increase the toxicity of pollutants such as ammonia.

A long-term debate over the construction of six power stations on the Hudson River culminated in the conclusion that the major impacts would be impingement and entrainment of fish (primarily eggs and larvae) in the cooling water intake system (Limburg et al. 1989). Mitigation of these impacts has taken the form of hatchery production and stocking of striped bass fingerlings in the Hudson River.

Water releases from hydroelectric dams to downstream reaches often coincide with peak electrical demands, which have little to do with the needs of the river biota. Peak demands occur daily (morning and evening) or seasonally (midwinter for heating and midsummer for air conditioning). Hydrographs in rivers such as the Columbia, which in the pristine condition were primarily influenced by spring runoff, are now predominantly driven by electrical power consumption, often with disastrous effects on early life stages of fish (Figure 19.7). Young salmonids become stranded on land due to dewatering when releases from dams are

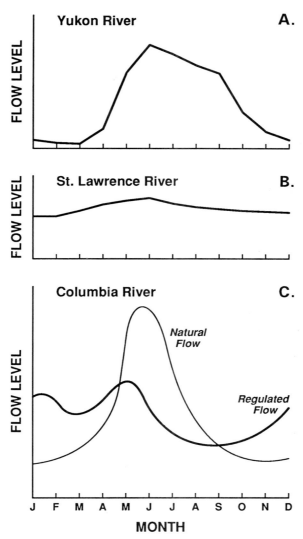

Figure 19.7 Effects of season (spring thaw), catchment characteristics, and storage reservoirs on discharge in large rivers: (A) Hydrograph (average annual discharge) of the Yukon River demonstrating the effect of season on discharge; this river is an extreme case because its drainage basin thaws over much of the year. (B) Hydrograph of the St. Lawrence River, another extreme case; fluctuations in discharge are reduced by the massive storage capacity of the Great Lakes, the river's source (Czaya 1981). (C) Generalized discharge pattern in the Columbia River under natural and regulated flow regimes (Ebel et al. 1989); regulated releases from mainstem and tributary storage reservoirs affect discharge in the Columbia River much as the Great Lakes do for the St. Lawrence River.

curtailed. Flows are completely halted during the night in some reaches of the Columbia River system, having effects on migratory fishes which orient to currents. These types of situations have stimulated interest in determining and establishing minimum instream flows to protect fish populations and channel morphology.

Water quality in hydropower dam releases is also a concern. Releases from 10 Tennessee Valley Authority dams failed to meet the Tennessee water quality standard of 5 mg/L dissolved oxygen, and five were at less than 1 mg/L (Voigtlander and Poppe 1989). Toxic metals, such as iron and manganese, which precipitate as oxides when dissolved oxygen is high, reenter solution when dissolved oxygen is low, also harming fish. High-head hydropower dams can also cause gas-embolism disease in fish, when the falling water entrains air. Gases from entrained air bubbles enter solution due to the increased hydrostatic pressure, and these gases become supersaturated when the water leaves the plunge basin.

There is little doubt that power plants can impact fish populations; however, they can also provide unique opportunities for enhancing river fish populations (Lewis 1987). Inexpensive electrical power available diurnally during periods of low electrical demand can be used for aeration of backwaters susceptible to oxygen depletion or for other beneficial activities. Power plant waste heat can be used to stimulate plankton blooms in sport fish rearing ponds and for accelerating growth of fingerlings for river stocking programs.

19.5.5 Pollution

Water pollution can be divided into five general classes: (1) eutrophication, (2) putrecible organic matter (sewage wastes) that causes dissolved oxygen depletion, (3) suspended solids, (4) toxic chemicals (ammonia, metals, acids and bases, biocides, detergents, petroleum hydrocarbons), and (5) thermal pollution. The first four categories have profound influences on large rivers, although water quality related to the first two has generally improved since the 1960s.

Decades of short-sighted land-use practices resulting in erosion and sedimentation, culminate today in the near total loss of extra-channel fish habitats in many rivers. Also, North American industry produces more than 55,000 different chemicals, many of which eventually find their way into rivers.

Heavy metals (e.g., mercury, selenium) and halogenated-hydrocarbon pesticides have reduced exploitation of several large-river fisheries. The discovery of PCBs in the Hudson River ecosystem below Hudson Falls forced closure of the freshwater fishery indefinitely (Limburg et al. 1989). The Food and Drug Administration detected PCB concentrations exceeding the action level in common carp from the Mississippi River, causing the destruction of 27,000 kg of presumably contaminated fillets. Despite the ban on sales and manufacture of PCBs since 1977, 60% of upper Mississippi River shovelnose sturgeon recently examined exceeded the action level of 2 mg PCB/kg. The long-term trend for PCBs appears to be on the decline, but just as the DDT contamination era was supplanted by PCBs, other toxic substances are on the increase.

The most significant pollutant in surface waters in North America is suspended solids. Suspended solids have had greater influence on fish diversity and abundance in the midwestern United States than any other factor. The source of most suspended solids is soil erosion, and soils become highly erodible when terrestrial vegetation is denuded due to agriculture, intensive forestry, or construction in the watershed.

Sewage treatment plants contribute suspended solids to streams (Figure 19.8), but primary environmental concerns related to sewage center around eutrophication, elevated ammonia, high biological oxygen demand, and disinfectants such as

Figure 19.8 Sewage treatment plant effluent. The turbid effluent visible near the riverbank resulted from storms that caused the plant to process water beyond its design capacity. Total ammonia was in excess of 50 mg/L in the plume. Catastrophic events such as this one as well as the cumulative effects of all industrial and municipal effluents should be taken into account when assessing pollution problems in rivers. (Photograph by L. Helfrich.)

chlorine which combine with ammonia to form toxic chloramines. Biologists have noted the near elimination of mollusks, such as mussels, for long distances downstream of sewage treatment plants. Sewage effluents can account for a significant portion of total discharge and of the allochthonous organic inputs (e.g., 24% of the total organic carbon inputs to the Hudson River; Limburg et al. 1989).

The threat of catastrophic pollution events (spills) is greater in large rivers than other aquatic systems. Large rivers attract industry to their shores because of the availability of water, inexpensive shipping, available hydroelectric power, and because these same factors promote urbanization, and provide a work force and support facilities. Consequently, there is a continuous threat of spills from industrial storage facilities and barges which transport the raw materials, products, and by-products of industry.

There is, however, cause to be encouraged by trends in water quality in some large rivers in recent decades. Many river reaches have followed a scenario similar to the Kanawha River, once considered one of the most polluted streams in the United States. Little information is available regarding the historical effects of pollution, but Addair (1944) noted that fishes were absent in the pool most severely affected by pollution. The biological oxygen demand was extremely high and dissolved oxygen ranged as low as 0 mg/L in some reaches as late as 1970, while ammonia concentrations and turbidity were elevated, lead and zinc were at toxic levels, and pH fluctuated from 4.2 to 9.5. A trend towards improved water quality in the Kanawha River began in the 1960s due to the enforcement of wastewater treatment requirements set by the Federal Water Pollution Control Administration, founded in 1963. Recent data indicate that only fecal coliforms and iron levels violate West Virginia standards. Average mainstem dissolved oxygen concentrations exceed 8 mg/L and dissolved oxygen minima during a recent 3-year study did not fall below 4 mg/L. The fish community now appears typical of large rivers in terms of standing crop and diversity.

19.5.6   Flow Diversions

River waters are diverted to supply irrigation for crop lands, municipalities, and to augment the needs of the electrical power industry. Flow diversions have also been used to scour and deepen harbors. The Colorado River is an example of the effects of flow diversions. A reduction in discharge is tantamount to a loss of habitat. Entire species can become eliminated when discharge is reduced below critical levels. Spawning migrations of anadromous fishes can also become jeopardized.

19.6   FISHERIES

19.6.1   Sport Fisheries

Sport fisheries are often underexploited in large rivers; even so, fishing is a leading river recreational activity. Angling is the leading recreational use of the Missouri River in Missouri (Weithman and Fleener 1988), and accounts for more than 35% of the total recreational activity in Pools 11 to 22 of the upper Mississippi River.

Major sport fishes of large rivers vary predictably across the climates of North America. The annual thermal cycle dictates what groups will be abundant, so the temperature altering effects of impoundments have led to deviations from the natural pattern. Coldwater species dominate the large Arctic rivers of Canada, but the commercial fishery reaps most of the harvest. Impoundment has led to an increase in predators of young salmonids (e.g., northern squawfish, bass, and walleye) in the Columbia River. The walleye is gaining importance in the Columbia River sport fishery, and other introduced fishes enter the creel, such as largemouth bass, smallmouth bass, yellow perch, crappie, and catfish (Ebel et al. 1989).

Warmwater species are important in rivers at lower latitudes and elevations. The Missouri River supports 13 sport species: channel catfish, freshwater drum, and common carp are frequently harvested. The sport fishery on the upper Mississippi River consists of about 30 species. Bluegill and crappie are often the most sought-after fishes in warmwater rivers, but a number of other fishes are important, such as channel catfish, bullheads, largemouth bass, freshwater drum, northern pike, and common carp. Walleye and sauger are important in the northern reaches of the river.

Season, river stage, flow, and habitat affect harvest rates and species harvested. The sport fisheries in unchannelized reaches of the Missouri River tend to be about 2 to 2.5 times more productive than they are in channelized reaches (Groen and Schmulbach 1978). Tailwaters below mainstem dams and spillways, main channel border dike fields, and backwaters generally produce the greatest yields in impounded rivers. Tailwater fisheries are generally good year-round especially for walleye and sauger, even when much of the river is ice covered. Backwaters with adequate depth and habitat diversity can provide year-round fishing.

Table 19.2 Major commercially harvested finfish of the upper Mississippi River and its tributaries.

Bowfin	Paddlefish	Buffalo
Bullheads	Catfish	Shad (*Dorosoma*)
Suckers (*Moxostoma*)	Northern pike	Pickerel spp.
Gar	Shovelnose sturgeon	Mooneye
Burbot	Lake whitefish	Quillback carpsucker
Common carp	Grass carp	Freshwater drum
White bass	Crappie	Yellow perch
Walleye		

19.6.2 Commercial Fisheries

The designation commercial fishery is more of a management concept than a biological one. Angler preferences, angler effort, market demand, and management agency philosophy are key factors determining whether commercial harvest is permitted. Highly sought game fish are routinely harvested commercially in large rivers. It seems inevitable that the number of species on commercial lists will decline. If the trend towards accelerated sportfishing pressure continues, recreational fishing demand for many river fishes will increase to the point where it will become uneconomical to permit their commercial harvest. The relative value of sport fish to local economies greatly outweighs their value as commercial fish, especially when angler cash outlays and resulting ''ripple effects'' are considered.

There are about 20 commercially harvested finfishes in the Mississippi River and its tributaries (Table 19.2). Some of these species support commercial fisheries in other large warmwater rivers of North America. Although data are not available in many cases, the untapped commercial fishery resource appears to be considerable in many large warmwater rivers.

Approximately 95% of the catch and 99% of the value of the upper Mississippi River commercial fishery consists of common carp, buffalo, catfishes, and freshwater drum. Total commercial landings in 1975 from the Mississippi River (including tributaries) totaled 4,651 tonnes (an average of 22 kg/hectare of river at lowflow) with an ex-vessel value of $1,942,000 and a processed value of $5,000,000 (Fremling et al. 1989). More than 30,000 kg of channel catfish and 223,781 kg of common carp were taken commercially from the Iowa, Missouri, and Nebraska portions of the Missouri River in 1976 and 1977 (Hesse 1982).

There is little doubt that anthropogenic factors greatly influence exploitation of commercial fisheries. The majority of the commercial catch consists of fishes that are largely benthophagic. Being benthophagic, commercial fishes consume prey that are in close contact with sediments, and they themselves are often in proximity to sediments which they disturb and mix. Many commercial species are at high risk regarding the uptake of contaminants that are associated with sediments. Thus, a major responsibility of management agencies is to monitor contaminant concentrations in commercial species. Chlordane and PCB contamination has recently forced management agencies to close the commercial harvest of certain species in rivers of the midwestern United States.

Human activities also dramatically affect the abundance of commercial species, as reflected in commercial catch. Commercial catch in the upper Mississippi River appears to be related to total water surface area in navigation pools, and commercial catch per unit area (CPUA) also appears to be positively correlated with the proportion of shallow marshy habitat, and negatively correlated with barge traffic. The commercial catch from the Illinois River was estimated at 77 to 200 kg/hectare in 1908, and CPUA was highest in reaches with the highest proportions of backwaters and lakes connected to the river (Richardson 1921). Catch per unit area declined to 45.6 kg/hectare by the 1950s and to 8.4 kg/hectare in the 1970s (Fremling et al. 1989). The decline in the commercial catch in the Illinois River during this century can be directly attributed to habitat loss.

Two groups of invertebrates, freshwater mussels and crayfish, are harvested in significant numbers from large rivers. Mussels were once taken in great numbers for the pearl button industry in the early part of the century in several eastern and midwestern U.S. rivers. This industry declined with the invention of plastic buttons and interest in the resource faded.

Today, the primary market for mussels is the Japanese cultured pearl industry, which began using mussel shells as seeds for pearl formation in the 1970s. Commercially important mussels are limited to species that have heavy shells and reach a fairly large size. Suitable pearl "blanks" can only be cut from the thick shells of these relatively large species. The commercial mussel fishery is currently highly profitable, with high-quality shells selling for as much as a dollar per pound—this puts considerable pressure on the resource. The majority of the harvest comes from the Mississippi and Tennessee river systems.

With the rekindling of the commercial mussel industry came renewed concern for mussels and mussel communities, also sparking interest in protection of endangered mussels. These concerns were heightened in the 1980s when numerous mussel die-offs were reported in many large rivers.

Crayfish are harvested for use as bait and food. The two primary species taken as food are the red swamp crawfish *Procambarus clarkii* and the white river crawfish *Procambarus acutus*. They are harvested from the lower Mississippi River and the Atchafalaya River basins to satisfy the growing demand for Cajun food delicacies. Harvests are greatest in high-water years when more of the floodplain is inundated, providing the decaying vegetation fundamental to the food web supporting crayfish production. Most of the harvest is in Louisiana (more than 20 million kg annually), where the unique topography results in huge overflow swamps.

19.7 FISHERIES MANAGEMENT

Fisheries management can be accomplished through manipulating (1) fish stocks and other biotic communities, (2) anglers, and (3) the physical and chemical environment. All three of these options are greatly affected by anthropogenic factors associated with other primary uses in most large North American rivers.

Large rivers remain among the few inland waters where sport-fish harvest can be significantly increased. However, it is generally agreed by fisheries managers that little can be done using standard population manipulation techniques. Conversely, harvest manipulation and habitat management can be effective and economical. A number of important river fisheries already depend primarily on fish culture and stocking, and this trend may increase in the future, unless natural habitats are restored.

19.7.1 Stock Assessment

The first step in managing a fishery, before any consideration should be given to manipulation, is to gather information about the fish population such as growth, abundance, recruitment, mortality, yield, exploitation, age and size structure, and habitat requirements. A multiple-gear approach is usually warranted in sampling fish communities in large rivers because of biases associated with various gear types and because of strong interactions between the environment and sampling efficiency. Commonly used river sampling tools include gill nets, trammel nets, seines, trawls (bottom and midwater), hoop nets, fyke nets, frame nets, larval fish nets, and electrofishing gear.

19.7.2 Fish Population Manipulations

Population reductions in large rivers are generally considered infeasible due to the magnitude of the resource and expense involved. When needed, such reductions are usually handled through the establishment of a commercial fishery. Managers can also enhance populations of large piscivorous fishes through stocking or catch restrictions to reduce forage species, such as shad and alewives, when they become too abundant.

Just as they receive all the point- and nonpoint-source pollution in their watersheds, large rivers are also the receptacle of downstream drift from every upstream minnow bucket, aquarium, hatchery, pond, and reservoir. Thus, the potential for species introductions is constantly present. Fortunately, these introductions often have little if any impact on large-river fish populations, although introduced warmwater and coolwater fishes have become serious threats to young salmonids in the Columbia River as previously noted. There is also concern that the history of the common carp in North America may be repeated by other carp species such as the grass carp (stocked for vegetation control in small impoundments). The grass carp, introduced originally in Arkansas, has found its way as far upstream in the Mississippi River as Pool 5a on the Minnesota–Wisconsin border. Natural reproduction has not been reported in the Upper Mississippi River, but there is evidence of reproduction in the Lower Mississippi, Arkansas, Red, Black, and Oachita rivers.

Reservoir stocking programs and species introductions can have dramatic effects on river fisheries due to spillway escapement, sometimes creating quality tailwater fisheries and sometimes causing impacts on endemic river fishes. The early assumption was that reservoirs would not support self-reproducing fish populations, so initial stocking focused on indigenous species. Today, more nonnative species, stocks, and hybrids are being introduced, such as striped bass and its hybrids; rainbow trout and brown trout; kokanee, cisco, rainbow smelt, alewife, and saugeye. Brown trout and rainbow trout introductions have been highly successful in Tennessee River mainstem reservoirs (Voigtlander and Poppe 1989). Striped bass and hybrid striped bass introductions, however, have resulted in high-profile trophy fisheries. The striped bass will not successfully reproduce in most of these reservoirs, but inability to reproduce is often desirable for an introduced species because it gives the manager more control over predator–prey balance.

Stocking may be the only answer if spawning has been disrupted, spawning grounds have been destroyed, exploitation is extensive, production of a given species is reduced, or a new species is desired. Stocking is now the mainstay in many salmonid fisheries because of dams that block spawning migrations, the degradation of spawning grounds, and other factors. In 1977, about 85% of the coho salmon harvested off the Columbia River coast came from hatchery production (Ebel et al. 1989). Stocking can also be used to reintroduce and enhance threatened and endangered species (Chapter 15). Missouri's paddlefish stocking program was stimulated by the construction of the Harry S. Truman Dam on the Osage River which inundated the spawning grounds of this species. Currently, however, stocking is being discouraged in many rivers due to the threat of disrupting the genetic integrity of wild populations.

19.7.3 Habitat Manipulations

Habitat manipulation can be a cost-effective method for enhancing river ecosystems and fish communities. However, most habitat manipulation projects on large rivers are complex and expensive. Input from biologists and engineers with a variety of specialties is required to ensure that a project has the desired effects, and that it endures beyond the next period of high water.

Habitat modification and enhancement is a relatively new science to large warmwater rivers, but a handbook of potential mitigation and enhancement techniques for large rivers has been compiled (Schnick et al. 1982). Most of the earlier work on habitat manipulation was directed at large coldwater rivers in the western United States to ensure passage and continued propagation of anadromous species (Chapter 23). The Upper Mississippi River System Master Plan initiated a major environmental management program in 1987, which includes habitat rehabilitation and enhancement projects. Planned projects include backwater dredging, dike and levee construction, island creation, bank stabilization, side channel openings and closures, wing and closing dike modification, aeration and water control systems, waterfowl nesting cover, acquisition of wildlife lands, and forest management. This environmental management program should generate information on habitat manipulation in large coolwater and warmwater rivers.

Modification of wing and closing dikes to reduce navigation channel maintenance impacts while improving fish habitat has been used extensively on the upper Mississippi River and Missouri River. Many designs have been tried including notched wing dikes, orientation of dikes to the current, rootless dikes, and use of vane dikes. The objective is to reduce aggradation and to encourage the river itself to develop aquatic habitats usable at various river stages (Hesse et al. 1989). A new river training technology, "bendway weirs," is currently under evaluation in the Mississippi River.

Blockage of fish movements in warmwater rivers has not been considered a major problem because spawning was not generally prevented. However, ongoing degradation of the Missouri River caused by upstream reservoirs and maintenance of the flood control and navigation projects may be threatening catfish spawning in major tributaries. Grade stabilization structures, installed to prevent bed degradation in Missouri River tributaries, have blocked catfish from reaching spawning areas. A project implemented on the Little Sioux River in Iowa to provide for catfish passage is currently being evaluated.

Young salmon are known to congregate at high densities during the winter in lentic components of river complexes, such as channel and extra-channel ponds, gravel pits, and cut-off channel meanders. These habitats, as compared to flowing channel habitats where temperatures fall to 0°C for extended periods of time, provide at least two conditions that probably promote overwinter survival: (1) warmer temperature due to stratification, and (2) lack of flow. This has led to calls for creation and enhancement of lentic habitats in salmon rivers. Lentic habitats are aggrading and disappearing in large, aging, modified rivers, also leading to calls for creating and enhancing these habitats.

Water level management for flood control and navigation projects often affects river fisheries. Public Law 697, passed in 1948 and known as the "Anti-Drawdown Law," prevents winter drawdown of slackwater navigation pools in the upper Mississippi River for flood control in the interest of preserving wintering habitat for fish. The law ordered the Army Corps of Engineers to maintain pools "as though navigation was carried on throughout the year."

Water level manipulation of any kind in developed areas rarely meets with favor from riparian property owners. Consequently, flow augmentation and water level manipulation on a large scale is rarely, if ever, practiced for large-river fisheries. However, control of water levels and flows on a smaller scale in isolated backwaters would be a useful tool for large-river fisheries and wildlife managers. Several of these types of flow manipulation projects have been completed with varying degrees of success on the upper Mississippi River.

Water releases from dams can be planned to promote beneficial effects in downstream reaches. Guidelines for planning releases can be developed from minimum instream flow studies of key river fishes. Water releases could also be used to simulate seasonal inundation of the floodplain (Ward and Stanford 1989), stimulating fish production and reproduction.

Advances have been made in developing techniques to predict the physical impacts of towboat traffic (Bhowmik et al. 1982; Chen et al. 1984). Reduction of navigation impacts on water quality may require improved regulations (including possible speed limits and establishment of protected areas), enforcement, and perhaps redesign of boat propellers.

Mainstem reservoirs can improve water quality in downstream reaches by trapping sediments. However, reservoir releases are often devoid of oxygen and toxic metals will enter release waters under these conditions. A program to alleviate these and other problems in reservoirs on the Tennessee River includes (1) reservoir destratification and hypolimnetic aeration, (2) hydroelectric turbine modification to increase dissolved oxygen, (3) aeration of release waters, and (4) weirs to automatically provide minimum instream flows (Voigtlander and Poppe 1989).

19.7.4 Species Manipulation

Overabundance of large forage species such as adult gizzard shad remains a problem in some areas. Introduction of a large predator, such as the striped bass, has been successful in some instances to convert this unused biomass into a trophy fishery. Some states are attempting to promote sport fishing for traditionally underexploited rough fish such as common carp, bullhead, and redhorse sucker through public educational efforts and urban fishing programs. Other states have developed pamphlets to enhance public awareness of river angling opportunities.

19.7.5 User Manipulation

Most state agencies have regarded fisheries in large warmwater rivers as limitless resources requiring little regulation. Consequently, more liberal, less restrictive regulations have been imposed. However, advances in angling technology, increases in available free time, organized fishing contests, and increased competition between anglers for use of inland waters have led to a rediscovery of large-river fishing.

There is a growing desire on the part of river managers to protect populations from overexploitation, resulting in more restrictive harvest regulations. Consequently, many inland fisheries regulations (including some creel limits and closed seasons) presently also apply to certain large-river reaches. For example, a 38-cm minimum length limit for largemouth bass was recently imposed on Pools 16–19 of the Upper Mississippi River.

Commercial fishing is a traditional practice on almost every large river, constituting a way of life for some individuals. The commercial fishery on the upper Mississippi River can be described as a limited-entry, territorial fishery. The commercial fisheries of the basin as a whole are being exploited at nearly optimal levels based on the observation that catch per fisher declined and yields remained stable during the 1950s to 1970s when effort increased substantially. Unfortunately, good catch records are not often kept for most large rivers, and the reporting procedures are usually voluntary, causing most harvest estimates to be low. However, even if underestimated, trend information can be obtained from these data if one is willing to assume that underestimates have been similarly made each year.

Commercial fishing for salmonids, on the other hand, is often highly regulated, with extensive information kept on catch. Catch can be allocated and partitioned into harvest by fishing gear types, and even into various user groups (e.g, Native Americans). Although there are high standing stocks of large salmonids in some large rivers, such as the Churchill and Mackenzie (Bodaly et al. 1989), they cannot tolerate much fishing pressure. Salmon in Arctic rivers are slow growing, but long lived (due to low exploitation), permitting them to attain large size.

Loss of habitat, unexplained die-offs, and overharvest threaten the continued existence of commercially harvested mussels. Mussel sanctuaries have now been established below all Cumberland River and Tennessee River dams in Tennessee, Kentucky, and Alabama. Parallel efforts are underway for some historically important mussel beds in the Upper Mississippi River.

19.8 BASINWIDE MANAGEMENT

Management of river fisheries requires an improved understanding of large river fisheries biology. This in turn requires collection of long-term data sets on population dynamics, movements, and habitat requirements. Because of the scale of these efforts, and the fact that rivers cross political boundaries, biologists should develop long-term management plans in cooperation with their peers in other states and agencies. This would make procurement of research funds more successful, extend the coverage of available funding, and eliminate much of the parochialism which presently exists in many large-river management activities.

The Hudson River Foundation provides $1,000,000 each year to support river

Figure 19.9 Mississippi River paper plant and hydroelectrical station, Sartell Dam, St. Cloud, Minnesota. Large rivers attract industries and municipalities to their banks because they provide water, inexpensive transportation, recreational opportunities, and other resources, such as hydroelectrical power. Anthropogenic impacts and periodically harsh natural conditions, such as ice-covered channels, low temperatures, and high discharge, combine to provide a challenging environment for river fishes. (Photograph by D. Logsdon.)

research and education; an excellent method of supporting and coordinating activities on that river. Biologists on the upper Mississippi River in Illinois, Iowa, Missouri, Minnesota, and Wisconsin have worked toward gear standardization and cooperative sampling efforts since 1943 through the upper Mississippi River Committee. Similar groups are emerging or in place on the lower Mississippi, the Missouri, and the Ohio rivers.

19.9 THE FUTURE

19.9.1 Multiple-Use Conflicts

Multiple-use conflicts on our large rivers will continue to increase as river uses increase (Figure 19.9). Fortunately, environmental awareness has come of age. The days of single-purpose exploitation have largely become a thing of the past, and will remain so as long as environmental regulations remain strong. It is imperative, however, that present and future river biologists and decision makers respect the environmental gains of the past 2 decades, and realize that they can easily be lost if taken for granted or misused.

19.9.2 Land Management and Sedimentation

Poor land management is the major source of riverine sedimentation, which annually destroys thousands of acres of prime spawning, rearing, and feeding habitat. Major programs are needed to influence land-use management decisions and offset these losses. The U.S. Food Security Act of 1986, if properly implemented, along with more stringent soil conservation measures, such as the "Sodbuster" program implemented with the 1985 Farm Bill, are the kinds of programs needed to effectively alter the course of past and ongoing soil erosion.

19.9.3 Water Quality

Contaminants, toxic spills, and suspended solids will probably be the greatest water quality problems in large rivers in the future. The ability of rivers to cleanse themselves and assimilate tremendous volumes of wastes are largely a function of natural fluviatile processes. Obviously, many human uses of rivers have reduced assimilation capacities, which can in turn diminish water quality. However, the outlook for river water quality in the future is still bright, if humans have the good sense to work with the river and protect it. For example, environmental legislation and tougher laws in Minnesota and Wisconsin have been influential in restoring water quality to reaches of the upper Mississippi River downstream from Minneapolis. The recurrence of the *Hexagenia* mayfly in recent years is proof.

19.9.4 Directions for Fish Management

Large U.S. rivers have been irreversibly altered from pristine conditions, and they are already artificially managed for other primary uses. This necessitates innovation and artificiality in managing biotic communities as well. River managers should consider the maintenance of a healthy ecosystem as their primary goal. Supplemental fish stocking, introductions, and habitat restoration and enhancement projects may be required. In general, fish management objectives in large rivers should be set according to the following criteria:

1. threatened and endangered species and remaining unique habitats should be preserved, and reintroductions of extirpated species should be contemplated;

2. fish communities should be managed to maximize species diversity, ensuring that available trophic niches are occupied;

3. resources such as commercial and sport fisheries should be managed for optimum sustained yield; and

4. habitat diversity should be maintained minimally at its current level through sound conservation practices, or (hopefully) increased through construction projects directed at restoration, enhancement, or creation of fish habitat and near natural flow regimes.

All indications point to significantly increased activity in monitoring, enhancing, and managing river habitats and fish populations in the future. The fisheries of large rivers can and should be managed as in other aquatic ecosystems.

19.10 REFERENCES

Addair, J. 1944. Fishes of the Kanawha River system and some factors which influence their distribution. Doctoral dissertation. Ohio State University, Columbus.

Alabaster, J. S., and R. Lloyd. 1980. Water quality criteria for freshwater fish. Butterworth, Stoneham, Massachusetts.

Bhowmik, N. G., M. Demissie, and C. Y. Guo. 1982. Waves generated by river traffic and wind on the Illinois and Mississippi rivers. University of Illinois, Water Resources Center, Illinois State Water Survey, UILU-WRC-82-0167 Research Report 167, Champaign.

Bodaly, R. A., J. D. Reist, D. M. Rosenberg, P. J. McCart, and R. E. Hecky. 1989. Fish and fisheries of the Mackenzie and Churchill river basins, northern Canada. Canadian Special Publication of Fisheries and Aquatic Sciences 106:128–144.

Chen, Y. H., D. B. Simons, R. Li, and S. S. Ellis. 1984. Investigation of effects of navigation traffic activities on hydrologic, hydraulic, and geomorphic characteristics in the upper Mississippi River system. Pages 229–324 in J. G. Wiener, R. V. Anderson and D. R. McConville, editors. Contaminants in the upper Mississippi River. Butterworth, Stoneham, Massachusetts.

Czaya, E. 1981. Rivers of the world. Cambridge University Press, New York.

Ebel, W. J., C. D. Becker, J. W. Mullan, and H. L. Raymond. 1989. The Columbia River—toward a holistic understanding. Canadian Special Publication of Fisheries and Aquatic Sciences 106:205–219.

Fremling, C. R., J. L. Rasmussen, R. E. Sparks, S. P. Cobb, C. F. Bryan, and T. O. Clafling. 1989. Mississippi River fisheries: a case history. Canadian Special Publication of Fisheries and Aquatic Sciences 106:309–351.

Groen, C. L., and J. C. Schmulbach. 1978. The sport fishery of the unchannelized and channelized middle Missouri River. Transactions of the American Fisheries Society 107:412–418.

Hesse, L. W. 1982. The Missouri River channel catfish. Nebraska Game and Parks Commission, Nebraska Technical Series 11, Lincoln.

Hesse, L. W., and six coauthors. 1989. Missouri River fishery resources in relation to past, present, and future stresses. Canadian Special Publication of Fisheries and Aquatic Sciences 106:352–371.

Karr, J. K., L. A. Toth, and D. R. Dudley. 1985. Fish communities of midwestern rivers: a history of degradation. BioScience 35:90–95.

Lewis, W. M. 1987. Production and stocking of fishes: observations concerning questions of potential interest to the power industry. Pages 1–9 in R. G. Otto and associates. Mechanism of compensating response of fish populations: workshop proceedings. Electric Power Research Institute EA5202, Project 1633, Palo Alto, California.

Limburg, K. E., S. A. Levin, and R. E. Brandt. 1989. Perspectives on management of the Hudson River ecosystem. Canadian Special Publication of Fisheries and Aquatic Sciences 106:265–291.

Nielsen, L. A., R. J. Sheehan, and D. J. Orth. 1986. Impacts of navigation on riverine fish production in the United States. Polish Archives of Hydrobiology 33:377–394.

Rasmussen, J. L. 1979. A compendium of fishery information on the upper Mississippi River. Upper Mississippi River Conservation Committee, Rock Island, Illinois.

Richardson, R. E. 1921. The small bottom and shore fauna of the middle and lower Illinois River and its connecting lakes, Chillicothe to Grafton: its valuation; its sources of food supply; and its relation to the fishery. Illinois Natural History Survey Bulletin 13:363–522.

Schnick, R. A., J. M. Morton, J. C. Mochalski, and J. T. Bailly. 1982. Mitigation and enhancement techniques for the upper Mississippi River system and other large river systems. U.S. Fish and Wildlife Service Resource Publication 149.

Sedell, J. R., and J. E. Richey. 1989. The river continuum concept: a basis for the expected ecosystem behavior of very large rivers? Canadian Special Publication of Fisheries and Aquatic Sciences 106:49–55.

Vannote, R. L., G. W. Winshall, K. W. Cummins, J. R. Sedell, and C. E. Cushing. 1980. The river continuum concept. Canadian Journal of Fisheries and Aquatic Sciences 37:130–137.

Voigtlander, C. W., and W. L. Poppe. 1989. The Tennessee River. Canadian Special Publication of Fisheries and Aquatic Sciences 106:372–384.

Ward, J. V., and J. A. Stanford. 1989. Riverine ecosystems: the influence of man on catchment dynamics and fish ecology. Canadian Special Publication of Fisheries and Aquatic Sciences 106:56–64.

Weithman, A. S., and G. G. Fleener. 1988. Recreational use along the Missouri River in Missouri. Pages 67–78 *in* N. G. Benson, editor. The Missouri River, the resources, and their uses and value. American Fisheries Society, North Central Division, Special Publication 8, Bethesda, Maryland.

Welcomme, R. L. 1979. Fisheries ecology of floodplain rivers. Longman Group, London.

Wright, T. D. 1982. Potential biological impacts of navigation traffic. U.S. Army Corps of Engineers, South Atlantic Division, Environmental and Water Quality Operation Studies Miscellaneous Paper E-82, Atlanta.

Chapter 20

Small Impoundments

STEPHEN A. FLICKINGER AND FRANK J. BULOW

20.1 INTRODUCTION

Small impoundments are ponds made by constructing dams or dikes to impound water from springs, streams, wells, direct precipitation, and surface runoff, or by digging depressions or pits to hold water (Dendy 1963). What constitutes a small impoundment varies regionally, but the designation generally includes impoundments ranging from less than 0.4 to more than 40 hectares. The Central States Pond Management Work Group defined a pond as an impoundment with a surface area of 0.2–2.4 hectares (Anderson 1978). Manageability is a key feature of most ponds that are constructed for fish production, and such ponds are typically furnished with a drain. Another common feature of ponds is that they are usually privately owned. Farm or ranch ponds are most commonly built for livestock water and irrigation (Figure 20.1).

By 1980, more than 2.1 million farm and ranch ponds had been built in the United States by private land owners (SCS 1982). Ponds are most numerous in the central and southeastern United States (Figure 20.2). The Canadian provinces have a large number and variety of natural ponds or potholes that are primarily the result of relatively recent glaciation. Thousands of small, often temporary "sloughs" in the southern prairie provinces are important to waterfowl and agriculture of the semiarid prairie. In addition, many small reservoirs have been constructed in Canada. Severe winter stagnation is a common limnological characteristic which impedes sustained fish production in smaller, northern-latitude impoundments. Alaskan pond management is very limited, but an active program does exist for stocking borrow pits along the highway system. There is a paucity of small impoundments in Mexico, and those that exist are used primarily to store water for livestock.

In many areas, farm ponds make an enormous contribution to sport fishing. Recent surveys indicated farm ponds supported approximately 40% of the fishing pressure in Georgia, while ponds and small lakes accounted for more than 23,000,000 fishing recreation days (35%) in Texas.

Development of farm pond management in North America has a long history. Regier (1962), Dendy (1963), Swingle (1970), and Bennett (1971) should be consulted for overviews. The present management philosophy for recreational pond fisheries is sustained or improved fishing quality and favorable benefit:cost ratios, the ultimate goal being high catch rates and above-average sizes. Many of

Figure 20.1 Farm ponds and other small impoundments are constructed for livestock water, irrigation, fish production, field and orchard spraying, domestic water supply, fire protection, energy conservation, wildlife habitat, recreation, and landscape improvement. Small impoundments are typically more easily managed for increased fish production than are large, multi-purpose reservoirs.

the principles and techniques developed for the management of farm ponds are being applied to small natural lakes, flood control impoundments, and impoundments created as a by-product of surface material or mineral deposit mining.

Although management techniques have been developed by fisheries biologists, it is usually the pond owner who is responsible for implementation. Literature and personal assistance are often available to the pond owner, and specific topics not covered in this chapter, such as fish parasites and muddy water, are typically addressed in booklets and extension literature. Guidance can often be obtained from Cooperative Extension Services, the Soil Conservation Service, or various fisheries agencies. In addition, pond management services are available from private consulting firms.

20.2 ECOLOGY

Current recommendations for pond construction, species and numbers of fish to stock, and subsequent management techniques are based on an understanding of basic ecological principles relating to trophic relationships and population dynamics. Carrying capacity when applied to pond fish populations may be defined as the maximum weight of a given species of fish that a pond will support during a stated interval of time; standing crop is the actual weight of a given species or complex of species present in a pond at a specific moment (Bennett 1971). The growing season for warmwater species may be 2 months in Manitoba, 3 months in Minnesota, 6 to 7 months in Illinois, and 11 to 12 months in Louisiana or Mexico. Consequently, new or reclaimed ponds initially stocked with largemouth bass and bluegills may reach carrying capacity in 1 year at southern latitudes, but may take two or more years at northern latitudes.

Fishes are consumers, being dependent upon the production of phytoplankton and a series of lower level consumers such as zooplankton, benthos, and other smaller fishes. Because the amount of usable energy is reduced with each transfer in the food chain, planktivorous fishes are expected to yield more biomass per unit area for a given level of primary production than are carnivores at the end of a long food chain. Standing crop tends to increase with number of species because

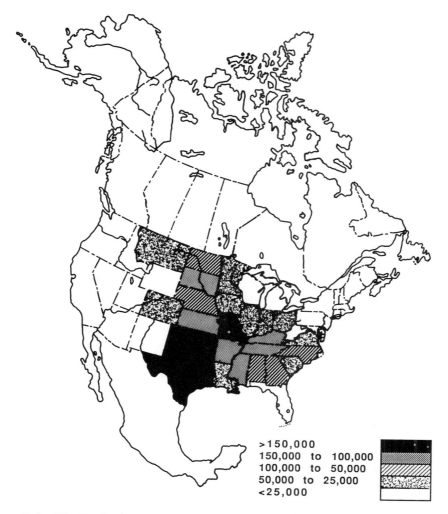

>150,000
150,000 to 100,000
100,000 to 50,000
50,000 to 25,000
<25,000

Figure 20.2   Distribution by states and provinces of the total number of ponds constructed in North America. (From Modde 1980.)

of greater use of available resources; however, simple species combinations are more manageable in producing fish considered useful in a sport fishery.

Relationships between carrying capacity, population density, and individual sizes have long been recognized. At a given carrying capacity, the average size of bluegills, for example, will increase in direct proportion to a decrease in population density (Bennett 1971). The carrying capacity of a pond may be increased by fertilization, feeding, or stocking different species.

20.3   HISTORY

20.3.1   Warmwater Ponds

The drought of the 1930s led to the Soil Conservation Service farm pond building program in the United States and the Prairie Farm Rehabilitation Act

Pond Construction Program in Canada. The demands for protein and inexpensive recreation during World War II intensified the interest in pond construction and management. The question of what species to stock initiated a search for a stocking combination that would establish balanced predator–prey populations and sustain quality fishing. Two schools of thought developed concerning farm pond stocking strategies, one centered in Alabama and one in the Midwest (Regier 1962). The former considers the bass–bluegill combination as optimal and the latter has sought other forage species to stock with bass.

Research at Auburn University established the largemouth bass and bluegill as a suitable stocking combination (Swingle and Smith 1938). The basic theory with this combination was that both bass and bluegills would provide fishing. The bluegill would convert invertebrate production into bluegill flesh and small bluegills would serve as food for the bass; bass predation would control excessive numbers of bluegills, and the few bluegills surviving would grow to a large average size because of low population density. Although Swingle experimented with alternative forage species, he found the bluegill to be most desirable because it met all of his criteria for a good forage fish for bass (Dillard and Novinger 1975): (1) in adult form it must not be too large for bass to eat; (2) in adult form it must not be too small or bass will eliminate their population; and (3) to be of continuing value as bass forage, the population must be reduced annually by angling or predation to stimulate spawning. After years of experimentation with fish size and stocking density, Swingle (1951) concluded that stocking fingerlings of each species was the most desirable technique. He obtained the highest average standing crop of harvestable bass by stocking 250 bass and 2,500 bluegill per hectare in fertilized ponds (Dillard and Novinger 1975).

A symposium on the management of farm ponds included evidence that stocking and management procedures applicable to Alabama did not necessarily apply elsewhere (Meehean 1952). The bass–bluegill combination frequently leads to an overpopulation of bluegills which interferes with successful reproduction of bass. This situation is more common in northern latitudes where interest in bluegill fishing is not as great as in the South. The shorter growing season and slower growth rate in northern regions mean that bass and bluegill may not reach harvestable size until 3 years after stocking. The faster-growing, southern-latitude bass also exert greater predatory control of bluegills. The northern-latitude bass and bluegill may require an additional year to reach sexual maturity than southern populations, and bluegills will spawn 1 year before bass spawn. These differences in growth and spawning, along with a high mortality rate within the population prior to the attainment of a harvestable-size fish, result in a lack of predictability for a given stocking. The larger, deeper (and less manageable) ponds necessary to escape winterkill in northern climates also contribute to lower success rate with the bass–bluegill combination. An additional contributory factor is the protection afforded to bluegills by excessive aquatic macrophytes which are often controlled by fertilization in southern but not in northern latitudes.

20.3.2 Coldwater Ponds

There are considerably fewer coldwater ponds than warmwater ponds, and there are no coldwater pond research counterparts to the works of Swingle, Bennett, and Regier. Rainbow, brook, cutthroat, and brown trout have been stocked in ponds, rainbow trout being the most common choice. Trout are usually not stocked in combination with other species; in fact other fish species, if present in the pond, are removed prior to trout stocking. Ponds of average fertility will produce enough natural food to support an average of 250 kg of trout per hectare, but 10 to 20-fold increases can be produced with supplemental feeding of pelleted foods (Marriage et al. 1971). Repeated stockings are necessary because trout reproduction seldom occurs in ponds.

20.3.3 Coolwater Ponds

The term coolwater fishes has been introduced to denote an intermediate category of water temperatures and species suited to those temperatures. Despite the validity of a coolwater concept, the few papers published on stocking these species in ponds have reported poor to limited success. More research is needed in this area. Coolwater species include the walleye, yellow perch, northern pike, striped bass, and white bass × striped bass hybrid. These species have usually been introduced to increase predator pressure on forage populations and provide an additional sport fish.

20.4 STOCKING STRATEGIES

Stocking recommendations vary according to purpose and region. Experimentation with alternatives to the standard bass–bluegill stocking combination (Dillard and Novinger 1975; Modde 1980) is ongoing. Dillard and Novinger (1975) stressed the need to develop a stocking strategy that will provide a desirable pond fishery in the shortest time possible within physical and fiscal limits. This philosophy dictates a diversity of stocking strategies rather than one standard.

Another school of thought concerning stocking is emerging. This is the replacement of the traditional standardized recommendations for each area with selected stocking combinations to most appropriately meet specific environmental situations and the angling preferences or other goals of the pond owner. Regier (1962) suggested that the kind of fish and fishing that the pond owner desires should be given greater consideration. Dillard and Novinger (1975) explored this concept in detail and suggested that pond owners could be presented with several stocking and management options even if they had to buy fish from private suppliers. Along these lines, Lewis and Heidinger (1978a) presented several stocking combinations to satisfy specific conditions and objectives, and Gabelhouse et al. (1982) presented five optional stocking and management strategies to meet specific objectives.

20.4.1 All-purpose Option

This option allows the harvest of largemouth bass, bluegill, and sometimes channel catfish of a variety of sizes. It employs a slot limit of 30–38 cm for largemouth bass 4 years after stocking. To reduce competition and allow bass to consistently grow over 38 cm, about 75 largemouth bass per hectare, in the range of 20–30 cm, should be harvested annually after the fourth year following bass fingerling stocking. Release of all 30–38-cm largemouth bass caught will ensure that at least 10% of the catchable-size bass survive to lengths of 38 cm and longer. A larger population of 30–38-cm bass will help reduce densities of intermediate-size bluegills and allow some bluegills to reach 20 cm. Bluegills and channel catfish can be harvested as desired, but catfish must be replaced with 20-cm or longer catfish. With this option, if anglers do not obey the slot limit, overharvest of largemouth bass can result in overpopulation of bluegills. On the other hand, if anglers release bass in the slot limit, but do not harvest 20–30-cm largemouth bass under the slot, overpopulation and stunting of bass can occur.

20.4.2   Harvest Quota Option

With some of the problems of size limits just mentioned above, quotas have been proposed as an alternative way to regulate harvest of largemouth bass. This option involves the harvest of a given number or weight of largemouth bass annually, regardless of size. Catch-and-release fishing is possible after the quota has been reached. Few or no largemouth bass should be harvested for the first 4 years, followed by the annual harvest of about 125 individuals or 22 kg of largemouth bass per hectare without regard for length. With this option, there is a tendency to overharvest large bass and underharvest small bass. Channel catfish harvest is unrestricted, but those harvested should be replaced with 20-cm or larger individuals. There is no limit for bluegill harvest, but various state recommendations with this option call for the harvest of 3–10 kg of bluegills for each kilogram of largemouth bass harvested. Although this plan is generally expected to keep the pond in balance, Gabelhouse et al. (1982) questioned its effectiveness. Obviously, very accurate harvest records would have to be kept. Application of quotas in management of small impoundments has greatly restricted harvest opportunities in some cases (Ming 1972), but has met with success in others (Powell 1975).

20.4.3   Panfish Option

This option emphasizes the harvest of big panfish instead of largemouth bass by imposing a 38-cm minimum length limit for largemouth bass. Few bass will grow beyond 38 cm because of overpopulation. The high density of 20–38-cm largemouth bass will reduce bluegill densities and allow growth of bluegills to 20 cm and longer. The panfish option can also involve channel catfish, black crappies, or black bullheads. Because crappies and bullheads tend to overpopulate, addition of these fishes requires release of nearly all largemouth bass and maintenance of water clarity beyond 46 cm. With adherence to the size restriction on harvest of largemouth bass, this approach to an option that emphasizes panfish is unlikely to fail. However, pond owners may be disappointed in the small size of largemouth bass caught.

An alternative approach to panfish fishing is to stock hybrid sunfish with largemouth bass. Several sunfish crosses produce offspring that are mostly male (Childers 1967; Lewis and Heidinger 1978b). By having such limited reproductive potential, hybrid sunfish do not overpopulate, and thereby they achieve large sizes. Hybrid sunfish also exhibit hybrid vigor and higher vulnerability to angling. Ellison and Heidinger (1976) reported that most pond owners who had tried hybrid sunfish and largemouth bass would do it again, but a majority of pond owners who had tried hybrid sunfish without largemouth bass would not stock them again. Hybrid sunfish are of little value in ponds containing other sunfish. Hybrids, or the parents to produce hybrids, must be restocked every few years.

20.4.4 Big Bass Option

The objective of this option is the consistent production of trophy largemouth bass without regard to bluegill size. Catch rate for largemouth bass will be low in this option, but the bass caught should be large. After 4 years, densities of 20–38-cm largemouth bass should be greatly reduced to allow for rapid growth of those remaining. With a pond of average fertility, 75 largemouth bass 20–30 cm long and about 12 largemouth bass 30–38 cm long per hectare should be harvested each year. Unless a trophy bass is caught, all bass over 38 cm should continue to be released.

This option includes the stocking of 50 adult gizzard shad per hectare 2 years after fingerling bass stocking. The gizzard shad increase the chances of producing trophy largemouth bass as they serve as food for large bass while bluegills serve as stable food for small bass. Gabelhouse et al. (1982) caution that this option is largely unevaluated and may be practical in only larger ponds because maintenance of bass over 1.5 kg may be limited to 25 per hectare. In addition to trophy bass for adult anglers, this option is expected to produce a high catch rate of small bluegills which may be of interest to children. With this option, overpopulation of bluegills may reduce recruitment of largemouth bass. The catchability of large bass may also be low because of their low density in the presence of abundant prey.

20.4.5 Catfish-only Option

This option is especially desirable for muddy ponds where sight-feeding largemouth bass and bluegills would not do well, or in ponds less than 0.2 hectares where overharvest of largemouth bass would be likely. The pond should be free of any structures that would provide the seclusion required for channel catfish spawning. Fathead minnows can be stocked as forage and harvest of channel catfish is unrestricted. Replacement stocking of channel catfish over 20 cm should take place during the cool of spring or fall, with the number harvested plus an additional 10% to account for natural mortality. A density of 250–500 channel catfish per hectare should be maintained in ponds receiving no supplemental feeding.

20.4.6 Bass-only Option

For those anglers who are not interested in panfish or for very shallow ponds where excessive aquatic vegetation provides too much cover for bluegills, black bass alone is a viable option. Bennett (1952) found that spotted, smallmouth, or largemouth bass, when stocked alone in Illinois ponds, were able to do well by feeding on crayfish, large aquatic insects, and their own young. Bennett (1971) reported that such populations produced more kilograms of bass per hectare than did populations where bass were stocked with other species. Bennett cautioned that it was necessary to stock several year-classes to prevent the development of a dominant year-class which might become stunted. Buck and Thoits (1970) also found that standing crops of largemouth bass or smallmouth bass stocked by themselves were comparable to those of bass in multiple-species populations; however, these monocultured bass grew more slowly. Swingle (1952) experimented with stocking bass alone, but considered this to be an inefficient use of available food resources of the pond, and this method resulted in about a third of the of angling pleasure as the bass–bluegill combination.

20.4.7 Trout Options

Because rainbow trout will not spawn in standing water, growth rate and ultimate size are easily controlled by stocking rates. In addition, rainbow trout readily accept formulated feeds if supplemental feeding is desired. However, many strains of hatchery-reared rainbow trout have become so domesticated that stocked trout do not live much beyond 3 years.

Interviews with pond owners have revealed that knowing what species and approximately what size trout are going to be caught reduces the anticipation of going fishing. Thought should be given to adding a few large fish or to stocking another species or color variation. The occasional catch of something different will provide excitement. Discarded brood stock become available for stocking from time to time. Brown trout generally live longer and grow larger than rainbow trout, but they can be cannibalistic and hard to catch. Several state and private trout hatcheries have yellow-colored strains of rainbow trout that would provide a surprise catch when stocked in low numbers.

20.5 STOCKING RATES

20.5.1 Warmwater Ponds

Stocking rates vary in different areas, but 250 largemouth bass and 2,500 bluegills (or bluegill–redear sunfish) per hectare remain the most common. Successive or split stocking usually involves stocking fingerling sunfish in the fall and fingerling bass the following spring; however, in some northern regions this sequence is reversed. Due to the slow growth of largemouth bass, simultaneously stocked bluegills become too big to be prey for the stocked bass. Consequently, by stocking bluegills ahead of largemouth bass, the bluegills reach sexual maturity by the time the bass are stocked. Bluegill spawning produces appropriate size prey for stocked fingerling largemouth bass.

When channel catfish are desired, they should be stocked at a rate of 250 fish per hectare with other species and up to 500 fish per hectare when alone. Periodic restocking is necessary; if adult largemouth bass are present, fingerling catfish longer than 20 cm should be stocked to avoid predation.

Ponds should not be stocked with fish caught elsewhere by anglers. In some states, it is illegal for anglers to transport live fish. In addition, stocking a few adults of sometimes erroneously identified species is not likely to produce a balanced community. Also, shipments of fish should be checked for unwanted species before stocking. Most warmwater fish-rearing facilities raise many species, and it is not uncommon for a few stray fish to find their way into a shipment of another species. Sometimes a shipment contains compatible species like fathead minnows mixed with largemouth bass, but other times the mixture is incompatible such as goldfish and channel catfish.

Although biologists often rely on stocking to achieve desired results, it is usually true that harvest has much more to do with a pond's success than does stocking.

20.5.2 Coldwater Ponds

Marriage et al. (1971) recommended fall stocking with 5–10-cm trout at 600 per hectare in western states and at 1,500 per hectare in eastern states. If larger fish are used, smaller numbers should be stocked. Fingerlings less than 10 cm are adequate for initial stocking, but subsequent stockings should be done with fish greater than 10 cm. Record keeping will aid in determining the need for more or fewer trout in subsequent stockings. Some owners of coldwater ponds prefer to stock in alternate years to reduce cost of transportation. However, such practice results in uneven fishing success.

20.6 REGULATING FISH GROWTH

20.6.1 Predation

In a predator–prey fish community, not only do the predators gain from eating prey, but also the prey benefit by having fewer individuals with which to compete for a limited food supply. There are many problems associated with maintaining a predator population in a pond. For some species, such as northern pike and muskellunge, the carrying capacity is often one adult or less per hectare. Most predators are highly prized game fish in North America and hence anglers actively seek the species in least abundance. Northern pike and largemouth bass are notorious for their emigration over spillways. While there are many desirable attributes of cover, too much cover, especially aquatic vegetation, makes it difficult for predators to capture prey. High turbidity has the same detrimental effect. Finally, some predators have rigid spawning requirements or do not recruit well in the forced interaction that occurs in the small area typical of ponds.

20.6.2 Harvest

As mentioned in many of the stocking options, regulation of largemouth bass harvest is important in the success of those options. It is common for 75% of adult largemouth bass to be caught in the first few days of fishing a new pond. If that many bass are harvested, the predator–prey balance will be ruined. Minimum size limits, if obeyed, will protect against overharvest. If largemouth bass recruitment is high, a minimum size limit will result in overpopulation and reduced growth. That condition is desirable in some management options but not in others. A protected range, or slot limit, would be a better choice when larger bass are desired. However, anglers must harvest smaller bass below the slot; otherwise the regulation would function like a minimum size limit.

20.6.3 No Spawning

For most stocking options, reproduction is desired and necessary; however, the use of hybrid sunfish is an example of reducing spawning potential. In the catfish-only option, no reproduction is desired because there would be no predation to control population size. Measures should be taken to eliminate spawning sites for channel catfish. Brook trout are the only trout species likely to spawn in a pond. Therefore, most trout ponds can be stocked with appropriate numbers of fish to achieve desired growth.

20.6.4 Supplemental Feeding

Use of formulated feeds is usually not recommended for fishing ponds because of potential dissolved oxygen depletion problems from heavy feeding and the fact that natural food production will support normal harvest rates. Nevertheless, supplemental feeding can be useful where fishing pressure is high or where large fish are desired. Channel catfish and trout are most commonly fed formulated feeds, but there is also interest in feeding bluegills, hybrid sunfish, and other species. Floating feed is advisable because the fish can be observed feeding and feeding rates can be appropriately adjusted. Schmittou (1969) found that supplemental feeding increased fishing success and total production in bass–bluegill ponds.

Feeding seasons, quantities, and frequencies vary with fish species, fish sizes, water temperature, water quality, weather conditions, and quality of feed. Seasonal feeding of channel catfish, for example, typically commences when water temperatures are over 15°C. They may be fed at 3% of their body weight per day with a maximum food input of 22 kg/hectare per day. The addition of large amounts of feed increases the danger of oxygen depletion and feeding is discontinued when surface dissolved oxygen levels drop below 5.0 mg/L. Otherwise a feeding program, once begun, should be continued or the fish will lose weight. Addition of feed increases the carrying capacity of the pond to a level which is dependent upon the feed. If the primary goal is to produce food fish, confining the fish to floating cages offers several advantages: (1) feeding activity and health of the fish can be more readily monitored, (2) it is economical to treat the fish for parasites and diseases, (3) nonseinable ponds can be used, and (4) harvest is rapid and easy.

20.6.5 Fertilization and Liming

In many southeastern states and other areas where soil fertility may be low, fertilizers are added to ponds to increase fish production by stimulating phytoplankton growth. Ponds on fertile watersheds, in well-managed pastures with high densities of cattle, in northern latitudes where winterkill is a problem, in areas where clay turbidity is a persistent problem, and ponds that receive very little fishing pressure generally should not be fertilized. Fertilization programs are used in Alabama, Arkansas, some parts of northern Florida, Georgia, Kentucky, Mississippi, North Carolina, South Carolina, Tennessee, Texas, and Virginia. With a few exceptions, fertilization and liming are generally not recommended elsewhere.

Swingle and Smith (1938) found that fertilized ponds in Alabama supported four to five times the weight of largemouth bass and bluegills as unfertilized ones. The shading effect of a phytoplankton bloom also aids in control of submerged aquatic macrophytes, and this beneficial turbidity increases fishing success. However, most fertilizers, especially the inorganics, are expensive, and the pond owner should determine whether the increased production justifies the expense. Bennett (1971) cautioned about the dangers of summerkill at any latitude and winterkill in northernmost states when abundant nutrients result in excessive plant growth.

Inorganic fertilizers are most commonly used as pond fertilizers in the United States. A common fertilizer is a granular 20–20–5. A 20–20–5 grade of fertilizer contains 20% nitrogen (as N), 20% phosphorus (as P_2O_5), and 5% potassium (as

---

Box 20.1   Standard Pond Fertilization Program for Southeastern United States (Boyd 1979).

1. In mid-February or early March (when water temperatures rise to 16-18°C) apply 45 kg/hectare of 20–20–5 fertilizer (equivalent amounts of other formulations are also used). Follow with two additional applications at 2-week intervals.

2. Make three more applications of 45 kg/hectare of 20–20–5 at 3-week intervals.

3. Continue applications of 45 kg/hectare of 20–20–5 at monthly intervals or whenever the water clears so that a Secchi disk or a white disk attached to a stick is visible to a depth of 45 cm.

4. Discontinue applications for the year by the last week in October.

Once a pond fertilization program is begun, it should be continued each year or stunted fish and the emergence of nuisance aquatic macrophytes may result. The standard fertilization rate of 8–12 applications may cost the pond owner $80–100 per hectare each year.

K_2O, also called potash). Secondary nutrients in fertilizers include calcium, magnesium, and sulfur. Minor or trace nutrients such as copper and zinc may also be present. In recent years, liquid formulations such as 10–30–4 or 10–34–0 have been found to be more economical and easier to apply.

Organic fertilizers include various animal manures, plants, and plant products, and can be inexpensive. Organic fertilizers have a low fertilizer grade, however, and large quantities are needed to supply equivalent quantities found in small amounts of inorganic fertilizers. Dissolved oxygen problems are sometimes created when large amounts of organics are used. Organic fertilizers are also of unpredictable nutrient grade, and they tend to encourage growth of macrophytic algae.

Boyd (1979) described the chemistry of fertilizers, relationships between chemical fertilization and fish production, and the experiments of Swingle and Smith which he modified into a standard fertilizer application procedure for the southeastern United States (Box 20.1). Additional accounts of early research on pond fertilization were presented by Dendy (1963), Swingle (1970), and Bennett (1971).

Swingle et al. (1965) found that old ponds that had been fertilized for many years had adequate supplies of nitrogen and potassium through nitrogen fixation by bacteria or algae and decomposition of bottom organic materials. For such ponds, they recommended the standard fertilization rate with phosphorus-only fertilizer using 45 kg/hectare of superphosphate (16–20%) or 20 kg/hectare of triple (or treble) superphosphate (44–54%). If a satisfactory bloom cannot be maintained, use of a complete fertilizer should be resumed.

Granular fertilizers applied to waters over 1.0 m deep may sink into the bottom mud, be locked into the bottom water, and be unavailable to phytoplankton.

Therefore, fertilizers should be broadcast over shallow water or, preferably, be placed on underwater platforms which are in the euphotic zone (Lawrence 1954). Platforms should be about 0.3 m underwater, and a platform with an area of 4 m^2 is adequate for 2–4 hectares of pond area (Boyd 1979). The mixing of liquid fertilizers in the prop wash of a boat motor is another alternative.

In certain regions, pond waters are characteristically low in hardness and may be very acidic. Such waters may not respond to fertilization unless some form of liming is used. Lime has been routinely used to increase fish production in ponds in many other nations and in areas of the southeastern United States where natural shortages of calcium result in low pH, hardness, or alkalinity. Liming has several effects on water quality and productivity including the following: (1) increases the pH of bottom mud and thereby increases the availability of phosphorous added in fertilizer; (2) increases benthos production by increasing nutrient availability; (3) increases alkalinity and thereby the availability of carbon dioxide for photosynthesis; (4) increases microbial activity through the increased pH; (5) buffers against pH shifts common in soft-water, eutrophic ponds; (6) increases total hardness through the addition of calcium and magnesium; and (7) may clear water of humic stains of vegetative origin which reduce sunlight penetration. The net effect is an increase in phytoplankton productivity which leads to increased food production.

Boyd (1979) described methods used in analysis of agricultural soils and pond muds to determine the lime requirement. Liming needs are also determined by total alkalinity and hardness measurements of water. Agricultural limestone, $CaCO_3$, or $CaMg(CO_3)_2$, is the material most commonly used in fish ponds. New ponds can be limed by spreading limestone over the pond bottom and disking it. Old ponds can be limed by spreading limestone over the entire pond surface (Figure 20.3). Addition of limestone or other liming materials is best accomplished during the winter. The length of time a treatment is effective will depend upon the rate of water loss through overflow and seepage.

20.7 BALANCE AND POPULATION ANALYSIS

The concept of balanced fish populations was proposed by Swingle (1950) and elaborated by Anderson (1973). Ponds are usually stocked with species combinations that are likely to attain balance. A balanced system is basically a dynamic one characterized by continual reproduction of predator and prey species, diverse size composition of prey species so that food is available for all sizes of predators, high growth rates of predators and prey, and an annual yield of harvestable-size fish in proportion to basic fertility.

Bennett (1971) refuted the concept of balance largely on the basis of ponds representing artificial ecosystems that cannot be expected to show great stability. More recently, Gabelhouse et al. (1982) and Gabelhouse (1984) proposed panfish and big bass management options that are out of balance but are useful. The all-purpose option does fit the concept of balance, and that is the management strategy that Swingle and Anderson had in mind.

No one has applied the concept of balance to coldwater pond fish populations. To be sure, lack of reproduction and no prey fish are major deviations from Swingle's (1950) thinking. However, the objective of sustained crops of harvest-

Figure 20.3 With a pontoon barge and pump sprayers, two tonnes of agricultural limestone can be applied in several minutes. Lime increases production in impoundments with acid bottom muds and soft water (less than 20 mg/L total alkalinity). Lime is especially important in making more phosphorus available for phytoplankton production.

able-size fish in proportion to pond fertility is equally valid for coldwater systems. Perhaps the relationship between zooplankton size and fishing quality for rainbow trout will emerge as an assessment of balance in coldwater ponds (Galbraith 1975).

20.7.1 Biomass Indices

In an attempt to better describe the conditions of balanced and unbalanced populations, Swingle (1950) analyzed biomass data from 55 balanced and 34 unbalanced ponds and computed certain biomass ratios or indices. Except for four ponds poisoned by rotenone, the population of each pond was determined by draining. The most commonly used biomass indices are the F:C ratio, the Y:C ratio, and the A_T value.

The F:C ratio is the total weight of all forage fish (F) divided by the total weight of all carnivorous fish (C). The F:C ratios from 3.0 to 6.0 were considered the most desirable in the balanced range (Figure 20.4).

The Y:C ratio is the total weight of all forage fish that are small enough to be eaten by the average-size individual in the C group divided by the C value. The most desirable range for Y:C ratios in balanced populations is between 1.0 and 3.0.

The A_T value (total availability value) is the percentage of the total weight of a fish population composed of fish of harvestable size. In calculating the A_T value, it is necessary to define minimum sizes suitable for harvest, and Swingle (1950) did this for common pond species (for example, sunfish 45 g, largemouth bass 180 g, and channel catfish 230 g). Swingle suggested a range of 60–85% as most desirable for balanced populations.

Although Swingle intended these biomass indices primarily for use in post-mortem evaluations of experimental stocking, thinning, or similar management

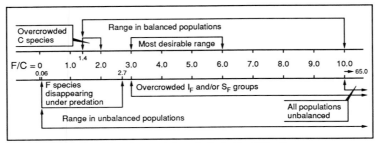

Conditions in populations indicated by F/C ratios.

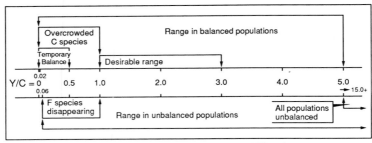

Conditions in populations indicated by Y/C ratios.

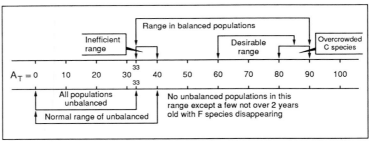

Conditions in fish populations indicated by various A_T values.

Figure 20.4 Conditions in fish populations indicated by F:C ratios, Y:C ratios, and A_T values. (From Swingle 1950.)

techniques, he did apply them to subsamples in ponds, rivers, and large impoundments (Swingle 1954). Subsequent application of these biomass indices to subsamples, such as cove rotenone studies, has been problematical because of sampling bias associated with the cove (i.e., overestimations of forage and young of all species) and unequal vulnerability of fish species to various other types of sampling gear. It appears that these indices are best suited for post-mortem evaluation of pond populations.

Swingle must have recognized these sampling problems, for in 1956 he presented a method of analysis that used a minnow seine to sample for evidence of reproduction and a larger seine to sample intermediate and harvestable-size fish (Figure 20.5). Swingle's (1956) description of possible conditions and their interpretation are summarized in Box 20.2. Because so much of the analysis is

Figure 20.5 The minnow seine method of pond analysis is a simple method for determining the state of balance of a population within a short time and without destruction of the population. The state of balance is deduced from the degree of success of reproduction of bass and bluegills, and the abundance of intermediate size (8–13 cm) bluegills.

based on presence or absence of young fish, sampling must be done after spawning has occurred.

In the same publication (Swingle 1956), a method of distinguishing balanced and unbalanced ponds from angler catch was outlined. Assuming that anglers have been fishing for both largemouth bass and bluegill, the following interpretations can be made: (1) in balanced ponds most bluegills caught will be larger than 15 cm, the average largemouth bass will be 0.5–1.0 kg, but smaller and larger ones also will be caught; (2) in unbalanced ponds the catch will be principally bluegill ranging from 7.5 to 12.5 cm, and the few largemouth bass caught will be larger than 1 kg (note the similarity to the big bass option—Section 20.4.4); and (3) in ponds crowded with largemouth bass (unbalanced condition), bluegills will average more than 150 g and largemouth bass will be less than 0.5 kg and in poor condition (note similarity to the panfish option—Section 20.4.3).

20.7.2 Length-Frequency Indices

It is difficult to sample different species and sizes of fish with equal effort to obtain the biomass figures needed in Swingle's (1950) pond analysis system. Consequently, Anderson (1976) introduced a different approach called proportional stock density (PSD). The PSD, and related length-frequency indices, can be calculated from subsamples of pond populations (Anderson and Gutreuter 1983). These indices reflect an interaction of rates of reproduction, growth, and mortality of the age groups present.

PSD is calculated by dividing the number of fish ≥ quality size by the number of fish ≥ stock size × 100 (Chapter 6). Quality size is the size most anglers like to catch and stock size is the size at or near which fish often reach sexual maturity

---

Box 20.2 Swingle's (1956) Method of Pond Analysis Based on Seining.

1. No young largemouth bass present:
   A. Many recently hatched bluegills; no or very few intermediate bluegills. (Temporary balance with bass overcrowded.)
   B. No recent hatch of bluegills; many intermediate bluegills. (Unbalanced population with overcrowded bluegills and insufficient bass.)
   C. No recent hatch bluegills; many intermediate bluegills; many tadpoles and/or minnows and/or crayfish. (Unbalanced population with overcrowded bluegills and very few bass.)
   D. No recent hatch of bluegills; few intermediate bluegills. (Unbalanced population, crowding due to species competitive with bluegills.)
   E. No recent hatch of bluegills; few intermediate bluegills; many intermediate fish of a species competitive with bluegills. (Unbalanced population due to crowding by competitive species.)
   F. No recent hatch of bluegills; no intermediate bluegills. (Unbalanced population; possible no fish present or water unsuitable for bass-bluegill reproduction.)

2. Young largemouth bass present:
   A. Many hatched bluegills; few intermediate bluegills. (Balanced population.)
   B. Many recently hatched bluegills; very few or no intermediate bluegills. (Balanced population with slightly crowded bass.)
   C. No recent hatch of bluegills; no intermediate bluegills. (Unbalanced population; bluegills prevented from spawning by low water temperature or salinity, etc.)
   D. No recent hatch of bluegills; few intermediate bluegills. (Temporary balance with possibility of imbalance developing due to a reduction of the food available to the bluegill or overcrowding by a species growing to a competitive size.)
   E. No recent hatch of bluegill; many intermediate bluegills. (Unbalanced population similar to 1.B., but less severely overcrowded.)

(Table 20.1). Sizes (stock and quality) are based on a percentage of world record length, and the sizes are standard for anyone using them (Anderson and Gutreuter 1983). In contrast, Swingle's (1950) notations of small (S), intermediate (I), and large (A) had enough latitude that different biologists could put different size fish into the categories. Such a practice hampers discussion of results from different geographical regions.

Proportional stock density is similar to A_T (Swingle 1950) in that both are a percentage of fish that are an attractive size to anglers. PSD, however, is based on length frequency instead of total weight. In addition, fish below stock size are not included in PSD; therefore, no special effort is needed to sample small fish. For balanced ponds, largemouth bass PSD should be between 40 and 60% and bluegill PSD should be between 20 and 40% (Anderson 1978). Sequential sampling for

Table 20.1 Proposed maximum total length (cm) for minimum stock, quality, preferred, memorable, and trophy sizes based on percentages of world record lengths. (From Anderson and Gutreuter 1983.)

Species	Size designation				
	Stock	Quality	Preferred	Memorable	Trophy
Largemouth bass	20	30	38	51	63
Smallmouth bass	18	28	35	43	51
Spotted bass	18	28	35	43	51
Walleye	25	38	51	63	76
Sauger	20	30	38	51	63
Muskellunge	51	76	97	107	127
Northern pike	35	53	71	86	112
Chain pickerel	25	38	51	63	76
Blue catfish	30	51	76	89	114
Channel catfish	28	41	61	71	91
Flathead catfish	28	41	61	71	91
Bluegill	8	15	20	25	30
Green sunfish	8	15	20	25	30
Pumpkinseed	8	15	20	25	30
Redear sunfish	10	18	23	28	33
Rock bass	10	18	23	28	33
Black crappie	13	20	25	30	38
White crappie	13	20	25	30	38
White bass	15	23	30	38	46
Yellow bass	10	18	23	28	33
White perch	13	20	25	30	38
Yellow perch	13	20	25	30	38
Black bullhead	15	23	30	38	46
Common carp	28	41	53	66	84
Freshwater drum	20	30	38	51	63
Bigmouth buffalo	28	41	53	66	84
Smallmouth buffalo	28	41	53	66	84
Gizzard shad	18	28	—	—	—

PSD may shorten the amount of time spent sampling and also allows calculation of confidence limits (Weithman et al. 1980; Gustafson 1988).

The relative stock density (RSD) is the proportion of fish of any designated size group in the stock size and larger portion of a population. Anderson and Gutreuter (1983) considered RSD to be a more sensitive indicator of potential fishing quality than PSD alone because RSD provides opportunity to divide a population into more than just stock and quality sizes. Gabelhouse (1984) described the basis of a five-cell analysis of fish lengths. Gabelhouse did not use RSD to assess balance, but he proposed desirable ranges in the five cells for different management options.

20.7.3 Abundance and Weight Indices

All of the indices presented so far in this section have no reference to abundance. Desirable ratios might exist with very low abundance, giving the impression that good fishing is available when catch rates probably would be poor. Relative weight (W_r) is the actual weight of a fish divided by a standard weight for the same length for that species times 100 (Anderson and Gutreuter 1983; Chapter 6). Equations for calculating standard weights have been developed for selected species (Table 20.2). Fish with W_r close to 100 are in balance with their food supply. Fish with values below 85 are underweight and may be too abundant for

Table 20.2 Equations for calculating standard weight for a given length to be used in relative weight (W_r) computations.

Species	Equation[a]	Source
Largemouth bass	$\log W_s = -5.316 + 3.191 \log L$	Wege and Anderson (1979)
Smallmouth bass	$\log W_s = -4.983 + 3.055 \log L$	Anderson (1980)
Bluegill	$\log W_s = -5.374 + 3.316 \log L$	Anderson (1980)
Green sunfish	$\log W_s = -4.8139 + 3.0558 \log L$	Gabelhouse[b]
White crappie	$\log W_s = -5.642 + 3.332 \log L$	Neumann and Murphy (1991)
Black crappie	$\log W_s = -5.168 + 3.345 \log L$	Neumann and Murphy (1991)
Channel catfish	$\log W_s = -5.649 + 3.243 \log L$	Anderson (1980)
Rainbow trout	$\log W_s = -5.194 + 3.098 \log L$	Anderson (1980)

[a]Total length is in millimeters and weight is in grams.
[b]D. W. Gabelhouse, Jr., Kansas Fish and Game Commission, personal communication.

their food supply. On the other end, fish with W_r above 105 are more plump than necessary, reflecting an overabundant food supply. In such a situation, the pond could support more fish without detrimental effects. Average W_r should not be calculated for an entire population. Different size individuals of the same species often have different food habits; consequently, fish of different lengths could have considerably different W_r.

Another measure of fish abundance that is sometimes used is the number of stock size and larger fish per hour of electrofishing. Some biologists break the total down to the Gabelhouse (1984) stock, quality, preferred, memorable, and trophy size categories. The general consensus among biologists using this expression is that 100 stock size and longer largemouth bass per hour of electrofishing is a dense bass population.

The progression of material presented in this section may lead readers to the conclusion that Swingle's concepts of balance have been outdated. This is not at all true, and there is no better discussion of the structure and dynamics of warmwater pond fish populations than Swingle's 1950 publication.

20.8 AQUATIC VEGETATION

Abundant vegetation creates unpleasant conditions for anglers, swimmers, and boaters. Aquatic plant respiration and decomposition depletes oxygen and can result in dead fish. Certain plant species can contribute to taste and odor problems in fish. There is a positive side, however, to aquatic vegetation. Primary production by aquatic plants, especially phytoplankton, is the means by which a portion of the sun's energy is fixed into forms available to other aquatic organisms. A variety of invertebrates and a few vertebrates feed on living and dead aquatic plants. Many aquatic organisms use plants for shelter and attachment. Isolated weed beds become fish attractors in the same way tires and concrete blocks serve as artificial reefs. Vegetation may need to be controlled through reduction, but control is not synonomous with elimination.

There are times when growth of aquatic vegetation should be encouraged. Fertilizing to promote phytoplankton increases fish production and discourages more noxious plants. If a pond is to be used for waterfowl, at least a portion of the pond should have shallow water to encourage growth of vascular plants; fertilization may or may not be needed. An interesting intentional plan for interspersing plants has been outlined by Belusz (1986). High and low spots in shallow areas of a pond will encourage and discourage, respectively, growth of aquatic plants.

Publications on control of aquatic vegetation list three general methods: mechanical, chemical, and biological. Mechanical methods are labor intensive and produce short-term results. When herbicides work, they give results quickly, but sometimes, such as with fall applications, they show no detectable effect. Hard waters also reduce or negate the effectiveness of herbicides, and some plant species are more difficult to kill than others. Three major disadvantages to the use of herbicides are (1) only a few compounds have been cleared, at least in the United States, for use on food fish (Schnick et al. 1989), (2) all approved herbicides except copper sulfate are expensive, and (3) decomposition of vast amounts of dead plants depletes dissolved oxygen. Fish pond publications, extension agents, and fisheries agencies should be consulted for herbicides recommended for a particular area.

Biological control of aquatic vegetation is receiving considerable attention largely because of environmental concerns over chemicals. However, biological control, specifically use of grass carp, has generated other environmental concerns. Use of grass carp is illegal in some states, and other states require the use of triploid (sterile) carp or have other restrictions. Swanson and Bergersen (1988) have written a grass carp stocking model that incorporates the major variables important to the success or failure of grass carp: daily temperature units, vegetation density and distribution, vegetation value, feeding preference, disturbance, management objectives, fish size, and genetic makeup. Because grass carp eat less as they get larger and will not spawn in standing water, it is necessary to periodically stock additional fish.

20.9   ASSESSMENT OF OLD PONDS

Fisheries biologists working on ponds today will be heavily involved with assessing old ponds and devising management plans that will make the best out of what someone unknowingly has obtained. Management strategies for old ponds will fall into the categories of do nothing, corrective stocking, or reclamation.

20.9.1   Do Nothing

The first question that a fisheries biologist should ask himself or herself after assessing a pond community is, What will happen if nothing is done? Both short- and long-term answers should be assessed. Sometimes nothing needs to be done, especially in the short-term. Sometimes nothing can be done to improve the condition. With limits on time and money, it is better to focus on ponds that can be managed relatively easily.

20.9.2   Corrective Stocking

Usually when pond fish communities are out of balance, there are too few predators. Through the use of formulated feeds, advanced sizes of largemouth bass and tiger muskellunge are available from hatcheries. In theory, stocking predators should help reduce abundant prey, but results have been variable (Bennett 1971; Boxrucker 1987). Success may be improved if standing crop is reduced below carrying capacity and enough fish are introduced to approach the carrying capacity for that species (Bennett 1971).

20.9.3   Reclamation

For older ponds that have unwanted species of fish, often the best remedy is to eliminate the fish community and start over. If the pond has a bottom drain, it is a simple matter to drain the pond. If any puddles remain, the fish can be poisoned. If a pond cannot be drained, the entire pond can be poisoned. Although such a procedure appears to destroy potential fishing, good fishing can occur in as little as 2 years after reclamation. Without reclamation, some older ponds never will produce good fishing.

20.10   CONSTRUCTING PONDS

Although construction of ponds has slowed in recent years, it is likely that fisheries biologists will be called upon to advise on construction of ponds. Biologists should focus their advice on aspects that will have an impact on future management.

The concept of multiple use that is more commonly applied to larger impoundments should also be considered for ponds. A pond that is intended for fishing, swimming, and duck hunting will need a few special provisions during construction, such as fish attractors, a sand beach, and a point or perhaps an island to best accommodate the three recreational activities. Nearness to the residence may also

Figure 20.6 Properly constructed spillway barriers prevent harvestable-size fish from escaping and require little maintenance. Certain species such as grass carp and juvenile largemouth bass are especially vulnerable to losses over emergency spillways.

Figure 20.7 Fishing piers can be used in conjunction with various fish attractors. In addition, they improve fishing opportunities for anglers lacking boats. Piers are especially helpful to handicapped and elderly anglers. Such structures can be provided with open and shaded areas.

vary for different uses. Livestock watering is a commonly desired use. Without carefully planned restricted access or a livestock water tank below the dam, livestock will ruin pond banks. In arid parts of North America, pond owners may want to use pond water for irrigation. Anything more than watering the lawn should be discouraged for impoundments of the size considered in this chapter.

Ponds can be too small as well as too large. Hooper (1970) found that 80% of warmwater ponds less than 0.1 hectare failed to achieve balance with multiple species. An option to consider for small ponds is single-species stocking and supplemental feeding such as is done with rainbow trout and channel catfish. Ponds are too large when owners cannot afford to buy the quantity of fish that should be stocked, or the amount of chemicals such as herbicides and fertilizers that may be necessary to maintain the pond. Pond owners should be aware of such potential expenses before a large pond is designed. They should also be advised of the minimum depths necessary to avoid winterkill problems at northern latitudes and higher elevations (SCS 1982).

Installation of a bottom drain should be emphasized. With a drain, fall drawdowns, reclamation, and renovation are simple matters. A considerable number of fish can be lost over emergency spillways (Louder 1958; Lewis et al. 1968). The spillway should be wide so that flows will be shallow, or a parallel bar barrier (Figure 20.6) should be installed (Powell and Spencer 1979).

Older extension bulletins on pond management stated that all brush and trees should be removed from a pond site because they provided too much cover for bluegills. Fish attractors are popular in fish management now (Johnson and Stein 1979; Wege and Anderson 1979), and newer pond bulletins are suggesting leaving or placing trees and brush in ponds (Figure 20.7). Also, leaving the pond bottom irregular provides fish habitat. Humps are better than dips so that no puddles remain if the pond has to be drained.

Shallow water of less than 1 m is undesirable from an aquatic plant management perspective. For safety and stability, a gradual slope of 3:1 is recommended, which means 1 m of depth will be reached about 3 m from shore. Plants can be prevented from growing in an area such as a swimming beach by a process called blanketing (Belusz 1986). Black plastic sheeting is laid over the area during construction, and then 15–20 cm of sand or fine gravel is placed over the plastic.

Before any pond is built, it may be necessary to obtain permits from several agencies. Employees of agencies such as the Soil Conservation Service (SCS) have the technical knowledge and experience to advise on actual construction and on application for federal cost-sharing funds. If a biologist wants to become more conversant about pond construction, a useful reference is the SCS Agriculture Handbook 590, *Ponds: Planning, Design, Construction* (SCS 1982).

20.11 SUMMARY

Although small impoundments are often constructed for purposes other than sport fish production, their great numbers and small sizes do provide convenient angling opportunities for millions of people of all ages. Unbalanced fish populations and poor fishing generally result when the pond owner lacks an understanding of basic fish management concepts or is unwilling to alter practices, such as direct livestock watering, which degrade fish habitat.

Fisheries biologists have too often tried to make an impoundment be all things to all people. Pond owner or angler desires and production capabilities of the impoundment should be used in establishing management objectives. Balance is best illustrated in ponds with relatively simple communities, but it is a valid concept regardless of impoundment size and community complexity. However, there are several useful options that differ from the traditional concept of a balanced bass–bluegill community.

The goal of fisheries management biologists should be to provide annual crops of harvestable-size fish commensurate with fertility of the impoundment. It is the responsibility of the pond owner, however, to implement management procedures designed to achieve fish management objectives. Successful angling opportunities can be provided in a multi-purpose pond with minimal impact on other uses, provided that it is properly constructed, stocked, and managed.

20.12 REFERENCES

Anderson, R. O. 1973. Application of theory and research to management of warmwater fish populations. Transactions of the American Fisheries Society 102:164–171.

Anderson, R. O. 1976. Management of small warm water impoundments. Fisheries (Bethesda) 1(6):5–7, 26–28.

Anderson, R. O. 1978. New approaches to recreational fishery management. Pages 73–78 *in* G. D. Novinger and J. G. Dillard, editors. New approaches to the management of small impoundments. American Fisheries Society, North Central Division, Special Publication 5, Bethesda, Maryland.

Anderson, R. O. 1980. Proportional stock density (PSD) and relative weight (W_r): interpretive indices for fish populations and communities. Pages 27–33 *in* S. Gloss and B. Shupp, editors. Practical fisheries management: more with less in the 1980's. American Fisheries Society, New York Chapter, Ithaca, New York. (Available from

New York Cooperative Fishery and Wildlife Research Unit, Cornell University, Ithaca.)

Anderson, R. O., and S. J. Gutreuter. 1983. Length, weight, and associated structural indices. Pages 283–300 in L. A. Nielsen and D. L. Johnson, editors. Fisheries techniques. American Fisheries Society, Bethesda, Maryland.

Belusz, L. C. 1986. Aquatic plant management in Missouri. Missouri Conservation Commission, Jefferson City.

Bennett, G. W. 1952. Pond management in Illinois. Journal of Wildlife Management 16:249–253.

Bennett, G. W. 1971. Management of lakes and ponds. Van Nostrand Reinhold, New York.

Boxrucker, J. 1987. Largemouth bass influence on size structure of crappie populations in small Oklahoma impoundments. North American Journal of Fisheries Management 7:273–278.

Boyd, C. E. 1979. Water quality in warmwater fish ponds. Alabama Agricultural Experiment Station, Auburn University, Auburn.

Buck, D. H., and C. F. Thoits III. 1970. Dynamics of one-species populations of fishes in ponds subjected to cropping and additional stocking. Illinois Natural History Survey Bulletin 30:69–165.

Childers, W. F. 1967. Hybridization of four species of sunfishes (Centrarchidae). Illinois Natural History Survey Bulletin 29:159–214.

Dendy, J. S. 1963. Farm ponds. Pages 595–620 in D. G. Frey, editor. Limnology in North America. University of Wisconsin Press, Madison.

Dillard, J. G., and G. D. Novinger. 1975. Stocking largemouth bass in small impoundments. Pages 459–474 in H. Clepper, editor. Black bass biology and management. Sport Fishing Institute, Washington, D.C.

Ellison, D. G., and R. C. Heidinger. 1978. Dynamics of hybrid sunfish in southern Illinois farm ponds. Proceedings of the Annual Conference Southeastern Association of Fish and Wildlife Agencies 30(1976):82–87.

Gabelhouse, D. W., Jr. 1984. A length-categorization system to assess fish stocks. North American Journal of Fisheries Management 4:273–285.

Gabelhouse, D. W., Jr., R. L. Hager, and H. E. Klaassen. 1982. Producing fish and wildlife from Kansas ponds. Kansas Fish and Game Commission, Pratt.

Galbraith, M. G. 1975. The use of large Daphnia as indices of fishing quality for rainbow trout in small lakes. Internationale Vereinigung für theoretische und angewandte Limnologie Verhandlungen 19:2485–2492.

Gustafson, K. A. 1988. Approximating confidence intervals for indices of fish population size structure. North American Journal of Fisheries Management 8:139–141.

Hooper, G. R. 1970. Results of stocking largemouth bass, bluegill, and redear sunfish in ponds less than 0.25 acre. Proceedings of the Annual Conference Southeastern Association of Game and Fish Commissioners 23(1969):474–479.

Johnson, D. L., and R. A. Stein, editors. 1979. Response of fish to habitat structure in standing water. American Fisheries Society, North Central Division, Special Publication 6, Bethesda, Maryland.

Lawrence, J. M. 1954. A new method of applying inorganic fertilizer to farm fish ponds. Progressive Fish-Culturist 16:176–178.

Lewis, W. M., and R. C. Heidinger. 1978a. Tailor-made fish populations for small lakes. Southern Illinois University, Fisheries Bulletin 5, Carbondale.

Lewis, W. M., and R. C. Heidinger. 1978b. Use of hybrid sunfishes in the management of small impoundments. Pages 104–108 in G. D. Novinger and J. G. Dillard, editors. New approaches to the management of small impoundments. American Fisheries Society, North Central Division, Special Publication 5, Bethesda, Maryland.

Lewis, W. M., R. Heidinger, and M. Konikoff. 1968. Loss of fishes over the drop box spillway of a lake. Transactions of the American Fisheries Society 97:492–494.

Louder, D. 1958. Escape of fish over spillways. Progressive Fish-Culturist 20:38–41.

Marriage, L. D., A. E. Borell, and P. M. Scheffer. 1971. Trout ponds for recreation. U.S. Department of Agriculture Farmers' Bulletin 2249.

Meehean, O. L. 1952. Problems of farm fish pond management. Journal of Wildlife Management 16:233–238.

Ming, A. 1972. Regulation of largemouth bass harvest with a quota. Pages 39–53 in J. L. Funk, editor. Symposium on overharvest and management of largemouth bass in small impoundments. American Fisheries Society, North Central Division, Special Publication 3, Bethesda, Maryland.

Modde, T. 1980. State stocking policies for small warmwater impoundments. Fisheries (Bethesda) 5(5):13–17.

Neumann, R. M., and B. R. Murphy. 1991. Evaluation of the relative weight (W_r) index for the assessment of white crappie and black crappie populations. North American Journal of Fisheries Management 11:543–555.

Powell, D. H. 1975. Management of largemouth bass in Alabama's state-owned public fishing lakes. Pages 386–390 in H. Clepper, editor. Black bass biology and management. Sport Fishing Institute, Washington, D.C.

Powell, D. H., and S. L. Spencer. 1979. Parallel-bar barrier prevents fish loss over spillways. Progressive Fish-Culturist 41:174–175.

Regier, H. A. 1962. On the evolution of bass-bluegill stocking policies and management recommendations. Progressive Fish-Culturist 24:99–111.

Schmittou, H. R. 1969. Some effects of supplemental feeding and controlled fishing in largemouth bass-bluegill populations. Proceedings of the Annual Conference Southeastern Association of Fish and Wildlife Agencies 22(1968):311–320.

Schnick, R. A., F. P. Meyer, and D. L. Gray. 1989. A guide to approved chemicals in fish production and fishery resource management. University of Arkansas, Cooperative Extension Service, Little Rock.

SCS (Soil Conservation Service). 1982. Ponds: planning, design, construction. U.S. Department of Agriculture, Agriculture Handbook 590, Washington, D.C.

Swanson, E. D., and E. P. Bergersen. 1988. Grass carp stocking model for coldwater lakes. North American Journal of Fisheries Management 8:284–291.

Swingle, H. S. 1950. Relationships and dynamics of balanced and unbalanced fish populations. Alabama Polytechnic Institute, Agricultural Experiment Station Bulletin 274, Auburn.

Swingle, H. S. 1951. Experiments with various rates of stocking bluegills, Lepomis macrochirus Rafinesque, and largemouth bass, Micropterus salmoides (Lacepède) in ponds. Transactions of the American Fisheries Society 80(1950):218–230.

Swingle, H. S. 1952. Farm pond investigations in Alabama. Journal of Wildlife Management 16:243–249.

Swingle, H. S. 1954. Fish populations in Alabama rivers and impoundments. Transactions of the American Fisheries Society 83(1953):47–57.

Swingle, H. S. 1956. Appraisal of methods of fish population study, part 4: determination of balance in farm fish ponds. Transactions of the North American Wildlife Conference 21:298–322.

Swingle, H. S. 1970. History of warmwater pond culture in the United States. American Fisheries Society Special Publication 7:95–105.

Swingle, H. S., B. C. Gooch, and H. R. Rabanal. 1965. Phosphate fertilization of ponds. Proceedings of the Annual Conference Southeastern Association of Game and Fish Commissioners 17(1963):213–218.

Swingle, H. S., and E. V. Smith. 1938. Fertilizers for increasing the natural food for fish in ponds. Transactions of the American Fisheries Society 68:126–135.

Wege, G. J., and R. O. Anderson. 1979. Influence of artificial structures on largemouth bass and bluegills in small ponds. Pages 59–69 in D. L. Johnson and R. A. Stein, editors. Response of fish to habitat structure in standing water. American Fisheries Society, North Central Division, Special Publication 6, Bethesda, Maryland.

Weithman, S. A., J. B. Reynolds, and D. E. Simpson. 1980. Assessment of structure of largemouth bass stocks by sequential sampling. Proceedings of the Annual Conference Southeastern Association of Fish and Wildlife Agencies 33(1979):415–424.

Chapter 21

Natural Lakes and Large Impoundments

DANIEL B. HAYES, WILLIAM W. TAYLOR, AND
EDWARD L. MILLS

21.1 INTRODUCTION

Natural and artificial lakes of 4 hectares or more constitute an important component of recreational fisheries. Within this set of water bodies there is a wide diversity of fisheries, from largemouth bass in Florida to rainbow trout in Washington, from walleye in South Dakota to chain pickerel in Maine, from introduced tilapias in Mexico to salmonids in Canada. Even though each of these species has different environmental requirements, the management problems and strategies for their solution are similar. The objectives of this chapter are to (1) provide a basis for understanding fish production in lakes, (2) outline typical fisheries management problems in lakes and some of their solutions, and (3) indicate current trends in lake management.

For our purposes, we will consider lakes as all bodies of water 4 hectares or larger in size (Noble 1980). Impoundments (also called reservoirs) are artificial lakes, and in this chapter we will focus on the management of impoundments 200 hectares or more in size.

21.1.1 Characteristics of Natural Lakes

In North America, most natural lakes occur in Canada and the northern third of the United States due to the fact that most natural lakes in North America were formed as a result of past glaciations. Glaciation had little effect on lake formation in the southern two-thirds of the United States and Mexico. In selected areas of this region, natural lakes were formed from a variety of processes including dissolution of limestone, movement of the earth's crust, wind action, river activity, and volcanic activity. Lakes formed by these various processes often differ in their morphometry and water chemistry; therefore, lakes are classified based on their method of formation (i.e., karst lakes, fault block lakes, oxbow lakes).

The physical and chemical nature of lakes in North America varies enormously. The size of lakes ranges from small kettle lakes scattered throughout Canada and the northern United States to Lake Superior which covers 82,414 km². Depths range from a meter or less up to 614 m in the case of Great Slave Lake, Canada. Salinity, alkalinity, and other chemical features of lakes also vary widely. These

493

chemical factors vary regionally due to climatic influences, lake formation processes, or watershed geochemistry. Even within a watershed, lakes show individual variation. For example, maximum summer lake temperature is partly a function of latitude: northern lakes are cooler than southern lakes. At a given latitude, however, lake depth and exposure to wind will determine the degree of thermal stratification and the range of temperatures found within a lake.

Despite the range of physicochemical conditions found in North American lakes, most lakes generally contain at least one fish species. The assemblage of fish species depends on the lake's limnological characteristics, its geographical location, and human effects through species introduction and management. Intentional and unintentional human alterations to lakes have occurred across the continent, altering the original limnology and fish communities of many lakes. Fish introductions have broken down many of the geographical barriers to fish dispersal, expanding the range of many species (Chapter 12). Disturbances to watersheds, including acid rain, eutrophication, and water diversions have altered the chemistry and hydrology of lakes in vast areas of North America. Due to the scope and magnitude of these disturbances, fisheries managers have been forced to approach lake management problems from a broader perspective. To successfully manage our continent's lake and reservoir fisheries requires an ecosystem approach; consideration must be given not only to internal processes within lakes, but to external processes affecting lake airsheds and watersheds. Furthermore, because of a human propensity to alter the world we live in, fisheries managers must not view lakes as fixed or stable ecosystems, but rather as entities functioning in an ever-changing environment.

21.1.2   Characteristics of Impoundments

North American impoundments provide hydroelectric power, flood control, and water storage as well as fishing and other recreational opportunities. In 1980, there were nearly 1,600 impoundments with surface areas greater than 200 hectares in the United States. Reservoirs are particularly important in providing fisheries for the southern two-thirds of the United States where there are relatively few natural lakes.

In many ways, the range of physicochemical factors in impoundments is similar to that of natural lakes, with both regional and local variation occurring among reservoirs. The physical morphology of impoundments is often a result of engineering concerns for effective dam building as defined by the objectives for the reservoir. Choice locations for impoundments are narrow river valleys where dams can be built providing the most water retaining capabilities for a given width of dam. Because of this, large impoundments are often deep with steep drop-offs associated with the old river valley. Through flooding of river channels upstream of dams, impoundments often have a large main body associated with the main river channel, but may have many smaller branches or bays associated with smaller tributaries. The steep depth gradient of impoundments and their high length:width ratio encourages the development of strong thermal and chemical gradients (Wunderlich 1971).

21.1.3   Differences Between Lakes and Impoundments

The defining characteristic of impoundments is that they are artificial. Because of this, they differ from natural lakes in several important ways. Physically, impoundments tend to show greater shoreline development because they typically occupy river valleys. Despite this, littoral areas in impoundments may be limited by relatively steep depth contours. This problem is exacerbated by rapid fluctuations in water level associated with power generation, flood control, or agricultural uses. When shoreline areas are dewatered for extended periods, use by fish and habitation by benthic invertebrates may be prevented.

Other physical characteristics typical of impoundments include high water flow-through rate, turbidity, and sediment input. Turbidity affects light penetration and can cause decreased primary production. Additionally, turbidity affects the ability of visually orienting predators to detect prey, thereby decreasing prey vulnerability, potentially leading to an unbalanced fishery. Sediment deposition decreases the abundance of benthic invertebrates (Aggus 1971) and can cause significant loss of fish eggs.

As in natural lakes, water chemistry in impoundments is strongly influenced by the geochemical nature of the watershed and basin morphometry. Due to the irregularity in basin morphometry of most reservoirs, relatively strong internal variation in nutrient concentrations, temperature, and other physicochemical factors often results (Wunderlich 1971).

When considering fish communities, it is important to consider the age of the water body. Impoundments are substantially younger than glaciated lakes, the majority being less than 100 years old. When a river is impounded, the majority of fish species present prior to impoundment are often displaced (Fitz 1968), leaving an impoverished fish community that may require a stocking program (Chapter 13). Fish communities in impoundments range from a simple to a highly diverse species mixture, and erratic year-class strength of the dominant species is often exhibited.

21.2 BIOLOGICAL AND ENVIRONMENTAL BASIS OF FISH PRODUCTION

There are a myriad of physical, chemical, and biological factors that determine the level of fish production in a given body of water (Figure 21.1). Most significant among these are food abundance and production, dissolved oxygen, temperature, lake water pH and acidity, and availability of reproductive sites.

21.2.1 Trophic State Model

The trophic state model is a widely recognized conceptual framework for describing the ontogeny of lentic systems, their productivity, and the process of eutrophication (Wetzel 1975). Within this framework, three major classes of lakes and impoundments are recognized: oligotrophic, mesotrophic, and eutrophic. Trophic state designations are based on nutrient loading and primary production, with oligotrophic lakes having the lowest levels of nutrient loading and primary production, eutrophic lakes having the highest, and mesotrophic lakes being somewhere in-between. In addition to nutrient inputs, climate plays an important role in determining trophic status of a lake. When lakes are compared globally, primary production is more closely linked to latitude and radiant energy inputs than it is to nutrients (Brylinsky and Mann 1973). For lakes at a similar latitude,

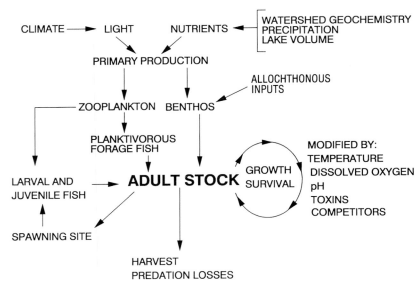

Figure 21.1 Biotic and abiotic factors influencing fish production in lakes and reservoirs.

or for a single lake over time, the importance of nutrient loading becomes more important.

Trophic state classification has been related to the productivity of all biological components of lake ecosystems. These relationships have been demonstrated primarily through correlational studies and further support the concept that nutrient inflow drives primary production which then determines zooplankton and benthic invertebrate production. Production of these food organisms in turn determines fish production. Some empirical relationships of particular interest to fisheries managers are between (1) fish yield and total phosphorus concentration; (2 and 3) fish yield or harvest and phytoplankton standing stock; and (4) fish yield and macrobenthos abundance:

(1)   Yield (kg/hectare) = 0.792 + 0.072 total P (μg/L)
      $r^2 = 0.84$ $n = 21$ (Hanson and Leggett 1982)
(2)   Yield (g dry wt/m/yr) = $-1.92 + 1.17$ log Chlorophyll a (mg/m^3)
      $r^2 = 0.84$ $n = 19$ (Oglesby 1977)
(3)   Harvest (kg/hectare) = $-1.8 + 2.7$ Chlorophyll a (mg/m^3)
      $r^2 = 0.83$ $n = 25$ (Jones and Hoyer 1982)
(4)   Yield (kg/hectare) = 1.293 + 0.012 macrobenthos (kg/hectare)
      $r^2 = 0.47$ $n = 26$ (Hanson and Leggett 1982)

A widely used index that is conceptually related to trophic state classification is the morphoedaphic index (MEI) which is the ratio of total dissolved solids to mean lake depth (Ryder 1982). In effect, these parameters integrate many of the limnological characteristics that affect lake and fish productivity. Total dissolved solids provides an indication of the amount of nutrient influx into a lake, which as indicated earlier, is one of the primary determinants of trophic status. Although the concentration of total dissolved solids is a less accurate indicator of nutrient availability than direct measurements, the relative ease of measuring total dissolved solids makes this attractive for use in an index such as the MEI. Mean lake depth provides an indication of lake morphometry. In general, deeper lakes are cooler and tend to have higher energy losses to deep-water sediments. Furthermore, deep lakes have more water volume than shallow lakes of the same area, and thus nutrient inputs are diluted to a greater degree (Ryder 1982). At a regional level, the MEI can be used to provide fisheries managers with a preliminary estimate of the amount of fish yield that can be expected from a given lake. When used on a continentwide or global basis, the MEI should be modified to account for latitudinal differences in climate and growing season.

For fisheries managers, the trophic state model and associated empirical regressions provide an index of lake fertility and potential fish production. Such simple indices are confounded, however, by factors such as pH or availability of reproductive sites which also influence the level of fish production in lakes. Another limitation of the trophic state model is that the species composition and size structure of fish populations in a given lake can not be directly estimated. Furthermore, trophic state classifications are sometimes erroneously equated with lake desirability. For example, eutrophic lakes are likely to be perceived as being less desirable by swimmers and boaters, but may provide productive largemouth bass fisheries. Fisheries managers must be aware of such conflicts in the uses of lakes and consider the spectrum of public demands on aquatic resources when constructing management plans.

21.2.2 Food Web Model

In the trophic state model, production and biomass at each trophic level is controlled by nutrients and primary production "from the bottom up." In contrast, in the food web model or consumer controlled model, fish predation determines the structure and abundance of prey items (Carpenter et al. 1985). By decreasing the abundance of planktivorous forage fish, zooplankton abundance and size-structure are affected. Alterations in the zooplankton community further leads to changes in algal and nutrient dynamics. Thus, fish predation has direct effects on prey which in turn affects other biological components of lake ecosystems. Such top-down (Mills and Forney 1988) control of trophic structure has generated theories of biomanipulation (Shapiro and Wright 1984) and cascading trophic interactions (Carpenter et al. 1985). Empirical relationships have been developed in support of the food web model, and include the correlation between fish mean size and zooplankton mean size, fish biomass and zooplankton biomass, and fish biomass and phytoplankton biomass (McQueen et al. 1986).

Fish introductions have provided additional evidence for top-down control of lake ecosystems. The classic example is the effect alewives had on the structure of the zooplankton communities in coastal ponds in Connecticut (Brooks and Dodson 1965). The results of their investigations showed that ponds with alewives had zooplankton communities dominated by small species such as *Bosmina longirostris, Ceriodaphnia lacustris, Cyclops bicuspidatus thomasi,* and *Tropocyclops prasinus.* In contrast to this, the zooplankton community where alewives were absent was predominantly large planters such as *Daphnia* spp., large *Diaptomus* spp., and *Mesocyclops edax.* In Crystal Lake, Connecticut, alewives were absent in 1942, and the zooplankton assemblage present was composed of large species, with a modal length of 0.79 mm. By 1964, alewives had become abundant in Crystal Lake, and consequently the zooplankton community had shifted to smaller species with a modal length of 0.29 mm.

While both top-down and bottom-up models have been used to describe food web dynamics, it is likely that the processes that each model focuses on occur simultaneously in aquatic systems. In the alewife example above, it is true that these fish had a dramatic influence on the zooplankton community. However, it can be argued that the ultimate biomass of the alewife population was determined by the rate of nutrient flux into these ponds. A current synthesis of top-down and bottom-up models proposes that predation controls community structure while competition and availability of nutrients limit maximum production at each trophic level (Mills and Forney 1988). A critical feature of both models and the synthesis is that productivity of different trophic levels is linked, and management actions imposed on one trophic level will have effects on other trophic levels. Understanding that all trophic levels in lakes are linked leads to the development of an ecosystem approach to lake management, with consideration given to anglers in addition to other lake users. Given the diverse demands being placed on our aquatic resources, an ecosystem approach is necessary to evaluate the consequences a given management prescription will have on food web interactions, trophic dynamics, and ultimately fish production.

The specific problem of maximizing fish abundance, fish yield, or the quality of fish produced depends critically on how production is packaged in a food web. If one considers "producer" versus "consumer" effects on food web dynamics, there are two alternative approaches to increasing fish yield (Figure 21.2). The first involves enlarging the food resource base without altering energy transfer efficiency among trophic levels. This approach amounts to increasing lake fertility to achieve peak biomass possible at all trophic levels. The second approach increases transfer efficiency without changing the size of the food resource base. This is most commonly achieved by maintaining an efficient and balanced suite of predator and prey species. Each of these approaches will be discussed in more detail through case history studies later in this chapter.

21.2.3 Biological Variability

Temporal and spatial variability of biotic components in ecological communities is a source of frustration for fisheries managers because it impedes predictability. Concern over variability in recruitment of fish is a problem which can only be faced through a better understanding of both short- and long-term factors influencing the recruitment process. Temporal gradients of predator and prey species in lakes and impoundments are often dynamic, and both the timing and location of management actions may influence the outcome of the fishery.

Community-level interactions and variability in fish recruitment can be illustrated from long-term records. One example is the Oneida Lake ecosystem which is a large eutrophic lake located in the state of New York. The fish community's impact on lake ecosystem structure and function has been the subject of investigation for nearly 30 years. The walleye is the dominant piscivorous fish, with age-0 yellow perch, white perch, and gizzard shad serving as the primary link in the transfer of energy from secondary to higher trophic levels (Mills and Forney 1988). Because cohorts of age-0 fish vary in initial size and are subject to different

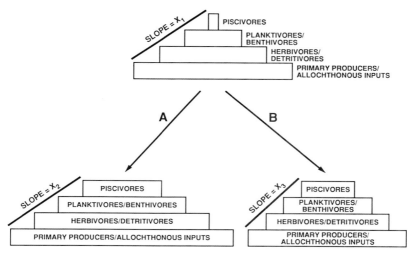

Figure 21.2 Alternate methods of increasing fish yield. Box size represents production at each trophic level. Slopes of dashed lines represent energy transfer efficiency. Method A involves enlarging the food resource base (e.g., fertilization) without changing the energy transfer efficiency among trophic levels ($X_1 = X_2$). Method B involves a food web manipulation (e.g., introduction of a predator) designed to increase energy transfer efficiency ($X_3 > X_1$) without an increase in food resource base. (After Wagner and Oglesby 1984.)

rates of predation, the biomass of these planktivores fluctuates both seasonally and annually (Mills and Forney 1988). In Oneida Lake, climate-induced variation in early survival coupled with predator-mediated depensatory mortality in late summer lead to rapid divergence in year-class abundance. Year-classes of yellow perch in mid-October have varied by as much as 100-fold over a 25-year period (Mills and Forney 1988). The effects of differences in abundance of age-0 fishes filters both up in the food web to walleye and down to zooplankton and phytoplankton. This is just one example of how stochastic events such as weather can lead to variable fish recruitment, with dramatic effects throughout the food web. The effect of stochastic events on lake ecosystems may in fact set inherent limits on the predictability of aquatic systems.

The problem of variable recruitment is often intensified in impoundments through the action of varying water levels. The reproductive success of many species including northern pike, largemouth bass, and common carp depends on the occurrence of high water levels and the types of substrate flooded in reservoirs (Aggus 1979). For example, largemouth bass reproduction is enhanced if shoreline vegetation is inundated (Von Geldern 1971). As such, dewatering of the littoral zone can significantly impact reproductive success and juvenile survival of largemouth bass. The timing and magnitude of water level fluctuations will affect species differentially, giving managers a tool for controlling undesirable species while promoting growth and survival of desired species (Chapter 14).

The ecological maturity of impoundments is also a source of recruitment variability since the fish species mixture usually has had little time to coevolve. New impoundments are particularly vulnerable to highly variable conditions

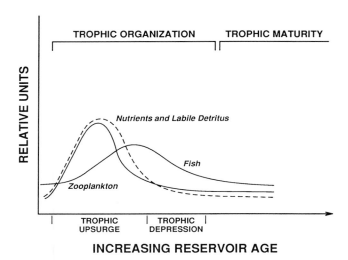

Figure 21.3 A generalized model of trophic dynamics as a reservoir matures and ages. (After Kimmel and Groeger 1986.)

which have a direct impact on fish populations. Kimmel and Groeger (1986) coined the terms trophic upsurge and trophic depression to describe the changes that occur in reservoir productivity with age. A generalized model of trophic dynamics as a reservoir matures is shown in Figure 21.3. The magnitude and variability of biological production is greatest in the early phase of an impoundment's existence because of the surplus of available nutrients and large quantities of organic material. Often this initial period of high productivity has produced false hopes for long-term fish production, but as reservoirs mature, fish populations usually stabilize (5–10 years) at a lower level of abundance. As reservoirs mature, productivity tends to decline as nutrients released through impoundment decline, and as competition reduces food availability (Kimmel and Groeger 1986). Because of the trophic simplicity of most reservoir systems, such water bodies continue to lack stability and are vulnerable to overfishing and environmental variation.

21.3 APPROACHES TO LAKE MANAGEMENT

The process of fisheries management is ideally goal-oriented, with a concrete set of objectives in place prior to taking any management action. One's goals and objectives are ultimately set by the public's demands (Chapter 7) but are limited by the capability of a lake to produce fish while maintaining water quality suitable for other uses. Determining the fishery potential of a lake is difficult due to the complexity of factors influencing fish production as described earlier. One approach is to use a limnological indicator associated with fish production such as the morphoedaphic index. Although this index provides an indication of the current level of fish production, it cannot predict what level of fish production a lake can achieve through appropriate habitat manipulation (i.e., fertilization, brush piles). Our abilities to predict the extent and magnitude of the responses to such habitat manipulations are limited due primarily to a lack of concrete mechanistic data. As such, determining a lake's potential is one example of where fisheries management is more of an art than a precise science.

21.3.1 Lake Assessment

The next step in management is to sample the key limnological and fishery characteristics of the lake in question in order to prescribe a management program to achieve the previously set goals. There are two general strategies for assessing fisheries. The first is to directly sample and measure the fish populations in terms of their growth, recruitment, and population size-structure. The key to successful direct assessment is obtaining a representative sample with sufficient sample sizes to provide statistically reliable estimates of fish population characteristics. Interpretation of the results from this type of assessment program depend on an understanding of the population ecology and biology of the species present in relation to their lake environment. Several indices are available for comparing fish population growth and size-structure between lakes. Rate of fish growth is one of the most sensitive parameters to ecosystem change and is perhaps best indexed by comparing length at age with statewide or regional averages. Difficulties emerge in this approach when growth stanzas in a fish's life history differ in their response to ecosystem stress or management activities. For example, in a population of stunted perch, age-0 perch may show above-average growth, while older individuals grow very slowly. Thus, depending on the life stage being examined, different conclusions regarding the general well-being of the population could be drawn. Fish growth can also be evaluated using condition factors or associated measures such as the relative weight (W_r) index (Wege and Anderson 1978). Higher values of W_r indicate that the fish in the sampled population have a higher weight at a given length than a broad regional mean. Interpretation of W_r is not always straightforward since W_r varies seasonally, especially during periods of gonadal development (see Chapter 6).

To index population size-structure, the proportional stock density (PSD) index was developed by Anderson (1976) based on research performed by Swingle (1950) concerning the question of balance in predator–prey systems. Basically, PSD indicates the proportion of the reproductive segment of a fish population that is of quality size to the angler (Chapter 6). Most successful applications of PSD have occurred in small ponds or reservoirs (see Chapter 20).

The second strategy for fishery assessment is to examine fishery impacted components of the ecosystem. These indirect assessment approaches (morphoedaphic index and regressions of fish yield on phosphorus, phytoplankton, and macrobenthos) provide information regarding fish production but do not require the direct sampling of fish populations.

From a fisheries management point of view, these indices may have limited value since they do not provide information regarding fish population size-structure. An indicator that has been successful in this respect is the size-structure of the zooplankton (Mills et al. 1987). This measure follows from work by Galbraith (1967), where he recognized the importance of large-bodied *Daphnia* to rainbow trout production in lakes. From his research, he suggested that a simple indicator of the probable success of rainbow trout stocking in lakes is the density of large *Daphnia*. This simple and inexpensive index stimulated the investigation of further relationships between the zooplankton and fish communities. Most recently, Mills and Schiavone (1982) and Mills et al. (1987) have shown that the

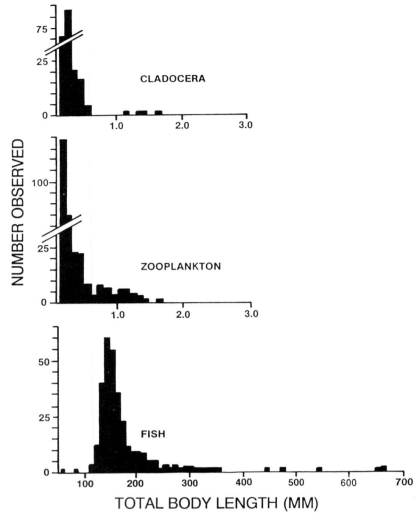

Figure 21.4 Illustration of a warmwater freshwater lake (Black Lake, New York) with a stunted panfish population, few predators, and small sized zooplankton. (After Mills and Schiavone 1982.)

mean body size of the crustacean zooplankton community is correlated with the mean body size of the fish community, particularly in lakes dominated by percids and centrarchids (Figures 21.4 and 21.5). Representative samples of zooplankton can be taken much more easily and quickly than samples of fish to provide an indication of the fish community size-structure within a lake. Data from Cana-darago Lake, a eutrophic lake with a warmwater fishery in New York State, is an excellent example of the use of zooplankton size-structure for indirect fishery assessment (Mills et al. 1987). Predators were stocked from 1977 to 1982 to shift a stunted panfish community toward a balanced predator–prey system. Coincident with the establishment of predatory fish and the suppression of stunted forage species, average zooplankton size increased and the shift toward larger

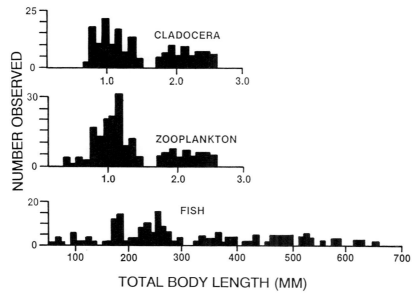

Figure 21.5 Illustration of a warmwater freshwater lake (Clear Lake, New York) with a balanced piscivore-panfish community and numerous large-bodied crustacean zooplankton. (After Mills and Schiavone 1982.)

zooplankton coincided with a change toward *Daphnia* pulex (Figure 21.6). The body-size approach to ecosystem analysis has been further generalized to all planktonic trophic levels (Sheldon et al. 1972, 1973), and is a step toward a holistic view of lake management.

21.3.2 Overview of Management Approaches

One classification scheme of management approaches arranges management prescriptions into three general classes: people management, population management, and habitat management. The goal of people management is to provide access opportunities to aquatic resources for all interested parties and to ensure resource availability and integrity over time. Secondary goals are to provide for specialized opportunities (e.g., quality fishing, catch-and-release) for various user groups when the resource permits. As discussed in Chapter 16, the most common forms of people regulation are length limits, bag limits, closed seasons, gear restrictions, and refugia.

Population management involves the direct manipulation or introduction of aquatic populations. Some examples include stocking of desired game species, rough fish removal, and thinning of overcrowded game species. As fisheries management arose from the fish culture discipline, it is not surprising that many of our historical management tools centered on stocking. In fact, a common stereotype shows the fisheries manager with a bucket of rotenone in one hand and a bucket of fish in the other. Early concepts of fisheries management focused on the idea that the failure of a lake to produce the desired fishery was a function of high exploitation combined with the inability of fish to reproductively sustain themselves (Chapters 12 and 13). These continue to be concerns but successful

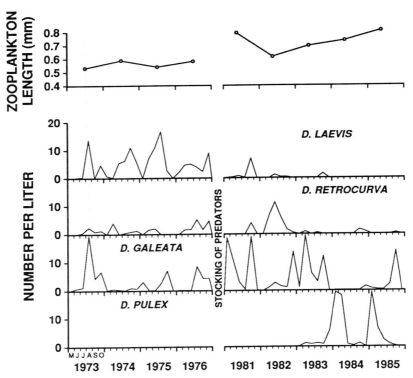

Figure 21.6 Mean size of crustacean zooplankton and average number of *Daphnia* spp. per liter in plankton samples from Canadarago Lake, New York, in May–October, 1973–1976 and 1981–1985. Predators were stocked from 1977 to 1982. (After Mills et al. 1987.)

management today relies on an ecosystem approach to management which integrates the people, aquatic population, and habitat components of a lake.

Habitat management in lakes has matured more slowly than people or population management, possibly due to the difficulty in directly observing the elements of fish habitat. There are two distinct categories of habitat manipulations used in lakes. One approach is to directly alter structure within lakes. Examples include aquatic macrophyte management, formation of spawning areas, or addition of brush as cover (Chapter 10). Perhaps the most extreme example is building impoundments where lotic fish habitats are converted to lentic fish habitats. Impoundments increase the total area of a stream, thereby producing larger potential fish production. The desirability of converting a stream to an impoundment depends critically on the value of the current uses of the stream and the amount of specialized opportunity that may be lost in relation to the benefits provided by the impoundment.

An alternative approach to lake habitat management is to increase the productivity of a fishery through manipulation of the physical and chemical attributes of a lake. Nutrient additions, phosphorus detergent bans, artificial aeration, and destratification are all examples of managing limnological characteristics (Chapters 8 and 10).

Table 21.1 Classes of lake fisheries management problems, their causes, and appropriate management prescriptions.

Management problem	Causes	Management prescriptions
Absence of desired species	Winterkill or summerkill	Decrease nutrient inputs
		Artificial aeration
	Low pH	Mitigative liming
		Stock acid-tolerant strains of fish
		Reduce emissions of sulphur and nitrogen-based acids
	Current fish species not desirable	Stock desired species
	Toxic chemical pollution	Discontinue inputs
Low abundance of desired species	Inadequate recruitment	Add new spawning sites
		Manipulate water levels
		Close angling season during spawning
		Add cover to increase young of the year survival
	High adult fishing mortality	Bag limits
		Length limits
	Fish dispersed over wide areas	Add cover to congregate fish
	Low system productivity	Fertilize
Slow growth	Intraspecific competition	Thin population with nets or chemicals
		Add or protect predators
		Fertilize lake
		Reduce weeds to increase predation
		Stock new prey species
		Stock sterile hybrids
		Stock triploids
	Interspecific competition	Remove with nets or chemicals
		Fertilize lake
		See above methods
	Low lake productivity	Fertilize lake
Poor population size structure	Slow growth	See above
	High adult mortality	Impose length limits
		Impose bag limits
		Stock catchable-size fish

21.3.3 Overview of Management Problems

Although the exact nature of the management problem encountered in a given lake will depend to a large degree on lake limnology and the fish assemblage present, there are several general classes of management obstacles that commonly occur. Organizing these problems into a logical scheme is useful in choosing among possible management prescriptions. In Table 21.1, we have provided a general classification of common management problems occurring in lakes, their causes, and possible management solutions.

21.4 CLASSES OF MANAGEMENT PROBLEMS

21.4.1 Absence of Desired Species

The absence of desired fish species is commonly a result of extremely harsh environmental conditions such as low oxygen concentrations or low lake pH.

Figure 21.7   The scenic beauty of many lakes in the northeastern United States belies the damage being done to their fisheries through acid precipitation and deposition.

Winterkill and summerkill occur most frequently in shallow eutrophic lakes where decomposition and algal respiration decrease oxygen. In stratified reservoirs, hypolimnetic oxygen is often depleted through the decomposition of organic matter associated with newly flooded land and allochthonous material brought in through tributary streams. Coldwater species such as rainbow trout are particularly sensitive to oxygen deficits since they cannot tolerate low hypolimnetic oxygen tensions and are unable to tolerate warm epilimnetic waters for prolonged periods of time. The thermal habitat of striped bass in southern reservoirs is similarly limited by low hypolimnetic oxygen. As a consequence, the viability of this species is significantly reduced in these reservoirs (Coutant 1987). In natural lakes, low oxygen levels may be mitigated by imposing restrictions on point and nonpoint sources of nutrients. A side benefit of this management scheme is to reduce algal blooms and macrophyte growth, thus increasing lake desirability for swimmers and boaters. In some cases nutrient reduction is not feasible and more intensive measures will need to be taken. One such action is the installation of artificial aeration or destratification devices. These devices are successful in increasing oxygen levels but do incur substantial installation and operation costs, which may be prohibitive for larger lakes (Cooke et al. 1986).

There is worldwide concern for the impact of acidic precipitation on lake pH (Figure 21.7). A long-term solution to lake acidification lies in reducing emissions of sulfur- and nitrogen-based acids. Short-term solutions include mitigative liming to increase buffering capacity and pH, and development and stocking of acid-tolerant strains of fishes. Both of these measures are effective for individual lakes, but the scope and magnitude of acidification makes treating all affected lakes impractical. Associated with low lake pH are increased heavy metal lability (especially mercury) and decreased productivity of lower trophic levels. These concerns seriously limit the feasibility of using acid-tolerant fish species to mitigate lake acidification.

Toxic chemical inputs from air and watersheds can also reduce or eliminate fish populations. Coal mine tailings are one example of a source of toxins that can

Figure 21.8 Slow growth or stunting is a problem commonly encountered in the management of yellow perch, bluegill, and other panfish. The challenge facing fisheries managers is to convert fisheries predominated by small fish into ones with abundant quality size individuals.

cause the loss of fish species in a lake or river. The effects of toxins may be more subtle than causing direct loss of a species. Contamination of fish by such chemicals as DDT, PCB, dioxin, mirex, and toxaphene may cause direct losses of fish and fish-eating birds and mammals. Perhaps more importantly, the presence of these contaminants affects human health.

The definition of "desired" species and the relative rank of desirability is a regional phenomenon and is dependent on local fishing traditions. An example of a species with variable desirability is the yellow perch (Figure 21.8). Throughout midwestern North America, yellow perch are a desirable panfish and are sought after by many anglers throughout the year. However, in the Owasco Lake basin (New York) the results of a survey of angler preferences rated yellow perch tied for seventh, with only 6% of anglers preferring that species (Duttweiler 1976). The results of this study highlight the importance of knowing angler preferences when forming management prescriptions. When the current fish community doesn't contain desirable species, a common approach is to stock the desired species. Spectacular results have occurred, such as the salmonid fishery that has been developed in the Great Lakes (Chapter 22) and Lake Ohae, South Dakota. Stocking failures have also occurred. A classic example is the introduction of peacock bass into Lake Gatun, Panama. The introduction of this desired sport fish significantly impacted the overall ecosystem structure and function. Within 2 years, predation by this fish reduced the abundance of all other fishes by 99%, and eventually eliminated all but one of the species present prior to introduction (Zaret 1979).

21.4.2 Low Abundance of Desired Species

Low fish abundance may occur for a variety of reasons including inadequate recruitment, overfishing, or low system productivity. In natural lakes, poor recruitment may occur due to limited spawning habitat, extreme environmental conditions, or low parental stock. In reservoirs, recruitment problems are often intensified due to the inherent variability of the physicochemical factors which often affect spawning success and survival of young fish. For example, when water levels are maintained above a normal summer stage, reproductive success of largemouth bass is enhanced resulting in a strong year-class (Martin et al. 1981; Rainwater and Houser 1982; Miranda et al. 1984). When water levels are lower than normal, or shoreline vegetation is dewatered quickly, reproductive success of largemouth bass is reduced. Thus, one way to increase largemouth bass abundance is to maintain high summer water levels at 2–3 year intervals. High abundances of undesirable fishes may be managed in a similar fashion by water level manipulation during the appropriate time period (Chapter 14). These management prescriptions are contingent, however, on competing uses for reservoir water during the summer months (i.e., irrigation and flood control).

For some species, the lack of suitable reproductive sites has a technological solution. The construction of spawning reefs in Brevort Lake, Michigan, has greatly enhanced walleye reproduction and substantially increased walleye harvest. Another case is Smith Mountain Lake, Virginia, where largemouth bass were observed nesting near tire reefs (Prince et al. 1986). In addition to acting as spawning sites, artificial reefs may serve to concentrate diffuse fish populations, thereby increasing angler success. Production of food items is another potential benefit of artificial structures by providing substrate for periphyton and invertebrates. The success of these habitat devices is contingent upon proper structural design and placement which necessitates consideration of the target species ecological requirements.

Low lake productivity has been implicated as being the cause of low fish abundance in certain lakes. However, low productivity in itself is not usually the sole reason for low fish abundance since low-productivity lakes in northern Canada support substantial populations of char. Low abundance more often results from harvest rates that are in excess of system productivity. Where this is a problem, lake fertilization may increase basal fish production resulting in greater harvest and higher standing stocks. This has been successfully applied in lakes in British Columbia where salmonid production has been significantly increased resulting in substantial economic benefits. However, adding excessive nutrients will have negative side effects including the occurrence of blue-green algal blooms, increased macrophyte growth, decreased water clarity and decreased hypolimnetic oxygen concentration. Due to the cost and potential for negative side effects of lake fertilization, managers generally regulate fish harvest in relationship to lake productivity rather than increase lake productivity directly.

Recreational angling rarely endangers the persistence of fish species, but decreases in population abundance through overfishing have been observed. Such overexploitation can occur surprisingly quickly. Eagle Lake, Maine, a 176-hectare lake in Acadia National Park was stocked with brook trout to provide an ice fishery. Prior to the opening of the season, an estimated 1,556 brook trout were present in the lake. A thorough creel census of anglers indicated that a minimum of 1,344 of these fish were caught in 742 angler trips by February 6 (Havey and Locke 1980). Despite the near total decimation of the stocked population, an additional 706 angling trips were taken during the winter season, yielding only 121 brook trout. Clearly these fish were highly vulnerable to anglers and, despite drastically reduced catch rates, were subjected to continued high fishing pressure. To reduce the impact of fishing on the population, regulations such as length limits or bag limits could have been imposed (Paragamian 1982). However, compliance with these regulations must be high (>80%) to achieve significant population improvement (Gigliotti and Taylor 1990).

21.4.3 Slow Growth of Desired Species

Slow fish growth is probably the most common fishery problem experienced and perhaps the most difficult to manage. Growth is highly plastic in fishes with many factors determining growth rates. Food and temperature are generally the primary determinants of fish growth. The amount of food available to a fish may depend on intraspecific competition, interspecific competition, lake productivity, and limnological characteristics of the lake such as turbidity. Food availability is further complicated by the presence of predators which influences habitat choice of vulnerable size classes. Habitats that provide protection against predators generally have less available food. Thus, fish using these habitats experience increased survival but at a cost of substantially reduced growth. For example, in weedy lakes young bluegills live almost entirely in weed beds (Werner et al. 1983). Predation rates are generally low and food resources in the weed beds often become depleted. This leads to poor growth but relatively high survival of young bluegill.

To effectively manage the growth of fish populations, one must be able to identify the key factors limiting growth. Because of the diverse set of circumstances that can lead to poor growth (Table 21.1), a number of options are available to managers to attack this problem. Where intraspecific competition or low lake productivity has led to slow growth, reductions in fish abundance have led to higher growth rates. The most desirable means of decreasing fish abundance is to increase predation pressure on the species in question. Yellow perch and bluegills are two species that show a tendency to grow very slowly or stunt when large predators are low in abundance. Increasing the appropriate predator populations by stocking can be an effective method of thinning the population and increasing growth (Figure 21.9). An appropriate predator must be large enough to consume the size prey available and be effective in consuming them in their habitats. As these predators are also game fish, the manager should limit their harvest through regulations in order to allow them to effectively control their prey species. When a balanced predator–prey system can not be achieved by stocking and regulations, other methods of population reduction might be needed. These include removal of all or part of the population using chemicals (i.e., rotenone or antimycin) or mechanical (i.e., nets or electrofishing) means. A successful application of fish removal through mechanical means was reported by Hanson et al. (1983). Removal of black crappie and black bullhead with fyke nets resulted a higher abundance of large individuals through increased growth rates. The effectiveness of fish removal projects depends on maintaining balanced population density over time. This is difficult since most fish have a high reproductive potential, and their populations can quickly rebound after artificial thinning (Rutledge and Barron 1972).

Interspecific competition may result in the depletion of food resources and reduced growth of desired species. Reducing the abundance of these competitors, especially if they are considered socially undesirable (Chapter 14), has led to enhanced sport fish populations. An excellent example is the removal of white suckers to improve yellow perch fisheries. In Big Bear Lake, Michigan, perch harvest increased from 500 per year to over 12,000 per year after a sustained

Figure 21.9 The tiger muskellunge, a hybrid between northern pike and muskellunge, is an example of a large predator whose population is maintained through stocking. These large predators provide a fishery in their own right while helping to maintain a balanced population in their prey.

sucker removal project (Schneider and Crowe 1980). Similarly, Johnson (1977) observed higher recruitment and growth of yellow perch after reducing sucker abundance in Wilson Lake, Minnesota. However, not all removals have resulted in improved fish populations (Holey et al. 1979).

When food resources are limited, an intuitive solution is to provide more food by introducing additional prey species (Chapter 12); however, complications often arise from introductions of forage species. Results of stocking gizzard shad and threadfin shad as prey for game species in southern reservoirs provide examples of both successes and failures in forage fish management. The addition of these fish has benefitted largemouth bass, walleye, and striped bass in many of these reservoirs (Noble 1981). However, in systems where predation pressure is insufficient to control numbers of gizzard shad, this species has a tendency to overpopulate at sizes too large for predators to exploit. Where this occurs, condition and reproductive success of largemouth bass and centrarchids may suffer (Noble 1981). Threadfin shad, which do not grow as large as gizzard shad, are often perceived as being a better forage species. Despite this advantage, threadfin shad may compete with small game fish, causing decreased growth and survival (Noble 1981). Thus, the manager needs to evaluate the overall impact of forage introductions on the fish community to ensure that the expected benefits will outweigh any negative interactions.

Food availability is not always the primary factor limiting growth rates. Temperature significantly affects growth, survival, and habitat choice of fish. The way that the temperature regime influences growth, however, is a function of the genetic composition of the population. For example, the northern strain of largemouth bass grows poorly at high temperatures, whereas the Florida strain grows rapidly at these warmer temperatures (Philipp et al. 1981). Consequently, by selecting fish species or strains whose growth optima match the thermal regime in a lake, managers can improve the growth efficiency and productivity of the fishery.

To assess the effects of temperature and food, bioenergetics models based on fish physiology have been developed (Kitchell et al. 1977). These models have been used primarily (1) to estimate growth from feeding rate and temperature, and (2) to estimate feeding rate from observed growth and temperature. This information can then be used by the manager to identify which prey are limiting fish growth and to determine the extent that temperature influences growth. When using bioenergetics models, it is important to note that the genetic composition of a fish strain affects the growth-temperature response and thermal tolerance portions of these models. As such, bioenergetics models whose parameters are obtained from one strain of fish may not accurately predict the growth of other strains. Because of this, managers may need to parameterize the model with data from the strain of fish they are managing.

21.4.4 Poor Population Size-Structure

Management practices to improve population size-structure typically focus on balancing predator–prey relationships (Chapter 20). It should be noted here that the desirable population size-structure varies by species and by region. For example, a yellow perch population with a majority of fish in the 23–30 cm size category would be considered highly desirable, while a walleye population with the same size structure would not. One of the reasons for poor population size-structure is slow growth, and the management prescriptions discussed in Section 21.4.3 would be appropriate for fishery improvement. However not all slow-growing populations exhibit a poor population size-structure. Even where growth rates are low, sizeable numbers of large fish can be maintained in a population if mortality rates are low enough. Examples of this are found in unexploited whitefish and char populations in Canada (Johnson 1976). Conversely, good growth rates do not ensure abundant large fish since excessive exploitation can reduce the numbers of individuals above the desired size. Thus, the size-structure of a fish population results from the interaction of both growth and survival rates. Where growth rates are low and mortality on recruited individuals high, poor population size-structure is virtually guaranteed.

Where macrophytes are overly abundant, undesirable size-structure often results for game fish. This is especially true for panfish such as bluegill which are forced to live in weed beds to escape predation. Where fish are crowded in weed beds, intraspecific competition for food is often severe and growth is often poor even though ample food resources for growth are available in the limnetic zone of the lake (Werner et al. 1983). Additionally, growth of their predators is depressed as they cannot effectively forage when dense weeds are present.

Aquatic weeds are often removed by mechanical or chemical means. However, if the limnological conditions resulting in heavy weed growth (i.e., eutrophication) are not treated, macrophytes may quickly return. Biological control of macrophytes with grass carp has been successfully used in some areas of the country, but the importation of this species is illegal in many states due to concerns regarding their compatibility with other fish and prey species (see Chapter 12).

A specialized aspect of population size-structure management occurs in trophy fisheries. Generally all anglers want to catch big fish, but they want to be able to keep some fish. In light of the discussion in Chapter 5, it is apparent that these two objectives cannot usually be met in a single lake. One way to provide both of these opportunities is to manage some lakes for trophy fisheries and other lakes for keeper fisheries. Anglers can then choose the type of fishing experience they desire most, and no single fishery is pushed to produce conflicting products. Maintenance of large numbers of trophy-size fish usually requires low mortality rates, and consequently bag limits are usually restricted.

21.5 FUTURE DIRECTIONS

In many ways, technological advances have increased our ability to gather, process, and synthesize information about lake ecosystems. A good example of an emerging technology is the use of hydroacoustics for fish population assessment. Through the use of hydroacoustic data, managers will be able to assess lakes more quickly and accurately than with traditional fisheries gear. This equipment not only allows us to gather traditional data on abundance and size-structure, but also expands the types of problems we can address. For instance, the spatial and temporal distribution of fishes is not easily determined with conventional gear, but is amenable to description with hydroacoustics techniques.

Another technological advance that is changing the nature of fisheries science is the computer. The use of computers for data processing is obvious to anyone doing statistical analysis of fisheries data. An intriguing, and as of yet under-developed aspect of computers is their utility as data-collection devices. Automated systems for measuring fish annular growth, length, and prey sizes are just a few examples where computers have improved data collection speed and accuracy. Integration of computers into fishery assessment will improve managers' abilities to understand and manage lake ecosystems.

The greatest conceptual challenge to fisheries science is to construct a model of lake ecosystem structure and function which allows for a holistic lake management approach. This is the first step in managing lakes as interacting systems rather than as disjointed parts. Demands for quality fish populations, the emphasis toward clean water, the importance of accountability in environmental issues, and the need for cost effectiveness are just some of the needs and management challenges in the future. Furthermore, alterations of the environment at continental or global scales, such as acid rain and global warming through increased CO_2, requires fisheries managers to take an ecosystem approach in managing fisheries. Traditionally, fisheries scientists have focused on understanding the dynamics of higher trophic levels while limnologists have concentrated on the physical and chemical processes in lakes and lower trophic levels. True ecosystem management depends on a thorough understanding of the abiotic and biotic factors controlling lake fish populations. Thus, for a successful holistic lake management paradigm to be built, limnologists and fisheries scientists must work together. In addition, with the realization that lake productivity is ultimately managed for human benefit, a greater understanding of the human dimensions (Chapter 7) aspect of lake management must ensue.

21.6 REFERENCES

Aggus, L. R. 1971. Summer benthos in newly flooded areas of Beaver Reservoir during the second and third years of filling 1965–1966. American Fisheries Society Special Publication 8:139–152.

Aggus, L. R. 1979. Effects of weather on freshwater fish predator–prey dynamics. Pages 47–56 in H. Clepper, editor. Predator–prey systems in fisheries management. Sport Fishing Institute, Washington, D.C.

Anderson, R. O. 1976. Management of small warmwater impoundments. Fisheries (Bethesda) 1(6):1:5–7.

Brooks, J. L., and S. I. Dodson. 1965. Predation, body size, and composition of plankton. Science (Washington, D.C.) 150:28–35.

Brylinsky, M., and K. H. Mann. 1973. An analysis of factors governing productivity in lakes and reservoirs. Limnology and Oceanography 18:1–14.

Carpenter, S. R., J. F. Kitchell, and J. R. Hodgson. 1985. Cascading trophic interactions and lake productivity. BioScience 35:634–639.

Cooke, G. D., E. B. Welsh, S. A. Peterson, and P. R. Newroth. 1986. Lake and reservoir restoration. Butterworth, Stoneham, Massachusetts.

Coutant, C. C. 1987. Poor reproductive success of striped bass from a reservoir with reduced summer habitat. Transactions of the American Fisheries Society 116:154–160.

Duttweiler, M. W. 1976. Use of questionnaire surveys in forming fishery management policy. Transactions of the American Fisheries Society 105:232–238.

Fitz, R. B. 1968. Fish habitat and population changes resulting from impoundment of Clinch River by Melton Hill Dam. Journal of the Tennessee Academy of Science 15:329–341.

Galbraith, M. G., Jr. 1967. Size-selective predation on Daphnia by rainbow trout and yellow perch. Transactions of the American Fisheries Society 96:1–10.

Gigliotti, L. M., and W. W. Taylor. 1990. The effect of illegal harvest on recreational fisheries. North American Journal of Fisheries Management 10:106–110.

Hanson, D. A., B. J. Belonger, and D. L. Schoenike. 1983. Evaluation of a mechanical population reduction of black crappie and black bullheads in a small Wisconsin lake. North American Journal of Fisheries Management 3:41–47.

Hanson, J. M., and W. C. Leggett. 1982. Empirical prediction of fish biomass and yield. Canadian Journal of Fisheries and Aquatic Sciences 39:257–263.

Havey, K. A., and D. O. Locke. 1980. Rapid exploitation of hatchery-reared brook trout by ice fishermen in a Maine lake. Transactions of the American Fisheries Society 109:282–286.

Holey, M., and six coauthors. 1979. Never give a sucker an even break. Fisheries (Bethesda) 4(1):2–6.

Johnson, F. H. 1977. Responses of walleye (Stizostedion vitreum vitreum) and yellow perch (Perca flavescens) populations to removal of white sucker (Catostomus commersoni) from a Minnesota lake, 1966. Journal of the Fisheries Research Board of Canada 34:1633–1642.

Johnson, L. 1976. Ecology of arctic populations of lake trout Salvelinus namaycush, lake whitefish, Coregonus clupeaformis, arctic char, S. alpinus, and associated species in unexploited lakes of the Canadian Northwest Territories. Journal of the Fisheries Research Board of Canada 33:2459–2488.

Jones, J. R., and M. V. Hoyer. 1982. Sportfish harvest predicted by summer chlorophyll-a concentration in midwestern lakes and reservoirs. Transactions of the American Fisheries Society 111:176–179.

Kimmel, G. L., and A. W. Groeger. 1986. Limnological and ecological changes associated with reservoir aging. Pages 103–109 in G. E. Hall and M. J. Van Den Avyle, editors. Reservoir fisheries management: strategies for the 80's. American Fisheries Society, Southern Division, Reservoir Committee, Bethesda, Maryland.

Kitchell, J. F., D. J. Stewart, and D. Weininger. 1977. Applications of a bioenergetic model to yellow perch (*Perca flavescens*) and walleye (*Stizostedion vitreum vitreum*). Journal of the Fisheries Research Board of Canada 34:1922–1935.

Martin, D. B., L. J. Mengel, J. F. Novotny, and C. H. Walburg. 1981. Spring and summer water levels in a Missouri River reservoir: effects on age-0 fish and zooplankton. Transactions of the American Fisheries Society 110:370–381.

McQueen, D. J., J. R. Post, and E. L. Mills. 1986. Trophic relationships in freshwater pelagic ecosystems. Canadian Journal of Fisheries and Aquatic Sciences 43:1571–1581.

Mills, E. L., and J. L. Forney. 1988. Trophic dynamics and development of freshwater pelagic food webs. Pages 11–29 in S. R. Carpenter, editor. Complex interactions in lake communities. Springer-Verlag, New York.

Mills, E. L., D. M. Green, and A. Schiavone, Jr. 1987. Use of zooplankton size to assess the community structure of fish populations in freshwater lakes. North American Journal of Fisheries Management 7:369–378.

Mills, E. L., and A. Schiavone, Jr. 1982. Evaluation of fish communities through assessment of zooplankton populations and measures of lake productivity. North American Journal of Fisheries Management 2:14–27.

Miranda, L. E., W. L. Shelton, and T. D. Bryce. 1984. Effects of water level manipulation on abundance, mortality, and growth of young-of-the-year largemouth bass in West Point Reservoir, Alabama–Georgia. North American Journal of Fisheries Management 4:314–320.

Noble, R. L. 1980. Management of lakes, reservoirs, and ponds. Pages 265–295 in R. T. Lackey and L. A. Nielsen, editors. Fisheries management. Blackwell Scientific Publications, Oxford, UK.

Noble, R. L. 1981. Management of forage fishes in impoundments of the southern United States. Transactions of the American Fisheries Society 110:738–750.

Oglesby, R. T. 1977. Relationships of fish yield to lake phytoplankton standing crop, production and morphoedaphic factors. Journal of the Fisheries Research Board of Canada 34:2271–2279.

Paragamian, V. L. 1982. Catch rates and harvest results under a 14.0-inch minimum length limit for largemouth bass in a new Iowa impoundment. North American Journal of Fisheries Management 2:224–231.

Philipp, D. P., W. F. Childers, and G. S. White. 1981. Management implications for different genetic stocks of largemouth bass (*Micropterus salmoides*) in the United States. Canadian Journal of Fisheries and Aquatic Sciences 38:1715–1723.

Prince, E. D., O. E. Maughan, and P. Brouha. 1986. Summary and update of the Smith Mountain Lake artificial reef project. Pages 401–430 in F. M. D'Itri, editor. Artificial reefs—marine and freshwater applications. Lewis Publishers, Chelsea, Michigan.

Rainwater, W. C., and A. Houser. 1982. Species composition and biomass of fish in selected coves in Beaver Lake, Arkansas, during the first 18 years of impoundment (1963–1980). North American Journal of Fisheries Management 2:316–325.

Rutledge, W. P., and J. C. Barron. 1972. The effects of the removal of stunted white crappie on the remaining crappie populations of Meridian State Park Lake. Texas Parks and Wildlife Department, Technical Series 12, Austin.

Ryder, R. A. 1982. The morphoedaphic index—use, abuse, and fundamental concepts. Transactions of the American Fisheries Society 111:154–164.

Schneider, J. C., and W. R. Crowe. 1980. Effect of sucker removal on fish and fishing at Big Bear Lake. Michigan Department of Natural Resources, Institute for Fisheries Research, Report 1887, Lansing.

Shapiro, J., and D. I. Wright. 1984. Lake restoration by biomanipulation. Freshwater Biology 14:371–383.

Sheldon, R. W., A. Prakash, and W. H. Sutcliffe, Jr. 1972. The size distribution of particles in the ocean. Limnology and Oceanography 17:327–340.

Sheldon, R. W., W. H. Sutcliffe, Jr., and A. Prakash. 1973. The production of particles in the surface waters of the ocean with particular reference to the Sargasso Sea. Limnology and Oceanography 18:719–733.

Swingle, H. S. 1950. Relationships and dynamics of balanced and unbalanced fish populations. Alabama Agricultural Experiment Station, Auburn University, Bulletin 274:1–74.

Von Geldern, C. E., Jr. 1971. Abundance and distribution of fingerling largemouth bass, *Micropterus salmoides*, as determined by electro-fishing at Lake Nacimiento, California. California Fish and Game 57:228–245.

Wagner, K. J., and R. T. Oglesby. 1984. Incompatibility of common lake management objectives. Pages 97–100 *in* Lake and reservoir management. U.S. Environmental Protection Agency, 440/5/84-001, Washington, D.C.

Wege, G. J., and R. O. Anderson. 1978. Relative weight (W_r): a new index of condition for largemouth bass. Pages 79–91 *in* G. Novinger and J. Dillard, editor. New approaches to the management of small impoundments. American Fisheries Society, North Central Division, Special Publication 5, Bethesda, Maryland.

Werner, E. E., J. F. Gilliam, D. J. Hall, and G. G. Mittelbach. 1983. An experimental test of the effects of predation risk on habitat use in fish. Ecology 64:1540–1548.

Wetzel, R. G. 1975. Limnology. Saunders, Philadelphia.

Wunderlich, W. O. 1971. The dynamics of density-stratified reservoirs. American Fisheries Society Special Publication 8:219–231.

Zaret, T. M. 1979. Predation in freshwater fish communities. Pages 135–143 *in* H. Clepper, editor. Predator–prey systems in fisheries management. Sport Fishing Institute, Washington, D.C.

Chapter 22

The Great Lakes Fisheries

DAVID J. JUDE AND JOSEPH LEACH

22.1 INTRODUCTION

The Laurentian Great Lakes make up the world's largest freshwater system and contain the world's largest freshwater fish resource. Because they are distributed over such a large area, the individual lakes differ in climatic, physical, and chemical characteristics, and consequently in productivity and fish species diversity (Figure 22.1; Table 22.1). Much has been written about changes in water quality and biota, particularly fish stocks, in the Great Lakes. Early accounts of explorers and fur traders present a picture of crystal-clear water and abundant fish. Historical changes in water quality and biota of the Great Lakes are related to cultural development. Increasing environmental degradation paralleled human progression through exploitation of fur, timber, agriculture, mineral, and energy resources. The degree of environmental alteration to the lakes from stresses related to cultural development has been and continues to be substantial (Table 22.2).

Smith (1970a) attributed deterioration of fish habitat and stocks partly to piecemeal management resulting from the division of the Great Lakes jurisdiction among two nations, eight states, and one province. Since management approaches have not been uniform throughout the lakes, waste disposal, fishing, and species introductions have been largely uncontrolled.

Recent initiatives in Great Lakes fisheries science have been directed toward rehabilitation. Smith (1970a) considered that the mechanisms for rehabilitation of Great Lakes fisheries had been put in place with ratification of the 1955 Convention of Great Lakes Fisheries and formation of the Great Lakes Fishery Commission (GLFC). The Commission undertook the task of sea lamprey control and coordination of research and regulatory efforts of management agencies. Early problems with water quality led to formation of the International Joint Commission (IJC) under the Boundary Waters Treaty of 1909. Water quality degradation in the 1950s and 1960s, particularly in the lower Great Lakes, spurred renewal of the IJC's activities.

Many changes have occurred in the Great Lakes fisheries since 1970. There has been an increase in the amount of information accruing from research and assessment efforts, in public perceptions and expectations of ecosystem quality in the basin, in attitudes of the public toward allocation of fish resources, and in institutional arrangements. A series of international symposia, supported largely by the GLFC, focused international knowledge on several key areas important to Great Lakes management (Loftus and Regier 1972; Colby 1977; Berst and Simon

517

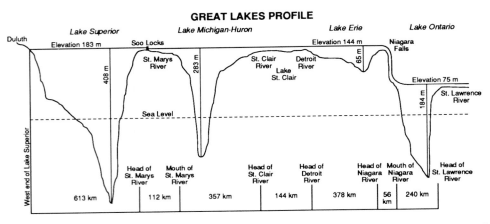

Figure 22.1 Map and profile of the Great Lakes. (Adapted from Michigan Department of Natural Resources Magazine.)

1981; Spangler et al. 1987). Reports of the U.S. Federal Water Pollution Control Administration (FWPCA 1968) and the IJC, and a report entitled "The Algal Bowl" (Vallentyne 1974) alerted the public and governments to the seriousness of pollution problems in the Great Lakes and the need for remedial efforts. Great Lakes Water Quality Agreements between the United States and Canada were signed in 1972 and 1978. The latter agreement incorporated the ecosystem approach toward Great Lakes management, a concept that was also supported by the GLFC. A major step in cooperation among fishery agencies occurred in 1980 with the signing of the Joint Strategic Plan for Management of Great Lakes Fisheries by all concerned parties. Clearly, the mechanisms for rehabilitation of Great Lakes fisheries have been enhanced and expanded in the past 2 decades.

In this chapter, changes in the fish communities of the Great Lakes, particularly

Table 22.1 Climatic, morphometric, and limnological characteristics of the Great Lakes. (Sources: Ryder 1972; Christie 1974; Dobson et al. 1974; Oglesby 1977; Matuszek 1978; Herdendorf 1982.) TDS = total dissolved solids, TP = total phosphorus, MEI = morphoedaphic index, and $\bar{Z}$ = mean depth.

Parameter	Lake Superior	Lake Michigan	Lake Huron	Lake Erie	Lake Ontario
Mean July temperature (°C)	16–18	18–21	18–24	21–24	21–24
Growing season (d)	140–160	160–200	160–240	220–240	200–220
Lake area (km²)	82,100	57,750	59,500	25,657	19,000
Watershed area (km²)	127,700	118,100	133,900	58,800	70,000
Mean depth (m)	149	85	59	18.5	86
Maximum depth (m)	407	282	229	64	245
Volume (km³)	12,230	4,920	3,537	483	1,637
Mean TDS (mg/L)	60	134	108	146	148
Mean TP (mg/m³)	3.3	6.6	4.4	17.8	19.4
Mean Chlorophyll a (mg/m³)	1.0	2.1	1.0	5.0	3.8
MEI (TDS/$\bar{Z}$)	0.40	1.58	1.83	7.89	1.72
No. of native fish species	67	114	99	114	112

since 1970, are outlined with explanations as to their most likely cause. Rehabilitation efforts with respect to responses of management agencies to resource crises are considered in detail with the obvious successes being highlighted. Finally, the major challenges that now face fisheries managers are discussed and strategies for avoiding future crises are suggested.

22.2 BACKGROUND AND OVERVIEW

22.2.1 Native and Introduced Fish Populations

The original species assemblage was similar in Lakes Superior, Michigan, Huron, and Ontario, but markedly different in Lake Erie. The most abundant fishes in Lake Erie were lake whitefish, lake herring, lake sturgeon, blue pike, walleye, sauger, yellow perch, freshwater drum, and channel catfish. Remaining lakes contained lake trout, lake whitefish, lake sturgeon, lake herring, several species of deepwater cisco, burbot, deepwater sculpin, slimy sculpin, emerald shiner, and (in Lake Ontario only) Atlantic salmon. Green Bay of Lake Michigan and Saginaw Bay of Lake Huron also contained large populations of walleye and yellow perch. In Lakes Superior, Michigan, and Huron, top predators were lake trout and burbot, of which lake trout were the more abundant; prey were mainly coregonines, including perhaps 11 species in Lake Michigan and somewhat fewer

Table 22.2 Degree of environmental alteration of the Great Lakes from anthropogenic impacts. (From Ryder 1972.)

Perturbation	Lake Superior	Lake Michigan	Lake Huron	Lake Erie	Lake Ontario
Fishing	Intensive	Intensive	Intensive	Intensive	Intensive
Introductions	Many	Many	Many	Many	Many
Pollution	Moderate	Moderate	Extensive	Extensive	Extensive
Other factors[a]	Many	Many	Many	Many	Many

[a] Other factors include alteration to habitat due to agriculture, forestry, mining, shipping, drainage, damming, diversions, canals, and shoreline restructuring.

in Lakes Huron and Superior (Koelz 1929), and sculpins. Lake trout became the dominant predator in Lake Ontario after 1850 when Atlantic salmon disappeared from that lake (Christie 1974). Top predators in Lake Erie were walleye and blue pike, except that lake trout may have been the dominant predator of the limnetic zone of eastern Lake Erie, at least until about 1890 (Regier and Hartman 1973); primary prey were probably yellow perch and freshwater drum. After settlers arrived, nearly all but the smallest native species became heavily fished. Perhaps the most valuable as human food have been lake trout, lake whitefish, walleye, and blue pike.

Most native fish populations in the Great Lakes suffered dramatic declines between the mid-1800s and the mid-1900s. For example, lake sturgeon had become scarce in all the lakes by 1900 and by the mid-1950s lake trout were extinct in Lakes Michigan, Ontario, and Erie, nearly extinct in Lake Huron, and at low levels of abundance in Lake Superior. Lake herring collapsed in lake Erie in the 1920s, followed by declines in lake whitefish, blue pike, and walleye. The declines in key predators, such as lake trout, resulted in an explosion of prey species and a constantly changing assemblage of fishes.

One result of these changes was the emergence of a large biomass of rainbow smelt and alewife, which provided the opportunity to stock salmon and trout. Consequently, fish assemblages now largely consist of pelagic species, a condition that probably did not exist 200 years ago when nearshore littoral communities were in dynamic equilibrium with pelagic offshore communities. Small, fecund pelagic species (alewife, rainbow smelt) now predominate over the larger species that were originally more closely associated with the benthic and coastal habitats of the lakes. Pacific salmon (coho, chinook, pink, and sockeye) have been introduced into all the lakes. Self-sustaining populations of pink salmon exist in all the Great Lakes and residual stocks of reproducing sockeye salmon persist in Lake Huron. White perch have recently invaded and prospered in Lakes Erie and Huron. In spite of major changes in species composition of the Lake Erie fish assemblages, fish yield (total biomass) has remained approximately the same as it was prior to the major cultural stresses (Regier and Hartman 1973). The present identity of most of the Great Lakes fish stocks is but faintly recognizable as being derived from the diverse communities of 200 years ago.

22.2.2 Factors Causing Changes

Changes in native Great Lakes fish stocks have been attributed to overfishing, degradation of habitat, and invasion of exotic fish species, particularly the sea lamprey, alewife, and rainbow smelt, from the Atlantic Ocean (Table 22.3). Lake Ontario was probably the earliest to be affected, followed by Lakes Erie, Huron, Michigan, and Superior. Lake Superior has had far less drastic changes than the other lakes.

Overfishing probably began during the mid-1800s, as settlement in the Great Lakes basin and the consequent demand for fresh fish increased. Increasing removal of stocks resulted not only from greater commercial fishing effort, but also from more efficient gear. As local stocks were depleted, the fishers moved farther away from home and continued the process. A good example of overfishing concerns the lake trout. During early periods of exploitation there existed many ecologically and genetically subspecific subpopulations of lake trout (Brown

Table 22.3 Summary of stresses affecting major fish stocks in the Great Lakes, approximate dates of population declines, and current status of native stocks. xx = major factor in decline; x = minor factor in decline; P = present; E = extinct; − = population has not recovered; + = population resurgence. SL = sea lamprey; RS = rainbow smelt; AL = alewife; WP = white perch.

Impacted fish species	Period of major decline	Exploitation	Introduced species SL	RS	AL	WP	Nutrient loading	Toxic substances	Alteration of spawning habitats	Current status of native stocks
Lake Superior										
Lake trout	1930s–1950s	x	xx					P	P	−
Lake whitefish	1890s–1950s	xx	xx					P	P	+
Lake herring	1940s–1970s	xx		x				P		+
Lake Michigan										
Lake trout	1940s–1950s	xx	xx					P		E
Lake whitefish	1890s–1950s	xx	x					P	x	+
Lake herring	1950s	P		x	xx			P	x	−
Coregonus spp.	1960s	P			xx			P		+
Yellow perch	1960s–1980s	xx			x					+
Lake Huron										
Lake trout	1930s–1950s	xx	xx					P		−
Lake whitefish	1930s & 1950s	x	x					P		+
Lake herring	1940s–1950s	P					x	x	x	−
Coregonus spp.	1960s	xx	x					P		+
Yellow perch	1970s	x				P		P		+
Walleye	1940s	xx					xx	P		+
Lake Erie										
Lake trout	late 1800s	xx							x	E
Lake whitefish	1950s	x		x			x	P	x	−
Lake herring	1920s	xx					x		x	−
Blue pike	1950s	xx		x			x	P		E
Sauger	1920s–1950s	xx					x	P	x	E
Walleye	1950s	xx		x			x	P	x	+
Yellow perch	1970s	xx		P		x	x	P		−
Lake Ontario										
Lake trout	1920s–1950s	xx	x					P	x	E
Lake whitefish	1920s–1960s	xx	x					P		−
Lake herring and *Coregonus* spp.	1940s–1950s	xx		x	x			P	x	−
Walleye	1960s					x	x	P	x	+

et al. 1981; Dehring et al. 1981). The most accessible, those of greatest commercial value, were depleted first, often to extinction, before less accessible stocks were influenced. Although commercial harvest figures remained fairly constant for a time, overall populations declined as the different subpopulations were successively depleted. Another prime example of fish that were overexploited were Lake Michigan deepwater ciscoes, which originally included seven species that differed in size at maturity. Use of decreasing mesh sizes as stocks were depleted sequentially decimated the young of the larger species (Smith 1964). This process and sea lamprey predation led to extirpation of all but the smallest species; highly prized species were all overfished in all the Great Lakes.

Loss and degradation of habitat in the Great Lakes began soon after settlers arrived and has become widespread. The consequences of habitat alteration were demonstrated early in Lake Ontario, where increased water temperature and sedimentation in spawning streams—a result of deforestation of watersheds—led to the extinction of Atlantic salmon by the mid-1800s (Christie 1974). By the late 1800s many tributary streams and estuaries of Lakes Superior, Michigan, and Huron were useless for spawning due to contamination by sawdust and other sawmill wastes. Habitat degradation has also resulted from agricultural runoff (causing excessive enrichment of waters), damming of spawning streams, pollution from industrial and domestic sources, and filling of marshes used as spawning and nursery grounds. About 60% of the 39 km^2 of marsh originally lining the nearshore areas of Green Bay has been lost to commercial development (Whillans 1982). These changes have set into motion the irreversible changes which precipitated the calamitous disruptions of Great Lakes ecosystems.

Perhaps the most notable of the exotic species to invade the Great Lakes has been the parasitic sea lamprey, which attaches itself to its host fish, rasps a hole, and sucks out its body juices, often killing it in the process. Although the sea lamprey preys on many species of fish, its greatest impact has been on lake trout. The negative impact of sea lamprey on lake trout may have been exacerbated by commercial fishing. The fishery tended to remove the larger trout, leaving the brunt of the sea lamprey attacks to be borne by smaller lake trout, which are less likely to survive such attacks (Christie 1972). Sea lampreys are believed to have entered Lake Ontario via the Hudson River and the Erie Canal. The date of their first arrival in lake Ontario is not known, but it was probably at least as early as 1819, when the Erie Canal was opened to Lake Ontario. The parasite probably entered Lake Erie via the Welland Canal and was first reported in Lake Erie in 1929; it spread to Lake Superior by 1946 (Figure 22.2). The impact of the sea lamprey on lake trout in Lake Ontario was not great until deforestation in the watershed was sufficient to provide (through resulting siltation of sea lamprey spawning streams) optimum substrate for larval sea lampreys which reside in streams up to 7 years. Thus, an environmental stress was first necessary in Lake Ontario before sea lampreys markedly affected the indigenous community. The sea lamprey never did well in Lake Erie because this lake lacks suitable cool, clean spawning tributaries. In Lakes Huron and Michigan, however, the sea lamprey, in concert with overfishing, reduced lake trout to extinction by the late 1940s. In Lake Superior, the sea lamprey was well on its way to exacting the same toll on lake trout as in Lake Huron and Michigan, but treatment of spawning streams with the selective lampricide TFM (3-trifluoromethyl-4-nitrophenol), under the direction of the GLFC, has reduced sea lamprey numbers and allowed native lake trout to rebound.

A second invader that has had adverse effects on native fish populations is the alewife. Loss of key predators allowed proliferation of alien species into disrupted fish communities. The alewife was first recorded in Lake Ontario in 1873 (Figure 22.2), then moved into Lake Huron by 1933, Lake Michigan by 1949, and Lake Superior by 1954 (Smith 1970b). The alewife was prevented from proliferating in Lake Ontario prior to the mid-1800s by cold temperatures (maintained by heavily forested watersheds) that prevented spawning, and predation from the well-established top predators, lake trout and Atlantic salmon. Alewife never attained great abundance in Lakes Erie and Superior. In Lake Erie, high winter mortality (resulting from the low winter water temperatures in that shallow lake) and stable predator populations are thought to have prevented the alewife from establishing

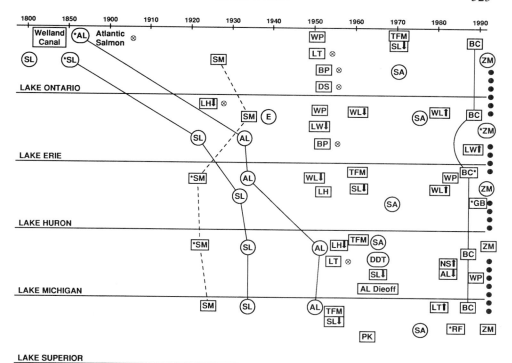

Figure 22.2 Temporal sequence of various critical events affecting fish populations in the Great Lakes. SL = sea lamprey, SM = rainbow smelt, WP = white perch, LW = lake whitefish, WL = walleye, AL = alewife, LH = lake herring, PK = pink salmon, SA = salmonines stocked, NS = native species (i.e., yellow perch, bloater), GB = Gobiidae (tubenose goby and round goby), DS = deepwater sculpin, WP = white perch, RF = ruffe, ZM = zebra mussel, BC = *Bythotrephes*, TFM = lamprey larvacide, E = eutrophication, ↓ = decline, ⊗ = extinction, ↑ = resurgence, circled species codes indicate major impact, species codes with squares designate them as more informational, * = first occurrence.

high populations (Smith 1968, 1970b); whereas, in Lake Superior, cold water in spring and summer inhibited reproduction (Smith 1972a; Christie 1974). The story was much different in Lakes Michigan and Huron. In the absence of terminal predators and a severely altered coregonine population, alewife became the most abundant species, often undergoing massive die-offs (Brown 1972). With increased stocking of salmonines and successful sea lamprey control that reduced salmonine mortality, large alewife die-offs have not been recorded in recent years and in fact alewives have suffered a decline in Lake Michigan (Jude and Tesar 1985; Eck and Wells 1987).

Alewives detrimentally affected other species by inhibiting reproduction. In Lake Michigan, alewives, through size-selective predation on zooplankton, caused a loss of larger individuals among the plankton (Wells 1970), which created bottlenecks in the food supply to the extent that recruitment of native species was affected (Crowder et al. 1987). Alewives are known to eat pelagic fish larvae (Kohler and Ney 1980; Brandt et al. 1987) and eggs (Wells 1980), which appear to be most susceptible to alewife predation. The emerald shiner, once extremely abundant in Lake Michigan, has large, 2-mm-diameter pelagic eggs (Auer 1982), pelagic larvae, a short life cycle, and was the first species to decline precipitously as the alewife entered and moved through Lake Michigan (Smith 1970b). Other species such as the deepwater cisco, lake herring, and rainbow smelt suffered declines suspected to be related to alewife predation on eggs and larvae (Smith 1970b). Deepwater sculpins and yellow perch, which both have pelagic larvae, were also reduced to low levels in Lake Michigan, then increased in the 1980s as alewife populations declined (Jude and Tesar 1985; Eck and Wells 1987). In Lake Ontario, O'Gorman et al. (1987) found that abundant adult alewives suppressed production of age-0 and yearling alewives, thereby depressing growth of young brown trout and coho salmon which fed on young alewives.

Alewives composed a large part of the diet of all salmonines in Lake Michigan (Jude et al. 1987) and Lake Ontario (Brandt 1986), probably because alewives are more pelagic, slower swimming, and in larger and denser schools than are native prey species. Heavy predation on alewives has led to an inverse relation between salmonine and alewife abundance. In addition, Pacific salmon eat mainly alewives because of the propensity of salmonines to feed in mid-water (Jude et al. 1987).

Rainbow smelt had colonized all five Great Lakes by the 1930s and have been implicated in adversely affecting many native species. At times they have been abundant in all the lakes and suffered huge die-offs in Lakes Michigan and Huron in the 1940s. They are believed to suppress recruitment of native coregonines either through food competition (Anderson and Smith 1971) or through predation on fry (Selgeby et al. 1978). However, predation was not believed to be an important factor in the decline of Lake Superior lake herring stocks, but predation or some other density-dependent factor was thought to maintain overexploited stocks at low levels (Selgeby 1982).

22.3   CHANGES IN FISH COMMUNITIES

22.3.1   Generalized Fish Population Responses

The importance of natural assemblages of fishes was stressed by Evans et al. (1987) regarding disruptions of Great Lakes fish communities and their potential for restoration. This view is also part of the current, strongly advocated ecosystem approach to Great Lakes rehabilitation. Fish communities among the oligotrophic lakes of the north temperate zone were similar and have responded similarly to perturbations. They have persisted rather well under moderate exploitation but have changed greatly under extreme disturbances. For example, loss of predators has been followed by major shifts in abundance of prey species and establishment of large stocks of introduced species, often small planktivores. These observations suggest structural conservatism and resistance to change among the original species assemblages, but stress the importance of the integrative influence of large piscivores on community structure. Whillans (1979) analyzed the changes in fish communities in three Great Lakes bays and found that fishes that survived perturbations were less reliant on specific pathways than the species they replaced. As a result, stressed communities develop looser structures. Therefore, as Christie et al. (1987a) have propounded, the best insurance against unpleasant surprises in the future is to manage in the direction of increased system maturity at every opportunity. This concept requires a movement away from artificial fisheries (stocking, genetic manipulation) toward restoration of native fish, control over factors which impinge on these goals (e.g,

Table 22.4 Community responses of an oligotrophic system under cultural stress. (From Ryder and Edwards 1985.)

1. Mean size of organisms decreases	9. Food webs shorten
2. Life span shortens	10. Introgression may occur
3. Pelagic organisms predominate	11. Reproduction may cease
4. Benthic organisms decline	12. Species rations change
5. Foragers increase	13. Number of species reduced
6. Top predators decrease	14. Mesotrophic forms increase
7. Eurybionts replace stenobionts	15. Production declines
8. Food webs decompose or simplify	

overfishing, habitat destruction, sea lamprey, pollution), and communication of these views to user groups, managers, and politicians.

Stresses, particularly cultural stresses, can lead to an array of ecosystem, community, and species responses, many of which may be predictable. Ryder and Edwards (1985) and Francis et al. (1979) discussed stress responses at various hierarchical levels but specifically those occurring within communities in oligotrophic systems (Table 22.4). Stress effects may be perceived first within specific hierarchical levels (e.g., species and populations) due to stress specificity: as bioaccumulation of toxicants in individual fish; as inordinate sensitivity of an organism; or through integral effects, either among component organisms, or between organisms and their abiotic environment (Ryder and Edwards 1985). The hierarchical level that first demonstrates readily recognizable signs of malaise may not necessarily be the level of greatest ultimate stress impact. Species with maximum growth or reproductive compensation, such as alewife and rainbow smelt, tend to have survival advantage under stress over species with minimal density-dependent response flexibility such as lake trout. Consequently, the species and community compensation characteristics set the stage, at least in part, for the sequential order of species and community demise in a culturally stressed system.

There has been a progressive depletion of many indigenous fish stocks in the Great Lakes due to one or more stresses. Stocks of lake sturgeon, deepwater cisco, lake trout, lake whitefish, lake herring, and walleye have collapsed and been replaced by other often less valuable species. Elimination of indigenous forms such as blue pike, Atlantic salmon, lake trout, and many others adapted to specific environments, represents a loss of irreplaceable genetic material.

Ecosystem response to stress may also be evaluated by the particle-size concept (Sheldon et al. 1972) whereby in marine pelagic communities, on a $\log_{10}$ size interval basis, organisms occur at approximately equal biomass, whether they are algae or whales. Kerr and Dickie (1984) extended this concept to ecosystems and Ryder et al. (1981) noted a similar characteristic community structure in the Great Lakes during early periods of stability. They speculated that particle-size changes due to cultural stresses in the Great Lakes would cause the large terminal predators and benthic feeders to disappear first. Ultimate survivors would likely be small opportunistic pelagic species. This theory was validated, because the lake sturgeon (large benthic feeder) and the lake trout (terminal predator) were highly susceptible to cultural stresses; whereas, alewife and rainbow smelt, two small pelagic species, tended to thrive through ecological opportunism, high fecundity, and short turnover times. Intermediate-size species such as lake herring, lake whitefish, and walleye were drastically reduced in an orderly sequence that was consistent with the particle-size-density theory. Even within species, the largest individuals tended to disappear first. Disruption in the particle-size relations of organisms, from whatever causal mechanism, may be an indicator of community malaise and is perhaps indicative of a dysfunction in the ecosystem as a whole. This theory was used to predict changes in the shape of the size spectrum of organisms in Lake Ontario after toxic chemical contamination (Borgmann and Whittle 1983) and to predict Lake Ontario production (Borgmann 1982).

Degradation of salmonine communities in response to fishing, cultural eutrophication, and introduced species results in similar changes in fish communities (guild shifts) no matter the perturbation (Christie et al. 1987a). This common stress response suggests that some fundamental principles of ecosystem organization govern community development and the response to strong environmental perturbations. Implicit within this approach is the understanding that degradation of aquatic ecosystems is at least a partially reversible process. An example of this reversibility is the Bay of Quinte improvement after nutrient loading reductions (Minns et al. 1986). Many important issues facing ongoing management of the Great Lakes fisheries relate to the concept of ecological succession in ecosystems and fish communities.

One generalized population response to stress is introgression, particularly among salmonines and coregonines. Before the 1800s, high levels of morphological and functional diversity of phenotypically plastic fish stocks, such as the lake trout and cisco complexes, allowed ready adaptation to the heterogeneous habitats of the lake basins. Losses of particular stocks identified as either genotypically or phenotypically distinct, through extinction processes or introgression, may be a symptom of more insidious environmental stresses. These changes are usually associated with intense harvesting, cultural eutrophication, and physical perturbation or restructuring of the nearshore zone (Regier et al. 1969). Steedman and Regier (1987) made a strong case for the importance of the nearshore zone as ''centers of organization'' which are often neglected, destroyed, or modified, as in the case of wetlands (Whillans 1982; Jude and Pappas 1992) and harbors. Smith (1964) documented introgression for Lake Michigan coregonines, in which six of seven species were lost as a result of overfishing and sea lamprey predation; only the least valuable and smallest of the species, the bloater, survived. Introgression was also cited as contributing to the demise of blue pike and sauger in Lake Erie (Regier et al. 1969), and in Lake Ontario, blackfin cisco, shortnose cisco, and kiyi were extirpated, leaving only the bloater, which eventually also disappeared (Christie 1972). This pattern—a progressive decline in abundance and eventual extinction of members of a congeneric group—is common.

In Lake Ontario, the deepwater sculpin became extinct and the slimy sculpin, a species with a more shallow water distribution, replaced it. In Lake Michigan, the spoonhead sculpin, a sensitive species, was thought to be extinct, but some were found recently (Potter and Fleischer 1992). The deepwater sculpin and slimy sculpin are the two more common remaining species. In all of these replacements, premium or highly sensitive species were replaced by a less valuable, more tolerant species. The collective human-induced changes in Great Lakes species composition demand a more basinwide effort to eliminate pollution and stronger initiatives to restore native fish communities.

22.3.2 Lake Superior Fish Community Dynamics

Lake Superior's highly oligotrophic state is reflected by its native fish community, which was once dominated by lake trout, burbot, lake whitefish, lake sturgeon, lake herring, suckers, sculpins, and several species of deepwater ciscoes. The community has been altered in recent decades by an intense fishery, predation or competition from introduced species (particularly sea lamprey and rainbow smelt), habitat loss, and water quality deterioration in some nearshore areas and tributaries. Lake Superior is unique in that it has not had the excessive nutrient loadings that the lower lakes have experienced. Lawrie and Rahrer (1972) and Lawrie (1978) summarized changes in native fish populations as follows: the introduced rainbow smelt had become abundant by the 1930s, and sea lamprey by 1950; alewives entered the lake in the 1950s, but never became abundant. The lake trout, which was being overfished in the 1930s and 1940s, declined in the 1950s as a result of added stress from sea lamprey predation. Lake whitefish populations have fluctuated since the late 1800s; sharp declines have been attributed to overfishing, spawning habitat degradation resulting from lumbering operations, and sea lamprey predation. Lake herring suffered steep declines in abundance between the 1930s and 1960s; these declines have been attributed to overfishing of spawning aggregations (Selgeby 1982), and expansion of rainbow smelt populations (Anderson and Smith 1971). Lake sturgeon and the deepwater ciscoes were also exploited commercially.

Recent trends show a shift in abundance from rainbow smelt, which has dominated prey populations since the 1960s, to lake herring, the historical dominant in the prey species (Selgeby 1985; MacCallum and Selgeby 1987). The rainbow smelt decline (more than 90% from 1978 to 1981) is attributed to salmonine predation; rainbow smelt stocks have increased recently (Selgeby 1985). The lake herring recovery from near extinction in the 1960s is significant because no other lake herring stocks in the Great Lakes have recovered after reaching such low levels, but reasons for the recovery are not obvious (MacCallum and Selgeby 1987). An immense year-class was produced in 1984 despite the buildup of large populations of predators. Stocks of lake whitefish have increased since the 1970s, probably due to sea lamprey control measures and favorable abiotic conditions. Recent harvests are the largest since the turn of the century.

In addition to lake trout, plantings of splake (lake trout × brook trout hybrid), rainbow trout, brown trout, brook trout, and chinook, sockeye, and coho salmon have been made in the past 2 decades. Also, pink salmon were inadvertently introduced in 1956. Natural reproduction by some of these species is believed to be large (MacCallum and Selgeby 1987), although lake trout continue to be the most important species in that respect. All contribute to sport fisheries in nearshore areas.

Some major controls have been put in place on the Lake Superior commercial fisheries, including quotas, banning of gill nets by Minnesota, sanctuaries, and sport-fishing-only zones (MacCallum and Selgeby 1987). Michigan has forced commercial fishers to use impoundment gear. However, tribal groups fish commercially, primarily with gill nets, throughout the lake.

Strategies for rehabilitation of stocks commenced in 1953 with attempts to control the sea lamprey, mainly by blocking spawning streams with electrical barriers. Control measures became much more effective in 1958 with the use of the lampricide TFM (see Section 22.4.3). Since large-scale stocking of lake trout began in 1958, abundance has increased in most areas, partly a result of improved natural reproduction believed to be derived mainly from native fish which never became extinct (Krueger et al. 1986). The long-range goal for lake trout rehabilitation calls for self-sustaining populations capable of yielding 1,810 tonnes annually. Various management actions in the past 25 years, such as reduction in exploitation, sea lamprey control, and heavy stocking of salmonines, have resulted in conspicuous reversals in downward trends of most native species.

22.3.3 Lake Michigan Fish Community Dynamics

In the past 150 years, native fish stocks in Lake Michigan have experienced huge fluctuations (Wells and McLain 1973). Until the 1940s, exploitation was the most important factor affecting the Lake Michigan fish community; the sea lamprey and alewife were most influential in later years (Eck and Wells 1987). Overfishing caused reductions in abundance of lake whitefish, lake trout, lake sturgeon, and the largest species of coregonines. Sea lamprey predation caused the extinction of lake trout and adversely affected the abundance of lake whitefish. The alewife has been accused of inhibiting reproduction of native species (Eck and Wells 1987), and along with rainbow smelt, caused the lake herring collapse in the late 1950s. More recently, cultural eutrophication and contaminants have damaged the habitat in southern Green Bay and affected populations of walleye and lake herring.

Many changes have occurred in native stocks. Burbot are now common in northern Lake Michigan, but scarce in the south; their recovery is attributed to sea lamprey control. Slimy sculpins increased from 1973 to 1976 but declined between 1976 and 1984, probably due to lake trout predation. Spoonhead sculpins (which were thought to be extinct in Lake Michigan), lake herring, and emerald shiners declined to extremely low levels of abundance before the 1970s and remain scarce (Eck and Wells 1987). Emerald shiners have shown some increases in the early 1980s.

The lake trout collapse in the 1950s, the rise in abundance of rainbow smelt and alewife in the early 1960s, and the implementation of sea lamprey control in the late 1960s created an opportunity for stocking top predators. Lake trout plantings commenced in 1965 and have continued at an annual rate in excess of 2 million yearlings. These fish grew well and contributed to an expanding sport fishery, but there is little evidence of natural reproduction. Planting of coho and chinook salmon began in 1966 and 1967, respectively, and these species, together with lake trout, rainbow trout, and brown trout, contribute to a sport fishery valued at over $150 million annually (Rakoczy and Rogers 1987).

Lake whitefish stocks have expanded since the late 1960s, probably due to reduced sea lamprey predation, and in the 1980s harvest reached levels of pre-lamprey years (Eck and Wells 1987). On the other hand, lake herring stocks have not recovered and contribute very little to the commercial fishery or the population of prey fish. Of the seven species of deepwater ciscoes originally present, only the smallest, the bloater, remained abundant by the early 1960s (Smith 1964). Bloater stocks declined in the 1960s and early 1970s due to interference with reproduction by alewives and an intensive fishery. Harvest restrictions (beginning in 1976) and diminished alewife populations have resulted in bloater recovery.

Since rainbow smelt became established in the 1930s, populations have fluctuated. Abundance was low in 1942–1943 due to winter mortality, and in the early 1960s, probably due to competition and predation from alewives. Rainbow smelt are now abundant and contribute to the diet of salmonines (Jude et al. 1987).

Yellow perch contributed between 450 and 1,350 tonnes annually to the commercial harvest during the first half of this century. Production increased briefly in the 1960s (to a peak of 2,700 tonnes) due to increased fishing intensity, then declined in the late 1960s. The primary cause of the decline was poor reproduction resulting from competition with, and predation on, larval yellow perch by alewives. Yellow perch populations remained low through the 1970s, but as alewives declined in the early 1980s, yellow perch populations rebounded, particularly in the southeastern part of the lake (Jude and Tesar 1985). Abundance has continued to expand and the yellow perch is now the most numerous species in the sport catch in Michigan's waters (Rakoczy and Rogers 1987).

22.3.4 Lake Huron Fish Community Dynamics

Victims of overfishing and predation by sea lampreys, native lake trout are extinct in Lake Huron, except for remnant stocks in Georgian Bay and the North Channel. Populations are maintained through stocking which began in 1973. Although the first round of stream treatments with lampricides was completed in 1969, control has not been as successful as in the other Great Lakes. There is concern that the St. Marys River, which is difficult to treat effectively with lampricide, is a refugium for ammocetes, larval sea lampreys (Eshenroder et al. 1987). Backcrosses (splake: three-quarters lake trout and one-quarter brook trout, F_n), which are much more abundant than pure lake trout in Canadian waters, are less vulnerable than are lake trout to attacks from sea lampreys because they are smaller and mature earlier. Limited reproduction from hybrids was first observed in 1985. The Lake Huron Management Plan for Lake Trout Rehabilitation calls for self-sustaining populations capable of providing annual harvests of about 900 tonnes by the year 2010.

Despite sea lamprey predation and an intensive fishery, lake whitefish populations have remained viable. Harvests increased in the 1980s due to increased pressure from commercial fisheries. Lake whitefish year-classes have been strong and prospects for future harvests are good, with the exception of southern Georgian Bay.

Lake herring, which was an important commercial species in the first 4 decades of this century, has not recovered from a population collapse that occurred in the 1940s and 1950s. The most probable cause was deterioration in water quality, since most lake herring spawned in Saginaw Bay. Lake herring were already overfished in the 1920s.

The deepwater cisco fishery provided stable catches until about 1940 and then declined to low levels, possibly due to the alewife invasion (Brown et al. 1987) and overfishing. Between 1958 and 1966, harvests increased and populations collapsed. The commercial fishery for deepwater cisco was closed in 1970 and by the late 1970s, stocks began to recover in association with predator restorations. Surveys in 1985 indicated deepwater ciscoes in Lake Huron are dominated by the smallest of the species, the bloater, which persisted as the larger species were reduced to low levels (or exterminated) by fishing and sea lamprey predation.

Although the rainbow smelt has been in Lake Huron at least since 1925, it has not been prominent in the commercial fishery. It has been an important prey species and is now the primary prey of salmonines. Abundance has remained high throughout the 1980s, whereas stocks of alewives have declined. Abundance of alewives has been declining since the early 1970s (Henderson and Brown 1985) and currently is much below pre-1970s levels due to a combination of cold winters and increased salmonine predation.

Yellow perch are harvested mainly from Saginaw Bay and the southern part of Lake Huron. Abundance in Saginaw Bay declined in the 1970s due to overfishing; the fishing-up sequence was more intense in the outer bay (Eshenroder 1977). Yields from Ontario waters increased gradually from 1900 to a peak in the 1960s (Spangler et al. 1977). Production of yellow perch in the 1980s has been relatively high in Saginaw Bay and elsewhere in Lake Huron.

Historically, the walleye fishery of Saginaw Bay was the second largest in the Great Lakes, occasionally exceeding 700 tonnes annually (Schneider and Leach 1977). The population collapsed in the 1940s probably due to degradation of spawning habitat and overfishing. Commercial fishing for walleye was banned in Saginaw Bay in 1969; with improved water quality, a successful sport fishery now exists which is supported largely by stocking. Large numbers of walleye larvae are produced in the Saginaw River, but few survive to contribute to the fishery (Jude 1992). The current commercial harvest, which is relatively stable, is mainly from the southern end of the lake.

22.3.5 Lake Erie Fish Community Dynamics

Lake Erie is considerably shallower and warmer than the other Great Lakes, has the highest turnover and sedimentation rate, and has undergone some of the most dramatic changes of any large lake. The lake has three distinct basins (western, central, and eastern) which range from oligotrophic in the east to eutrophic in the west. Most of the deterioration of Lake Erie's environment and fish populations has occurred in the last century. Major sources of stress in order of impact have been exploitation of fish, cultural eutrophication, introduced species, habitat destruction, siltation, and inputs of toxic substances. These changes resulted in dissolved oxygen depletion, major decline of an important fish food organism (*Hexagenia*), and extinctions.

Lake Erie provides an example of sequential degradation of fish communities due to multiple stresses of cultural origin (Regier and Hartman 1973). Fish stocks underwent these changes before they occurred in the upper Great Lakes. Despite dramatic changes, commercial fish yield from Lake Erie has generally been higher than all the other Great Lakes combined (Baldwin et al. 1979). Warmwater species such as channel catfish, emerald shiner, white bass, gizzard shad, alewife, spottail shiner, and common carp are maintaining healthy populations. However, other populations declined starting with lake trout which were moderately abundant in the eastern basin (Applegate and VanMeter 1970) and lake sturgeon, and progressed through lake whitefish, lake herring, sauger, blue pike, walleye, and yellow perch. By 1970, the exotic and low-valued rainbow smelt was the predominant species over much of Lake Erie. Rainbow smelt have been in Lake Erie since 1935 and fished commercially since the 1950s by an efficient trawl fishery developed in Ontario waters, mainly in the eastern and central basins. The yellow perch has been the most valuable commercial species since the 1950s (Hartman 1988). Annual production declined from a peak of more than 15,000 tonnes in 1969 to about 3,000 tonnes in the early 1980s, and has since held steady at around 5,000 tonnes, mostly from Ontario waters.

Populations of lake trout, lake herring, lake whitefish, and longjaw cisco all declined to low levels or extinction. The commercial fishery had reduced native lake trout by 1900 and habitat deterioration (mainly eutrophication) virtually eliminated them by the 1930s (Hartman 1972). Stocking efforts in the eastern basin have not established a self-sustaining population. Sea lamprey predation appears to be preventing survival of lake trout over 3 years of age. Rehabilitation of lake trout will require limiting total annual mortality to 40% or less, which will require sustained sea lamprey control measures and new sea lamprey control initiatives by the GLFC.

Commercial production of lake whitefish was irregular during the first half of this century due to occasional strong year-classes, but the stock collapsed in the 1950s due to degradation of summer and spawning habitat and exploitation (Hartman 1988). A remnant population remains in the eastern part of the lake and spawns regularly, mainly in the western basin. There are indications that the population could recover. On the other hand, lake herring remains commercially extinct. This species was the most abundant commercial species in the early decades, but their populations collapsed in 1925, probably due to excessive fishing pressure during spawning migrations (Regier et al. 1969).

Alewives entered Lake Erie around 1931 via the Welland Canal from Lake Ontario (Smith 1970a). Predators and low water temperatures in winter have limited their abundance (Hartman 1972).

Harvests of blue pike, another very important commercial species, fluctuated widely before the fishery collapsed by 1958. Reasons for the loss of the blue pike, which is now considered extinct, are not clear but an intensive fishery, loss of summer habitat, and competition and predation from rainbow smelt have been suggested (Regier et al. 1969; Leach and Nepszy 1976). Introgressive hybridization with walleyes may have caused final disappearance of the species. Native sauger may have experienced the same fate. Because the fishery declined gradually over a long period, environmental deterioration, particularly pollution, siltation, and dam construction, are considered (along with increasing fishing pressure) to have caused the demise of native sauger. The state of Ohio began stocking sauger in the western basin in the 1970s, but a self-sustaining population has yet to become established.

Walleyes in western Lake Erie have undergone fluctuations in abundance. Declines in the 1950s and 1960s have been attributed to overexploitation, interactions with invading rainbow smelt and nutrient loading (Schneider and Leach 1977). Nutrient loadings degraded spawning and summer habitats and adversely affected benthic organisms, such as the burrowing mayfly (Regier et al. 1969). Since 1970, western basin walleyes have responded favorably to rehabilitation measures, especially limitations on harvest and reductions in phosphorus loadings and have increased to or exceeded former abundance levels.

White perch, an East Coast estuarine species, was first reported in 1953, but did not become established until the 1970s (Boileau 1985). It has expanded its range to include Saginaw Bay and Green Bay. In Saginaw Bay, this species suffers high overwinter mortalities. In Lake Erie it is most prevalent in the western end where it contributes to the commercial fishery.

22.3.6 Lake Ontario Fish Community Dynamics

Lake Ontario is unique among the Great Lakes because of its high level of nutrient loading (Beeton 1965), its connection to the Atlantic Ocean, and its native fish assemblage. Its early fish community differed from the upper Great Lakes by the presence of Atlantic salmon and American eels (Ryder and Edwards 1985). Major changes to Lake Ontario's stocks due to cultural interventions preceded those of the other Great Lakes by several decades. There have been several extinctions in Lake Ontario including Atlantic salmon around 1900, lake trout, blue pike, deepwater sculpin (Brandt 1986), and four coregonines due to over-fishing and possibly rainbow smelt predation. There have been drastic declines in catches of lake whitefish and lake sturgeon because of overfishing (Christie 1972). Lake herring and coregonine stocks have not recovered from declines that occurred in the 1940s and 1950s, but lake whitefish populations in the east end (Bay of Quinte) appear to be in a state of recovery (Christie et al. 1987b). Introduced species include alewife, white perch, common carp, and rainbow smelt. This fish potpourri is complicated by the stocking of several species of salmon and trout. Currently, fish stocks are in an unpredictable and continual state of flux.

Lake trout rehabilitation in Lake Ontario is hampered by low survival rates of adults and lack of natural reproduction. There is evidence that sportfishing mortality may be excessive. Lack of substantial reproductive success (Krueger et al. 1986) despite 10 years of sea lamprey control indicates that spawning habitat for lake trout may be degraded.

The Lake Ontario alewife–salmonine interactions appear to be following the same trajectory as in Lake Michigan, but there are several differences. In Lake Michigan it is generally believed that salmonine predation (Stewart et al. 1981) and possibly cold winters (Eck and Brown 1985) caused a precipitous decline in alewife populations (Jude and Tesar 1985). Large-bodied *Daphnia* appeared in Lake Michigan after the alewife decline (Evans and Jude 1986), but similar changes in zooplankton have not yet been observed in Lake Ontario (Johannsson 1987). The alewife decline in Lake Michigan was followed by a dramatic increase in yellow perch, bloater, and to some degree rainbow smelt populations (Jude and Tesar 1985). These species may now compose a higher proportion of salmonine diets in Lake Michigan. In Lake Ontario, bloaters and deepwater sculpins are extinct, so it is not clear what species may respond to an alewife decline. O'Gorman et al. (1987) believe that yellow perch, white perch, and the remnant stocks of lake herring may increase if alewives decline. Impetus is thus given to reestablishment of the endemic fauna, especially bloaters and deepwater sculpins, should the alewife population decline dramatically. The New York Department of Environmental Conservation has a policy of maintaining a reasonable population of alewives to serve as prey for salmonines (Brandt 1986).

Currently, more salmon are stocked in Lake Ontario on a surface-area basis than have gone into Lake Michigan; yet alewife populations in Lake Ontario have been affected more by weather than by predation. There is, however, some evidence that the first predation-induced alewife declines may be occurring.

Anglers are harvesting large numbers of salmonines (O'Gorman et al. 1987). Because of phosphorus controls, the Lake Ontario carrying capacity for alewife will probably be reduced. Therefore, Lake Ontario is poised on the threshold of change. An alewife die-off in 1976–1977 was followed by increased growth of age-0 alewife and a strong yellow perch year-class. Rainbow smelt populations have increased, apparently buffered by alewife, while slimy sculpin numbers declined due to juvenile lake trout predation. An alewife population decline will have far-reaching effects throughout the lake and has the potential to impact the sport fishery as well. In fact, during 1993, in response to lower abundance and poor growth of alewives and rainbow smelt, substantial stocking reductions (>50%) were made for chinook salmon and lake trout.

22.4 MAJOR FISH MANAGEMENT ACTIONS

22.4.1 Recognition of the Overfishing Problem

Overfishing has altered most Great Lakes fish communities. Resilience, competitive advantage, and resistance of fish to stress can all be impacted by the progressive removal of the largest members of a population. Early exploitation by fishers and other perturbations (e.g., pollution, habitat modification, nutrients, and thermal loads) impacted the nearshore zone first. The overfishing problem and the inability of fisheries managers to detect or control it was exemplified by the lake trout fisheries (Brown et al. 1981). Lake trout production was a composite of many stocks, which represented many ages and sizes of mature and immature fish. This gave the population stability and resilience during times of adverse density-independent effects (bad weather) and overfishing. Stability of the fishery was maintained because commercial fishers were able to move sequentially from depleted to unexploited stocks (Lawrie and Rahrer 1972). As a result, there was little evidence in total catch data of fluctuations which would result from marked variations in year-class strength. Although some stocks became greatly depleted or extinct, there was no indication of general depletion (Lawrie 1978). Smith (1968) recognized this problem and provided the basis for future management efforts to curb and control harvest. Efforts to control harvest have resulted in resurgence of bloater in Lakes Michigan and Superior, walleye in Lake Erie, and lake herring in Lake Superior.

22.4.2 Alewife Invasion

The exploding alewife population in Lakes Ontario, Michigan, and Huron caused damage to native stocks (e.g., yellow perch and emerald shiner) and changed zooplankton size and diversity (Wells 1970). Alewives plugged intakes and littered beaches, causing economic and aesthetic havoc among residents and businesses which depended on tourists. The population explosion was successfully thwarted by restoration of top predators. Tanner (1988) detailed the state of Michigan's plan to both control the alewife using large predacious fish and create a sport fishery, thereby relegating commercial fishing to a much lesser role. In 1964–1965, one million coho salmon were obtained from Oregon and raised to smolts; 850,000 fish were released into tributaries of Lakes Michigan and Superior in 1966. Due to an alewife population explosion, a massive dieoff occurred in 1967 (Brown 1972), so additional stocking of chinook salmon and other salmonines was promulgated. The number of salmonines stocked into the Great Lakes is now over 30 million annually. Stocking salmon to control alewives can be viewed as an end in itself with the additional economic and recreational benefits of a world-renowned sport fishery. It can also be viewed as a stopgap measure along a road to rehabilitation of the ecosystem. We are seeing the revival of native species and a substantial decline in alewife populations in Lakes Michigan and Huron.

22.4.3 Sea Lamprey Control

Fisheries management agencies in the Great Lakes were challenged in the late 1940s to 1960s by degraded fish populations and habitats. Populations of lake trout and other large species collapsed sequentially in the lakes due to sea lamprey predation and overfishing; the alewife invasion followed and conditions were further exacerbated by habitat destruction. The Great Lakes Fishery Commission, a binational agency of the governments of the United States and Canada, was established in 1955 primarily to develop ways to control sea lamprey (Fetterolf 1980, 1988). Vernon Applegate was a pivotal figure in applying scientific principles to a study of sea lamprey biology (Applegate 1950), and subsequently testing thousands of chemicals to apply to the identified vulnerable life history stage, the ammocete. Thus TFM was discovered (Applegate et al. 1961). An adult sea lamprey kills many fish (up to 18 kg) during its life (Fetterolf 1988). Partial sea lamprey control using selective toxicants has been accomplished since 1958 on Lake Superior, 1960 on Lake Michigan, 1966 on Lake Huron, and 1971 on Lake Ontario. Control of sea lamprey is expensive; over $7.5 million was spent in 1988. However, the combined worth of the sport and commercial fisheries is $2–4 billion (Fetterolf 1988).

22.4.4 Resource Allocation Problems

Conflicts have developed between commercial fishing and sport fishing interests (Berkes et al. 1983). Activities of one group can adversely impact those of the other. Each group is opposed to allocation decisions which appear to be unfavorable to its specific interests. Currently, sport fisheries in most of the Great Lakes have economic values that far exceed commercial fisheries. In an economic analysis of the allocation of Lake Michigan alewife to commercial (harvest for pet food) or sport fishery (prey for salmonines) interests, the commercial value was near 0, while the marginal net social value was $4.10 per salmonine, strongly suggesting alewives should not be harvested by commercial fishers (Samples and Bishop 1987). However, there is definitely a place for regulated commercial fisheries in the Great Lakes (Figure 22.3). There are many species, such as the deepwater ciscos, which are not exploited by anglers, and others such as rainbow smelt and yellow perch, which are underharvested in some lakes.

22.4.5 Eutrophication Abeyance

Each of the Great Lakes has experienced cultural eutrophication over the last 150 years; Lake Erie experienced very high levels while Lake Superior experienced low levels (Loftus and Regier 1972). Eutrophication still poses an environmental problem. Fish community response to nutrient enrichment can take two courses (Evans et al. 1987). If food abundance changes but habitat quality does not, increased growth rates, earlier maturation, and increased production are

Figure 22.3 Commercial fishing has influenced the composition of fish communities in the Great Lakes. Commercial harvests are currently regulated to avoid such problems and to enable sustained harvest. Shown here is a trap net being serviced by a commercial fishing boat.

expected. If nutrient enrichment is severe, habitat is degraded and reduced survival of sensitive species results in a decline in abundance (Leach et al. 1977). If the most sensitive species is also a top piscivore, dramatic secondary shifts in dominance toward increased abundance of small prey species and possibly some benthivores would be expected. Further ramifications would include a decline in large zooplankton and an increase in algae. Overfishing would cause a similar sequence of changes. Minns et al. (1986) concluded that nutrient additions in the Bay of Quinte set the general level of production, but to a major degree, species interactions may explain observed changes in the structure of the fish community. Responses to cultural enrichment have included increased algal populations, hypolimnial dissolved oxygen depletion, and dramatic changes in fish communities (Ryder et al. 1981). In Lake Erie, nutrient-induced dissolved oxygen depletion in the central basin hypolimnion was cited (Hartman 1972) as contributing to the ultimate biological extinction of lake trout. Excessive phosphorus inputs result in reduction in size of lake trout stocks by limiting the amount of suitable habitat, and in some situations it can lead to extinction. In Lake Erie, reduced loadings of phosphorus have caused ambient concentrations of phosphorus in the west and central basins of the lake to decline 35% and 33% respectively between 1970 and 1980 (Great Lakes Water Quality Board 1981). However, anoxia in central basin bottom waters still occurs in most years. Lake Ontario shows some of the signs of degradation evident in Lake Erie. However, Lake Ontario has an additional increasing concentration of hazardous chemicals, particularly those originating from the Niagara River (Allan et al. 1983).

It is estimated that projected reductions in municipal phosphorus loads will decrease fish yield by 15% or more in lower Green Bay, main Lake Michigan,

Saginaw Bay, western and central Lake Erie, and Lake Ontario (Sullivan et al. 1981). Decreased fish yield from reduced phosphorus loads to the lakes is based on reduction in phytoplankton growth. These reductions should result in some increase in macrophytes because of improved transparency, especially in harbors, embayments, and nearshore areas of the lower lakes. Fish species associated with these habitats should therefore increase in abundance.

22.4.6 Cascading Trophic Interactions

Heightened predator–prey interactions, whether caused by an increase in the number of predators and their impact on the prey or by a decrease in the larger members of the community as in overfishing, can cascade through the trophic food web (Carpenter et al. 1985). When the largest terminal predators are harvested, prey fishes such as coregonines may grow beyond optimal size for the next largest predators that were not harvested. Species like lake herring, when released from intense predation, may become superabundant and overgraze zooplankton. If these impacts are intense enough lake herring may essentially reach a predation refugium and become increasingly abundant. An overly large and abundant prey species may then overgraze the zooplankton food supply which would lead to inefficient transfer of energy to the terminal predator. The reverse of this case occurred in Lake Michigan where increasing numbers of stocked salmon overgrazed alewife populations (Stewart et al. 1981) and this effect, in concert with a series of cold winters (Eck and Brown 1985), reduced alewife populations in 1983 to their lowest levels since first entering the lake. Reduced densities of alewives released predation on large zooplankton and a dramatic increase in large-bodied zooplankton occurred (Evans and Jude 1986). Competing species such as yellow perch, bloater, and rainbow smelt also increased in abundance (Jude and Tesar 1985). Large-bodied zooplankton cleared the water column more efficiently and water clarity increased (Scavia et al. 1986). Consequently, the maintenance of appropriate relative size relations and numbers between any two trophic levels is important in ensuring a stable fishery and maintaining the community at steady state (Ryder and Edwards 1985).

22.5 FISH POPULATION RESTORATIONS AND REHABILITATIONS

22.5.1 Ecosystem Rebirth

Christie et al. (1987a) are optimistic about the future of the Great Lakes and the opportunities for providing valuable fisheries. Realization of these opportunities will depend upon accurate assessment of ecosystem health and development of management measures that will maintain or enhance it.

In Lake Ontario, the 1970s decade saw implementation of sea lamprey control, a quota system for commercial fishers, intensive salmonine stocking programs, and reduced emissions of mercury, DDT, PCBs, and phosphorus (Christie et al. 1987b). The changes wrought by these actions have all been in the direction of a return toward earlier system status, but as in the case of Lake Superior, considerable effort and money must be committed to maintain the current reversals, and more needs to be done.

22.5.2 Eutrophication Control

Nutrient input to the Great Lakes has been reduced, most dramatically in Lake Erie. Improved water quality there has assisted in the walleye resurgence and improved beaches, all resulting in increased tourist revenues and resource use. In Lake Michigan, total phosphorus declined over the 1970s and early 1980s, especially in 1977 and 1982 (Scavia et al. 1986). The former was due to reduced phosphorus regeneration from the sediments during an unusually cold winter of extensive ice cover, and the latter was attributed to a salmonine-induced increase in abundance of large-bodied *Daphnia* which cleared the water column. Similar decreases in phosphorus in Lake Ontario have not shown similar effects, but the alewife and rainbow smelt populations have declined substantially.

22.5.3 The Strategic Fish Management Plan

A continuing problem in the Great Lakes has been the fragmented control and responsibility for management of the fisheries (Francis 1987). In 1978 the IJC, along with the GLFC, nurtured into general acceptance the ecosystem approach to Great Lakes problems, recognizing that any impact may affect a Great Lake, connecting channel, or even the entire basin. As part of this approach the GLFC invoked the concept of a strong, practical Great Lakes fisheries management plan to ensure that the public's fishery resources receive full recognition and consideration in any plans for development in the Great Lakes basin (Great Lakes Fishery Commission 1980; Fetterolf 1988). The process was begun in 1978 through the Council of Lake Committees. To ensure success, the effort was well funded and commitments were secured from high-ranking fishery agency officials in Canada and the United States. These officials and their staffs were involved from the beginning, and the plan was to be strategic in scope and form a nucleus from which all efforts could be coordinated.

Managers, with the help of the GLFC and the IJC, have made progress since 1955 in rehabilitating Great Lakes fish communities and habitat. The sea lamprey and alewife are largely under control, and catch restrictions have restored some predator populations. The annual regional economic impact has been estimated at $270 million for commercial fisheries and from $2 to 4 billion with 55 million angler days for recreational fisheries (Fetterolf 1988).

22.5.4 Toxicant Discharge Control

A major effort has been underway to control toxic substances input to the Great Lakes because of the high accumulation rates in bottom-dwelling and top predator fishes. Many toxic compounds, including DDT and PBCs, have been banned or restricted in use. A decline in toxic contaminants in fishes has been the general response (D'Itri 1988). Studies are now underway to determine bioaccumulation pathways from the sediments through the food chain, to determine sources of these anthropogenic substances, and their rates of entry to and release from the Great Lakes. Large quantities of toxic substances are now coming into the Great Lakes via the air from outside the Great Lakes basin.

22.5.5 Salmonine Stocking Programs

The first stocking of Pacific salmon occurred in 1966 when over a half million coho salmon were placed into Lake Michigan. Returns of over 30% in spawning

runs and catches were more than expected. Salmon have done well in all the upper
Great Lakes except Lakes Superior and Erie where alewives are relatively scarce,
which indicates the importance of this prey fish. This also forewarns of problems
in lakes such as Lake Michigan where alewife stocks have declined. Response of
salmonines to changing prey fish populations is still being evaluated (Jude et al.
1987).

Kitchell and Hewett (1987) assessed the potential role of sterile chinook salmon
(triploids) stocked into Lake Michigan to provide a trophy fishery. These fish can
potentially reach 23 kg (50 pounds) or more. At current growth rates fish should
begin to appear in the fishery about 5 years after stocking and they will consume
1.5 times as much forage as an equivalent stocking of standard chinook salmon.
About 0.3% of sterile chinook salmon stocked would appear in the trophy fishery.
Two drawbacks of these fish include the possible negative impacts of genetic
manipulation and increased contaminant buildup, making these fish hazardous to
consume.

22.5.6   Walleye Resurgences

Declines in walleye stocks in the Great Lakes due to cultural perturbations have
been reviewed by Schneider and Leach (1977). Some stocks have responded
favorably to efforts to reduce sources of stress. One of the best examples of
success has been the walleye resurgence in western Lake Erie (Hatch et al. 1987),
which was accomplished mainly because the fishery was closed due to mercury
contamination (Muth and Wolfert 1986). A charter-boat fishery has been established and the stock has extended its range into the central basin. Future fisheries
management efforts will be directed at increasing walleye harvest, which should
increase walleye growth, decrease age at maturity, and bring prey and predator
stocks more nearly into balance, thereby promoting greater yields.

A walleye stock in the St. Louis River, Lake Superior, has been restored
because of paper mill waste diversions (MacCallum and Selgeby 1987). Many of
the fish caught were old individuals that accumulated in the river because their
tainted flesh was repugnant to anglers.

Saginaw Bay once supported annual commercial yields of about 700 tonnes
(Schneider and Leach 1977), but stocks were reduced by 1950 due to a deterio-
rating environment, exploitation, and a moderate amount of sea lamprey predation. Walleyes were reared and stocked around Saginaw Bay by the Michigan
Department of Natural Resources which resulted in a spectacular sport fishery in
the 1980s (Rakoczy and Rogers 1987). A ban on commercial fishing for walleyes
and limits on yellow perch are now in effect in the bay. In each of these cases,
major impediments to stock revivals were removed, first through improvements in
wastewater treatment and, in the case of Saginaw Bay, stocking to initiate and
enhance natural reproduction.

22.5.7   Recovery of Native Stocks

There have been many positive results of the steps taken to curb overfishing,
control expanding introduced fish populations, reverse eutrophication, reduce or
eliminate toxic substances, and maintain the sea lamprey at low levels. Lake
whitefish are now thriving in Lakes Superior, Huron, and Michigan. In Lake
Superior, native lake trout are reproducing at a near record rate, lake herring have
been resurrected from very low levels, and bloater populations have been

restored. In Lake Michigan, alewife populations have declined, but still provide considerable prey for salmonines. In response to the alewife decline, yellow perch and bloater populations have rebounded to all-time high levels. The once depressed deepwater sculpin has recovered, while emerald shiners are showing signs of a modest return. In Lake Huron, bloaters have recovered moderately and walleyes currently provide renowned sport fisheries in Saginaw Bay (through stocking) and Lake Erie (through natural reproduction).

22.6   AVOIDING FUTURE CRISES

22.6.1   Public Expectations

Fisheries managers must be cognizant of what Gale (1987) called "resource miracles and rising expectations" of the fishing public. He argued that dismal conditions (e.g., degraded water quality of Lake Erie and dead alewives on Lake Michigan beaches) will not stimulate revolt if those conditions are consistent with public expectation. Instead, the revolution comes when conditions improve and there is a gap between actual improvement and rapidly increasing public expectations. Accelerated public expectations could force managers into an impossible situation, where available resources cannot catch up with escalating expectations. In 1987, which was an unusually warm year, fishers had more difficulty catching salmonines in Lake Michigan and there were complaints about the paucity of large fish. There was even some evidence that fishers shifted effort from chinook salmon—which were difficult to catch—to lake trout (Rakoczy and Rogers 1987). This in turn will further inhibit lake trout restoration efforts in Lake Michigan. Another aspect of public expectations is the perception that through technology we can provide artificial fisheries or "fixes" by stocking more or different strains of fish or other massive and costly fishery rehabilitation techniques. Again, we lose sight of underlying causes and the long-term goal of restoring ecosystem integrity which has many benefits, including self-reproducing populations, clean water, and fish communities which can act as a "miner's canary" for environmental quality for the humans who inhabit the basin. The challenge for fisheries managers is to keep expectations in line with biological and ecological possibilities for improvement, aggressively try new techniques, build new constituencies (involve environmental groups and agencies concerned with water quality and public health), and begin new research initiatives. Fisheries managers must plan for a future of reduced expectations now.

22.6.2   Competing Strategies: Top-Down Versus Bottom-Up

In Lake Ontario, the phosphorus management strategy (reducing phosphorus loading) and fish management strategy (stocking of salmonines) have been successful (Great Lakes Science Advisory Board 1988). These two strategies may promote system instability if they are not managed effectively. Alewife is the main food for salmonines and can also control zooplankton size, and hence algal populations. Alewives can also depress native fish stocks through predation and competition. By decreasing phosphorus loadings and increasing salmonine predation pressure, we may be negatively impacting the alewife, which in turn may allow native stocks to increase along with increases in zooplankton sizes. Winter severity may also negatively impact the alewife, further exacerbating these complex interactions. Thus Lake Ontario is on the threshold of major changes, which will most likely be keyed on the stochastic influence of weather on alewife mortality. Christie et al. (1987b) maintain that bottom-up forces (nutrient-dominated pathways and interactions) prevail in the early stages of rehabilitation and top-down forces (predation impacts on lower levels) dominate at later stages of the maturation cycle. Vertical vectoring of energy may be inhibited in Lake Ontario because of the loss of the deepwater coregonine complex, so consideration should be given to restoration of this component of the fish community to promote increased ecosystem efficiency and energy transfer among trophic levels.

22.6.3 Global Warming

Changes that may occur in species and abundances of fish induced by future global warming may be dramatic (Meisner et al. 1987). Predicted warming of air temperatures of 3.2–4.8°C could cause a shrinkage of extant populations of salmonines and coregonines through reduction in preferred thermal habitat, while range extensions are expected for more warmwater species. Spawning and nursery areas may be reduced along with yields.

22.6.4 Institutional Arrangements

There is an exceedingly complex system of institutional arrangements which bear on all aspects of the planning and management of the Great Lakes ecosystem (Francis 1987). These complexities have hampered effective management of the Great Lakes (Smith 1970a), contributed to legacies of extinct species, contaminated aquatic organisms, lost wetlands, and irrevocably changed some of the most complex and fragile ecosystems in North America. Francis (1987) stressed that rehabilitation measures must take into account two things for effective design: first, that some areas are local and require local governmental input and responsibility, and second, that basin- or lakewide problems (contamination of fish) require governmental inputs at a higher level. The IJC and the GLFC are international leaders in these efforts. The IJC is currently spearheading efforts to clean contaminated sites throughout the Great Lakes, especially at the 43 areas of concern (Hartig and Thomas 1988). The GLFC provides initiatives in sea lamprey control and rehabilitation of native stocks, especially lake trout. Ironically, as water quality of tributary streams improves, more habitat is available for spawning sea lampreys.

Establishment of a salmonine sport fishery in some of the Great Lakes and partial rehabilitation of the lake trout fishery, especially in Lake Superior, are notable achievements. The present put-grow-and-take sport fishery gives management agencies flexibility in managing the system (Smith 1972b), though at greater cost than would a self-sustaining natural predator–prey system, which is more consistent with the ecosystem approach (Christie et al. 1987a). However, issues remain which are intractable with existing approaches. The Joint Strategic Plan for Management of the Great Lakes Fisheries (Great Lakes Fishery Commission 1980) provides some mechanisms for overcoming past obstacles. These mechanisms operate within the framework of lake committees, which act as a forum for agency representatives to represent their own interests and negotiate consensus decisions regarding joint concerns. This process has worked very well in the past, especially in planning strategies to control sea lamprey. Much more could be done since the mechanisms for united management are in place. For example, a unified fish contaminants advisory has recently been initiated for Lake Michigan; a similar plan is needed for controlling the number and species of salmonines stocked, based on current forage fish estimates for all the lakes.

22.6.5   Sea Lamprey Conundrums

The sea lamprey replaced the lake trout and burbot as the dominant terminal predator in the Great Lakes. Chemical sea lamprey control has been successful in allowing reestablishment of top predators including salmon and trout (Eshenroder 1987). In fact, burbot populations have increased dramatically in some parts of Lakes Michigan, Huron, and Ontario, and may require some control to reduce competition with salmonines. However, lake trout have not been restored to their former stock diversity. According to Smith (1973), stocking will likely have to be continued indefinitely to maintain lake trout as long as current circumstances of low levels of sea lamprey predation and overfishing by commercial fishers, anglers, and Native American interests continue. Average sea lamprey length has doubled over the past 2 decades in direct correspondence with salmonine stocking rates in Lake Michigan and probably the other Great Lakes as well. As a result, potential mortality imposed by a fixed number of sea lamprey may have increased as much as sixfold for small salmonines (J. F. Kitchell, University of Wisconsin, Madison, personal communication). In addition, ammocetes may reside in soft sediments off the mouth of rivers where they are too costly to treat, TFM may lose its potency, or resistant strains of sea lamprey may develop. Accordingly, alternatives to control sea lamprey (e.g., sterilization of males) are being implemented (Hansen and Manion 1980).

22.6.6   Toxic Substances: Sublethal Effects, Bioaccumulation

Currently, toxic contamination of fish is one of the most serious problems facing Great Lakes fisheries and environmental agencies. Identification of contaminants in several important fish species has caused disruption of commercial and sport fishing in some lakes. Despite some successes in toxic contaminant reduction, the IJC has identified 43 areas of concern in the basin (Hartig and Thomas 1988). These areas are identified mainly because of serious problems with fish and sediment contamination and nutrient inputs. These problems impact the fisheries, since the areas of concern are in tributaries, connecting channels, embayments, and nearshore areas, some of which are prime nursery, spawning, and fishing grounds.

D'Itri (1988) noted that between 1965 and 1980, PCBs, total DDT, and mercury concentrations in lake trout, bloater, coho salmon, and chinook salmon decreased significantly throughout the Great Lakes. These declines reflect more stringent controls on point discharges of contaminants into the Great Lakes since the 1970s. Since 1980, however, there appears to be a leveling off or even slight increase in contaminant burdens. Declines observed were probably due to easily controlled inputs, and now more intractable sources, such as the atmosphere, spurious industrial and domestic discharges, runoff, landfills, and resuspension of contaminated sediments continue to plague Great Lakes organisms.

Xenobiotic substances never before documented in aquatic systems are now common in some Lake Michigan biota. For example, Willford et al. (1981) demonstrated that water from Lake Michigan was contaminated to such a degree that survival of young lake trout carrying maternally inherited residues was impeded. Hesselberg and Seelye (1982) demonstrated that adult lake trout from Lake Michigan contained some 167 identified organic chemicals as compared with eight identified compounds in hatchery brood fish reared in well water.

Currently we appear to be at a plateau in the decline of toxic substances in fish in the Great Lakes. We now must deal with these anthropogenic substances in the various compartments of the ecosystem, especially sediments and fish (Camanzo et al. 1987). Large amounts continue to enter the ecosystem from the air, being carried in from far beyond the watershed of the Great Lakes basin (Eisenreich et al. 1981). Contamination of premium sport and commercial fishes persists and confounds management efforts to enhance stocks that may have long-term impacts on human health. Tanner (1988) maintains that fisheries managers in the Great Lakes should seriously consider their goals of restoring the long-lived lake trout and brown trout because of their known accumulation of toxic substances. However, these fish are bellwethers of ecosystem health and as long as they remain contaminated, we need to redouble our efforts to control xenobiotic substance entry into the Great Lakes. Bioaccumulation through food webs continues to be a potential source of disruption despite successes in decreasing the input of toxic chemicals to the system.

22.6.7   Introduced Species

Emery (1985) reviewed 34 species of fish that were introduced into the Great Lakes basin between 1819 and 1974 and documented their distribution and present status. Only about half of the introduced species were successful. Mills et al. (1993) provide a recent summary. The sea lamprey, alewife, and white perch invaded the Great Lakes through canals built to link the Great Lakes for shipping. Many species of animals and plants have arrived here through the ballast water of foreign vessels. Because of heavy traffic in ocean-going vessels in the Great Lakes, the potential for introduction of exotic biota with release of ballast water is high. The GLFC is currently pursuing this issue with Canada and promoting the exchange of ballast water at sea before ships enter the Great Lakes.

Goldfish entered via hobbyists no longer enamored with their pets and *Eurytemora affinis,* an invertebrate, occurs in all the Great Lakes, but has had little impact (Anderson and Clayton 1959). A more recent invader, *Bythrotrephes cederstroemi,* is an epilimnial zooplankton species from Europe, which preys on smaller zooplankton (Evans 1988). It is now found in all the Great Lakes where it has the potential of disrupting zooplankton communities and adversely affecting bloater recruitment (Warren and Lehman 1988). It has been found in the stomachs of alewife and yellow perch in Lake Michigan, and in yellow perch and coho salmon in Lake Erie. The zebra mussel *Dreissena polymorpha,* recently found in high densities in Lake St. Clair and Lake Erie, has spread through all the Great Lakes and far outside the basin. It has a planktonic veliger stage, which attaches itself to solid substrates (Figure 22.4). This organism is a nuisance in Europe where it fouls water intakes (Stanczykowska 1978) and has become one in North America as well. The recent discovery of another closely related form, the "quagga mussel," portends even more dire consequences, since this form tolerates colder water temperatures and occurs in deeper waters than the zebra mussel. Although impacts on fisheries have not yet been measured, managers are

Figure 22.4 Zebra mussels have been introduced to the Great Lakes and are of concern to fisheries managers, as well as municipal and industrial users of water.

concerned that the invader could affect fish production in two ways. The fouling of spawning reefs could affect reproduction of shoal spawners if the habitat is altered sufficiently to reduce egg deposition and survival. Preliminary results from observations on walleyes in the western basin of Lake Erie indicated that reproduction and survival in 1990 were satisfactory. Secondly, the shifting of organic matter from pelagic to benthic areas through filtering by zebra mussels could disrupt the food web that links primary production to larval fish. The three-spine stickleback is another exotic present in Lake Ontario, and it recently was collected in the St. Marys River, northern Lakes Michigan and Huron (Stedman and Bowen 1985), and most recently, the St. Clair River. The ruffe has reached high densities in the St. Louis River, a Lake Superior tributary (Pratt 1988). In Europe it feeds on fish eggs and zooplankton and has impacted stocks of yellow perch and trout-perch in the St. Louis harbor area, probably because of its ability to feed at night. The most recent piscine invaders to become established in the Great Lakes are the tubenose and round gobies. They were first discovered in the St. Clair River during 1990 and have increased in abundance and spread throughout the river (Jude et al. 1992). They are believed to have entered prior to 1990—possibly with zebra mussels—in ballast water of foreign tankers from the Black or Caspian seas. The tubenose goby is an endangered species in Russia, while the round goby reaches lengths up to 400 mm and is harvested commercially. They are benthic, hearty fishes, and are expected to act as prey for walleyes and displace some of the darters, sculpins, and percids that currently occupy benthic niches in the St. Clair River.

Introduced species have at one time or another dominated the fish communities of all the Great Lakes. The sea lamprey and Pacific salmon are dominant top predators in Lakes Ontario, Michigan, and Huron. White perch, pink salmon, ruffe, three-spine sticklebacks, and gobies are recent invaders, ample evidence that we have not established well-integrated fish communities which would resist expansion of alien species.

22.6.8   Lake Trout Rehabilitation

Native lake trout are essentially extinct in all the Great Lakes except Lake Superior. They are victims of many human interventions on their stocks and habitat. Populations in the other lakes are currently maintained through stocking which began in 1958. To date, limited natural reproduction has been noted and many agencies continue striving to solve the rehabilitation problem.

Ultra-oligotrophic conditions optimize lake trout reproduction, while early mesotrophic conditions are best for growth (Ryder and Edwards 1985). In the Great Lakes, Lake Superior may be viewed as the lake with the best conditions for lake trout reproduction, while the central or eastern basins of Lake Erie likely provided maximal growth rates.

Factors that may inhibit lake trout reproduction include: contaminants, deterioration of incubation environment, the hatchery experience disrupting reproductive biology, inappropriate stocking sites and hence homing to the wrong spawning areas (Horrall 1981), incorrect genetic strains, and insufficient numbers of spawners because of high fishing mortality (Eshenroder et al. 1984). Scattered reproduction is occurring in Lakes Michigan, Huron, and Ontario, while substantial reproduction is occurring in Lake Superior, where native stocks never went extinct. Krueger et al. (1986) believe that most reproduction is by wild, not hatchery, fish. Rehabilitation of lake trout invariably involves stocking of hatchery-reared individuals. Selection in hatcheries is usually for maximum production of eggs and survivability, which sometimes results in diminished genetic diversity. Failure to determine causes for poor success with natural reproduction, and despite almost 25 years of intensive stocking of yearlings, points to the difficulty in rehabilitating lake trout. Simple restocking without restoration of the water quality, co-occurring native species, and reduction in the stresses of overfishing and sea lamprey predation, has not worked in lakes below Lake Superior.

22.7   CONCLUSIONS

The Great Lakes represent a vast aquatic ecosystem which has been used and abused by the humans who inhabit the basin. The dark forces of habitat destruction, overexploitation, and introduced species have made massive changes to the fish and fabric of the ecosystem. For example, most native lake trout are extinct along with many other species. Attempts to rehabilitate have met with mixed success; hatchery fish lack genetic diversity, incubation habitat may have deteriorated, and fishers are unwilling to reduce harvests to allow more spawner escapement.

The resource has been altered for human use, often to the detriment of the lakes. Water is used for drinking, industrial applications, and dilution of domestic wastes. Commercial and sport fishers ply these waters for premium species. Ships move large quantities of goods throughout the system and recreational boats abound in nearshore waters. Demands for more fish, more harbors, more water (within and outside the basin) for irrigation, industry, and dilution of human wastes continue. We have not been good stewards. We have allowed nutrient input because it was expedient and inexpensive; we allowed harbor development and shoreline and wetlands destruction for economic growth. We have allowed toxic contamination to assault the Great Lakes from the land, tributaries, and the air. We now know that dilution is not the solution, that toxic substances will accumulate somewhere, and that the premium fishes and humans are the terminal receptors of these xenobiotic substances. The changes in fish habitat and fish populations were incremental, insidious, and sometimes irrevocable. No other large freshwater resource has undergone the amount of change in the last 40 years as have the Great Lakes. We have been in a reactive mode over most of the history of Great Lakes changes, doing postmortems on ecosystem ills, seldom learning, less often applying lessons of past failures. However, there is hope. Reconstituting the Great Lakes to their former diversity and elegance is impossible, but some compromise, using the ecosystem approach as a template and retaining as many of the native endemic species and genetic diversity along with some of the suite of modern predators, may prove to be the best we can do with current levels of political will and technology. We have made improvements; however, conditions had to deteriorate to grossly unacceptable levels before progress in rehabilitation was made. Nutrient enrichment has been lessened and overfishing curbed. We have slowed down toxic substance entry into the Great Lakes, and attained some semblance of control over the pestiferous alewife and the parasitic sea lamprey.

Anglers now enjoy fishing for a variety of native and exotic predators, from Atlantic salmon, through the various trout species, to walleye. Signs of successful rehabilitation are present throughout the Great Lakes. We are in an era of recovery; we will continue to need the economic and political will of the people and governments in the Great Lakes basin to maintain the Great Lakes and improve them for future residents, so they too can catch uncontaminated yellow perch and chinook salmon from piers, swim and enjoy clear waters on beaches, and watch the eternal, ever-varying sunrises and sunsets on these jeweled inland seas.

22.8 REFERENCES

Allan, R., A. Murdock, and A. Sudar. 1983. An introduction to the Niagara River/Lake Ontario pollution problem. Journal of Great Lakes Research 9:249–273.

Anderson, D., and D. Clayton. 1959. Plankton in Lake Ontario. Ontario Department of Lands and Forests, Division of Research, Phytoplankton Section, Phytoplankton Research Note 1, Toronto.

Anderson, E. D., and L. L. Smith, Jr. 1971. Factors affecting abundance of lake herring (*Coregonus artedii* Lesueur) in western Lake Superior. Transactions of the American Fisheries Society 100:691–707.

Applegate, V. C. 1950. Natural history of the sea lamprey (*Petromyzon marinus*) in Michigan. U.S. Fish and Wildlife Service Special Scientific Report Fisheries 55:1–237.

Applegate, V. C., J. H. Howell, J. W. Moffett, B. G. H. Johnson, and M. A. Smith. 1961. Use of 3-trifluormethyl-4-nitrophenol as a selective sea lamprey larvicide. Great Lakes Fishery Commission Technical Report 1.

Applegate, V. C., and H. D. VanMeter. 1970. A brief history of commercial fishing in Lake Erie. U.S. Fish and Wildlife Service Bureau of Commercial Fisheries Fishery Leaflet 630:1–28.

Auer, N. A. 1982. Identification of larval fishes of the Great Lakes Basin with emphasis on the Lake Michigan drainage. Great Lakes Fishery Commission, Special Publication 82-3, Ann Arbor, Michigan.

Baldwin, N. S., R. Saalfeld, M. Ross, and H. J. Buettner. 1979. Commercial fish production in the Great Lakes 1967–1977. Great Lakes Fishery Commission Technical Report 3.

Beeton, A. M. 1965. Eutrophication of the St. Lawrence Great Lakes. Limnology and Oceanography 10:240–254.

Berkes, F., J. Dennis, P. Hayes, R. Morris, and D. Pocock. 1983. Ontario's Great Lakes fisheries: managing the user groups. Brock University, Institute of Urban and Environmental Studies, Working Paper 19, St. Catharines, Ontario.

Berst, A. H., and R. C. Simon. 1981. Introduction to the proceedings of the 1980 Stock Concept International Symposium (STOCS). Canadian Journal of Fisheries and Aquatic Sciences 38:1457–1458.

Boileau, M. G. 1985. The expansion of white perch, *Morone americana,* in the lower Great Lakes. Fisheries (Bethesda) 10(1):6–10.

Borgmann, U. 1982. Particle-size-conversion efficiency and total animal production in pelagic ecosystems. Canadian Journal of Fisheries and Aquatic Sciences 39:668–674.

Borgmann, U., and D. M. Whittle. 1983. Particle-size-conversion efficiency and contaminant concentrations in Lake Ontario biota. Canadian Journal of Fisheries and Aquatic Sciences 40:328–336.

Brandt, S. 1986. Food of trout and salmon in Lake Ontario. Journal of Great Lakes Research 12:200–205.

Brandt, S. 1986. Disappearance of the deepwater sculpin (*Myoxocephalus thompsoni*) from Lake Ontario: the keystone predator hypothesis. Journal of Great Lakes Research 12:18–24.

Brandt, S. B., D. M. Mason, D. B. MacNeill, T. Coates, and J. E. Gannon. 1987. Predation by alewives on larvae of yellow perch in Lake Ontario. Transactions of the American Fisheries Society 116:641–645.

Brown, E. H., Jr. 1972. Population biology of alewives, *Alosa pseudoharengus,* in Lake Michigan, 1949–1970. Journal of the Fisheries Research Board of Canada 29:477–500.

Brown, E. H., Jr., R. L. Argyle, N. R. Payne, and M. E. Holey. 1987. Yield and dynamics of destabilized chub (*Coregonus* spp.) populations in Lakes Michigan and Huron, 1950–84. Canadian Journal of Fisheries and Aquatic Sciences 44 (Supplement 2):371–383.

Brown, E. H., Jr., G. W. Eck, N. R. Foster, R. M. Horrall, and C. E. Coberly. 1981. Historical evidence for discrete stocks of lake trout (*Salvelinus namaycush*) in Lake Michigan. Canadian Journal of Fisheries and Aquatic Sciences 38:1747–1758.

Camanzo, J., C. Rice, D. Jude, and R. Rossmann. 1987. Organic priority pollutants in nearshore fish from 14 Lake Michigan tributaries and embayments, 1983. Journal of Great Lakes Research 13:296–309.

Carpenter, S., J. F. Kitchell, and J. Hodgson. 1985. Cascading trophic interactions and lake productivity—fish predation and herbivory can regulate lake ecosystems. BioScience 35:634–639.

Christie, W. J. 1972. Lake Ontario: effects of exploitation, introductions, and eutrophication on the salmonid community. Journal of the Fisheries Research Board of Canada 29:913–929.

Christie, W. J. 1974. Changes in the fish species composition of the Great Lakes. Journal of the Fisheries Research Board of Canada 31:827–854.

Christie, W. J., K. A. Scott, P. G. Sly, and R. H. Strus. 1987b. Recent changes in the aquatic food web of eastern Lake Ontario. Canadian Journal of Fisheries and Aquatic Sciences 44(Supplement 2):37–52.

Christie, W. J., and six coauthors. 1987a. A perspective on Great Lakes fish community rehabilitation. Canadian Journal of Fisheries and Aquatic Sciences 44(Supplement 2):486–499.

Colby, P. J. 1977. Introduction to the proceedings of the 1976 Percid International Symposium (PERCIS). Journal of the Fisheries Research Board of Canada 34:1447–1449.

Crowder, L. B., M. E. McDonald, and J. A. Rice. 1987. Understanding recruitment of Lake Michigan fishes: the importance of size-based interactions between fish and zooplankton. Canadian Journal of Fisheries and Aquatic Sciences 44(Supplement 2):141–147.

Dehring, T. R., A. F. Brown, C. H. Daugherty, and S. R. Phelps. 1981. Survey of the genetic variation among eastern Lake Superior lake trout (*Salvelinus namaycush*). Canadian Journal of Fisheries and Aquatic Sciences 38:1738–1746.

D'Itri, F. 1988. Contaminants in selected fishes from the upper Great Lakes. Pages 51–84 *in* Schmidtke (1988).

Dobson, H. F. H., M. Gilbertson, and P. G. Sly. 1974. A summary and comparison of nutrients and related water quality in Lakes Erie, Ontario, Huron, and Superior. Canadian Journal of Fisheries and Aquatic Sciences 31:731–738.

Eck, G. W., and E. H. Brown, Jr. 1985. Lake Michigan's capacity to support lake trout (*Salvelinus namaycush*) and other salmonines: an estimate based on the status of prey populations in the 1970s. Canadian Journal of Fisheries and Aquatic Sciences 42:449–454.

Eck, G. W., and L. Wells. 1987. Recent changes in Lake Michigan's fish community and their probable causes, with emphasis on the role of alewife (*Alosa pseudoharengus*). Canadian Journal of Fisheries and Aquatic Sciences 44(Supplement 2):53–60.

Eisenreich, S., B. Looney, and J. Thornton. 1981. Airborne organic contaminants in the Great Lakes ecosystem. Environmental Science and Technology 15:30–38.

Emery, L. 1985. Review of fish species introduced into the Great Lakes, 1819–1974. Great Lakes Fishery Commission, Technical Report 45.

Eshenroder, R. L. 1977. Effects of intensified fishing, species changes, and spring water temperatures on yellow perch, *Perca flavescens,* in Saginaw Bay. Journal of the Fisheries Research Board of Canada 34:1830–1838.

Eshenroder, R. L. 1987. Socioeconomic aspects of lake trout rehabilitation in the Great Lakes. Transactions of the American Fisheries Society 116:309–313.

Eshenroder, R. L., T. P. Poe, and C. Olver. 1984. Strategies for rehabilitation of lake trout in the Great Lakes: proceedings of a conference on lake trout research. Great Lakes Fishery Commission Technical Report 40.

Eshenroder, R. L., and seven coauthors. 1987. St. Marys River Sea Lamprey Task Force report to Great Lakes Fishery Commission, Ann Arbor, Michigan.

Evans, D. O., B. A. Henderson, N. J. Bax, T. R. Marshall, R. T. Oglesby, and W. J. Christie. 1987. Concepts and methods of community ecology applied to freshwater fisheries management. Canadian Journal of Fisheries and Aquatic Sciences 44(Supplement 2):448–470.

Evans, M. E. 1988. *Bythotrephes cederstroemi*: its new appearance in Lake Michigan. Journal of Great Lakes Research 14:234–240.

Evans, M. S., and D. J. Jude. 1986. Recent shifts in *Daphnia* community structure in southeastern Lake Michigan: a comparison of the inshore and offshore regions. Limnology and Oceanography 31:56–67.

Fetterolf, C. M., Jr. 1980. Why a Great Lakes Fishery Commission and why a Sea Lamprey International Symposium. Canadian Journal of Fisheries and Aquatic Sciences 37:1588–1593.

Fetterolf, C. M., Jr. 1988. Uniting habitat quality and fishery programs in the Great Lakes. Pages 85–100 *in* Schmidtke (1988).

Francis, G. 1987. Toward understanding Great Lakes "organizational ecosystems." Journal of Great Lakes Research 13:233.

Francis, G. R., J. J. Magnuson, H. A. Regier, and D. R. Talhelm. 1979. Rehabilitating Great Lakes ecosystems. Great Lakes Fishery Commission Technical Report 37.

FWPCA (Federal Water Pollution Control Administration). 1968. Lake Erie environmental summary 1963-64. U.S. Department of the Interior, Federal Water Pollution Control Administration, Great Lakes Region, Cleveland Program Office, Cleveland, Ohio.

Gale, R. P. 1987. Resource miracles and rising expectations: a challenge to fishery managers. Fisheries (Bethesda) 12(5):8–13.

Great Lakes Fishery Commission (GLFC). 1980. A joint strategic plan for management of Great Lakes fisheries. GLFC, Ann Arbor, Michigan.

Great Lakes Science Advisory Board. 1988. Rehabilitation of Lake Ontario: the role of nutrient reduction and food web dynamics. International Joint Commission, Windsor, Ontario.

Great Lakes Water Quality Board. 1981. Report on Great Lakes water quality. Appendix: Great Lakes surveillance. International Joint Commission, Windsor, Ontario.

Hansen, L. H., and P. J. Manion. 1980. Sterility method of pest control and its potential role in an integrated sea lamprey (Petromyzon marinus) control program. Canadian Journal of Fisheries and Aquatic Sciences 37:2108–2117.

Hartig, J. H., and R. L. Thomas. 1988. Development of plans to restore degraded areas in the Great Lakes. Environmental Management 132:327–347.

Hartman, W. L. 1972. Lake Erie: effects of exploitation, environmental changes, and new species on the fishery resources. Journal of the Fisheries Research Board of Canada 29:899–912.

Hartman, W. L. 1988. Historical changes in the major fish resources of the Great Lakes. Pages 103–132 in M. E. Evans, editor. Toxic contaminants and ecosystem health: a Great Lakes focus. Wiley, New York.

Hatch, R. W., S. J. Nepszy, K. M. Muth, and C. T. Baker. 1987. Dynamics of the recovery of the western Lake Erie walleye (Stizostedion vitreum vitreum) stock. Canadian Journal of Fisheries and Aquatic Sciences 44(Supplement 2):15–22.

Henderson, B. A., and E. H. Brown, Jr. 1985. Effects of abundance and water temperature on recruitment and growth of alewife (Alosa pseudoharengus) near South Bay, Lake Huron, 1954–82. Canadian Journal of Fisheries and Aquatic Sciences 42:1608–1613.

Herdendorf, C. E. 1982. Large lakes of the world. Journal of Great Lakes Research 8:379–412.

Hesselberg, R. J., and J. G. Seelye. 1982. Identification of organic compounds in Great Lakes fishes by gas chromatography/mass spectrometry: 1977. U.S. Fish and Wildlife Service, Great Lakes Fishery Laboratory, Administration Report 82-1, Ann Arbor, Michigan.

Horrall, R. M. 1981. Behavioral stock-isolating mechanisms in Great Lakes fishes with special reference to homing and site imprinting. Canadian Journal of Fisheries and Aquatic Sciences 38:1481–1496.

Johannsson, O. E. 1987. Comparison of Lake Ontario zooplankton communities between 1967 and 1985: before and after implementation of salmonine stocking and phosphorus control. Journal of Great Lakes Research 13:328–339.

Jude, D. J. 1992. Evidence for natural reproduction by stocked walleyes in the Saginaw River tributary system, Michigan. North American Journal of Fisheries Management 12:386–395.

Jude, D. J., and J. Pappas. 1992. Fish utilization of Great Lakes Coastal Wetlands. Journal of Great Lakes Research 18:651–672.

Jude, D. J., R. H. Reider, and G. R. Smith. 1992. Establishment of Gobiidae in the Great Lakes basin. Canadian Journal of Fisheries and Aquatic Sciences 49:416–421.

Jude, D. J., and F. J. Tesar. 1985. Recent changes in the inshore forage fish of Lake Michigan. Canadian Journal of Fisheries and Aquatic Sciences 42:1154–1157.

Jude, D. J., F. J. Tesar, S. F. DeBoe, and T. J. Miller. 1987. Diet and selection of major prey species by Lake Michigan salmonines, 1973–1982. Transactions of the American Fisheries Society 116:677–691.

Kerr, S. R., and L. M. Dickie. 1984. Measuring the health of aquatic ecosystems. Pages

279–284 *in* V. W. Cairns, P. V. Hodson, and J. O. Nriagu, editors. Contaminant effects on fisheries. Wiley, New York.

Kitchell, J. F., and S. W. Hewett. 1987. Forecasting forage demand and yield of sterile chinook salmon (*Oncorhynchus tshawytscha*) in Lake Michigan. Canadian Journal of Fisheries and Aquatic Sciences 44 (Supplement 2):384–389.

Koelz, W. 1929. Coregonid fishes of the Great Lakes. U.S. Bureau of Fisheries Bulletin 43:1–643.

Kohler, C. C., and J. J. Ney. 1980. Piscivority in a landlocked alewife (*Alosa pseudoharengus*) population. Canadian Journal of Fisheries and Aquatic Sciences 37:1314–1317.

Krueger, C. C., B. L. Swanson, and J. H. Selgeby. 1986. Evaluation of hatchery-reared lake trout for reestablishment of populations in the Apostle Islands region of Lake Superior, 1960–84. Pages 93–107 *in* R. H. Stroud, editor. Fish culture in fisheries management. American Fisheries Society, Fish Culture Section and Fisheries Management Section, Bethesda, Maryland.

Lawrie, A. H. 1978. The fish community of Lake Superior. Journal of Great Lakes Research 4:513–549.

Lawrie, A. H., and J. F. Rahrer. 1972. Lake Superior: effects of exploitation and introductions on the salmonid community. Journal of the Fisheries Research Board of Canada 29:765–776.

Leach, J. H., M. G. Johnson, J. R. M. Kelso, J. Hartman, W. Numann, and B. Entz. 1977. Responses of percid fishes and their habitats to eutrophication. Journal of the Fisheries Research Board of Canada 34:1964–1971.

Leach, J. H., and S. J. Nepszy. 1976. The fish community in Lake Erie. Journal of the Fisheries Research Board of Canada 33:622–638.

Loftus, K. H., and H. A. Regier. 1972. Introduction to the proceedings of the 1971 symposium on salmonid communities in oligotrophic lakes. Journal of the Fisheries Research Board of Canada 29:613–616.

MacCallum, W. R., and J. H. Selgeby. 1987. Lake Superior revisited 1984. Canadian Journal of Fisheries and Aquatic Sciences 44(Supplement 2):23–36.

Matuszek, J. E. 1978. Empirical predictions of fish yields of North American lakes. Transactions of the American Fisheries Society 107:385–394.

Meisner, J. D., J. L. Goodier, H. A. Regier, B. J. Shuter, and W. J. Christie. 1987. An assessment of the effects of climate warming on Great Lakes basin fishes. Journal of Great Lakes Research 13:340–352.

Mills, E. L., J. H. Leach, J. T. Carlton, and C. L. Secor. 1993. Exotic species in the Great Lakes: a history of biotic crises and anthropogenic introductions. Journal of Great Lakes Research 19:1–54.

Minns, E. K., D. A. Hurley, and K. H. Nicholls, editors. 1986. Project Quinte: point source phosphorus control and ecosystem response in the Bay of Quinte, Lake Ontario. Canadian Special Publication of Fisheries and Aquatic Sciences 86.

Muth, K. M., and D. R. Wolfert. 1986. Changes in growth and maturity of walleyes associated with stock rehabilitation in western Lake Erie, 1964–1983. North American Journal of Fisheries Management 6:168–175.

Oglesby, R. T. 1977. Relationships of fish yield to lake phytoplankton standing crop, production and morphoedaphic factors. Journal of the Fisheries Research Board of Canada 34:2271–2279.

O'Gorman, R., R. A. Bergstedt, and T. H. Eckert. 1987. Prey fish dynamics and salmonine predator growth in Lake Ontario, 1978–84. Canadian Journal of Fisheries and Aquatic Sciences 44(Supplement 2):390–403.

Potter, R. L., and G. W. Fleischer. 1992. Reappearance of spoonhead sculpin (*Cottus ricei*) in Lake Michigan. Journal of Great Lakes Research 18:755–758.

Pratt, D. 1988. Distribution and population status of the ruffe (*Gynocephalus cernua*) in the St. Louis estuary and Lake Superior. Great Lakes Fishery Commission Research Completion Report, Ann Arbor, Michigan.

Rakoczy, G. P., and R. D. Rogers. 1987. Sportfishing catch and effort from the Michigan waters of Lakes Michigan, Huron, and Erie, and their important tributary streams, April 1, 1986-March 31, 1987. Michigan Department of Natural Resources, Fisheries Division, Fisheries Technical Report 87-6A, Lansing.

Regier, H., and W. L. Hartman. 1973. Lake Erie's fish community: 150 years of cultural stresses. Science (Washington, D.C.) 180:1248–1255.

Regier, H. A., V. C. Applegate, and R. A. Ryder. 1969. The ecology and management of the walleye in western Lake Erie. Great Lakes Fishery Commission Technical Report 15.

Ryder, R. A. 1972. The limnology and fishes of oligotrophic glacial lakes in North America (about 1800 A.D.). Journal of the Fisheries Research Board of Canada 29:617–628.

Ryder, R. A., and C. J. Edwards, editors. 1985. A conceptual approach for the application of biological indicators of ecosystem quality in the Great Lakes basin. International Joint Commission, Windsor, Ontario.

Ryder, R. A., S. R. Kerr, W. W. Taylor, and P. A. Larkin. 1981. Community consequences of fish stock diversity. Canadian Journal of Fisheries and Aquatic Sciences 38:1856–1866.

Samples, K., and R. Bishop. 1987. An economic analysis of integrated fisheries management: the case of the Lake Michigan alewife and salmonine fisheries. Journal of Great Lakes Research 8:593–602.

Scavia, D., G. L. Fahnenstiel, M. S. Evans, D. J. Jude, and J. T. Lehman. 1986. Influence of salmonid predation and weather on long-term water quality trends in Lake Michigan. Canadian Journal of Fisheries and Aquatic Sciences 43:435–443.

Schmidtke, N. W. 1988. Toxic contamination in large lakes, volume 2. Impact of toxic contaminants on fisheries management. Lewis Publishers, Chelsea, Michigan.

Schneider, J. C., and J. H. Leach. 1977. Walleye (*Stizostedion vitreum vitreum*) fluctuations in the Great Lakes and possible causes, 1800–1975. Journal of the Fisheries Research Board of Canada 34:1878–1899.

Selgeby, J. H. 1982. Decline of lake herring (*Coregonus artedii*) in Lake Superior: an analysis of the Wisconsin herring fishery, 1936–78. Canadian Journal of Fisheries and Aquatic Sciences 39:554–563.

Selgeby, J. H. 1985. Population trends of lake herring (*Coregonus artedii*) and rainbow smelt (*Osmerus mordax*) in U.S. waters of Lake Superior, 1968–1984. Pages 1–12 *in* R. L. Eshenroder, editor. Great Lakes Fishery Commission Technical Report 853.

Selgeby, J. H., W. R. MacCullum, and D. V. Swedberg. 1978. Predation by rainbow smelt (*Osmerus mordax*) on lake herring (*Coregonus artedii*) in western Lake Superior. Journal of the Fisheries Research Board of Canada 35:1457–1463.

Sheldon, R. W., A. Prakash, and W. H. Sutcliffe, Jr. 1972. The size distribution of particles in the ocean. Limnology and Oceanography 17:327–340.

Smith, S. H. 1964. Status of the deepwater cisco population of Lake Michigan. Transactions of the American Fisheries Society 93:155–163.

Smith, S. H. 1968. Species succession and fishery exploitation in the Great Lakes. Journal of the Fisheries Research Board of Canada 25:667–693.

Smith, S. H. 1970a. Trends in fishery management of the Great Lakes. Pages 107–114 *in* N. G. Benson, editor. A century of fisheries in North America. American Fisheries Society Special Publication 7.

Smith, S. H. 1970b. Species interactions of the alewife in the Great Lakes. Transactions of the American Fisheries Society 99:754–765.

Smith, S. H. 1972a. Factors in ecologic succession in oligotrophic fish communities of the Laurentian Great Lakes. Journal of the Fisheries Research Board of Canada 29:717–730.

Smith, S. H. 1972b. The future of salmonid communities in the Laurentian Great Lakes. Journal of the Fisheries Research Board of Canada 29:951–957.

Smith, S. H. 1973. Application of theory and research in fishery management of the Laurentian Great Lakes. Transactions of the American Fisheries Society 102:156–163.

Spangler, G. R., K. H. Loftus, and W. J. Christie. 1987. Introduction to the International Symposium on stock assessment and yield prediction (ASPY). Canadian Journal of Fisheries and Aquatic Sciences 44(Supplement 2):7–9.

Spangler, G. R., N. R. Payne, and G. K. Winterton. 1977. Percids in the Canadian waters of Lake Huron. Journal of the Fisheries Research Board of Canada 34:1839–1848.

Stanczykowska, A. 1978. Occurrence and dynamics of *Dreissena polymorpha* (Pall.) (Bivalvia). Internationale Vereinigung für theoretische und angewandte Limnologie Verhandlungen 20:2431–2434.

Stedman, R. M., and C. A. Bowen II. 1985. Introduction and spread of the threespine stickleback (*Gasterosteus aculeatus*) in Lakes Huron and Michigan. Journal of Great Lakes Research 11:508–511.

Steedman, R. J., and H. A. Regier. 1987. Ecosystem science for the Great Lakes: perspectives on degradative and rehabilitative transformations. Canadian Journal of Fisheries and Aquatic Sciences 44(Supplement 2):95–103.

Stewart, D. J., J. F. Kitchell, and L. B. Crowder. 1981. Forage fishes and their salmonid predators in Lake Michigan. Transactions of the American Fisheries Society 110:751–763.

Sullivan, R., T. Heidtke, J. Hall, and W. Sonzogni. 1981. Potential impact of changes in Great Lakes water quality on fisheries. Great Lakes Basin Commission, Contribution 26, Ann Arbor, Michigan.

Tanner, H. 1988. Restocking of Great Lakes fishes and reduction of environmental contaminants 1960–1980. Pages 209–288 *in* Schmidtke (1988).

Vallentyne, J. R. 1974. The algal bowl. Lakes and man. Canada Department of the Environment, Miscellaneous Special Publication 22, Ottawa.

Warren, G. J., and J. T. Lehman. 1988. Young-of-the-year *Coregonus hoyi* in Lake Michigan: prey selection and influence on the zooplankton community. Journal of Great Lakes Research 14:420–426.

Wells, L. 1970. Effects of alewife predation on zooplankton populations in Lake Michigan. Limnology and Oceanography 15:556–565.

Wells, L. 1980. Food of alewives, yellow perch, spottail shiners, trout-perch, and slimy and fourhorn sculpins in southeastern Lake Michigan. U.S. Fish and Wildlife Service Technical Papers 98.

Wells, L., and A. L. McLain. 1973. Lake Michigan: man's effects on native fish stocks and other biota. Great Lakes Fishery Commission Technical Report 20.

Whillans, T. H. 1979. Historic transformations of fish communities in three Great Lakes bays. Journal of Great Lakes Research 5:195–215.

Whillans, T. 1982. Changes in marsh area along the Canadian shore of Lake Ontario. Journal of Great Lakes Research 8:570–577.

Willford, W. A., and eight coauthors. 1981. Introduction and summary. Pages 1–7 *in* Chlorinated hydrocarbons as a factor in the reproduction and survival of the lake trout (*Salvelinus namaycush*) in Lake Michigan. U.S. Fish and Wildlife Service Technical Papers 105.

Chapter 23

Anadromous Stocks

JOHN R. MORING

23.1 INTRODUCTION

Effective management of anadromous fishes (fishes that are born in freshwater, travel to the sea to grow, then return to freshwater to spawn) is often much more complicated than management of inland species. This is because anadromous stocks often cross state, provincial, foreign, and international waters. Movements between jurisdictions involve additional management agencies. By definition, anadromous species move between the freshwater and marine domains (see Box 23.1). Even if such passage is limited, the management of the species may involve more than one principal agency. Sound management of anadromous sport fisheries is far from easy (Figure 23.1).

Following a series of court decisions in the 1970s and 1980s, Indian tribes in many areas obtained or reaffirmed legal jurisdiction of fish, guaranteed portions of the catch, or both. The Boldt decision (*United States* v. *Washington*) of 1974 ordered the catch of steelhead and Pacific salmon in western Washington to be evenly divided between the state and Indian tribes. The state of Washington and 21 Indian tribes determine catch and allocation of these resources; a court-established Fisheries Advisory Board handles disputes (Clark 1985). Similar decisions by Judge Belloni (*United States* v. *Oregon,* 1969 and 1974) affect four tribes and the states of Washington and Oregon in the management of salmon and steelhead along the Columbia River basin (Marsh and Johnson 1985). Joint management of anadromous fisheries can sometimes be a complicated task. Principal fisheries issues for tribes in the United States and Canada have been reviewed in detail by Busiahn (1984).

Fisheries managers responsible for anadromous species must often deal with both sport and commercial interests, address the concerns of the two interests, ensure that tribal interests are protected, yet still protect the fishes. In making decisions, managers rely on all the tools available: research, fish culture, regulation, habitat maintenance and restoration, and monitoring.

23.2 THE ROLE OF FISH CULTURE

23.2.1 Historical Perspective

Fish culture (and subsequent stocking) is a useful management tool, but not the only tool. No amount of stocking can restore or enhance a fishery unless the underlying causes of depleted fish populations are understood and addressed.

Figure 23.1   Steelhead anglers on the Little North Folk of the Santiam River, Oregon. (Photograph by John Moring.)

These impacts include overharvest, habitat destruction, pollution, dams and other obstructions, as well as other direct and indirect by-products of man. For example, the rapid decline of striped bass populations along the Atlantic coast is a result of several factors, including habitat destruction and pollution. Overfishing

---

Box 23.1   Principal Anadromous Fish Species of North America with Commercial or Sport Fisheries Importance or Endangered Status.

Striped bass	Sea-run brown trout	Sockeye salmon
Atlantic salmon	Sea-run brook trout	Chinook salmon
American shad	Sea-run cutthroat trout	Chum salmon
Alewife	Sea-run Dolly Varden	Coho salmon
Blueback herring	Sea-run Arctic char	Pink salmon
Steelhead	Rainbow smelt	Eulachon
Shortnose sturgeon	Atlantic sturgeon	White sturgeon
Green sturgeon		

has also been a significant factor. Until each of those impacts are addressed, stocking will never bring about true restoration.

Though techniques for culturing trout were brought to North America from Europe in 1853, the culture of anadromous species effectively began in the 1860s (Moring 1986). Initially, these efforts concentrated on Atlantic salmon and American shad, two species that were declining in abundance in New England during that era. Livingstone Stone, a fish culture pioneer, raised Atlantic salmon eggs from the Miramichi River, Canada, and sold the resultant progeny to state fisheries agencies in Massachusetts, Vermont, and New York (Bowen 1970). New Hampshire imported salmon eggs from Ontario as early as 1866, and Stone's Canadian partner, J. Goodfellow, sold salmon eggs to Vermont. Thus, using fish culture as a tool in anadromous fish enhancement was viewed as a viable management option over 120 years ago.

Efforts to revive declining runs of Atlantic salmon and American shad have continued sporadically since that time, while other anadromous species were being cultured, transported, and stocked widely. Rainbow smelt and alewives were distributed to many states during the last 3 decades of the 19th Century. Almost 200 million shad eggs were hatched in five northeastern states between 1866 and 1871 (Bowen 1970), and the first Atlantic salmon hatchery, at Craig Brook (then known as Craig's Brook), Maine, was constructed in 1871 (Figure 23.2). These Atlantic salmon eggs supplied Maine, Massachusetts, and Connecticut in an ultimately futile attempt to reverse the effects of dams and water pollution on fish populations. In the early 1870s alone, over 6 million Atlantic salmon were released in northeastern streams as far south as New Jersey (Bowen 1970).

The culture of anadromous stocks on the Pacific coast also began in the 1870s. A landmark of that era was the operation of the Crooks Creek egg taking station on the McCloud River, California. During the 1870s, 51 million eggs of chinook salmon were collected and distributed to public and private hatcheries in the West and East. After 2 decades of declining Pacific salmon runs to the Columbia River, the state of Washington initiated its first hatcheries (though quite small) in 1894 and 1895. By 1902, 15 hatcheries were operating in Washington.

Alaska's first hatchery was constructed in 1891 and canneries operated numerous private hatcheries that released 450 million smolts of sockeye and coho salmon by 1906 (McNeil 1980). Oregon's first hatchery began operating in 1877, British Columbia's in 1884, and the Maritimes' in the 1880s. Over 4.5 million chinook and sockeye salmon fry were stocked in British Columbia by early 1887 (Mowat 1889).

Most efforts of the U.S. Fish Commission in the 1870s were directed at stocking Atlantic salmon, American shad, chinook salmon, and a few other species in new or previously exploited waters. Some of these early introductions of anadromous fishes were successful. Hundreds of millions of American shad were stocked from state hatcheries and were introduced to California in a series of trips beginning in 1871. By 1880, 654,000 shad fry had been stocked in the Sacramento River (Bowen 1970) and runs have since been established in California, Oregon, and Washington, and some fish that occasionally stray to British Columbia as well. Striped bass were initially introduced to San Francisco Bay with shipments of 132 fish in 1879 and 300 fish in 1882; a commercial fishery developed within 10 years. Striped bass were reported from San Diego to Oregon as early as 1887, and

Figure 23.2 Transporting Atlantic salmon to Craig's Brook Hatchery (now Craig Brook National Fish Hatchery) in the late 1800s. (Photograph courtesy of the Fish and Wildlife Service.)

commercial and sport fisheries are now established in California and southern Oregon.

The value of producing anadromous fishes in hatcheries is substantial. It is estimated by some that more than one-third of the fish caught by anglers are hatchery-stocked fishes. Commercial catches in the United States in 1982 included over 270 million kg of Pacific salmon and 1 million kg of striped bass, a significant portion of which originated from hatcheries. Almost all of the Atlantic salmon returning to U.S. waters are of hatchery origin, and hatcheries are taking an increasingly important role in the production of steelhead and coho salmon.

23.2.2 Current and Future Directions

Modern culturists rely on the accumulated historical experience for raising Atlantic and Pacific salmon and steelhead that extends back over a century. Even species with less culture history in North America (e.g., white sturgeon, short-nose sturgeon, and rainbow smelt) can now be cultured with a degree of confidence. We know how to artificially spawn, raise, and stock anadromous fishes. The challenge to fish culturists is how to do it better; how to refine these culture techniques so that growth is faster, survival is higher, and fish are healthier.

A continuing concern among fish culturists and managers is the role of fish genetics. Hatchery-reared fish, particularly of highly inbred stocks, can be radically changed—morphologically, physiologically, and behaviorally—from their wild ancestors. The more that cultured stocks are inbred or are moved geographically, the less adaptable they become. Many studies have documented differences between wild and hatchery fish. The role of fish culture is to produce high-quality fish that can compete favorably with wild fish. The development of new and innovative techniques will continue to improve culture operations, reduce costs, or expand production capabilities.

23.3 FISH PASSAGE

23.3.1 Upstream Fish Passage

Under natural conditions, anadromous fishes have always experienced some mortality during upstream migration to spawning areas, but with dams or other artificial obstructions, the percentage increases. One of the principal causes of the elimination of Atlantic salmon from most rivers in the United States was the construction of dams without passage facilities. In the six major rivers undergoing restoration of Atlantic salmon, there are still 29 dams without fish passage facilities (Stolte 1986). When such passage is available, mortality can vary with the size of the dam (particularly height), type of fish ladder or lift, length of fish ladder, water flow, location upstream, and other factors. Passage efficiency can also vary with species; few fishways, such as at Bonneville on the Columbia River, pass sturgeon yet still pass Pacific salmon, shad, and steelhead.

Upstream passage mortalities, even with modern fish ladders, can be significant. Swartz (1985) reported a 4% loss of migrating steelhead at each of eight dams on the mainstem Columbia and Snake rivers. When properly designed fish passage facilities are present, losses can be minimal; only 0.1% of adult steelhead died while passing through the short ladders and traps at Foster and Green Peter dams in Oregon (Wagner and Ingram 1973).

23.3.2 Downstream Fish Passage

A rule of thumb for many years has been that 10% of salmonid smolts will be lost at each dam during their downstream migration. Fish mortalities can be high when fish pass through turbines or undergo stress caused by pressure changes. Although some hydro developers believe mortality can be reduced to 8% by technical improvements, a recent report by the Electric Power Research Institute analyzed 64 studies of fish passage and concluded that it will be unlikely that most fish passage facilities will reduce the site mortality below 10%. In some locations, even a 10% value may be conservative (Table 23.1).

In general, the extent of mortality will vary with each facility, each type of turbine, and each type of attraction or deflector system in use. There are also other considerations. Dams with poorly placed downstream passage facilities can cause delays in migration, and these delays can result in significant impacts. For example, at Foster Reservoir, migrating steelhead smolts were delayed by up to 2 weeks, making them susceptible to anglers seeking legal-sized rainbow trout. An estimated 19,500 to 22,800 of these steelhead smolts were taken annually in this relatively small, 494-hectare reservoir (Bley and Moring 1988).

Downstream migrating adult fish are also subject to high mortality at dams because many dams do not have proper adult passage facilities, and survival of

Table 23.1 Representative downstream passage mortalities reported by researchers.

Location	Species/stage	Mortality (%)	Source
Merrimack River, New Hampshire–Massachusetts	Atlantic salmon smolts	17	Unpublished[a]
Five Maine rivers	Atlantic salmon smolts	11–23	Unpublished[a]
Connecticut River, Massachusetts–Connecticut	Atlantic salmon smolts	12–14	Stier and Kynard (1986)
St. Croix River, Maine–New Brunswick	Atlantic salmon smolts	52	Watt (1986)
Penobscot River, Maine	Atlantic salmon smolts	9–11	Unpublished[a]
Columbia River, Washington–Oregon	Steelhead smolts at each dam	20–25	Swartz (1985)
S. Santiam River, Oregon	Steelhead and chinook salmon smolts	7–61	Wagner and Ingram (1973)
Connecticut River, Massachusetts–Connecticut	Juvenile American shad and blueback herring	62–82	Taylor and Kynard (1985)

[a]Unpublished data from the U.S. Fish and Wildlife Service.

adults passing through turbines is low. An estimate made using mark-and-recapture of adult American shad in the Susquehanna River was 50% mortality (Whitney 1961).

There is some innovative research on improving downstream survival at dams. Biologists are exploring ways of diverting smolts or adults to passage facilities and away from turbines using such features as inclined plane screens, angled drum screens, acoustic transducers, bar racks, and lights. Adult American shad, for example, are being repelled away from undesirable areas by strobe lights on the Connecticut River. Juvenile shad can now be passed at 90 to 100% efficiency with the proper flow velocities (Mecum 1980). A recent development has been the use of electric fields to guide juvenile and spent adult American shad past Holyoke Dam. Testing is continuing.

23.3.3 Trucking and Barging

One management technique for reducing the mortalities associated with fish passage at dams is to capture fish and transport them by truck or barge dams, releasing smolts farther downstream or adults farther upstream. Maine will soon be trucking American shad and alewives past a series of dams on the Kennebec River. In another example, Pacific salmon and steelhead trout are routinely captured at Lower Granite and Little Goose dams on the Snake River, Idaho, transported 460 km downstream to below Bonneville Dam on the Columbia River, and released. The advantage is the avoidance of successive 10% mortalities passing each dam.

Some biologists, however, believe that by bypassing much of the river on the journey, the homing instinct of returning adults will not be as intense. Experiments by Bjornn and Ringe (1984) seem to support this supposition. While transported fish had higher survival to the mouth of the Columbia River, adult returns to the upper Columbia River were highest for fish that had migrated the entire river distance as smolts. Additional studies seem to show higher stress levels in transported smolts (Congleton et al. 1984). Examination of Atlantic salmon in Norway has indicated significant behavior changes and straying in fish without a complete downstream migration experience (Jonsson et al. 1990).

Although trucking and barging operations are much more costly and labor intensive than normal releases, they can significantly improve the percentage of smolts reaching the lower part of a river or the number of adults reaching spawning grounds. In that respect, they can be effective management tools.

23.4 STOCK IDENTIFICATION AND SELECTION

23.4.1 Genetic Selection

With increasing use of stocking as a management tool, genetic selection—the development of selected traits—is a role taken on by management biologists, fish culturists, and geneticists. Fish managers and culturists may wish to select for growth, food conversion, survival, disease resistance, timing of return, age at maturity, fecundity, timing of spawning, or other characteristics.

In its basic form, genetic selection can be strictly selective breeding—the accentuation of certain traits in fish by selecting for those traits over generations of artificial breeding. The classic example of such selective breeding is the "super trout" or "Donaldson trout" developed over years by Lauren Donaldson at the University of Washington. Actually a rainbow trout × steelhead hybrid, these fish have been selected for such traits as weight, fecundity, and age at maturity. Large rainbow trout were crossed with steelhead to bring out the high meat content and rapid weight gain of large rainbow trout to the resulting hybrid.

After 36 generations (years), 2-year-old Donaldson trout averaged 4.5 kg compared to 40–50 g for wild rainbow trout. All females matured in the second year compared to some females in the fourth year and all females in the fifth year, and fecundity averaged 18,000 eggs in the third year compared to 400–500 eggs in the fifth year in the wild (Donaldson 1968, and additional data from Donaldson).

Genetic selection for other characteristics, such as resistance to disease in chinook salmon and resistance to low levels of pH in Atlantic salmon have also been useful to management. There is substantial genetic variability in freshwater growth characteristics of coho salmon, and biologists are selecting salmon for appropriate size at release (Iwamoto et al. 1982; Mahnken et al. 1982). Still other traits are being explored with long-term benefits. Scientists are isolating and transferring the antifreeze protein gene in winter flounder to Atlantic salmon in an attempt to produce fish able to withstand extremely low winter water temperatures for cage culture and other applications.

Yet, genetic selection is much more than simply selective breeding. It is a whole concept of planned genetic exploration. Hershberger and Iwamoto (1985) have pointed out that some less heritable traits may take considerable time to develop and that inbreeding can be a problem. Random mating as opposed to assortive (or selected) mating is often suggested. A fish may be large and produce many eggs, but is it of efficient form to survive in the wild and return?

There is the possibility of losing unique site-adapted characteristics (Thorpe et al. 1981). Some of these characteristics can be helpful to management. For example, studies by the Canadian Department of Fisheries and Oceans Biological Station at St. Andrews, New Brunswick, and the nearby North American Salmon Research Center have shown that genetic selection for certain strains can decrease the proportion of grilse in runs of Atlantic salmon. Grilse are small salmon returning after only 1 year at sea and form large proportions of many Canadian runs. By increasing the proportion of adult fish in the run, the size and potential catch can be increased, and preferred sizes will be available to anglers and commercial fishers.

23.4.2   Mixing and Straying of Stocks

A concern among geneticists and fisheries managers is that, with increased hatchery output to enhance existing anadromous fish populations, there will be greatly increased chances for straying and mixing of different stocks. Ricker (1972) and Saunders' (1981) definitions of the word stock are still widely used and refer to individuals of a species that spawn in a particular stream and do not interbreed with other such groups spawning in a different location or the same location at a different time. Many fish geneticists assert that each stream has its own stock of salmon, steelhead, shad, striped bass, or other species, and some larger rivers may have more than one stock.

Ricker (1972) concluded that most differences between stocks were due to environmental or genetic factors, and perhaps as many as 10,000 distinct stocks of Pacific salmon may exist along the north Pacific rim. Saunders and Bailey (1980) conservatively estimated 2,000 stocks of Atlantic salmon along the North Atlantic rim. Each of these stocks has unique characteristics that separate them from other stocks, such as fecundity, growth rates, egg size, survival, age and size at maturity, resistance to disease and pH, and timing of migrations. (Saunders 1981 reviewed the extensive literature on Atlantic salmon.) When two stocks are mixed at spawning, whether by chance or intentionally, these unique characteristics can be lost or certainly diluted in the genetic components. It is common to periodically introduce wild fishes to hatchery brood stocks to ensure the continuance of preferred, native traits in the gene pool. Stock diversity and preservation of unique stock characteristics are the management goals because they provide agencies with the most management options.

Most anadromous fish species exhibit homing to natal streams or river systems. This trait has been observed at least as far back as 1635, when Izaak Walton reported homing of Atlantic salmon in *The Compleat Angler*:

> Much of this has been observed by tying a Ribband or some known tape or thred, in the tail of some young Salmons, which have been taken in Wiers as they have swimm'd toward the salt water, and then by taking a part of them again with the known mark at the same place at their return from the Sea . . . which has inclined many to think that every Salmon usually returns to the same River in which it was bred.

The degree of such homing behavior depends on the species, but homing can be a mechanism for preserving genetic homogeneity and ensuring spawners return to appropriate spawning areas that have proven their value by producing fish that have completed a life cycle. Conversely, poor homing ability can be a mechanism to promote genetic heterogeneity and ensure that populations will not be seriously impacted by an environmental or biological catastrophe in one localized area. The progeny may still be returning to spawn, through it may be in some other stream.

How extensive is straying among anadromous fishes? It is an extremely difficult value to derive without adequate surveys or recovery systems in place because it takes the recovery of tags from all potential streams to give an accurate estimate. The precise nature of homing has been shown for Atlantic and Pacific salmon, and reviews by Harden-Jones (1968), Hasler and Scholz (1983), and others conclude that, based on artificially imprinted fish and tag returns for all salmonids, average straying accounts for a maximum of 5% of returning fish. There are, however, exceptions, with straying rates up to 45% reported for coho salmon and up to 4% for steelhead. Straying concerns fisheries managers because of the possibility of unplanned genetic mixing of wild and cultured stocks or populations.

23.4.3 In-System Versus Out-System Rearing

It is well known among geneticists and fish managers that survival is normally highest when using a native or local stock compared to using a genetic group from another river system or a foreign country (Reisenbichler and McIntyre 1986). That is why management agencies stocking striped bass into Chesapeake Bay waters try to use local stocks, rather than those from the Hudson River. Agencies stocking Pacific salmon into the Snake River of Idaho will use local, or at least upper Columbia River stocks, rather than those from Alaska or the Fraser River. There are, however, new refinements to this in-system versus out-system rearing. Buchanan and Moring (1986) noticed higher recycling of adult summer steelhead from groups reared out of the Santiam River system (Oregon), then released there, compared to groups reared and released within the Santiam system. Recycling refers to adult fish returning to a dam, entering the trap, being released into the reservoir forebay, then passing through the turbines in an attempt to move back downstream. This trait is being studied, but it may be a by-product of rearing conditions in the hatchery (the out-system fish were reared at constant temperature).

Managers are more concerned with the effect of in-system versus out-system rearing on homing of anadromous fishes and the accuracy of data relating to return survival. It is a common practice for management agencies to obtain gametes from returning spawners on one river, raise the fish in a hatchery, then distribute the fry or smolts to several rivers or river systems. However, biologists in Oregon have recently documented significant homing of steelhead stocked in one system back to the river of origin. Thus, the in-system versus out-system rearing procedures may have several long-term implications for management.

23.4.4 Effects of Aquaculture Operations

Commercial aquaculture operations in North America use two rearing procedures that can affect management of natural and hatchery-supplemented stocks. Cage culture operations are theoretically self-contained, but escapes almost always occur and the percentages are not trivial. Shrinkage or unknown losses of chinook salmon in 23 cages in Puget Sound ranged from 2.5 to 46.5%, without rips or holes in netting. When rips do occur, large numbers of fish can be released into a local area. Eventually, adults will return to nearby streams to spawn with native stocks. Worldwide, it is estimated that 10 to 30% of cage-reared salmonids disappear during rearing due to predation or escape (Moring 1989).

With ocean ranching, a portion of returning adults will not enter the collection facility of the commercial company, but will stray into local streams or into other streams along the migration route. If fishes are not strongly attracted to the return facility, the portion spawning elsewhere will be higher. Hasler and Scholz (1983) artificially imprinted coho salmon in hatcheries using the chemical morpholine.

Nine percent of the salmon returned to a stream where morpholine had been introduced, while 5% of the fish strayed to another stream.

The concern among fisheries managers is that occasionally these fishes used in aquaculture are of different stocks, from other river systems, other states and provinces, and even from other countries. For example, one commercial operation in Oregon imported chum salmon eggs from the former Soviet Union, and another ranching operation in Maine imported Pacific salmon from Alaska and Japan. A commercial cage operation in eastern Maine has received Atlantic salmon from Scotland as well as from North America. Genetic mixing of these stocks with native runs is an ongoing concern (Berg 1981). A detailed analysis by Thorpe et al. (1981) indicates these concerns are not without foundation.

23.5 STOCK MANIPULATION

Stock manipulation includes many techniques for enhancing and (rarely) reducing runs. But, the essence of stock–recruitment and production–yield models is maintaining runs necessary for continuation of populations while allocating that portion of the run surplus to those needs. In such models, the term stock, as defined by Gulland (1969), is sometimes different from the genetic definition and may include several populations or a part of a population managed with similar production characteristics.

Although maximum sustained yield (MSY) is a widely known concept—the maximum yield that can be produced year after year—it is also widely criticized in actual application, generally on economic grounds: the economic costs increase rapidly as the MSY approaches. Other approaches to managing stocks, such as maximum economic return, are often promoted, though population dynamics biologists differ on a universal concept. The theories and applications of fish population dynamics are complex and are discussed in Chapter 5.

23.5.1 Allocation of Catch

One of the difficult tasks of a fisheries manager is the allocation of catch and escapement of anadromous fishes. How many fish can be harvested during the course of the run and still provide sufficient escapement of potential spawners? To answer that question, managers need reliable information on numbers of adults ascending rivers, sport catch, commercial catch, and predictive factors on run size, such as smolt indices from previous years. In most cases, all of this information is not available to management agencies. As a result, stock manipulation is often an art. If one is too conservative on harvest allotments, commercial fisheries may suffer, particularly if run size is ultimately large. If one is too liberal on harvest allotments, commercial fisheries may boom in one season, but may be depressed in a subsequent season due to poor escapement and production of fish.

23.5.2 Regional Management Councils

In 1976, the Fishery Conservation and Management Act (FCMA) established a series of regional fisheries councils in the United States to manage coastal fisheries from the 3-mile limit out to 200 miles offshore. Although foreign vessels can fish for certain species within that zone under authorization of the councils, most anadromous species are taken strictly by home-based fisheries. These regional councils prepare fisheries management plans and coordinate plans with those of nearby councils when migratory species are involved.

Pacific salmon are considered a special case under the FCMA because stocks are far-reaching and have been protected under certain other treaties and agreements covering the high seas, such as the International North Pacific Fisheries Commission (United States, Canada, Japan). American shad, alewives, and striped bass are generally not governed by the FCMA as they are restricted to inshore coastal waters, within the 3-mile limit.

23.6 DISEASE RESEARCH AND CONTROL

Fish diseases affect management in many ways, including reduced production of fish available for stocking. Diseases are natural and kill unknown numbers of fish in the wild. It is in the hatchery that mortalities are visible. Diseases can spread quickly through hatchery facilities because fish there are held at densities much higher than in nature. Disease prevention and control affect management decisions in many areas.

1. Fish subjected to disease are stressed, may not feed, and may exhibit reduced growth rates, thus affecting release size and rearing costs.
2. Fish subjected to disease and therapeutic control may exhibit delayed mortality from stress after release.
3. Management decisions may expose fish to disease zones during their downstream migration.

As an example of this latter impact, biologists have long released steelhead smolts into the South Santiam River in Oregon. In the 1970s, however, biologists documented delays of up to 2 weeks in downstream migration caused by Foster Dam. In some years, smolts eventually reached the lower Willamette River when water temperatures had increased. This exposed them to outbreaks of *Ceratomyxa shasta,* and biologists believe that some of the variability in adult returns may be due to the effects of this disease on smolts.

The key aspect of effective disease management, including diseases promoted by nutritional deficiencies (Snieszko 1972), is prevention and early detection. It is to the mutual benefit of fisheries managers and culturists to prevent any disease outbreaks and minimize mortalities when outbreaks do occur. Most states, provinces, and federal agencies have disease policies that relate to importation of eggs and hatched fish; monitoring of hatchery lots; preventative measures; disinfection of hatchery trucks, raceways, nets, and equipment; detection procedures; and control measures. Maine, for example, allows only the importation of eggs that have been certified by a reputable pathologist as being disease free (a misnomer, as no tests can actually detect the presence of all diseases). But the state prohibits the importation of live fish. The law was contested to the U.S. Supreme Court in 1986, but it was upheld as a means of disease prevention.

When treatment for a disease outbreak is necessary, the prognosis is much higher when treatment is started early. However, there are few legal medicines available to culturists. Over time, various strains of pathogens and bacteria become resistant to tetracycline, sulfa, and other treatment compounds. Only a few such drugs are certified for use under the 1913 Federal Virus-Serum-Toxin Act, and the certification process is long (at least 2 years) and involved. It is to the advantage of managers and culturists to support continual disease research and drug certification efforts.

23.7 HABITAT STUDIES

Habitat is probably the most important factor affecting natural production of anadromous fishes. If spawning gravel is limited for salmonids in a stream or lake, egg yield will be limited and smolt and adult numbers will be affected. If nursery habitat (rearing area for juvenile fish) is limited, the carrying capacity of a stream for fry will be reduced. The quantity and quality of habitat defines the limits of a population within which food and environmental factors are influential.

Other chapters of this text discuss habitat issues in great detail, but some aspects are worth noting as they pertain to anadromous fishes. A few examples will relate how habitat availability and alteration can affect management decisions.

One of the causes of declines in anadromous fish populations has been habitat destruction. Urban development and pollutants have severely reduced the amount of acceptable spawning and nursery habitat available for striped bass in the Chesapeake Bay drainage. Part of the decline in many stocks of coho salmon near urbanized areas of Washington has apparently been due to the destruction of small streams. Coho salmon characteristically inhabit small streams: they seek them for spawning and the juveniles need them for foraging and protection. Similar declines in populations of sea-run cutthroat trout have occurred near population centers of British Columbia. The province has embarked on a large stocking program, hoping that hatchery fish can absorb demand from anglers while wild runs are allowed to increase through habitat improvement.

Logging and other land-use activities can have a profound impact on habitat of anadromous fishes. While the timber industry is extremely important in several coastal states and provinces, the value of anadromous salmonids is almost as great. Everest (1977) estimated fisheries as second in value only to timber in national forests of southwestern Oregon, and even higher in value in some watersheds. The environmental and biological consequences of logging activities on aquatic resources can be severe and include changes in water temperature, suspended sediment, dissolved oxygen, streamflow, and insect and fish populations (Moring et al. 1985). Nevertheless, timber management and anadromous fisheries management can coexist, particularly when riparian (streamside) vegetation is preserved. Maintenance of riparian vegetation is a key in the protection of instream habitat of anadromous fishes. The value of riparian zones has been recognized by federal management agencies, and some states, such as Oregon, provide property tax incentives to landowners who preserve and manage riparian zones for fisheries and wildlife.

23.8 RELEASE PROCEDURES

Since the earliest days of fish culture involving anadromous species, biologists have been attempting to refine their release procedures to determine how big fish should be at release and when they should be released.

In the 19th Century, hatchery managers kept meticulous records of everything. They drew subjective conclusions as to the proper time to release smolts, but it was not until fisheries managers began to tag large numbers of fish and experiment with different release times, sizes, and locations that quantitative information became available. That process continues throughout North America today, yet surprisingly little information is available (Moring 1986). The problem is the time span for most experiments. A group of marked fish may not return as adults for several years, and several replications are needed for each treatment group to provide statistically meaningful data. So, making a decision on efficient size and time of release for an anadromous stock could easily require 5 to 10 years of effort. These studies need to be well designed because the return rate of released smolts to adults may be 1% or less with many salmonid groups. The window of opportunity, when fish are osmotically ready for migration may be only a matter of weeks.

At one time, many steelhead were released as fry. Later experiments, however, concluded that ocean survival could be increased with increased size at release (Larson and Ward 1955). Some recent experimental releases in Oregon have indicated, for the Skamania stock of summer steelhead, that a 2-cm increase in size at release can result in up to twice the previous return rate for adults.

Sheppard (1972) summarized several time and size-at-release studies and concluded that smolt-to-returning adult survival is about 2.5% for releases of small, 1-year-old steelhead, 6% for 2-year-old steelhead, and 18% for 3-year-old steelhead. Studies in the 1950s and 1960s in the Alsea River, Oregon, showed the highest rates of return were for steelhead of the largest release size (12.1/kg), and the location of release affected survival as well (Wagner 1969). Most of Oregon's steelhead production today relies on releasing even larger smolts: 170–180 mm in length and 8.8–13.2/kg in weight. The success rate drops rapidly when steelhead are stocked below 160–170 mm in size, as scientists are finding such smaller fish may not migrate at all. The optimal time of stocking is now considered to be mid-April to mid-June; those fish stocked later will often not migrate.

23.9 PUBLIC INVOLVEMENT PROGRAMS

A growing trend with many state agencies is the interaction of the public with the management of anadromous species. Generally, this involves the active participation of a sports organization or limited number of individuals with selected, localized projects. For example, California has appropriated over $900,000 annually since 1981 to the Northcoast Cooperative Salmon and Steelhead Project (also known as the Bosco-Keene Program or AB951 Program). Funds are provided to nonprofit groups to rehabilitate Pacific salmon and steelhead streams on the northern coast of California.

A similar program in Oregon, the Salmon and Trout Enhancement Program, employs a full-time biologist to work with nonprofit groups to improve salmonid populations in streams. Projects to date have concentrated on incubating salmonid eggs in instream boxes, rehabilitating streambanks by planting willows, restoring riparian vegetation, creating spawning areas, and stocking steelhead trout fry. The Saco Salmon Club in Maine is currently engaged in a cooperative project with the University of New England to assist ongoing restoration efforts for Atlantic salmon. This privately funded effort will soon be incubating 100,000 eggs and releasing fry in the Saco River.

A related program in Washington since the mid-1970s has involved sports groups, Indian tribes, power companies, and private corporations in rearing Pacific salmon. Over 10 million salmon were released in this program in 1983 alone. Financial arrangements vary from donation of eggs by the state and rearing costs by the private groups, to the outright purchase of live fish, reared to smolt size or beyond, often in conjunction with delayed-release programs to improve sport fishing in Puget Sound. A similar, though smaller scale cooperative program has also been conducted with Pacific salmon in San Francisco Bay, and the initial success of the University of California-Davis culture studies with white sturgeon came with the assistance of local anglers in the San Francisco Bay area.

Taken individually, these public involvement projects and others have only minor effects on the total management of anadromous species. But, collectively (and certainly from a public relations standpoint), the effect can be significant, and some management agencies have felt the programs worthy of staffing full-time biologists to work with private groups and individuals.

23.10   MISCELLANEOUS TECHNIQUES

23.10.1   Delayed-Release Programs

Delayed release is simply a management technique where anadromous salmon smolts are held past their normal migration times, then subsequently released into saltwater areas. The result is localized populations of salmon and improved inshore sport fisheries for immature fishes. The most successful program to date has been in Puget Sound.

The idea was first proposed by biologists in Washington and Oregon in the 1950s and 1960s, after considering the relationship between catches of coho salmon, the distance travelled between hatcheries and saltwater, and the timing of releases (Allen 1966). Coho salmon released from hatcheries moved rapidly out of Puget Sound and the Strait of Juan de Fuca, then northward along the west coast of Vancouver Island. Commercial catches of those fishes from British Columbia were ten times higher than sport catches in Puget Sound (Novotny 1980). However, if salmon are delayed in their release past the normal migration time, many will remain in Puget Sound, contributing to a sport fishery. Buckley and Haw (1978) have since found that delayed-release coho salmon contributed 21 times more to the Puget Sound sport fishery compared to identical lots released at normal migration times (Figure 23.3).

In 1969 and 1980, the Washington Department of Fisheries (WDF) began experiments with delayed-release coho salmon at their Minter Creek Hatchery (Haw and Bergman 1972). Later delayed releases were made after transferring smolts to saltwater cages, then releasing them. The released coho salmon tended to remain within the confines of Puget Sound, and the program was expanded to chinook salmon in the early 1970s (Abbott and Salo 1973; Moring 1976). Similar procedures have also resulted in improved localized fisheries in Alaskan waters (Novotny 1975).

From the beginning, the delayed-release program in Puget Sound reaped benefits for fish managers. More than 14% of the 785 chinook salmon tagged and

Figure 23.3 "Blackmouth" or immature chinook salmon caught in Puget Sound, as part of the delayed-release program. (Photograph by Frank Haw.)

released by Abbott and Salo (1973) were ultimately recovered by anglers in or near the Sound, and 1.5% of the 24,500 chinook salmon tagged by Moring (1976) were recovered by anglers over a 2-year period. Returns from some lots of delayed-release coho salmon have been as high as 28% (Novotny 1980).

Federal and state studies also indicate that "sport reared" salmon, held in various locations, will ultimately provide fish for those localized fisheries. For example, 84% of the tagged fish recovered from a release made in southern Puget Sound were from the same portion of the Sound (Moring 1976). Delayed release has been shown to be an effective management technique for improving local sport fisheries for Pacific salmon, and there is some evidence that it may also be successful with sea-run cutthroat trout.

23.10.2   Control of Predators

It seems obvious that the elimination or reduction of potential predators of anadromous fishes may result in higher production, survival, and catches of the preferred species. Harris (1973), for example, reported a fivefold increase in Atlantic salmon fry-to-smolt survival when trout and eel predators were controlled. The impacts of predator control were also documented in the classic study of sockeye salmon in Cultus Lake, British Columbia, in the 1930s where over 20,000 potential predators were removed from the lake during 7 years of gillnetting. Survival to migration of sockeye salmon increased from a pretreatment average of 1.8 to 7.8% (or 9.0% from plants of eyed eggs).

Removing predators from a body of water can be labor intensive and costly despite the improvements in freshwater survival of anadromous fishes. Sometimes such predation is only significant during a brief period of the year, particularly during a vulnerable life stage. As examples, Foerster (1968) reported predation on sockeye salmon by prickly sculpins during periods when newly hatched fry are emerging from the gravel, when fingerlings move into shallow areas prior to migrating, and following releases from a hatchery. Although northern squawfish are known to be a predator on anadromous salmonids, Buchanan et al. (1981) suggested the numbers consumed are inflated in areas below dam spillways or following releases from hatcheries.

Similarly, aquatic birds are widely known to be predators on anadromous fishes, but particularly when the fish are confined in hatchery raceways or floating cages (Moring 1982), or when smolts are migrating seaward in surface waters. Double-crested cormorants *Phalacrocorax auritus* can be significant predators on migrating Atlantic salmon smolts in spring, with as many as 75 to 80% of the smolts being eaten in local areas during this period (Elson 1962). These birds are federally protected in the United States. Population numbers in Maine have increased from zero in 1930 to an estimated 25,000 nesting pairs today.

Fisheries managers have employed two techniques to help reduce the extent of this predation by cormorants. First, they have attempted to shift, as much as possible, the release times of hatchery smolts so that runs of anadromous American shad and alewives are available to cormorants as prey. Second, in the mid-1980s, biologists with the Maine Atlantic Sea Run Salmon Commission, Penobscot Indian Nation, and U.S. Department of Agriculture, conducted a cormorant removal and harassment program along the Penobscot River during the spring smolt migration; decreased bird population numbers resulted during this critical period (Moring 1987).

Other techniques used to dissuade predators include overhead bird nets or covered raceways at fish hatcheries, the use of harassing cannons near fish holding ponds, and electric fences around hatcheries or floating cages to prevent entry by terrestrial predators. Ocean cage culture operations sometimes use double nets to enclose rearing fish but growers almost always check and remove mortalities daily to prevent rips in netting from spiny dogfish. Ocean ranchers have found that barging salmon smolts offshore can sometimes avoid heavy mortality from fish and marine mammals in bays and estuaries near salmon release sites.

23.10.3   Interagency Involvement

Effective management of anadromous fishes often requires joint interagency decisions as well as intraagency decisions. When different agencies can work together in managing stocks, potential conflicts can sometimes be avoided. For example, Indian tribes in the Columbia River basin interact with state agencies in implementing tribal fisheries so that state agencies can sometimes adjust their management decisions and avoid possible future conflicts. Members of the Columbia Basin Fish and Wildlife Authority include all involved state and federal agencies and four lower Columbia River basin Indian tribes. Members confer on major management issues and recently completed a Columbia River Fish Management Plan to direct future allocation of catch and set management goals.

The restoration programs for Atlantic salmon in New England involve numerous agencies, even on a single river system. A Technical Advisory Committee makes recommendations to Maine's Atlantic Sea Run Salmon Commission on stocking rates and locations, disposition of eggs, fish culture activities, tagging, and other aspects of restoration policy for Maine rivers. Voting members include representatives from the Fish and Wildlife Service, the three state fisheries agencies, and the University of Maine. Regular participants also include the Penobscot Indian Nation and the Atlantic Salmon Federation. On the Connecticut River, a Policy and Technical Committee includes representatives from the states of Connecticut, Massachusetts, New Hampshire, and Vermont, along with the Fish and Wildlife Service and the National Marine Fisheries Service.

Another example is the centralized processing of coded wire tag information on salmon and steelhead along the Pacific coast. Hundreds of thousands of these tags (small pieces of binary-coded wire embedded in the snout of fish) are applied annually by agencies along the Pacific coast. The tags are recovered in fish harvested in nonhome states and British Columbia, as well as other countries. The coastal agencies have formed a central receiving house for tag information to maximize the recovery information ultimately returned to tagging agencies. Similarly, Atlantic salmon tagging information is now coordinated through the National Marine Fisheries Service in Woods Hole, Massachusetts, and there are plans by the North American Salmon Conservation Organization to coordinate worldwide Atlantic salmon tag data at one location.

23.11 RESTORATION AND ENHANCEMENT PROGRAMS

23.11.1 Striped Bass

Enhancement and restoration of striped bass is probably the highest national priority for any species of anadromous fish, particularly in the Chesapeake Bay region. Though runs in the Hudson River are increasing, it has been estimated that perhaps 90% of the striped bass along the east coast of the Atlantic originate from tributaries of the Chesapeake Bay. Because of declining populations in recent years, it is thought that the Chesapeake stock is now quite low (though how low is unknown), with the Hudson and the Roanoke (North Carolina and Virginia) stocks now being about equal in contribution (Deuel 1987).

At one time, striped bass populations were endemic from the Sabine River, Texas, along the Gulf of Mexico coast, and up the Atlantic coast to the St. Lawrence River. Today, however, most of the Gulf stocks are virtually extinct, and spawning stocks north of the Hudson River are essentially gone. Spawning populations from Florida to North Carolina are low, but apparently stable (a few may even be increasing), while the important stocks north of North Carolina are undergoing dramatic declines. Chesapeake Bay and Hudson River populations are

Box 23.2   The Decline of the Striped Bass Harvest in Atlantic
            Waters. (Figures summarized from Alperin 1987 and
            Deuel 1987.)

1. Average harvest 1958–1976   4.3 million kg
   Peak harvest 1973   6.7 million kg
   Harvest in 1983   0.8 million kg
   Harvest in 1985   0.5 million kg
2. The decline in the Maryland harvest alone was 94% between 1973 and
   1983.
3. The 1981 Atlantic hatch was the lowest of any season in the 28 years of
   monitoring.
4. Sport catches also declined from North Carolina to Maine: 2 million fish
   in 1979; 300,000 fish in 1985.

fished from North Carolina to Maine, but environmental factors, particularly water flow and water temperature, along with pollution, habitat degradation, and overfishing, have apparently been responsible for the severe decline of stocks in the past 20 years (see Box 23.2).

Striped bass, introduced from the East Coast, are found on the Pacific coast, from Baja California to British Columbia, but are concentrated in the Sacramento–San Joaquin River system of California and some smaller coastal runs, most notably the Coos and Umpqua rivers of Oregon. Pacific coast runs have declined to 25% of their former level, and causes of this decline are somewhat similar to those on the Atlantic coast: low numbers of spawners, pollution, loss of juveniles to water diversions, and reduced plankton densities (Stevens et al. 1985).

The emphasis in restoration and enhancement has been on the Atlantic coastal stocks, and catch restrictions (minimum lengths), sales restrictions, and moratoriums on fishing are now imposed in almost every coastal state. As a consequence of the Atlantic Striped Bass Conservation Act and amendments to the Anadromous Fish Conservation Act, the Atlantic States Marine Fisheries Commission has influenced coastal states (through financial and regulatory pressure) to reduce harvests of striped bass by 55%. Most states today have some type of minimum size limit for commercial and recreational fisheries. In seven states with limited populations, the striped bass is declared to be a sport fish, with commercial fishing banned, and there are even fishing moratoriums in some states.

The management philosophy for improving striped bass populations along the Atlantic coast involves monitoring, regulation, habitat improvement, research, and stocking (see Box 23.3). An important tool of managers is an effective monitoring program whereby any successes of other management procedures can be assessed. The Maryland-Chesapeake Bay monitoring of age-0 striped bass has been conducted since the mid-1950s. Similar sampling programs for juveniles now occur in five other states and the District of Columbia, and adults are being monitored in six states and the District of Columbia.

The imposition of minimum size limits and moratoriums are designed to reduce

Box 23.3 Annual Stocking of Anadromous Striped Bass in the Eastern United States in 1983 and 1984.

Maryland and Virginia, Chesapeake Bay	240,000
New York and New Jersey	337,000
Maine	46,000
Georgia and North Carolina	155,000–175,000
Gulf states	600,000–700,000
Eleven coastal states from federal hatcheries	3.8 million

one source of mortality—fishing—while other management measures are taking effect. One of these is habitat improvement. Six states are involved with projects dealing with defining and reducing the impacts of pollutants on striped bass. In particular, state and federal researchers have found that striped bass are very susceptible to arsenic, copper, cadmium, aluminum, and malathion, and there are closures or advisories against catching or eating PCB-laden striped bass in New York, New Jersey, and Rhode Island. Other researchers have shown that chlorine (used to treat sewage effluent, among other things) reduces zooplankton densities and, hence, food for juvenile striped bass. Applied research continues on the role of these potential toxicants and measures needed to clean up the sources, on feeding and nutritional requirements of striped bass, and on disease control and stock identification. In the latter area, considerable research has been conducted on genetic identification of stocks through electrophoretic or mitochondrial DNA techniques, but more recent successes have involved identifying stocks by scale shape.

As these other measures take effect, it is hoped that stocking will show increasing success. As one example, Maine has had a limited program of stocking striped bass since 1982, with almost 200,000 striped bass stocked in the Kennebec and Androscoggin rivers. For the past several years, juvenile bass have been captured by seine in Merrymetting Bay, probably the first such natural spawning since populations were destroyed in the 19th Century.

23.11.2 Atlantic Salmon

The restoration of Atlantic salmon in the United States relies almost entirely on fish stocking. By the 1860s, runs of salmon in many industrialized areas of the northeastern United States were depleted or eliminated entirely by a combination of dams and pollution. Later, as stocks declined, commercial fishing offshore and in streams helped speed the extinction. Only runs in eastern Canada and a few small streams in eastern and central Maine maintained wild stocks, even though numerous runs in eastern Canada have suffered the same consequences of overharvest, pollution (including acid precipitation), and dams.

Restoration was not able to proceed in any area of New England until progress was made on cleaning up water quality in rivers and providing fish passage around and over dams. Current restoration programs are centered in four regions: Maine (primarily the Penobscot River, but a dozen other rivers as well), the Merrimack

River, the Connecticut River, and, to a lesser extent, the Pawcatuck River in Rhode Island. In each case, a multi-agency group directs the restoration effort, with the federal government involved in fish culture and state agencies involved with monitoring and management.

Though Maine began a restoration program in 1948, it was the Model River Program for the Penobscot River (1965) that was the principal thrust for subsequent programs (Connecticut River in 1967, and Merrimack River in 1969) involving fish passage and stocking. Currently, there are 19 state, federal, and private rearing, holding, and kelt rejuvenation facilities in New England.

Full restoration of the Penobscot River will require a consistent annual run of at least 6,000 adults, unsupported by stocking (Moring 1986). The Penobscot, the most successful Atlantic salmon restoration river, receives 31% of the New England stocked salmon, yet is far from being restored. The run has averaged over 3,000 salmon over the past 11 years, including a record return of 4,529 in 1986. Although the proportion of the run originating from naturally constructed redds in the wild is steadily increasing, full restoration will not occur until the 21st Century, and may have to always be supplemented by some level of stocking.

Stocking programs in the maritime provinces totalled over 1.1 million in 1982; 71% of the salmon were placed in New Brunswick waters and 19% were placed in Nova Scotia. Wild runs, though declining in several areas, are still the mainstay of much of the fisheries in the Maritimes.

Part of the problem in restoring runs to U.S. streams has been the interception of fish on the high seas, primarily as they pass through Canadian fisheries. Some estimates place this interception figure at half of all U.S. fish. To identify U.S. stocks, there have been some promising studies with scale shape as a distinguishing characteristic and onsite monitoring programs at local Canadian and Greenland canneries.

Another constraint to restoration is the source of eggs. With the native runs to most New England rivers eliminated, the unique set of adaptive genetic characteristics of those populations has also been lost. The source of eggs for restoration of Maine rivers, typically now called Penobscot, was probably a combination of Canadian stocks and possibly some eastern Maine stocks that were spared (those populations have now been diluted by stocking). As a result, it has taken considerable time to create the new Penobscot strain that has shown satisfactory, if not spectacular return rates. The restoration of southern New England rivers will take even longer as those runs have relied on Penobscot salmon eggs as a source. The genetic and geographic origin of those eggs is even more removed, and it will take considerable time to create new Merrimack, Pawcatuck, and Connecticut stocks adapted for those waters. Nevertheless, successes are small but inspiring. Over the past several years, salmon have returned to the Pawcatuck, Merrimack, and White rivers—in each case for the first time in almost a century.

23.11.3 Pacific Salmon

Pacific salmon are the most valuable of North America's anadromous fishes. Recreational fisheries from northern California to Alaska are multimillion dollar industries and sport catches of Pacific salmon introduced in New England waters have been notable in the past. Although not an anadromous situation, the Pacific salmon fishery in the Great Lakes is estimated by some to exceed $1 billion. Thus, efficient management of this resource is of critical importance.

Table 23.2 Recent stockings of Pacific salmon, exluding the Great Lakes.

State or province	Year	Number
British Columbia	1982	449 million
Washington	1983	328 million
Oregon	1983	58 million
California	1983–1984	40 million
New Hampshire	1983	<0.5 million
Alaska	current	a

[a]20 hatcheries with a production capacity of almost 1 billion eggs and a goal of adding 47 million salmon to total runs.

Stocking of hatchery-reared fish is an essential, costly component of anadromous fish management in Pacific coastal states and British Columbia. The numbers are significant (Table 23.2).

Despite these large stocking programs, Pacific salmon runs in some areas of the Pacific coast have been declining rapidly, particularly in the last decade. This is a difficult problem for fish management. An analysis of population trends by Konkel and McIntyre (1987) concluded that south-central Alaska has the most increasing populations of salmon, while the Columbia River basin has the most declining populations. Three times more coho and chum salmon are declining than increasing along the Pacific coast. Populations in Alaska have been generally increasing for sockeye, pink, and chinook salmon, and populations in the Pacific Northwest have been either stable (pink salmon) or generally declining (sockeye salmon). Despite these declines, quotas or proposals for outright bans on commercial catches of coho salmon have met with heated opposition. Part of the decline has probably been due to an extended El Niño event in the 1980s that may have reduced ocean nutrients and food supplies.

On the positive side, some catches, such as in parts of Alaska, have increased in recent years (Konkel and McIntyre 1987), and improvements in water quality in major rivers, such as Oregon's Willamette, have led to increased runs in conjunction with stocking. The runs of chinook salmon to Oregon waters broke several records in 1987, including record commercial catches. There is little that management agencies can do about conditions in the ocean until salmon return to coastal waters. Hatchery stocking is an important management option, but the effect of large numbers of hatchery fish on supplies of ocean foods is still being debated.

There are, however, some other management options that can improve the picture, including more precise data on time and size at release, habitat improvement and protection, increased survival and returns via genetic selection, disease control, and regulations. Other options include expanding rearing capabilities, delayed-release (Section 23.10.1) and interaction with commercial ocean ranching facilities. In the latter case, commercial fishers and, to a lesser extent, sport anglers, can benefit from increased catches of privately released salmon, though other potential conflicts can exist.

23.11.4 Steelhead

With few exceptions, runs of steelhead have not been totally eliminated but, even in Alaska, have disappeared near population centers (Barnhart 1975). Stocking to supplement runs is a vital component of management programs, even though the commercial harvest has been greatly reduced over the past 60 years. Only one-third of the steelhead catch in Oregon consists of wild fish; the rest exist as a result of stocking. In most states and British Columbia, the species is managed as a sport fish, and increasing pressure from anglers has put pressure on management agencies to increase run sizes.

Steelhead have been transplanted across North America, but the anadromous subspecies is cultured primarily in northern California, Oregon, Washington, Idaho, British Columbia, and, to a lesser extent, Alaska. Over 28 million fish are stocked annually. Smolts accounted for over 70% of the 12.5 million steelhead released in Washington in 1983. Other state, federal, and provincial agencies are primarily releasing fish larger than fry size.

Stocking is a critical management tool in many waters. At least 80% of the run on the Alsea River, Oregon, is of hatchery origin. Where runs are supplemented by stocking, these hatchery fish are critical: from 75 to 80% of the steelhead in hatchery-supported runs in Oregon are of hatchery origin. To improve runs in southeastern and south-central Alaska, steelhead brood stocks have been established at five hatcheries with a combined capacity of over 200,000. Adult fish have been returning for several years.

23.11.5 Other Anadromous Salmonids

Sea-run brook trout occur in limited numbers along the northeastern coast of North America, and the subspecies is managed primarily as a sport fish. Some work on sea ranching has been conducted by the Matamek Research Station (Quebec) of Woods Hole Oceanographic Institution, but the principal management effort today is probably in Massachusetts. Sea-run populations were established long before the arrival of the Pilgrims. The runs were almost eliminated in Massachusetts, with populations in less than two dozen streams (concentrated on Cape Cod) remaining today (Bergin 1985). In order to protect these fragile populations, management efforts have been to stock brown trout to absorb angler pressure. Approximately 15,000 to 25,000 yearling brown trout are stocked annually in coastal streams of Massachusetts.

Connecticut has a limited stocking program for sea-run brown trout, derived from eggs obtained from Europe, and brown trout fisheries have been created in the estuaries of the Royal, St. George, and Ogunquit rivers of Maine by stocking nonanadromous brown trout directly in or near estuaries, and by directly stocking spring yearlings in streams. These brown trout are forced into an anadromous state by stocking in saline areas. They grow rapidly and popular fall fisheries have developed for fish up to 1.4 kg in weight.

Sea-run cutthroat trout are found from northern California to Alaska and, throughout the range, the species is avidly sought by anglers. Hatchery production for limited to moderate stocking programs is centered in Oregon, Washington, and British Columbia, and fisheries managers in these areas employ different strategies for stocking the subspecies. Oregon seeks to protect wild populations of sea-run cutthroat trout, while stocking trout in major coastal river systems to meet angler demand. About 900,000 cutthroat trout are stocked annually to provide spring, summer, and fall fisheries at four different life stages: a legal-size trout fishery prior to fish moving into the estuary in the spring, a fishery for larger trout in the estuary in summer, a fishery on returning spawners in the fall, and a limited fishery on downstream-migrating kelts in the following spring.

In 1983–1984, Washington stocked over 329,000 sea-run cutthroat trout—two-thirds of them migrant fish. A study in 1972 indicated that only 0.1% of cutthroat trout smolts stocked in Hood Canal (Puget Sound) were ultimately taken by anglers fishing in the Canal (Hisata 1973). By using delayed-release techniques that have been successful with Pacific salmon (see Section 23.10.1), cutthroat trout were held in floating cages in Puget Sound, then later released back into Hood Canal; the angler harvest rose to 9% (Johnston and Mercer 1976).

British Columbia has attempted to reverse a declining trend in cutthroat trout populations near urban areas. This Salmonid Enhancement Program hopes to eventually stock sufficient cutthroat trout juveniles to provide an additional 19,000 adults in coastal streams annually (Moring 1986).

Arctic char have both landlocked and anadromous runs in North America, and the species is managed as a sport fish in Alaska and several provinces of Canada. Commercial fisheries exist in the maritime provinces, where they are primarily taken by gill net. However, fisheries managers in Labrador have seen a rapid decline in the size of commercially caught Arctic char. Char processed in 1987 averaged 1 to 2 kg, compared to fish of 3.5 to 4.5 kg in the early 1980s. This may indicate a problem with overexploitation. Arctic char are also being reared commercially in hatcheries and floating cages along the coast of New Brunswick, and may become an important species for aquaculture in the 1990s. If population enhancement measures are needed to improve coastal populations of char, the culture techniques being developed may be effective management tools as well.

A related char, the Dolly Varden, is distributed from northern California to Alaska, and eastward into the Yukon and Northwest Territories, Alberta, Montana, Nevada, and Idaho, but anadromous runs exist from Washington northward. The species is a popular sport fish, particularly in areas without substantial populations of other trout. At one time, there was a 2.5 cent bounty on Dolly Varden in Alaska, and some $20,000 per year was spent on bounties in the 1930s (McPhail and Lindsey 1970). Like most bounties, this one was ill-conceived, and removing these predators probably had little major effect on salmon and trout populations. Although there has been some commercial processing of Dolly Varden in the 20th Century, it was only as a by-product of other fisheries, and today the species is managed exclusively as a sport fish.

23.11.6 American Shad, Alewives, and Blueback Herring

The American shad was probably the first species of anadromous fish to be aided by North American fish cultural programs because, along with Atlantic salmon, it was the first anadromous species to be severely effected by human impact. Over 200 million eggs were artificially hatched in northeastern states between 1866 and 1871 (Bowen 1970), but today restoration and enhancement efforts primarily involve transplantation of adults. For example, New Hampshire and Rhode Island have imported mature shad from the Connecticut River to stock in state waters, and Maine transfers shad from the Narraguagus River (Maine) and the Merrimack River (New Hampshire) to stock in Maine rivers that have limited or no shad populations (Moring 1986). Similar restoration programs are also underway in Massachusetts and Pennsylvania.

American shad are found from the Gulf of Mexico northward to New Brunswick, but it is in the mid- and north-Atlantic regions where American shad populations have declined due to dams, pollution, and overfishing. It is estimated that, in Maine, only 5% of the habitat formerly accessible to American shad can now be reached. For several years, there have been commercial and sport angling restrictions in some areas, but the annual U.S. harvest has stayed between 1 and 4 million tonnes since 1965. As a consequence, enhancement and restoration of runs in the northeastern United States has become a priority issue with management agencies. Most states have at least a moderate program, but major restoration efforts are underway in the Susquehanna and Schuylkill rivers of Pennsylvania. In the Susquehanna, dams have long blocked runs of shad to the upper river, and the fishery deteriorated. Downstream bay catches in Maryland dropped from about 900,000 kg in 1969 to 9,000 kg in 1980 (Moring 1986). Today, states are in the midst of a 10-year restoration program involving fish passage, hatchery culture, and research. In 1984, 7.6 million American shad were released in Pennsylvania alone.

Two related anadromous clupeids are also being stocked in limited enhancement programs. Sea-run alewives and blueback herring are commercially important along the Atlantic coast, with over 5,000 tonnes of the two species harvested annually (Mullen et al. 1986). While 90% of the catch in Maine is used for lobster bait, the two species are also valued for human consumption, fish meal, and highly prized roe. New Hampshire is currently improving fish passage to increase runs of alewives and blueback herring, and Massachusetts, Rhode Island, and Maine have ongoing stocking programs. The state of Maine has plans for an ambitious introduction program, stocking sea-run alewives in numerous lakes, following a legislative mandate to restore this species to its former range. Prior to implementing the full program, three state agencies and the University of Maine are conducting a long-term study on the impacts of stocking sea-run alewives on freshwater fish populations.

23.11.7 Other Species

Sturgeon have long been cultured in the former Soviet Union, though work in the United States since the mid-1970s has largely been concentrated in California and South Carolina. The ultimate management goal has been to increase population numbers by culturing sturgeon in hatchery situations for later stocking in rivers, bays, and estuaries. Researchers at the University of California-Davis have conducted numerous studies on white sturgeon and have been stocking fingerlings in San Francisco Bay since 1980 (Buddington and Doroshov 1984; Doroshov and Lutes 1984).

The endangered shortnose sturgeon has been cultured for over 75 years. In the early 1980s, efforts by the South Carolina Wildlife and Marine Resources Department and the Fish and Wildlife Service resulted in the culture and later release of juvenile shortnose and Atlantic sturgeon. Because the shortnose sturgeon is on the federal list of endangered species, and Atlantic sturgeon numbers are comparatively low, fish stocking and harvest restrictions may be the most effective management tools.

Researchers on the Pacific and Atlantic coasts have experimented with the culture and release of anadromous smelts. Taken as a group, these fishes are popular with anglers and are particularly valuable as forage species for freshwater, anadromous, and marine game fishes. On the Pacific coast, several species have been reared experimentally, though the eulachon may hold the most promise for future management options. Work with the rainbow smelt, an anadromous and freshwater species, has been successful from two standpoints: the culture of fish in tanks to commercial baitfish size, and the stocking of eyed-eggs in natural bodies of water (Akielaszek et al. 1985).

23.12 SUMMARY

Management of anadromous species can be highly complex, involving the input of many agencies, municipalities, and groups, as well as biological and political concerns. Fortunately, fisheries managers have numerous tools at their disposal. Fish culture has been an important component of management plans since the mid-1800s, but refinements in diet, disease control, growth, and survival are still needed. A critical concern for anadromous species is fish passage: maximizing survival for upstream and downstream-migrating fish. Computerized modelling exercises suggest that new techniques for diverting downstream fry, smolts, and spent adults may have a significant effect on survival, overall run numbers, and production.

Genetic selection and improvements in stream habitat are also options available to fisheries managers to improve survival of wild and hatchery-produced fish. Further management options include some of the more innovative techniques, such as delayed release programs to promote localized sport fisheries.

Fisheries managers must then decide how resources should be allocated. Along with this comes questions such as What is the optimal size of fish and time at release? How can the public become involved in habitat improvement and other restoration and enhancement programs? How can fisheries be improved through interagency and other types of cooperative management? These are the challenges in anadromous fish management.

23.13 REFERENCES

Abbott, R. R., and E. O. Salo. 1973. Contribution to the recreational fishery of pen-reared chinook salmon released as yearlings. Pages 34–35 in D. E. Rogers, editor. 1972 Research in fisheries. University of Washington, College of Fisheries, Seattle.

Akielaszek, J. J., J. R. Moring, S. R. Chapman, and J. H. Dearborn. 1985. Experimental culture of young rainbow smelt Osmerus mordax. Transactions of the American Fisheries Society 114:596–603.

Allen, G. H. 1966. Ocean migration and distribution of fin-marked coho salmon. Journal of the Fisheries Research Board of Canada 23:1043–1061.

Alperin, I. M. 1987. Management of migratory Atlantic coast striped bass: an historical perspective. Fisheries (Bethesda) 12(6):2–3.

Barnhart, R. A. 1975. Pacific slope steelhead trout management. Pages 7–11 in W. King, editor. Wild Trout Management Symposium. Trout Unlimited, Vienna, Virginia.

Berg, E. R. 1981. Management of Pacific Ocean salmon ranching: a problem of federalism in the coastal zone. Coastal Zone Management Journal 9:41–76.

Bergin, J. D. 1985. Massachusetts coastal trout management. Pages 137–142 in F. Richardson and R. H. Hamre, editors. Wild Trout III. Federation of Fly Fishers and Trout Unlimited, Vienna, Virginia.

Bjornn, T. C., and R. R. Ringe. 1984. Homing of hatchery salmon and steelhead allowed a short-distance voluntary migration before transport to the lower Columbia River.

University of Idaho, Idaho Cooperative Fishery Research Unit, Technical Report 84-1, Moscow.

Bley, P. W., and J. R. Moring. 1988. Freshwater and ocean survival of Atlantic salmon and steelhead. U.S. Fish and Wildlife Service Biological Report 88(9).

Bowen, J. T. 1970. A history of fish culture as related to the development of fishery programs. American Fisheries Society Special Publication 7:71–93.

Buchanan, D. V., R. M. Hooton, and J. R. Moring. 1981. Northern squawfish (*Ptychocheilus oregonensis*) predation on juvenile salmonids in sections of the Willamette River basin, Oregon. Canadian Journal of Fisheries and Aquatic Sciences 38:360–364.

Buchanan, D. V., and J. R. Moring. 1986. Management problems with recycling of adult summer steelhead trout at Foster Reservoir, Oregon. Pages 191–200 *in* Stroud (1986).

Buckley, R., and F. Haw. 1978. Enhancement of Puget Sound populations of resident coho salmon, *Oncorhynchus kisutch* (Walbaum). Canadian Fisheries and Marine Service Technical Report 759:93–103.

Buddington, R. K., and S. I. Doroshov. 1984. Feeding trials with hatchery produced white sturgeon juveniles (*Acipenser transmontanus*). Aquaculture 36:237–243.

Busiahn, T. R. 1984. An introduction to native peoples' fisheries issues in North America. Fisheries (Bethesda) 9(5):8–11.

Clark, W. G. 1985. Fishing in a sea of court orders: Puget Sound salmon management 10 years after the Boldt decision. North American Journal of Fisheries Management 5:417–434.

Congleton, J. L., T. C. Bjornn, C. A. Robertson, J. L. Irving, and R. R. Ringe. 1984. Evaluating the effects of stress on the viability of chinook salmon smolts transported from the Snake River to the Columbia River estuary. University of Idaho, Idaho Cooperative Fishery Research Unit, Technical Report 84-4, Moscow.

Deuel, D. 1987. The role of the National Marine Fisheries Service in the Emergency Striped Bass Research Study. Fisheries (Bethesda) 12(6):7–8.

Donaldson, L. R. 1968. Selective breeding of salmonid fishes. Pages 65–74 *in* W. J. McNeil, editor. Marine aquiculture. Oregon State University Press, Corvallis.

Doroshov, S. I., and P. B. Lutes. 1984. Preliminary data on the induction of ovulation in white sturgeon (*Acipenser transmontanus* Richardson). Aquaculture 38:231–227.

Elson, P. F. 1962. Predator–prey relationships between fish-eating birds and Atlantic salmon. Fisheries Research Board of Canada, Bulletin 133.

Everest, F. H. 1977. Evaluation of fisheries for anadromous salmonids produced on western national forests. Fisheries (Bethesda) 2(2):8–9,11,34–36.

Foerster, R. E. 1968. The sockeye salmon, *Oncorhynchus nerka*. Fisheries Research Board of Canada Bulletin 162.

Gulland, J. A. 1969. Manual of methods for fish stock assessment, part 1. FAO (Food and Agriculture Organization of the United Nations) Manuals in Fisheries Science M4.

Harden-Jones, F. R. 1968. Fish migration. Edward Arnold, London.

Harris, G. S. 1973. Rearing smolts in mountain lakes to supplement salmon stocks. International Atlantic Salmon Foundation Special Publication Series 4:237–252.

Hasler, A. D., and A. T. Scholz. 1983. Olfactory imprinting and homing in salmon. Springer-Verlag, Berlin.

Haw, F., and P. K. Bergman. 1972. A salmon angling program for the Puget Sound regions. Washington Department of Fisheries, Information Booklet 2, Olympia.

Hershberger, W. K., and R. N. Iwamoto. 1985. Systematic genetic selection and breeding in salmonid culture and enhancement programs. NOAA (National Oceanic and Atmospheric Administration) Technical Report NMFS (National Marine Fisheries Service) 27.

Hisata, J. S. 1973. Evaluation of stocking hatchery reared sea-run cutthroat trout in streams of Hood Canal. Washington State Department of Game, Job Completion Reports AFS-44-1 and AFS- 44-2, Olympia.

Iwamoto, R. N., A. M. Saxton, and W. K. Hershberger. 1982. Genetic estimates for length and weight of coho salmon (*Oncorhynchus kisutch*). Journal of Heredity 73:187–191.

Johnston, J. M., and S. P. Mercer. 1976. Sea-run cutthroat in saltwater pens: broodstock development and extended juvenile rearing (with a life history compendium). Washington State Department of Game, Fisheries Research Report, Olympia.

Jonsson, B., N. Jonsson, and L. P. Hansen. 1990. Does juvenile experience affect migration and spawning of adult Atlantic salmon? Behavioral Ecology and Sociobiology 26:225–230.

Konkel, G. W., and J. D. McIntyre. 1987. Trends in spawning populations of Pacific anadromous salmonids. U.S. Fish and Wildlife Service Fish and Wildlife Technical Report 9.

Larson, R. W., and J. H. Ward. 1955. Management of steelhead trout in the state of Washington. Transactions of the American Fisheries Society 84:261–274.

Mahnken, C., E. Prentice, W. Waknitz, G. Monan, C. Sims, and J. Williams. 1982. The application of recent smoltification research to public hatchery releases: an assessment of size/time requirements for Columbia River hatchery coho salmon (*Oncorhynchus kisutch*). Aquaculture 28:251–268.

Marsh, J. H., and J. H. Johnson. 1985. The role of Stevens Treaty tribes in the management of anadromous fish runs in the Columbia River. Fisheries (Bethesda) 10(4):2–5.

McNeil, W. J. 1980. Salmon ranching in Alaska. Pages 13–27 *in* J. E. Thorpe, editor. Salmon ranching. Academic Press, New York.

McPhail, J. D., and C. C. Lindsey. 1970. Freshwater fishes of northwestern Canada and Alaska. Fisheries Research Board of Canada Bulletin 173.

Mecum, W. L. 1980. The efficiency of various bypass configurations for juvenile striped bass, *Morone saxatilis,* and American shad, *Alosa sapidissima.* California Department of Fish and Game, Anadromous Fisheries Branch, Administrative Report 80-12, Sacramento.

Moring, J. R. 1976. Contributions of delayed-release, pen-cultured chinook salmon to the sport fishery in Puget Sound, Washington. Progressive Fish-Culturist 38:36–39.

Moring, J. R. 1982. Fin erosion and culture-related injuries of chinook salmon raised in floating net pens. Progressive Fish-Culturist 44:189–191.

Moring, J. R. 1986. Stocking anadromous species to restore or enhance fisheries. Pages 59–74 *in* Stroud (1986).

Moring, J. R. 1987. Restoration of Atlantic salmon to Maine: overcoming physical and biological problems in the estuary. Pages 3129–3140 *in* O. T. Magoon, H. Converse, D. Miner, L. T. Tobin, D. Clark, and G. Domurat, editors. Coastal Zone '87. American Society of Civil Engineers, New York.

Moring, J. R. 1989. Documentation of unaccounted-for losses of chinook salmon from saltwater cages. Progressive Fish-Culturist 51:173–176.

Moring, J. R., G. C. Garman, and D. M. Mullen. 1985. The value of riparian zones for protecting aquatic systems: general concerns and recent studies in Maine. U.S. Forest Service General Technical Report RM-120:315–319.

Mowat, T. 1889. Statistics of the fisheries of the province of British Columbia for 1886. U.S. Fish Commission Bulletin 7(1887):24–25.

Mullen, D. M., C. W. Fay, and J. R. Moring. 1986. Species profiles: life histories and environmental requirements of coastal fishes and invertebrates (North Atlantic)—alewife/blueback herring. U.S. Fish and Wildlife Service Biological Report 82(11.56).

Novotny, A. J. 1975. Net-pen culture of Pacific salmon in marine waters. U.S. National Marine Fisheries Service Marine Fisheries Review 37(1):36–47.

Novotny, A. J. 1980. Delayed release of salmon. Pages 325–369 *in* J. E. Thorpe, editor. Salmon ranching. Academic Press, New York.

Reisenbichler, R. R., and J. D. McIntyre. 1986. Requirements for integrating natural and artificial production of anadromous salmonids in the Pacific northwest. Pages 365–374 *in* Stroud (1986).

Ricker, W. E. 1972. Heredity and environmental factors affecting certain salmonid populations. Pages 27–160 *in* R. C. Simon and P. A. Larkin, editors. The stock concept in Pacific salmon. H.R. MacMillan Lectures in Fisheries, University of British Columbia, Vancouver.

Saunders, R. L. 1981. Atlantic salmon (*Salmo salar*) stocks and management implications in the Canadian Atlantic provinces and New England, USA. Canadian Journal of Fisheries and Aquatic Sciences 38:1612–1625.

Saunders, R. L, and J. K. Bailey. 1980. The role of genetics in Atlantic salmon management. Pages 182–200 *in* A. E. J. Went, editor. Atlantic salmon: its future. Fishing News Books, Farnham, UK.

Sheppard, D. 1972. The present status of the steelhead trout stocks along the Pacific coast. Pages 519–556 *in* D. H. Rosenberg, editor. A review of the oceanography and renewable resources of the northern Gulf of Alaska. University of Alaska, Institute of Marine Science Report R72-23 and Sea Grant Report 73-3, Fairbanks.

Snieszko, S. F. 1972. Nutritional fish diseases. Pages 404–437 *in* J. E. Halver, editor. Fish nutrition. Academic Press, New York.

Stevens, D. E., D. W. Kohlhorst, L. W. Miller, and D. W. Kelley. 1985. The decline of striped bass in the Sacramento–San Joaquin estuary, California. Transactions of the American Fisheries Society 114:12–30.

Stier, D. J., and B. Kynard. 1986. Use of radio telemetry to determine the mortality of Atlantic salmon smolts passed through a 17-MW Kaplan turbine at a low-head hydroelectric dam. Transactions of the American Fisheries Society 115:771–775.

Stolte, L. W. 1986. Atlantic salmon. Pages 696–713 *in* R. L. DiSilvestro, editor. Audubon Wildlife Report. National Audubon Society, New York.

Stroud, R. H., editor. 1986. Fish culture in fisheries management. American Fisheries Society, Fish Culture Section and Fisheries Management Section, Bethesda, Maryland.

Swartz, D. 1985. Critical steelhead habitat and habitat problems in Oregon. Pages 105–108 *in* F. D. Van Hulle, editor. Alaska Steelhead Workshop, 1985. Alaska Department of Fish and Game, Division of Sport Fish, Juneau.

Taylor, R. E., and B. Kynard. 1985. Mortality of juvenile American shad and blueback herring passed through a low-head Kaplan hydroelectric turbine. Transactions of the American Fisheries Society 114:430–435.

Thorpe, J. E, and eight coauthors. 1981. Assessing and managing man's impact on fish genetic resources. Canadian Journal of Fisheries and Aquatic Sciences 38:1899–1907.

Wagner, E., and P. Ingram. 1973. Evaluation of fish facilities and passage at Foster and Green Peter dams on the South Santiam River drainage in Oregon. Fish Commission of Oregon, Final Report to U.S. Army Corps of Engineers, Contract DACW57-68-C-0013, Portland.

Wagner, H. H. 1969. Effect of stocking location of juvenile steelhead trout, *Salmo gairdnerii*, on adult catch. Transactions of the American Fisheries Society 98:27–34.

Watt, W. D. 1986. Report to the St. Croix River Biological Working Party on statistical analysis of St. Croix River tag return data for information on downstream migration mortality of Atlantic salmon smolts. Department of Fisheries and Oceans, Fisheries Research Branch, Halifax, Nova Scotia.

Whitney, R. R. 1961. The Susquehanna fishery study, 1957–1960. A report of a study on the desirability and feasibility of passing fish at Conowingo Dam. Maryland Department of Resource Education, Contribution 169, Solomons.

Symbols and Abbreviations

The following symbols and other abbreviations are used in this book.

°C	degrees Celsius	min	minute
cm	centimeter	N	sample size
d	day	P	probability
e	base of natural logarithm (2.71828 . . .)	pH	negative log of hydrogen ion activity
e.g.	(exempli gratia) for example	R	multivariate correlation or regression coefficient
et al.	(et aliae) and others	r	simple correlation or regression coefficient
etc.	etcetera		
F	filial generation	s	second
g	gram	sp.	species (singular)
h	hour	spp.	species (plural)
i.e.	(id est) that is	subsp.	subspecies
k	kilo (10^3, as a prefix)	U.S.	United States (adjective)
kg	kilogram	μ	micro (10^{-6}, as a prefix)
km	kilometer	°	degree (temperature as a prefix, angular as a suffix)
L	liter		
m	meter (as a suffix or by itself)		
mg	milligram	%	percent (per hundred)
mm	millimeter	‰	per mille (per thousand)

Index